Universitext

Universitext

Universitext is a series of textbooks that presents material from a wide variety of mathematical disciplines at master's level and beyond. The books, often well class-tested by their author, may have an informal, personal, even experimental approach to their subject matter. Some of the most successful and established books in the series have evolved through several editions, always following the evolution of teaching curricula, into very polished texts.

Thus as research topics trickle down into graduate-level teaching, first textbooks written for new, cutting-edge courses may make their way into *Universitext*.

For further volumes:
www.springer.com/series/223

Friedrich Sauvigny

Partial Differential Equations 1

Foundations and Integral Representations

With Consideration of Lectures
by E. Heinz

Second Revised and Enlarged Edition

Prof. Dr. Friedrich Sauvigny
Mathematical Institute, LS Analysis
Brandenburgian Technical University
Cottbus, Germany

ISSN 0172-5939 ISSN 2191-6675 (electronic)
Universitext
ISBN 978-1-4471-2980-6 ISBN 978-1-4471-2981-3 (eBook)
DOI 10.1007/978-1-4471-2981-3
Springer London Heidelberg New York Dordrecht

Library of Congress Control Number: 2012936042

Mathematics Subject Classification: 28A20, 30-01, 30G20, 31-01, 31B20, 32A26, 33C55, 35-01, 35A01, 35A02, 35A08, 35B08, 35B50, 35J08, 35K05, 46-01

Based on a previous edition of the Work:
Partial Differential Equations 1 by Friedrich Sauvigny
© Springer Heidelberg 2006

Printed on acid-free paper

Springer is part of Springer Science+Business Media (www.springer.com)

Preface – Volume 1

Partial differential equations appear in both physics and geometry. Within mathematics they unite the areas of complex analysis, differential geometry and calculus of variations. The investigation of partial differential equations has contributed substantially to the development of functional analysis. Although a relatively uniform treatment of ordinary differential equations is possible, multiple and quite diverse methods are available for partial differential equations. With this two-volume textbook we intend to present the entire domain PARTIAL DIFFERENTIAL EQUATIONS – so rich in theories and applications – to students on the intermediate level.

We presuppose a basic knowledge of the analysis, as it is conveyed in the beautiful lectures [Hi1] and [Hi2] by S. Hildebrandt or in our lecture notes [S1] and [S2] or in W. Rudin's influential textbook [Ru]. For the convenience of the reader we further develop foundations from the analysis in a form adequate to the theory of partial differential equations. Therefore, this textbook can be used for a course extending over several semesters. We have intended to present the theory in the same form we know from books on complex analysis or differential geometry. In our opinion, gaining a deep understanding of the subject replaces the need for exercises, which are implicitly present in our text and may be supplemented from other books. By excluding exercises, we instead focus on presenting a complete and self-contained theory.

A survey of all the topics is provided by the table of contents, which naturally reflects the interests of the author. For advanced readers, each chapter may be studied independently from the others. In selecting the topics of our lectures and consequently for our textbooks, I tried to follow the advice of one of the first great scientists at the University of Göttingen, namely G.C. Lichtenberg: *Teach the students h o w to think and not always w h a t to think!* When I was a student at Göttingen, I admired the commemorative plaques throughout the city in honor of many great physicists and mathematicians. In this spirit, I attribute the results and theorems in our compendium to the persons who – to the best of my knowledge – created them.

The original version of this textbook, *Friedrich Sauvigny: Partielle Differentialgleichungen der Geometrie und der Physik – Grundlagen und Integraldarstellungen – Unter Berücksichtigung der Vorlesungen von E. Heinz*, was first published in 2004 by *Springer-Verlag*. A translated and expanded version of this monograph followed in 2006 as *Springer-Universitext*, namely *Friedrich Sauvigny: Partial Differential Equations 1*, and we are now presenting a second edition of this textbook.

In Chapter 1 we treat the *Differentiation and Integration on Manifolds*, where we use the improper Riemannian integral. After the Weierstraß approximation theorem in Section 1, we introduce differential forms in Section 2 as functionals on surfaces – parallel to [Ru]. The calculus rules for differential forms are immediately derived from the determinant laws and the transformation formula for multiple integrals. With the partition of unity and an adequate approximation we prove the Stokes integral theorem for manifolds in Section 4, which may possess singular boundaries of capacity zero besides their regular boundaries. In Section 5 we especially obtain the Gaussian integral theorem for singular domains as in [H1], which is indispensable for the theory of partial differential equations. After the discussion of contour integrals in Section 6, we shall follow [GL] in Section 7 and represent A. Weil's proof of the Poincaré lemma. In Section 8 we shall explicitly construct the $*$-operator for certain differential forms in order to define the Beltrami operators. Finally, we represent the Laplace operator in n-dimensional spherical coordinates.

In Chapter 2 we shall constructively supply the *Foundations of Functional Analysis.* Having presented Daniell's integral in Section 1, we shall continue the Riemannian integral to the Lebesgue integral in Section 2. The latter is distinguished by convergence theorems for pointwise convergent sequences of functions. We deduce the theories of Lebesgue measurable sets and functions in a natural way; see the Sections 3 and 4. In Section 5 we compare Lebesgue's with Riemann's integral. Then we consider Banach and Hilbert spaces in Section 6, and in Section 7 we present the Lebesgue spaces $L^p(X)$ as classical Banach spaces. Especially important are the selection theorems with respect to almost everywhere convergence due to H. Lebesgue and with respect to weak convergence due to D. Hilbert. Following ideas of J. v. Neumann we investigate bounded linear functionals on $L^p(X)$ in Section 8. For this Chapter 1 have profited from a seminar on functional analysis, offered to us as students by Professor Dr. E. Heinz in Göttingen.

In Chapter 3 we shall study topological properties of mappings in $\mathbb{R}^n$ and solve nonlinear systems of equations. In this context we utilize Brouwer's degree of mapping, for which E. Heinz has given an ingenious integral representation (compare [H8]). Besides the fundamental properties of the degree of mapping, we obtain the classical results of topology. For instance, the theorems of Poincaré on spherical vector-fields and of Jordan-Brouwer on topological spheres in $\mathbb{R}^n$ appear. The case $n = 2$ reduces to the theory of the winding number. In this chapter we essentially follow the first part of the lecture on fixed point theorems [H4] by E. Heinz.

In Chapter 4 we develop the theory of holomorphic functions in one and several complex variables. Since we utilize the Stokes integral theorem, we easily attain the well-known theorems from the classical theory of functions in the Sections 2 and 3. In the subsequent paragraphs we also study solutions of the inhomogeneous Cauchy-Riemann differential equation, which has been completely investigated by L. Bers and I. N. Vekua (see [V]). In Section 6 we assemble statements on pseudoholomorphic functions, which are similar to holomorphic functions as far as the behavior at their zeroes is concerned. In Section 7 we prove the Riemannian mapping theorem with an extremal method due to Koebe and investigate in Section 8 the boundary behavior of conformal mappings. Furthermore, we consider the discontinuous behavior of Cauchy's integral across the boundary in Section 9 and solve a Dirichlet problem for plane harmonic mappings. In this chapter we have profited from the beautiful lecture [Gr] on complex analysis by H. Grauert.

Chapter 5 is devoted to the *Potential Theory in* $\mathbb{R}^n$. With the aid of the Gaussian integral theorem we investigate Poisson's differential equation in Section 1 and Section 2, and we establish an analyticity theorem. With Perron's method we solve the Dirichlet problem for Laplace's equation in Section 3. Starting with Poisson's integral representation, we develop the theory of spherical harmonic functions in $\mathbb{R}^n$; see Section 4 and Section 5. This theory was founded by Legendre, and we owe this elegant representation to G. Herglotz. In this chapter as well, I was able to profit decisively from the lecture [H2] on partial differential equations by my academic teacher, Professor Dr. E. Heinz in Göttingen.

In Chapter 6 we consider *Linear Partial Differential Equations in* $\mathbb{R}^n$. We prove the maximum principle for elliptic differential equations in Section 1 and apply this central tool on quasilinear, elliptic differential equations in Section 2 (compare the lecture [H6]). In Section 3 we turn to the heat equation and present the parabolic maximum-minimum principle. Then in Section 4, we study characteristic surfaces and establish an energy estimate for the wave equation. In Section 5 we solve the Cauchy initial value problem of the wave equation in $\mathbb{R}^n$ for the dimensions $n = 1, 3, 2$. With the aid of Abel's integral equation we solve this problem for all $n \geq 2$ in Section 6 (compare the lecture [H5]). Then we consider the inhomogeneous wave equation and an initial-boundary-value problem in Section 7. For parabolic and hyperbolic equations we recommend the textbooks [GuLe] and [J]. Finally, we classify the linear partial differential equations of second order in Section 8. We discover the Lorentz transformations as invariant transformations for the wave equation (compare [G]).

With Chapters 5 and 6, we intend to give a geometrically oriented introduction into the theory of partial differential equations, without assuming prior functional analytic knowledge.

In this second edition of our monograph *Partial Differential Equations 1*, we have carefully revised Volume 1 and added Section 9 on the Boundary Behavior of Cauchy's Integral in Chapter 4. We shall present a revised version of our book *Partial Differential Equations 2* as well, where we shall add a new chapter on *Boundary Value Problems from Differential Geometry.* The topics of the new Chapter 13 are listed in the table of contents for our enlarged second edition of Volume 2.

We follow the Total Order Code of the Universitext series, however, adapt this to the present extensive contents. Since we see our books as an entity, we count the chapters throughout our two volumes from 1-13. In each section individually, we count the equations and refer to them by a single number; when we refer to an equation in another section of the same chapter, say the m-th section, we have to add *Section* m; when we refer to an equation in the m-th section of another chapter, say the l-th chapter, we have to add *Section* m *in Chapter* l.

We assemble *definitions, theorems, propositions, examples* to the expression *environment*, which is borrowed from the underlying TEX-file. Individually in each section, we count these environments consecutively by the number n, and denote the n-th environment within the m-th section by *Environment* $m.n$. Thus we have attributed a pair of integers $m.n$ to all definitions, theorems, propositions, and examples, which is unique within each chapter and easy to find. Referring to these environments throughout both books, we proceed as described above for the equations.

We add Figure 1.1 – Figure 1.9 to our Volume 1 and Figure 2.1 – Figure 2.11 to our Volume 2, which mostly represent portraits of mathematicians. This small photo collection of some scientists, who have contributed to the theory of Partial Differential Equations, already shows that our area is situated in the center of modern mathematics and possesses profound interrelations with geometry and physics.

This textbook PARTIAL DIFFERENTIAL EQUATIONS has been developed from lectures that I have been giving in the Brandenburgische Technische Universität at Cottbus from the winter semester 1992/93 to the present semester. The monograph, in part, builds upon the lectures (see [H1] – [H6]) of Professor Dr. Dr.h.c. E. Heinz, whom I was fortunate to know as his student in Göttingen from 1971 to 1978 and as postdoctoral researcher in his *Oberseminar* from 1983 to 1989. As an assistant in Aachen from 1978 to 1983, I very much appreciated the elegant lecture cycles of Professor Dr. G. Hellwig (see [He1] – [He3]). Since my research fellowship at the University of Bonn in 1989/90, an intensive scientific collaboration with Professor Dr. Dr.h.c.mult. S. Hildebrandt has developed, which continues to this day (see [DHS] and [DHT2] with the list of references therein). All three of these excellent representatives of mathematics will forever have my sincere gratitude and my deep respect!

Here I gratefully acknowledge the valuable and profound advice of Priv.-Doz. Dr. Frank Müller (Universität Duisburg-Essen) for the original edition *Partielle Differentialgleichungen* as well as the indispensable and excellent assistance of Dipl.-Math. Michael Hilschenz (BTU Cottbus) for the present edition *Partial Differential Equations 1*. Furthermore, my sincere thanks are devoted to Mrs. C. Prescott (Berlin) improving the English style of this second edition. Moreover, I would like to thank cordially Herr Clemens Heine (Heidelberg) and Mr. Jörg Sixt (London) as well as Mrs. Lauren Stoney (London) and all the other members of Springer for their helpful collaboration and great confidence.

Cottbus, February 2012 *Friedrich Sauvigny*

Contents – Volume 1: Foundations and Integral Representations

Chapter 1 Differentiation and Integration on Manifolds .. 1

1 The Weierstraß Approximation Theorem 2
2 Parameter-invariant Integrals and Differential Forms 12
3 The Exterior Derivative of Differential Forms 23
4 The Stokes Integral Theorem for Manifolds 30
5 The Integral Theorems of Gauß and Stokes 39
6 Curvilinear Integrals 56
7 The Lemma of Poincaré 67
8 Co-derivatives and the Laplace-Beltrami Operator 72
9 Some Historical Notices to Chapter 1 89

Chapter 2 Foundations of Functional Analysis 91

1 Daniell's Integral with Examples 91
2 Extension of Daniell's Integral to Lebesgue's Integral 96
3 Measurable Sets 109
4 Measurable Functions 121
5 Riemann's and Lebesgue's Integral on Rectangles 134
6 Banach and Hilbert Spaces 140
7 The Lebesgue Spaces $L^p(X)$ 151
8 Bounded Linear Functionals on $L^p(X)$ and Weak Convergence 161
9 Some Historical Notices to Chapter 2 172

Chapter 3 Brouwer's Degree of Mapping 175

1 The Winding Number 175
2 The Degree of Mapping in $\mathbb{R}^n$ 184
3 Topological Existence Theorems 194
4 The Index of a Mapping 196
5 The Product Theorem 205
6 Theorems of Jordan-Brouwer 211

Chapter 4 Generalized Analytic Functions 215

1 The Cauchy-Riemann Differential Equation 215
2 Holomorphic Functions in $\mathbb{C}^n$ 219
3 Geometric Behavior of Holomorphic Functions in $\mathbb{C}$ 233
4 Isolated Singularities and the General Residue Theorem 242
5 The Inhomogeneous Cauchy-Riemann Differential Equation . . 255
6 Pseudoholomorphic Functions 266
7 Conformal Mappings 270
8 Boundary Behavior of Conformal Mappings 286
9 Behavior of Cauchy's Integral across the Boundary 296
10 Some Historical Notices to Chapter 4 303

Chapter 5 Potential Theory and Spherical Harmonics . . . 305

1 Poisson's Differential Equation in $\mathbb{R}^n$ 305
2 Poisson's Integral Formula with Applications 317
3 Dirichlet's Problem for the Laplace Equation in $\mathbb{R}^n$ 329
4 Theory of Spherical Harmonics in 2 Variables: Fourier Series . 342
5 Theory of Spherical Harmonics in n Variables 347

Chapter 6 Linear Partial Differential Equations in $\mathbb{R}^n$ 363

1 The Maximum Principle for Elliptic Differential Equations . . . 363
2 Quasilinear Elliptic Differential Equations 373
3 The Heat Equation 378
4 Characteristic Surfaces and an Energy Estimate 392
5 The Wave Equation in $\mathbb{R}^n$ for $n = 1, 3, 2$ 403
6 The Wave Equation in $\mathbb{R}^n$ for $n \geq 2$ 411
7 The Inhomogeneous Wave Equation and an Initial-boundary-value Problem 422
8 Classification, Transformation and Reduction of Partial Differential Equations 427
9 Some Historical Notices to the Chapters 5 and 6 436

References 439

Index 443

Contents – Volume 2: Functional Analytic Methods and Differential Geometric Applications

Chapter 7 Operators in Banach Spaces

1 Fixed Point Theorems
2 The Leray-Schauder Degree of Mapping
3 Fundamental Properties for the Degree of Mapping
4 Linear Operators in Banach Spaces
5 Some Historical Notices to the Chapters 3 and 7

Chapter 8 Linear Operators in Hilbert Spaces

1 Various Eigenvalue Problems
2 Singular Integral Equations
3 The Abstract Hilbert Space
4 Bounded Linear Operators in Hilbert Spaces
5 Unitary Operators
6 Completely Continuous Operators in Hilbert spaces
7 Spectral Theory for Completely Continuous Hermitian Operators
8 The Sturm-Liouville Eigenvalue Problem
9 Weyl's Eigenvalue Problem for the Laplace Operator
10 Some Historical Notices to Chapter 8

Chapter 9 Linear Elliptic Differential Equations

1 The Differential Equation $\Delta\phi(x,y) + p(x,y)\phi_x(x,y) + q(x,y)\phi_y(x,y) = r(x,y)$
2 The Schwarzian Integral Formula
3 The Riemann-Hilbert Boundary Value Problem
4 Potential-theoretic Estimates
5 Schauder's Continuity Method
6 Existence and Regularity Theorems
7 The Schauder Estimates
8 Some Historical Notices to Chapter 9

Chapter 10 Weak Solutions of Elliptic Differential Equations

1 Sobolev Spaces
2 Embedding and Compactness
3 Existence of Weak Solutions
4 Boundedness of Weak Solutions
5 Hölder Continuity of Weak Solutions
6 Weak Potential-theoretic Estimates
7 Boundary Behavior of Weak Solutions
8 Equations in Divergence Form
9 Green's Function for Elliptic Operators
10 Spectral Theory of the Laplace-Beltrami Operator
11 Some Historical Notices to Chapter 10

Chapter 11 Nonlinear Partial Differential Equations

1 The Fundamental Forms and Curvatures of a Surface
2 Two-dimensional Parametric Integrals
3 Quasilinear Hyperbolic Differential Equations and Systems of Second Order (Characteristic Parameters)
4 Cauchy's Initial Value Problem for Quasilinear Hyperbolic Differential Equations and Systems of Second Order
5 Riemann's Integration Method
6 Bernstein's Analyticity Theorem
7 Some Historical Notices to Chapter 11

Chapter 12 Nonlinear Elliptic Systems

1 Maximum Principles for the H-surface System
2 Gradient Estimates for Nonlinear Elliptic Systems
3 Global Estimates for Nonlinear Systems
4 The Dirichlet Problem for Nonlinear Elliptic Systems
5 Distortion Estimates for Plane Elliptic Systems
6 A Curvature Estimate for Minimal Surfaces
7 Global Estimates for Conformal Mappings with respect to Riemannian Metrics
8 Introduction of Conformal Parameters into a Riemannian Metric
9 The Uniformization Method for Quasilinear Elliptic Differential Equations
10 Some Historical Notices to Chapter 12

Chapter 13 Boundary Value Problems from Differential Geometry

1 The Dirichlet Problem for Graphs of Prescribed Mean Curvature
2 Winding Staircases and Stable Minimal Graphs on a Riemann Surface
3 Mixed Boundary Value Problem for Minimal Graphs over the Etale Plane
4 Solution of Plateau's Problem with Constant Mean Curvature
5 Closed Surfaces of Constant Mean Curvature
6 One-to-one Solutions of Lewy-Heinz Systems
7 On Weyl's Embedding Problem and Elliptic Monge-Ampère Equations
8 Some Historical Notices to Chapter 13

Chapter 1

Differentiation and Integration on Manifolds

In this chapter we lay the foundations for our treatise on partial differential equations. A detailed description for the contents of Chapter 1 is given in the Introduction to Volume 1 above. At first, we fix some familiar notations used throughout the two volumes of our textbook.

By the symbol $\mathbb{R}^n$ we denote the n-dimensional Euclidean space with the points $x = (x_1, \ldots, x_n)$ where $x_i \in \mathbb{R}$, and we define their modulus

$$|x| = \Big(\sum_{i=1}^{n} x_i^2\Big)^{\frac{1}{2}}.$$

In general, we denote open subsets in $\mathbb{R}^n$ by the symbol Ω. By the symbol $\overline{M}$ we indicate the topological closure and by $\overset{\circ}{M}$ the open kernel of a set $M \subset \mathbb{R}^n$. In the sequel, we shall use the following linear spaces of functions:

- $C^0(\Omega)$ ...... continuous functions on Ω
- $C^k(\Omega)$ ...... k-times continuously differentiable functions on Ω
- $C_0^k(\Omega)$ ...... k-times continuously differentiable functions f on Ω with the compact support $\operatorname{supp} f = \overline{\{x \in \Omega : f(x) \neq 0\}} \subset \Omega$
- $C^k(\overline{\Omega})$ ...... k-times continuously differentiable functions on Ω, whose derivatives up to the order k can be continuously extended onto the closure $\overline{\Omega}$
- $C_0^k(\Omega \cup \Theta)$.. k-times continuously differentiable functions f on Ω, whose derivatives up to the order k can be extended onto the closure $\overline{\Omega}$ continuously with the property $\operatorname{supp} f \subset \Omega \cup \Theta$
- $C_*^*(*, K)$ ... space of functions as above with values in $K = \mathbb{R}^n$ or $K = \mathbb{C}$.

Finally, we utilize the notations

- ∇u .... gradient $(u_{x_1}, \ldots, u_{x_n})$ of a function $u = u(x_1, \ldots, x_n) \in C^1(\mathbb{R}^n)$

F. Sauvigny, *Partial Differential Equations 1*, Universitext,
DOI 10.1007/978-1-4471-2981-3_1, © Springer-Verlag London 2012

Δu .... Laplace operator $\sum_{i=1}^{n} u_{x_i x_i}$ of a function $u \in C^2(\mathbb{R}^n)$

J_f .... functional determinant or *Jacobian* of a function $f : \mathbb{R}^n \to \mathbb{R}^n \in C^1(\mathbb{R}^n, \mathbb{R}^n)$.

1 The Weierstraß Approximation Theorem

Let $\Omega \subset \mathbb{R}^n$ with $n \in \mathbb{N}$ denote an open set and $f(x) \in C^k(\Omega)$ with $k \in \mathbb{N} \cup \{0\} =: \mathbb{N}_0$ a k-times continuously differentiable function. We intend to prove the following statement:
There exists a sequence of polynomials $p_m(x)$, $x \in \mathbb{R}^n$ for $m = 1, 2, \dots$ which converges on each compact subset $C \subset \Omega$ uniformly towards the function $f(x)$. Furthermore, all partial derivatives up to the order k of the polynomials p_m converge uniformly on C towards the corresponding derivatives of the function f. The coefficients of the polynomials p_m depend on the approximation, in general. If this were not the case, the function

$$f(x) = \begin{cases} \exp\left(-\dfrac{1}{x^2}\right), & x > 0 \\ 0, & x \le 0 \end{cases}$$

could be expanded into a power series. However, this leads to the evident contradiction:

$$0 \equiv \sum_{k=0}^{\infty} \frac{f^{(k)}(0)}{k!} x^k .$$

In the following Proposition, we introduce a *mollifier* which enables us to smooth systematically integrable functions.

Proposition 1.1. *We consider the following function to each $\varepsilon > 0$, namely*

$$\begin{aligned} K_\varepsilon(z) &:= \frac{1}{\sqrt{\pi\varepsilon}^{\,n}} \exp\left(-\frac{|z|^2}{\varepsilon}\right) \\ &= \frac{1}{\sqrt{\pi\varepsilon}^{\,n}} \exp\left(-\frac{1}{\varepsilon}\left(z_1^2 + \ldots + z_n^2\right)\right), \qquad z \in \mathbb{R}^n. \end{aligned}$$

Then this function $K_\varepsilon = K_\varepsilon(z)$ possesses the following properties:

1. *We have $K_\varepsilon(z) > 0$ for all $z \in \mathbb{R}^n$;*
2. *The condition $\int_{\mathbb{R}^n} K_\varepsilon(z)\, dz = 1$ holds true;*
3. *For each $\delta > 0$ we observe: $\lim_{\varepsilon \to 0+} \int_{|z| \ge \delta} K_\varepsilon(z)\, dz = 0$.*

Proof:

1. The exponential function is positive, and the statement is obvious.
2. We substitute $z = \sqrt{\varepsilon}x$ with $dz = \sqrt{\varepsilon}^n\, dx$ and calculate

$$\int\limits_{\mathbb{R}^n} K_\varepsilon(z)\, dz = \frac{1}{\sqrt{\pi\varepsilon}^n} \int\limits_{\mathbb{R}^n} \exp\left(-\frac{|z|^2}{\varepsilon}\right) dz$$

$$= \frac{1}{\sqrt{\pi}^n} \int\limits_{\mathbb{R}^n} \exp\Big(-|x|^2\Big)\, dx = \left(\frac{1}{\sqrt{\pi}} \int\limits_{-\infty}^{+\infty} \exp\Big(-t^2\Big)\, dt\right)^n = 1.$$

3. We utilize the substitution from part 2 of our proof and obtain

$$\int\limits_{|z|\geq\delta} K_\varepsilon(z)\, dz = \frac{1}{\sqrt{\pi}^n} \int\limits_{|x|\geq\delta/\sqrt{\varepsilon}} \exp\Big(-|x|^2\Big)\, dx \longrightarrow 0 \qquad \text{for} \quad \varepsilon \to 0+.$$

q.e.d.

Proposition 1.2. *Let us consider $f(x) \in C_0^0(\mathbb{R}^n)$ and additionally the function*

$$f_\varepsilon(x) := \int\limits_{\mathbb{R}^n} K_\varepsilon(y-x) f(y)\, dy, \qquad x \in \mathbb{R}^n$$

for $\varepsilon > 0$. Then we infer

$$\sup_{x\in\mathbb{R}^n} |f_\varepsilon(x) - f(x)| \longrightarrow 0 \quad \textit{for} \quad \varepsilon \to 0+,$$

and consequently the functions $f_\varepsilon(x)$ converge uniformly on the space $\mathbb{R}^n$ towards the function $f(x)$.

Proof: On account of its compact support, the function $f(x)$ is uniformly continuous on the space $\mathbb{R}^n$. The number $\eta > 0$ being given, we find a number $\delta = \delta(\eta) > 0$ such that

$$x, y \in \mathbb{R}^n, \ |x-y| \leq \delta \implies |f(x) - f(y)| \leq \eta.$$

Since f is bounded, we find a quantity $\varepsilon_0 = \varepsilon_0(\eta) > 0$ satisfying

$$2 \sup_{y\in\mathbb{R}^n} |f(y)| \int\limits_{|y-x|\geq\delta} K_\varepsilon(y-x)\, dy \leq \eta \quad \text{for all} \quad 0 < \varepsilon < \varepsilon_0.$$

We note that

$$|f_\varepsilon(x) - f(x)| = \Big| \int\limits_{\mathbb{R}^n} K_\varepsilon(y-x)\, f(y)\, dy - f(x) \int\limits_{\mathbb{R}^n} K_\varepsilon(y-x)\, dy \Big|$$

$$\leq \Big| \int\limits_{|y-x|\leq\delta} K_\varepsilon(y-x)\, \{f(y) - f(x)\}\, dy \Big|$$

$$+ \Big| \int\limits_{|y-x|\geq\delta} K_\varepsilon(y-x)\, \{f(y) - f(x)\}\, dy \Big|,$$

and we arrive at the following estimate for all points $x \in \mathbb{R}^n$ and all numbers $0 < \varepsilon < \varepsilon_0$, namely

$$\begin{aligned} |f_\varepsilon(x) - f(x)| &\leq \int\limits_{|y-x|\leq\delta} K_\varepsilon(y-x)\,|f(y) - f(x)|\,dy \\ &\quad + \int\limits_{|y-x|\geq\delta} K_\varepsilon(y-x)\,\{|f(y)| + |f(x)|\}\,dy \\ &\leq \eta + 2\sup_{y\in\mathbb{R}^n}|f(y)| \int\limits_{|y-x|\geq\delta} K_\varepsilon(y-x)\,dy \leq 2\eta. \end{aligned}$$

We summarize our considerations to

$$\sup_{x\in\mathbb{R}^n} |f_\varepsilon(x) - f(x)| \longrightarrow 0 \qquad \text{for} \quad \varepsilon \to 0+\,.$$

q.e.d.

In the sequel, we need

Proposition 1.3. (Partial integration in $\mathbb{R}^n$)
When the functions $f(x) \in C_0^1(\mathbb{R}^n)$ and $g(x) \in C^1(\mathbb{R}^n)$ are given, we infer

$$\int\limits_{\mathbb{R}^n} g(x)\frac{\partial}{\partial x_i} f(x)\,dx = -\int\limits_{\mathbb{R}^n} f(x)\frac{\partial}{\partial x_i} g(x)\,dx \quad \textit{for} \quad i = 1,\ldots,n.$$

Proof: On account of the property $f(x) \in C_0^1(\mathbb{R}^n)$, we find a radius $r > 0$ such that $f(x) = 0$ and $f(x)g(x) = 0$ is correct for all points $x \in \mathbb{R}^n$ with $|x_j| \geq r$ for one index $j \in \{1,\ldots,n\}$ at least. The fundamental theorem of differential- and integral-calculus yields

$$\begin{aligned} &\int\limits_{\mathbb{R}^n} \frac{\partial}{\partial x_i}\Big\{f(x)g(x)\Big\}\,dx \\ &= \int\limits_{-r}^{+r} \cdots \int\limits_{-r}^{+r} \left(\int\limits_{-r}^{+r} \frac{\partial}{\partial x_i}\Big\{f(x)g(x)\Big\}\,dx_i \right) dx_1 \ldots dx_{i-1}dx_{i+1}\ldots dx_n = 0. \end{aligned}$$

This implies

$$0 = \int\limits_{\mathbb{R}^n} \frac{\partial}{\partial x_i}\Big\{f(x)g(x)\Big\}\,dx = \int\limits_{\mathbb{R}^n} g(x)\frac{\partial}{\partial x_i} f(x)\,dx + \int\limits_{\mathbb{R}^n} f(x)\frac{\partial}{\partial x_i} g(x)\,dx.$$

q.e.d.

Proposition 1.4. *Let the function $f(x) \in C_0^k(\mathbb{R}^n, \mathbb{C})$ with $k \in \mathbb{N}_0$ be given. Then we have a sequence of polynomials with complex coefficients*

$$p_m(x) = \sum_{j_1,\dots,j_n=0}^{N(m)} c^{(m)}_{j_1\dots j_n} x_1^{j_1} \dots x_n^{j_n} \quad \text{for} \quad m = 1, 2, \dots$$

such that the limit relations

$$D^\alpha p_m(x) \longrightarrow D^\alpha f(x) \qquad \text{for} \quad m \to \infty, \quad |\alpha| \le k$$

are satisfied uniformly in each ball $B_R := \{x \in \mathbb{R}^n : |x| \le R\}$ *with the radius* $0 < R < +\infty$. *Here we define the differential operator* D^α *with* $\alpha = (\alpha_1, \dots, \alpha_n)$ *by*

$$D^\alpha := \frac{\partial^{|\alpha|}}{\partial x_1^{\alpha_1} \dots \partial x_n^{\alpha_n}}, \qquad |\alpha| := \alpha_1 + \dots + \alpha_n,$$

where $\alpha_1, \dots, \alpha_n \ge 0$ *represent nonnegative integers.*

Proof: We differentiate the function $f_\varepsilon(x)$ with respect to the variables x_i, and together with Proposition 1.3 we see

$$\begin{aligned}
\frac{\partial}{\partial x_i} f_\varepsilon(x) &= \int\limits_{\mathbb{R}^n} \left\{ \frac{\partial}{\partial x_i} K_\varepsilon(y - x) \right\} f(y)\, dy \\
&= -\int\limits_{\mathbb{R}^n} \left\{ \frac{\partial}{\partial y_i} K_\varepsilon(y - x) \right\} f(y)\, dy \\
&= \int\limits_{\mathbb{R}^n} K_\varepsilon(y - x) \frac{\partial}{\partial y_i} f(y)\, dy
\end{aligned}$$

for $i = 1, \dots, n$. By repeated application of this device, we arrive at

$$D^\alpha f_\varepsilon(x) = \int\limits_{\mathbb{R}^n} K_\varepsilon(y - x) D^\alpha f(y)\, dy, \quad |\alpha| \le k.$$

Here we note that $D^\alpha f(y) \in C_0^0(\mathbb{R}^n)$ holds true. Due to Proposition 1.2, the family of functions $D^\alpha f_\varepsilon(x)$ converges uniformly on the space $\mathbb{R}^n$ towards $D^\alpha f(x)$ - for all $|\alpha| \le k$ - when $\varepsilon \to 0+$ holds true. Now we choose the radius $R > 0$ such that $\operatorname{supp} f \subset B_R$ is valid. Taking the number $\varepsilon > 0$ as fixed, we consider the power series

$$K_\varepsilon(z) = \frac{1}{\sqrt{\pi\varepsilon}^n} \exp\left(-\frac{|z|^2}{\varepsilon}\right) = \frac{1}{\sqrt{\pi\varepsilon}^n} \sum_{j=0}^\infty \frac{1}{j!} \left(-\frac{|z|^2}{\varepsilon}\right)^j,$$

which converges uniformly in B_{2R}. Therefore, each number $\varepsilon > 0$ possesses an index $N_0 = N_0(\varepsilon, R)$ such that the polynomial

$$P_{\varepsilon,R}(z) := \frac{1}{\sqrt{\pi\varepsilon}^n} \sum_{j=0}^{N_0(\varepsilon,R)} \frac{1}{j!} \left(-\frac{z_1^2 + \dots + z_n^2}{\varepsilon}\right)^j$$

is subject to the following estimate:

$$\sup_{|z|\le 2R} |K_\varepsilon(z) - P_{\varepsilon,R}(z)| \le \varepsilon.$$

With the expression

$$\widetilde{f}_{\varepsilon,R}(x) := \int_{\mathbb{R}^n} P_{\varepsilon,R}(y-x) f(y)\,dy$$

we obtain a polynomial in the variables $x_1, \ldots, x_n$ - for each $\varepsilon > 0$. Furthermore, we deduce

$$D^\alpha \widetilde{f}_{\varepsilon,R}(x) = \int_{\mathbb{R}^n} P_{\varepsilon,R}(y-x) D^\alpha f(y)\,dy \quad \text{for all} \quad x \in \mathbb{R}^n, \quad |\alpha| \le k.$$

Now we arrive at the subsequent estimate for all $|\alpha| \le k$ and $|x| \le R$, namely

$$\begin{aligned}
|D^\alpha f_\varepsilon(x) - D^\alpha \widetilde{f}_{\varepsilon,R}(x)| &= \Big| \int_{|y|\le R} \Big\{K_\varepsilon(y-x) - P_{\varepsilon,R}(y-x)\Big\} D^\alpha f(y)\,dy \Big| \\
&\le \int_{|y|\le R} |K_\varepsilon(y-x) - P_{\varepsilon,R}(y-x)||D^\alpha f(y)|\,dy \\
&\le \varepsilon \int_{|y|\le R} |D^\alpha f(y)|\,dy.
\end{aligned}$$

Therefore, the polynomials $D^\alpha \widetilde{f}_{\varepsilon,R}(x)$ converge uniformly on B_R towards the derivatives $D^\alpha f(x)$. Choosing the null-sequence $\varepsilon = \frac{1}{m}$ with $m = 1, 2, \ldots$, we obtain an approximating sequence of polynomials $p_{m,R}(x) := \widetilde{f}_{\frac{1}{m},R}(x)$ in B_R, which is still dependent on the radius R. We take $r = 1, 2, \ldots$ and find polynomials $p_r = p_{m_r,r}$ satisfying

$$\sup_{x\in B_r} |D^\alpha p_r(x) - D^\alpha f(x)| \le \frac{1}{r} \quad \text{for all} \quad |\alpha| \le k.$$

The sequence p_r satisfies all the properties stated above. q.e.d.

We are now prepared to prove the fundamental

Theorem 1.5. (The Weierstraß approximation theorem)
Let $\Omega \subset \mathbb{R}^n$ denote an open set and $f(x) \in C^k(\Omega, \mathbb{C})$ a function with the degree of regularity $k \in \mathbb{N}_0$. Then we have a sequence of polynomials with complex coefficients of the degree $N(m) \in \mathbb{N}_0$, namely

$$f_m(x) = \sum_{j_1,\ldots,j_n=0}^{N(m)} c^{(m)}_{j_1\ldots j_n} x_1^{j_1} \cdot \ldots \cdot x_n^{j_n}, \qquad x \in \mathbb{R}^n, \quad m = 1, 2, \ldots,$$

such that the limit relations

$$D^\alpha f_m(x) \longrightarrow D^\alpha f(x) \qquad \text{for} \quad m \to \infty, \quad |\alpha| \le k$$

are satisfied uniformly on each compact set $C \subset \Omega$.

Proof: We consider a sequence $\Omega_1 \subset \Omega_2 \subset \ldots \subset \Omega$ of bounded open sets exhausting Ω. Here we have $\overline{\Omega_j} \subset \Omega_{j+1}$ for all indices j. Via the partition of unity (compare Theorem 1.8), we construct a sequence of functions $\phi_j(x) \in C_0^\infty(\Omega)$ satisfying $0 \le \phi_j(x) \le 1$, $x \in \Omega$ and $\phi_j(x) = 1$ on $\overline{\Omega}_j$ for $j = 1, 2, \ldots$. Then we observe the sequence of functions

$$f_j(x) := \begin{cases} f(x)\phi_j(x), & x \in \Omega \\ 0, & x \in \mathbb{R}^n \setminus \Omega \end{cases}$$

with the following properties:

$$f_j(x) \in C_0^k(\mathbb{R}^n) \quad \text{and} \quad D^\alpha f_j(x) = D^\alpha f(x), \qquad x \in \Omega_j, \quad |\alpha| \le k.$$

Due to Proposition 1.4, we find a polynomial $p_j(x)$ to each function $f_j(x)$ satisfying

$$\sup_{x \in \Omega_j} |D^\alpha p_j(x) - D^\alpha f_j(x)| = \sup_{x \in \Omega_j} |D^\alpha p_j(x) - D^\alpha f(x)| \le \frac{1}{j}, \qquad |\alpha| \le k,$$

since Ω_j is bounded. For a compact set $C \subset \Omega$ being given arbitrarily, we find an index $j_0 = j_0(C) \in \mathbb{N}$ such that the inclusion $C \subset \Omega_j$ for all $j \ge j_0(C)$ is correct. This implies

$$\sup_{x \in C} |D^\alpha p_j(x) - D^\alpha f(x)| \le \frac{1}{j}, \qquad j \ge j_0(C), \quad |\alpha| \le k.$$

When we consider the transition to the limit $j \to \infty$, we arrive at the statement

$$\sup_{x \in C} |D^\alpha p_j(x) - D^\alpha f(x)| \longrightarrow 0$$

for all $|\alpha| \le k$ and all compact subsets $C \subset \Omega$. q.e.d.

Theorem 1.5 above provides a uniform approximation by polynomials in the interior of the domain for the respective function. Continuous functions defined on compact sets can be uniformly approximated up to the boundary of the domain. Here we need the following

Theorem 1.6. (Tietze's extension theorem)
Let $C \subset \mathbb{R}^n$ *denote a compact set and* $f(x) \in C^0(C, \mathbb{C})$ *a continuous function defined on* C. *Then we have a continuous extension of* f *onto the whole space* $\mathbb{R}^n$ *which means: There exists a function* $g(x) \in C^0(\mathbb{R}^n, \mathbb{C})$ *satisfying*

$$f(x) = g(x) \qquad \text{for all points} \quad x \in C.$$

Proof:

1. We take $x \in \mathbb{R}^n$ and define the function

$$d(x) := \min_{y \in C} |y - x|,$$

which measures the distance of the point x to the set C. Since C is compact, we find to each point $x \in \mathbb{R}^n$ a point $\overline{y} \in C$ satisfying $|\overline{y} - x| = d(x)$. When $x_1, x_2 \in \mathbb{R}^n$ are chosen, we infer the following inequality for $\overline{y}_2 \in C$ with $|\overline{y}_2 - x_2| = d(x_2)$, namely

$$\begin{aligned} d(x_1) - d(x_2) &= \inf_{y \in C} \Big(|x_1 - y|) - |x_2 - \overline{y}_2| \Big) \\ &\leq |x_1 - \overline{y}_2| - |x_2 - \overline{y}_2| \\ &\leq |x_1 - x_2|. \end{aligned}$$

Interchanging the points x_1 and x_2, we obtain an analogous inequality and infer

$$|d(x_1) - d(x_2)| \leq |x_1 - x_2| \qquad \text{for all points} \quad x_1, x_2 \in \mathbb{R}^n.$$

In particular, the distance $d : \mathbb{R}^n \to \mathbb{R}$ represents a continuous function.

2. For $x \notin C$ and $a \in \mathbb{R}^n$, we consider the function

$$\varrho(x, a) := \max \left\{ 2 - \frac{|x - a|}{d(x)}, 0 \right\}.$$

The point a being fixed, the arguments above tell us that the function $\varrho(x, a)$ is continuous in $\mathbb{R}^n \setminus C$. Furthermore, we observe $0 \leq \varrho(x, a) \leq 2$ as well as

$$\begin{aligned} \varrho(x, a) &= 0 \qquad \text{for} \quad |a - x| \geq 2d(x), \\ \varrho(x, a) &\geq \frac{1}{2} \qquad \text{for} \quad |a - x| \leq \frac{3}{2}\, d(x). \end{aligned}$$

3. With $\left\{a^{(k)}\right\} \subset C$ let us choose a sequence of points which is dense in C. Since the function $f(x) : C \to \mathbb{C}$ is bounded, the series below

$$\sum_{k=1}^{\infty} 2^{-k} \varrho\left(x, a^{(k)}\right) f\left(a^{(k)}\right) \quad \text{and} \quad \sum_{k=1}^{\infty} 2^{-k} \varrho\left(x, a^{(k)}\right)$$

converge uniformly for all $x \in \mathbb{R}^n \setminus C$, and represent continuous functions in the variable x there. Furthermore, we observe

$$\sum_{k=1}^{\infty} 2^{-k} \varrho\left(x, a^{(k)}\right) > 0 \quad \text{for} \quad x \in \mathbb{R}^n \setminus C,$$

since each point $x \in \mathbb{R}^n \setminus C$ possesses at least one index k with $\varrho(x, a^{(k)}) > 0$. Therefore, the function

$$h(x) := \frac{\sum\limits_{k=1}^{\infty} 2^{-k} \varrho\Big(x, a^{(k)}\Big) f\Big(a^{(k)}\Big)}{\sum\limits_{k=1}^{\infty} 2^{-k} \varrho\Big(x, a^{(k)}\Big)} = \sum_{k=1}^{\infty} \varrho_k(x) f\Big(a^{(k)}\Big), \qquad x \in \mathbb{R}^n \setminus C,$$

is continuous. Here we have set

$$\varrho_k(x) := \frac{2^{-k} \varrho\Big(x, a^{(k)}\Big)}{\sum\limits_{k=1}^{\infty} 2^{-k} \varrho\Big(x, a^{(k)}\Big)} \quad \text{for} \quad x \in \mathbb{R}^n \setminus C.$$

We have the identity

$$\sum_{k=1}^{\infty} \varrho_k(x) \equiv 1, \quad x \in \mathbb{R}^n \setminus C.$$

4. Now we define the function

$$g(x) := \begin{cases} f(x),\ x \in C \\ h(x),\ x \in \mathbb{R}^n \setminus C \end{cases}.$$

We have still to show the continuity of g on ∂C. We have the following estimate for $z \in C$ and $x \notin C$:

$$\begin{aligned} |h(x) - f(z)| &= \left| \sum_{k=1}^{\infty} \varrho_k(x) \Big\{ f\Big(a^{(k)}\Big) - f(z) \Big\} \right| \\ &\leq \sum_{k:|a^{(k)}-x| \leq 2d(x)} \varrho_k(x) \left| f\Big(a^{(k)}\Big) - f(z) \right| \\ &\leq \sup_{a \in C\,:\,|a-x| \leq 2d(x)} |f(a) - f(z)| \\ &\leq \sup_{a \in C\,:\,|a-z| \leq 2d(x)+|x-z|} |f(a) - f(z)| \\ &\leq \sup_{a \in C\,:\,|a-z| \leq 3|x-z|} |f(a) - f(z)|. \end{aligned}$$

Since the function $f : C \to \mathbb{C}$ is uniformly continuous, we infer

$$\lim_{\substack{x \to z \\ x \notin C}} h(x) = f(z) \quad \text{for} \quad z \in \partial C \quad \text{and} \quad x \notin C.$$

q.e.d.

The assumption of compactness for the subset C is decisive in the theorem above. The function $f(x) = \sin(1/x)$, $x \in (0, \infty)$ namely cannot be continuously extended into the origin 0.

Theorem 1.5 and Theorem 1.6 together yield

Theorem 1.7. *Let $f(x) \in C^0(C, \mathbb{C})$ denote a continuous function on the compact set $C \subset \mathbb{R}^n$. To each quantity $\varepsilon > 0$, we then find a polynomial $p_\varepsilon(x)$ with the property*

$$|p_\varepsilon(x) - f(x)| \leq \varepsilon \quad \textit{for all points} \quad x \in C.$$

We shall construct smoothing functions which turn out to be extremely valuable in the sequel. At first, we easily show that the function

$$\psi(t) := \begin{cases} \exp\left(-\dfrac{1}{t}\right), & \text{if } t > 0 \\ 0, & \text{if } t \leq 0 \end{cases} \tag{1}$$

belongs to the regularity class $C^\infty(\mathbb{R})$. We take $R > 0$ arbitrarily and consider the function

$$\varphi_R(x) := \psi\Big(|x|^2 - R^2\Big), \qquad x \in \mathbb{R}^n. \tag{2}$$

Then we observe $\varphi_R \in C^\infty(\mathbb{R}^n, \mathbb{R})$. We have $\varphi_R(x) > 0$ if $|x| > R$ holds true, $\varphi_R(x) = 0$ if $|x| \leq R$ holds true, and therefore

$$\operatorname{supp}(\varphi_R) = \Big\{x \in \mathbb{R}^n \ : \ |x| \geq R\Big\}.$$

Furthermore, we develop the following function out of $\psi(t)$, namely

$$\varrho = \varrho(t) : \mathbb{R} \to \mathbb{R} \ \in C^\infty(\mathbb{R}) \qquad \text{via} \quad t \mapsto \varrho(t) := \psi(1-t)\psi(1+t). \tag{3}$$

This function is symmetric, which means $\varrho(-t) = \varrho(t)$ for all $t \in \mathbb{R}$. Furthermore, we see $\varrho(t) > 0$ for all $t \in (-1, 1)$, $\varrho(t) = 0$ for all else, and consequently

$$\operatorname{supp}(\varrho) = [-1, 1].$$

Finally, we define the following ball for $\xi \in \mathbb{R}^n$ and $\varepsilon > 0$, namely

$$B_\varepsilon(\xi) := \Big\{x \in \mathbb{R}^n \ : \ |x - \xi| \leq \varepsilon\Big\} \tag{4}$$

as well as the functions

$$\varphi_{\xi,\varepsilon}(x) := \varrho\left(\frac{|x-\xi|^2}{\varepsilon^2}\right), \quad x \in \mathbb{R}^n. \tag{5}$$

Then the regularity property $\varphi_{\xi,\varepsilon} \in C^\infty(\mathbb{R}^n, \mathbb{R})$ is valid, and we deduce $\varphi_{\xi,\varepsilon}(x) > 0$ for all $x \in \overset{\circ}{B}_\varepsilon(\xi)$ as well as $\varphi_{\xi,\varepsilon}(x) = 0$ if $|x - \xi| \geq \varepsilon$ holds true. This implies

$$\operatorname{supp}(\varphi_{\xi,\varepsilon}) = B_\varepsilon(\xi).$$

A fundamental principle of proof is presented in the next

Theorem 1.8. (Partition of unity)
Let $K \subset \mathbb{R}^n$ denote a compact set, and to each point $x \in K$ the symbol $\mathcal{O}_x \subset \mathbb{R}^n$ indicates an open set with $x \in \mathcal{O}_x$. Then we can select finitely many points $x^{(1)}, x^{(2)}, \ldots, x^{(m)} \in K$ with the associate number $m \in \mathbb{N}$ such that the covering

$$K \subset \bigcup_{\mu=1}^{m} \mathcal{O}_{x^{(\mu)}}$$

holds true. Furthermore, we find functions $\chi_\mu = \chi_\mu(x) : \mathcal{O}_{x^{(\mu)}} \to [0, +\infty)$ satisfying $\chi_\mu \in C_0^\infty(\mathcal{O}_{x^{(\mu)}})$ for $\mu = 1, \ldots, m$ such that the function

$$\chi(x) := \sum_{\mu=1}^{m} \chi_\mu(x), \qquad x \in \mathbb{R}^n \tag{6}$$

has the following properties:

(a) The regularity $\chi \in C_0^\infty(\mathbb{R}^n)$ holds true.
(b) We have $\chi(x) = 1$ for all $x \in K$.
(c) The inequality $0 \le \chi(x) \le 1$ is valid for all $x \in \mathbb{R}^n$.

Proof:

1. Since the set $K \subset \mathbb{R}^n$ is compact, we find a radius $R > 0$ such that $K \subset B := B_R(0)$ holds true. To each point $x \in B$ we now choose an open ball $\overset{\circ}{B}_{\varepsilon_x}(x)$ of radius $\varepsilon_x > 0$ such that $B_{\varepsilon_x}(x) \subset \mathcal{O}_x$ for $x \in K$ and $B_{\varepsilon_x}(x) \subset \mathbb{R}^n \setminus K$ for $x \in B \setminus K$ is satisfied. The system of sets $\left\{\overset{\circ}{B}_{\varepsilon_x}(x)\right\}_{x \in B}$ yields an open covering of the compact set B. According to the Heine-Borel covering theorem, finitely many open sets suffice to cover B, let us say

$$\overset{\circ}{B}_{\varepsilon_1}(x^{(1)}), \overset{\circ}{B}_{\varepsilon_2}(x^{(2)}), \ldots, \overset{\circ}{B}_{\varepsilon_m}(x^{(m)}), \overset{\circ}{B}_{\varepsilon_{m+1}}(x^{(m+1)}), \ldots \overset{\circ}{B}_{\varepsilon_{m+M}}(x^{(m+M)}) .$$

Here we observe $x^{(\mu)} \in K$ for $\mu = 1, 2, \ldots, m$ and $x^{(\mu)} \in B \setminus K$ for $\mu = m+1, \ldots, m+M$, defining $\varepsilon_\mu := \varepsilon_{x^{(\mu)}}$ for $\mu = 1, \ldots, m+M$.
With the aid of the function from (5), we now consider the nonnegative functions $\varphi_\mu(x) := \varphi_{x^{(\mu)}, \varepsilon_\mu}(x)$. We note that the following regularity properties hold true: $\varphi_\mu \in C_0^\infty(\mathcal{O}_{x^{(\mu)}})$ for $\mu = 1, \ldots, m$ and $\varphi_\mu \in C_0^\infty(\mathbb{R}^n \setminus K)$ for $\mu = m+1, \ldots, m+M$, respectively. Furthermore, we define $\varphi_{m+M+1}(x) := \varphi_R(x)$, where we introduced φ_R already in (2). Obviously, we arrive at the statement

$$\sum_{\mu=1}^{m+M+1} \varphi_\mu(x) > 0 \qquad \text{for all} \quad x \in \mathbb{R}^n.$$

2. Now we define the functions χ_μ due to

$$\chi_\mu(x) := \Big[\sum_{\mu=1}^{m+M+1} \varphi_\mu(x)\Big]^{-1} \varphi_\mu(x), \quad x \in \mathbb{R}^n$$

for $\mu = 1, \ldots, m + M + 1$. The functions χ_μ and φ_μ belong to the same classes of regularity, and we observe additionally

$$\sum_{\mu=1}^{m+M+1} \chi_\mu(x) = \Big[\sum_{\mu=1}^{m+M+1} \varphi_\mu(x)\Big]^{-1} \sum_{\mu=1}^{m+M+1} \varphi_\mu(x) \equiv 1 \quad \text{for all} \quad x \in \mathbb{R}^n.$$

The properties (a), (b), and (c) of the function $\chi(x) = \sum_{\mu=1}^{m} \chi_\mu(x)$ are directly inferred from the construction above. q.e.d.

Definition 1.9. *We name the functions $\chi_1, \chi_2, \ldots, \chi_m$ from Theorem 1.8 a* partition of unity subordinate to the open covering $\{\mathcal{O}_x\}_{x \in K}$ of the compact set K.

2 Parameter-invariant Integrals and Differential Forms

In the basic lectures of analysis the following fundamental result is established.

Theorem 2.1. (Transformation formula for multiple integrals)
Let $\Omega, \Theta \subset \mathbb{R}^n$ denote two open sets, where we take $n \in \mathbb{N}$. Furthermore, let $y = (y_1(x_1, \ldots, x_n), \ldots, y_n(x_1, \ldots, x_n)) : \Omega \to \Theta$ denote a bijective mapping of the class $C^1(\Omega, \mathbb{R}^n)$ satisfying

$$J_y(x) := det\Big(\frac{\partial y_i(x)}{\partial x_j}\Big)_{i,j=1,\ldots,n} \neq 0 \quad \textit{for all} \quad x \in \Omega.$$

Let the function $f = f(y) : \Theta \to \mathbb{R} \in C^0(\Theta)$ be given with the property

$$\int_\Theta |f(y)|\, dy < +\infty$$

for the improper Riemannian integral of $|f|$. Then we have the transformation formula

$$\int_\Theta f(y)\, dy = \int_\Omega f(y(x))\, |J_y(x)|\, dx.$$

In the sequel, we shall integrate differential forms over m-dimensional surfaces in $\mathbb{R}^n$.

Definition 2.2. *Let the open set $T \subset \mathbb{R}^m$ with $m \in \mathbb{N}$ constitute the* parameter domain. *Furthermore, the symbol*

$$X(t) = \begin{pmatrix} x_1(t_1, \ldots, t_m) \\ \vdots \\ x_n(t_1, \ldots, t_m) \end{pmatrix} : T \longrightarrow \mathbb{R}^n \in C^k(T, \mathbb{R}^n)$$

represents a mapping - with $k, n \in \mathbb{N}$ and $m \leq n$ - whose functional matrix

$$\partial X(t) = \Big(X_{t_1}(t), \ldots, X_{t_m}(t)\Big), \quad t \in T$$

has the rank m for all $t \in T$. Then we call X a parametrized regular surface *with the* parametric representation $X(t) : T \to \mathbb{R}^n$.
When $X : T \to \mathbb{R}^n$ and $\widetilde{X} : \widetilde{T} \to \mathbb{R}^n$ are two parametric representations, we call them equivalent *if there exists a topological mapping*

$$t = t(s) = \Big(t_1(s_1, \ldots, s_m), \ldots, t_m(s_1, \ldots, s_m)\Big) : \widetilde{T} \longrightarrow T \in C^k(\widetilde{T}, T)$$

with the following properties:

1. $J(s) := \dfrac{\partial(t_1, \ldots, t_m)}{\partial(s_1, \ldots, s_m)}(s) = \begin{vmatrix} \frac{\partial t_1}{\partial s_1}(s) & \ldots & \frac{\partial t_1}{\partial s_m}(s) \\ \vdots & & \vdots \\ \frac{\partial t_m}{\partial s_1}(s) & \ldots & \frac{\partial t_m}{\partial s_m}(s) \end{vmatrix} > 0 \qquad$ *for all* $s \in \widetilde{T}$;
2. $\widetilde{X}(s) = X\Big(t(s)\Big)$ *for all* $s \in \widetilde{T}$.

We say that $\widetilde{X}$ originates from X by an orientation-preserving reparametrization. *The equivalence class $[X]$ consisting of all those parametric representations which are equivalent to X is named an* open, oriented, m-dimensional, regular surface of the class C^k in $\mathbb{R}^n$. *We name a surface* embedded in the space $\mathbb{R}^n$ *if additionally the mapping $X : T \to \mathbb{R}^n$ is injective.*

Example 2.3. (Curves in $\mathbb{R}^n$)
On the interval $T = (a, b) \subset \mathbb{R}$ we consider the mapping

$$X = X(t) = \Big(x_1(t), \ldots, x_n(t)\Big) \in C^1(T, \mathbb{R}^n), \qquad t \in T$$

satisfying

$$|X'(t)| = \sqrt{\{x_1'(t)\}^2 + \ldots + \{x_n'(t)\}^2} > 0 \quad \text{for all} \quad t \in T.$$

Then the integral

$$L(X) = \int_a^b |X'(t)|\, dt$$

determines the *arc length of the curve* $X = X(t)$.

Example 2.4. (Classical surfaces in $\mathbb{R}^3$)
When $T \subset \mathbb{R}^2$ denotes an open parameter domain, we consider the Gaussian surface representation

$$X(u,v) = \Big(x(u,v), y(u,v), z(u,v)\Big) \,:\, T \longrightarrow \mathbb{R}^3 \in C^1(T,\mathbb{R}^3).$$

The vector in the direction of the *normal* to the surface is given by

$$\begin{aligned} X_u \wedge X_v &= \left(\frac{\partial(y,z)}{\partial(u,v)}, \frac{\partial(z,x)}{\partial(u,v)}, \frac{\partial(x,y)}{\partial(u,v)}\right) \\ &= (y_u z_v - z_u y_v, z_u x_v - x_u z_v, x_u y_v - x_v y_u). \end{aligned}$$

The *unit normal vector* to the surface X is defined by the formula

$$N(u,v) := \frac{X_u \wedge X_v}{|X_u \wedge X_v|},$$

and we note that

$$|N(u,v)| = 1, \quad N(u,v) \cdot X_u(u,v) = N(u,v) \cdot X_v(u,v) = 0 \quad \text{for all} \quad (u,v) \in T.$$

Via the integral

$$A(X) := \iint\limits_T |X_u \wedge X_v| \, dudv$$

we determine the *area of the surface* $X = X(u,v)$. We evaluate

$$|X_u \wedge X_v|^2 = (X_u \wedge X_v) \cdot (X_u \wedge X_v) = |X_u|^2 |X_v|^2 - (X_u \cdot X_v)^2$$

such that

$$A(X) = \iint\limits_T \sqrt{|X_u|^2 |X_v|^2 - (X_u \cdot X_v)^2} \, dudv$$

follows.

Example 2.5. (Hypersurfaces in $\mathbb{R}^n$)
Let $X : T \to \mathbb{R}^n$ denote a regular surface - defined on the parameter domain $T \subset \mathbb{R}^{n-1}$. The $(n-1)$ vectors $X_{t_1}, \ldots, X_{t_{n-1}}$ are linearly independent for all $t \in T$; and they span the *tangential space to the surface* at the point $X(t) \in \mathbb{R}^n$. Now we shall construct the *unit normal vector* $\nu(t) \in \mathbb{R}^n$. Therefore, we require

$$|\nu(t)| = 1 \quad \text{and} \quad \nu(t) \cdot X_{t_k}(t) = 0 \quad \text{for all} \quad k = 1, \ldots, n-1$$

as well as

$$\det\Big(X_{t_1}(t), \ldots, X_{t_{n-1}}(t), \nu(t)\Big) > 0 \quad \text{for all} \quad t \in T.$$

Consequently, the vectors $X_{t_1}, \ldots, X_{t_{n-1}}$ and ν constitute a positive-oriented n-frame. In this context we define the functions

$$D_i(t) := (-1)^{n+i} \frac{\partial(x_1, x_2, \ldots, x_{i-1}, x_{i+1}, \ldots, x_n)}{\partial(t_1, \ldots, t_{n-1})}, \quad i = 1, \ldots, n.$$

Then we obtain the identity

$$\begin{vmatrix} \frac{\partial x_1}{\partial t_1} & \cdots & \frac{\partial x_n}{\partial t_1} \\ \vdots & & \vdots \\ \frac{\partial x_1}{\partial t_{n-1}} & \cdots & \frac{\partial x_n}{\partial t_{n-1}} \\ \lambda_1 & \cdots & \lambda_n \end{vmatrix} = \sum_{i=1}^{n} \lambda_i D_i \qquad \text{for all} \quad \lambda_1, \ldots, \lambda_n \in \mathbb{R}.$$

Now we introduce the *unit normal vector*

$$\nu(t) = \Big(\nu_1(t), \ldots, \nu_n(t)\Big) = \frac{1}{\sqrt{\sum_{j=1}^{n} (D_j(t))^2}} \Big(D_1(t), \ldots, D_n(t)\Big), \qquad t \in T.$$

Evidently, the equation $|\nu(t)| = 1$ holds true and we calculate

$$\sum_{i=1}^{n} D_i \frac{\partial x_i}{\partial t_j} = \begin{vmatrix} \frac{\partial x_1}{\partial t_1} & \cdots & \frac{\partial x_n}{\partial t_1} \\ \vdots & & \vdots \\ \frac{\partial x_1}{\partial t_{n-1}} & \cdots & \frac{\partial x_n}{\partial t_{n-1}} \\ \frac{\partial x_1}{\partial t_j} & \cdots & \frac{\partial x_n}{\partial t_j} \end{vmatrix} = 0, \qquad 1 \leq j \leq n-1.$$

This implies the orthogonality relation $X_{t_j}(t) \cdot \nu(t) = 0$ for all $t \in T$ and $j = 1, \ldots, n-1$. The *surface element of the hypersurface in* $\mathbb{R}^n$ is given by

$$\begin{aligned} d\sigma &:= \begin{vmatrix} \frac{\partial x_1}{\partial t_1} & \cdots & \frac{\partial x_n}{\partial t_1} \\ \vdots & & \vdots \\ \frac{\partial x_1}{\partial t_{n-1}} & \cdots & \frac{\partial x_n}{\partial t_{n-1}} \\ \nu_1 & \cdots & \nu_n \end{vmatrix} dt_1 \ldots dt_{n-1} \\ &= \sum_{j=1}^{n} \nu_j D_j \, dt_1 \ldots dt_{n-1} \\ &= \sqrt{\sum_{j=1}^{n} (D_j(t))^2} \, dt_1 \ldots dt_{n-1}. \end{aligned}$$

Consequently, the surface area of X is determined by the improper integral

$$A(X) := \int_T \sqrt{\sum_{j=1}^{n} (D_j(t))^2}\, dt.$$

Example 2.6. An open set $\Omega \subset \mathbb{R}^n$ can be seen as a surface in $\mathbb{R}^n$ - via the mapping

$$X(t) := t, \quad \text{with} \quad t \in T \quad \text{and} \quad T := \Omega \subset \mathbb{R}^n.$$

Example 2.7. (An m-dimensional surface in $\mathbb{R}^n$)
Let $X(t) : T \to \mathbb{R}^n$ denote a surface with $T \subset \mathbb{R}^m$ as its parameter domain and the dimensions $1 \le m \le n$. By the symbols

$$g_{ij}(t) := X_{t_i} \cdot X_{t_j} \quad \text{for} \quad i, j = 1, \ldots, m$$

we define the *metric tensor* of the surface X. Furthermore, we call

$$g(t) := \det \Big(g_{ij}(t) \Big)_{i,j=1,\ldots,m}$$

its *Gramian determinant.* We complete the system $\{X_{t_i}\}_{i=1,\ldots,m}$ in $\mathbb{R}^n$ at each point $X(t)$ by the vectors ξ_j with $j = 1, \ldots, n-m$ such that the following properties are valid:

(a) We have $\xi_j \cdot \xi_k = \delta_{jk}$ for all $j, k = 1, \ldots, n-m$;
(b) The relations $X_{t_i} \cdot \xi_j = 0$ for $i = 1, \ldots, m$ and $j = 1, \ldots, n-m$ hold true;
(c) The condition $\det \Big(X_{t_1}, \ldots, X_{t_m}, \xi_1, \ldots, \xi_{n-m} \Big) > 0$ is correct.

Then we determine the surface element as follows:

$$\begin{aligned} d\sigma(t) &= \det \Big(X_{t_1}, \ldots, X_{t_m}, \xi_1, \ldots, \xi_{n-m} \Big)\, dt_1 \ldots dt_m \\ &= \sqrt{\det \Big\{ (X_{t_1}, \ldots, \xi_{n-m})^t \circ (X_{t_1}, \ldots, \xi_{n-m}) \Big\}}\, dt_1 \ldots dt_m \\ &= \sqrt{\det \Big(g_{ij}(t) \Big)_{i,j=1,\ldots,m}}\, dt_1 \ldots dt_m \\ &= \sqrt{g(t)}\, dt_1 \ldots dt_m. \end{aligned}$$

In order to evaluate our surface element via the Jacobi matrix $\partial X(t)$, we need the following

Proposition 2.8. *Let A and B denote two $n \times m$-matrices, where $m \le n$ holds true. For the numbers $1 \le i_1 < \ldots < i_m \le n$, let $A_{i_1 \ldots i_m}$ define the matrix consisting of those rows with the indices $i_1, \ldots, i_m$ from the matrix A.*

Correspondingly, we define the submatrices of the matrix B. Then we have the identity

$$det(A^t \circ B) = \sum_{1 \le i_1 < \ldots < i_m \le n} \det A_{i_1 \ldots i_m} \det B_{i_1 \ldots i_m}.$$

Proof: We fix A and show that the identity above holds true for all matrices B.

1. When we consider the unit vectors $e_1, \ldots, e_n$ as columns in $\mathbb{R}^n$, the formula above holds true for all $B = (e_{j_1}, \ldots, e_{j_m})$ with $j_1, \ldots, j_m \in \{1, \ldots, n\}$, at first.
2. When the formula above holds true for the matrix $B = (b_1, \ldots, b_m)$, this remains true for the matrix $B' = (b_1, \ldots, \lambda b_i, \ldots, b_m)$.
3. When we have our formula for the matrices $B' = (b_1, \ldots, b_i', \ldots, b_m)$ and $B'' = (b_1, \ldots, b_i'', \ldots, b_m)$, this remains true for the matrix $B = (b_1, \ldots, b_i' + b_i'', \ldots, b_m)$.

q.e.d.

Corollary: Given the $n \times m$-matrix A, we have the identity

$$\det (A^t \circ A) = \sum_{1 \le i_1 < \ldots < i_m \le n} (\det A_{i_1 \ldots i_m})^2.$$

We write the metric tensor in the form

$$\Big(g_{ij}(t)\Big)_{i,j=1,\ldots,m} = \partial X(t)^t \circ \partial X(t)$$

with the functional matrix $\partial X(t) = \Big(X_{t_1}(t), \ldots, X_{t_m}(t)\Big)$, and we deduce

$$\begin{aligned} g(t) &= \det \Big(g_{ij}(t)\Big)_{i,j=1,\ldots,m} \\ &= \sum_{1 \le i_1 < \ldots < i_m \le n} \left(\frac{\partial(x_{i_1}, \ldots, x_{i_m})}{\partial(t_1, \ldots, t_m)}(t)\right)^2. \end{aligned}$$

Therefore, the *surface element* satisfies

$$\begin{aligned} d\sigma(t) &= \sqrt{g(t)}\, dt_1 \ldots dt_m \\ &= \sqrt{\sum_{1 \le i_1 < \ldots < i_m \le n} \left(\frac{\partial(x_{i_1}, \ldots, x_{i_m})}{\partial(t_1, \ldots, t_m)}(t)\right)^2}\, dt_1 \ldots dt_m. \end{aligned}$$

Definition 2.9. *The* surface area *of an open, oriented, m-dimensional, regular C^1-surface in $\mathbb{R}^n$ with the parametric representation $X(t) : T \to \mathbb{R}^n$ is given by the improper Riemannian integral*

$$A(X) := \int\limits_T \sqrt{\sum_{1 \le i_1 < \ldots < i_m \le n} \left(\frac{\partial(x_{i_1}, \ldots, x_{i_m})}{\partial(t_1, \ldots, t_m)} \right)^2} \, dt_1 \ldots dt_m.$$

Here the parameter domain $T \subset \mathbb{R}^m$ is open and the dimensions $1 \le m \le n$ are prescribed. If $A(X) < +\infty$ is valid, the surface $[X]$ possesses finite area.

Remarks:

1. With the aid of the transformation formula for multiple integrals, we immediately verify that the value of our surface area is independent of the parametric representation.
2. In the case $m = 1$, we obtain by $A(X)$ the arc length of the curve $X : T \to \mathbb{R}^n$. The case $m = 2$ and $n = 3$ reduces to the classical area of a surface X in $\mathbb{R}^3$. In the case $m = n - 1$ we evaluate the area of hypersurfaces in $\mathbb{R}^n$.

In physics and geometry, we often meet with integrals which only depend on the m-dimensional surface and which are independent of their parametric representation. In this way, we are invited to consider integrals over so-called differential forms.

Definition 2.10. *On the open set $\mathcal{O} \subset \mathbb{R}^n$, let the functions $a_{i_1 \ldots i_m} \in C^k(\mathcal{O})$ with $i_1, \ldots, i_m \in \{1, \ldots, n\}$ and $1 \le m \le n$ be given; where $k \in \mathbb{N}_0$ holds true. Now we define the set*

$$\mathcal{F} := \Big\{ X \mid X : T \to \mathbb{R}^n \text{ is a regular, oriented, } m\text{-dimensional surface with finite area such that } X(T) \subset\subset \mathcal{O} \Big\}.$$

By a differential form of the degree m in the class $C^k(\mathcal{O})$, *namely*

$$\omega := \sum_{i_1, \ldots, i_m = 1}^{n} a_{i_1 \ldots i_m}(x) \, dx_{i_1} \wedge \ldots \wedge dx_{i_m},$$

*or briefly an m-*form *of the class $C^k(\mathcal{O})$, we comprehend the function $\omega : \mathcal{F} \to \mathbb{R}$ defined as follows:*

$$\omega(X) := \int\limits_T \sum_{i_1, \ldots, i_m = 1}^{n} a_{i_1 \ldots i_m}(X(t)) \frac{\partial(x_{i_1}, \ldots, x_{i_m})}{\partial(t_1, \ldots, t_m)} \, dt_1 \ldots dt_m, \qquad X \in \mathcal{F}.$$

Remark:

1. We abbreviate $A \subset\subset \mathcal{O}$, if the set $\overline{A} \subset \mathbb{R}^n$ is compact and $\overline{A} \subset \mathcal{O}$ holds true.
2. Since the coefficient functions $a_{i_1 \ldots i_m}(X(t))$, $t \in T$ are bounded and the surface has finite area, the integral above converges absolutely.

3. When two differential symbols

$$\omega = \sum_{i_1,\ldots,i_m=1}^{n} a_{i_1\ldots i_m}(x)\, dx_{i_1} \wedge \ldots \wedge dx_{i_m}$$

and

$$\widetilde{\omega} = \sum_{i_1,\ldots,i_m=1}^{n} \widetilde{a}_{i_1\ldots i_m}(x)\, dx_{i_1} \wedge \ldots \wedge dx_{i_m}$$

are given, we introduce an equivalence relation between them as follows:

$$\omega \sim \widetilde{\omega} \quad \Longleftrightarrow \quad \omega(X) = \widetilde{\omega}(X) \quad \text{for all} \quad X \in \mathcal{F}.$$

Therefore, we comprehend a differential form as an *equivalence class* of differential symbols, where we choose a representative to characterize this differential form.

4. When $X, \widetilde{X} \in \mathcal{F}$ are two equivalent representations of the surface $[X]$, we observe

$$\begin{aligned}
\omega(\widetilde{X}) &= \int\limits_{\widetilde{T}} \sum_{i_1,\ldots,i_m=1}^{n} a_{i_1\ldots i_m}\Big(\widetilde{X}(s)\Big) \frac{\partial(\widetilde{x}_{i_1},\ldots,\widetilde{x}_{i_m})}{\partial(s_1,\ldots,s_m)}\, ds_1\ldots ds_m \\
&= \int\limits_{\widetilde{T}} \sum_{i_1,\ldots,i_m=1}^{n} a_{i_1\ldots i_m}\Big(X(t(s))\Big) \frac{\partial(x_{i_1},\ldots,x_{i_m})}{\partial(t_1,\ldots,t_m)} \frac{\partial(t_1,\ldots,t_m)}{\partial(s_1,\ldots,s_m)}\, ds_1\ldots ds_m \\
&= \int\limits_{T} \sum_{i_1,\ldots,i_m=1}^{n} a_{i_1\ldots i_m}\Big(X(t)\Big) \frac{\partial(x_{i_1},\ldots,x_{i_m})}{\partial(t_1,\ldots,t_m)}\, dt_1\ldots dt_m \\
&= \omega(X).
\end{aligned}$$

Therefore, ω is a mapping which is defined on the equivalence classes of the oriented surfaces $[X]$ with $X \in \mathcal{F}$.

5. An orientation-reversing parametric transformation $t = t(s)$ with $J(s) < 0$, $s \in \widetilde{T}$ induces the change of sign: $\omega(\widetilde{X}) = -\omega(X)$.

Definition 2.11. *A* 0-form *of the class $C^k(\mathcal{O})$ is simply a function $f(x) \in C^k(\mathcal{O})$ and more precisely*

$$\omega = f(x), \qquad x \in \mathcal{O}.$$

When $1 \le m \le n$ is fixed, we name

$$\beta^m := dx_{i_1} \wedge \ldots \wedge dx_{i_m}, \qquad 1 \le i_1,\ldots,i_m \le n$$

a basic m-form.

Definition 2.12. *Let $\omega, \omega_1, \omega_2$ represent three m-forms of the class $C^0(\mathcal{O})$ and choose $c \in \mathbb{R}$. Then we define the differential forms $c\omega$ and $\omega_1 + \omega_2$ by the prescription*

$$(c\omega)(X) := c\omega(X) \qquad \text{for all} \quad X \in \mathcal{F}$$

and

$$(\omega_1 + \omega_2)(X) := \omega_1(X) + \omega_2(X) \qquad \text{for all} \quad X \in \mathcal{F}$$

respectively.

The m-dimensional differential forms constitute a vector space with the null-element

$$o(X) = 0 \qquad \text{for all} \quad X \in \mathcal{F}.$$

Definition 2.13. (Exterior product of differential forms)
Let the differential forms

$$\omega_1 = \sum_{1 \le i_1, \dots, i_l \le n} a_{i_1 \dots i_l}(x)\, dx_{i_1} \wedge \dots \wedge dx_{i_l}$$

of degree l and

$$\omega_2 = \sum_{1 \le j_1, \dots, j_m \le n} b_{j_1 \dots j_m}(x)\, dx_{j_1} \wedge \dots \wedge dx_{j_m}$$

of degree m in the class $C^k(\mathcal{O})$ with $k \in \mathbb{N}_0$ be given. Then we define the exterior product *of ω_1 and ω_2 as the $(l+m)$-form*

$$\omega = \omega_1 \wedge \omega_2 := \sum_{1 \le i_1, \dots, i_l, j_1, \dots, j_m \le n} a_{i_1 \dots i_l}(x) b_{j_1 \dots j_m}(x)\, dx_{i_1} \wedge \dots \wedge dx_{i_l} \wedge dx_{j_1} \wedge \dots \wedge dx_{j_m}$$

of the class $C^k(\mathcal{O})$.

Remarks:

1. Arbitrary differential forms $\omega_1, \omega_2, \omega_3$ are subject to the associative law
$$(\omega_1 \wedge \omega_2) \wedge \omega_3 = \omega_1 \wedge (\omega_2 \wedge \omega_3).$$
2. When two l-forms ω_1, ω_2 and one m-form ω_3 are given, we have the distributive law
$$(\omega_1 + \omega_2) \wedge \omega_3 = \omega_1 \wedge \omega_3 + \omega_2 \wedge \omega_3.$$
3. The alternating character of the determinant reveals
$$dx_{i_1} \wedge \dots \wedge dx_{i_l} = \operatorname{sign}(\pi)\, dx_{i_{\pi(1)}} \wedge \dots \wedge dx_{i_{\pi(l)}}.$$
Here the symbol $\pi : \{1, \dots, l\} \to \{1, \dots, l\}$ denotes a permutation with $\operatorname{sign}(\pi)$ as its sign.

4. In particular, when the two indices i_{j_1} and i_{j_2} coincide, we deduce

$$dx_{i_1} \wedge \ldots \wedge dx_{i_l} = 0.$$

Therefore, each m-form in $\mathbb{R}^n$ with the degree $m > n$ vanishes identically.

5. An l-form ω_1 and an m-form ω_2 are subject to the commutator relation

$$\omega_1 \wedge \omega_2 = (-1)^{lm} \omega_2 \wedge \omega_1.$$

Therefore, the exterior product is not commutative.

6. We can represent each m-form in the following way:

$$\omega = \sum_{1 \le i_1 < \ldots < i_m \le n} a_{i_1 \ldots i_m}(x)\, dx_{i_1} \wedge \ldots \wedge dx_{i_m}.$$

The basic m-forms $dx_{i_1} \wedge \ldots \wedge dx_{i_m}$, $1 \le i_1 < \ldots < i_m \le n$ constitute a basis for the space of all differential forms, with coefficient functions in the class $C^k(\mathcal{O})$, where $k \in \mathbb{N}_0$ holds true.

Definition 2.14. *Let the symbol*

$$\omega = \sum_{1 \le i_1 < \ldots < i_m \le n} a_{i_1 \ldots i_m}(x)\, dx_{i_1} \wedge \ldots \wedge dx_{i_m}, \quad x \in \mathcal{O}$$

denote a continuous differential form on the open set $\mathcal{O} \subset \mathbb{R}^n$, with $1 \le m \le n$ being fixed. Then we define the improper Riemannian integral of the differential form ω *over the surface $[X] \subset \mathcal{O}$ via*

$$\int_{[X]} \omega := \int_T \sum_{1 \le i_1 < \ldots < i_m \le n} a_{i_1 \ldots i_m}\Big(X(t)\Big) \frac{\partial(x_{i_1}, \ldots, x_{i_m})}{\partial(t_1, \ldots, t_m)}\, dt_1 \ldots dt_m,$$

if ω is absolutely integrable over X *and consequently*

$$\int_{[X]} |\omega| := \int_T \Big| \sum_{1 \le i_1 < \ldots < i_m \le n} a_{i_1 \ldots i_m}\Big(X(t)\Big) \frac{\partial(x_{i_1}, \ldots, x_{i_m})}{\partial(t_1, \ldots, t_m)} \Big|\, dt_1 \ldots dt_m$$
$$< +\infty$$

is satisfied.

Remark: With the aid of the transformation formula, we show that these integrals are independent of the choice of the representatives for the surface. Therefore, we are allowed to write

$$\int_{[X]} |\omega| = \int_X |\omega|, \quad \int_{[X]} \omega = \int_X \omega.$$

Example 2.15. (Curvilinear integrals)
Let $a(x) = \Big(a_1(x_1,\dots,x_n),\dots,a_n(x_1,\dots,x_n)\Big)$ denote a continuous vector-field and

$$\omega = \sum_{i=1}^{n} a_i(x)\,dx_i$$

the associate 1-form or *Pfaffian form*. Furthermore, let

$$X(t) = \Big(x_1(t),\dots,x_n(t)\Big) \; : \; T \to \mathbb{R}^n \in C^1(T)$$

represent a regular C^1-curve defined on the parameter interval $T = (a,b)$. Then we observe

$$\int_X \omega = \int_a^b \left(\sum_{i=1}^{n} a_i\Big(X(t)\Big)x_i'(t)\right)\,dt.$$

We shall investigate curvilinear integrals in Section 6 more intensively.

Example 2.16. (Surface integrals)
Let the continuous vector-field $a(x) = \Big(a_1(x_1,\dots,x_n),\dots,a_n(x_1,\dots,x_n)\Big)$ with the associate $(n-1)$-form

$$\omega = \sum_{i=1}^{n} a_i(x)(-1)^{n+i}\,dx_1 \wedge \ldots \wedge dx_{i-1} \wedge dx_{i+1} \wedge \ldots \wedge dx_n$$

be given. Furthermore, let $X(t_1,\dots,t_{n-1}) \; : \; T \to \mathbb{R}^n$ represent a regular C^1-surface. Then we observe

$$\begin{aligned}\int_X \omega &= \int_T \sum_{i=1}^{n} a_i\Big(X(t)\Big)(-1)^{n+i}\,\frac{\partial(x_1,\dots,x_{i-1},x_{i+1},\dots,x_n)}{\partial(t_1,\dots,t_{n-1})}\,dt_1\ldots dt_{n-1}\\ &= \int_T \left(\sum_{i=1}^{n} a_i\Big(X(t)\Big)D_i(t)\right)\,dt_1\ldots dt_{n-1}\\ &= \int_T \{a(X(t))\cdot\nu(t)\}\,d\sigma(t).\end{aligned}$$

This surface integral will be studied more intensively in Section 5, when we prove the Gaussian integral theorem.

Example 2.17. (Domain integrals)
Let us consider the continuous function $f = f(x_1,\dots,x_n)$ with the associate n-form

$$\omega = f(x)\,dx_1 \wedge \ldots \wedge dx_n.$$

Furthermore, $X = X(t) : T \to \mathbb{R}^n$ represents a regular C^1-surface. Then we infer the identity

$$\int_X \omega = \int_T f\Big(X(t)\Big) \frac{\partial(x_1, \ldots, x_n)}{\partial(t_1, \ldots, t_n)} \, dt_1 \ldots dt_n.$$

This parameter-invariant integral is well-suited for transformations of the domain.

3 The Exterior Derivative of Differential Forms

We begin with the fundamental

Definition 3.1. *For a 0-form $f(x)$ of the class $C^1(\mathcal{O})$, we define the* exterior derivative *as its differential*

$$df(x) = \sum_{i=1}^{n} f_{x_i}(x) \, dx_i, \qquad x \in \mathcal{O}.$$

When

$$\omega = \sum_{1 \le i_1 < \ldots < i_m \le n} a_{i_1 \ldots i_m}(x) \, dx_{i_1} \wedge \ldots \wedge dx_{i_m}$$

represents an m-form of the class $C^1(\mathcal{O})$, we define its exterior derivative *as the $(m+1)$-form*

$$d\omega := \sum_{1 \le i_1 < \ldots < i_m \le n} \Big(da_{i_1 \ldots i_m}(x)\Big) \wedge dx_{i_1} \wedge \ldots \wedge dx_{i_m}.$$

Remarks:

1. When ω_1 and ω_2 are two m-forms in $\mathbb{R}^n$ and $\alpha_1, \alpha_2 \in \mathbb{R}$ are given, we have the identity

$$d(\alpha_1 \omega_1 + \alpha_2 \omega_2) = \alpha_1 d\omega_1 + \alpha_2 d\omega_2.$$

   Therefore, the differential operator d constitutes a linear operator.
2. When λ denotes an l-form and ω an m-form of the class $C^1(\mathcal{O})$, we infer the *product rule*

$$d(\omega \wedge \lambda) = (d\omega) \wedge \lambda + (-1)^m \omega \wedge d\lambda.$$

We shall prove only the last statement. Here it suffices to consider the situation

$$\omega = f(x)\beta^m, \quad \lambda = g(x)\beta^l,$$

where β^m and β^l are basic forms of the order m and l, respectively. Now we deduce

$$\omega \wedge \lambda = f(x)g(x)\beta^m \wedge \beta^l$$

and, moreover,

$$\begin{aligned} d(\omega \wedge \lambda) &= d\Big(f(x)g(x)\Big) \wedge \beta^m \wedge \beta^l \\ &= \Big(g(x)df(x) + f(x)dg(x)\Big) \wedge \beta^m \wedge \beta^l \\ &= d\omega \wedge \lambda + (-1)^m \omega \wedge d\lambda. \end{aligned}$$

Example 3.2. Taking the function $f(x) \in C^1(\mathcal{O})$, we can integrate immediately the differential form df over curves. With the curve

$$X(t) = \Big(x_1(t), \ldots, x_n(t)\Big) \in C^1([a,b], \mathbb{R}^n)$$

being given, we calculate

$$\begin{aligned} \int_X df &= \int_a^b \sum_{i=1}^n f_{x_i}\Big(X(t)\Big)\dot{x}_i(t)\,dt \\ &= \int_a^b \frac{d}{dt} f\Big(X(t)\Big)\,dt \\ &= f\Big(X(b)\Big) - f\Big(X(a)\Big). \end{aligned}$$

Example 3.3. We consider the Pfaffian form

$$\omega = \sum_{i=1}^n a_i(x)\,dx_i$$

of the class $C^1(\mathcal{O})$ and determine its exterior derivative as follows:

$$\begin{aligned} d\omega &= \sum_{j=1}^n da_j(x) \wedge dx_j = \sum_{i,j=1}^n \frac{\partial a_j}{\partial x_i}\,dx_i \wedge dx_j \\ &= \sum_{1 \le i < j \le n} \left(\frac{\partial a_j}{\partial x_i} - \frac{\partial a_i}{\partial x_j}\right) dx_i \wedge dx_j. \end{aligned}$$

Obviously, the identity $d\omega = 0$ holds true if and only if the functional matrix $\left(\frac{\partial a_i}{\partial x_j}\right)_{i,j=1,\ldots,n}$ is symmetric. In the case $n = 3$, we evaluate

$$
\begin{aligned}
d\omega &= \left(\frac{\partial a_2}{\partial x_1} - \frac{\partial a_1}{\partial x_2}\right) dx_1 \wedge dx_2 + \left(\frac{\partial a_3}{\partial x_1} - \frac{\partial a_1}{\partial x_3}\right) dx_1 \wedge dx_3 \\
&\quad + \left(\frac{\partial a_3}{\partial x_2} - \frac{\partial a_2}{\partial x_3}\right) dx_2 \wedge dx_3 \\
&= b_1(x)\, dx_2 \wedge dx_3 + b_2(x)\, dx_3 \wedge dx_1 + b_3(x)\, dx_1 \wedge dx_2.
\end{aligned}
$$

Here we have defined the vector-field

$$
\begin{aligned}
\Big(b_1(x), b_2(x), b_3(x)\Big) &= \left(\frac{\partial a_3}{\partial x_2} - \frac{\partial a_2}{\partial x_3}, \frac{\partial a_1}{\partial x_3} - \frac{\partial a_3}{\partial x_1}, \frac{\partial a_2}{\partial x_1} - \frac{\partial a_1}{\partial x_2}\right) \\
&= \nabla \wedge (a_1, a_2, a_3)(x) \ =: \ \operatorname{rot} a(x),
\end{aligned}
$$

where $\nabla := \left(\frac{\partial}{\partial x_1}, \frac{\partial}{\partial x_2}, \frac{\partial}{\partial x_3}\right)$ denotes the *nabla-operator*. Integration of this differential form $d\omega$ over surfaces in $\mathbb{R}^3$ will be possible by the *classical Stokes integral theorem.*

Definition 3.4. *We name*

$$
rot\, a(x) = \left(\frac{\partial a_3}{\partial x_2} - \frac{\partial a_2}{\partial x_3}, \frac{\partial a_1}{\partial x_3} - \frac{\partial a_3}{\partial x_1}, \frac{\partial a_2}{\partial x_1} - \frac{\partial a_1}{\partial x_2}\right)
$$

the rotation of the vector-field $a(x) = \Big(a_1(x), a_2(x), a_3(x)\Big) \in C^1(\mathcal{O}, \mathbb{R}^3)$.

Example 3.5. Now we consider a specific $(n-1)$-form in $\mathbb{R}^n$, namely

$$
\omega = \sum_{i=1}^{n} a_i(x)(-1)^{i+1}\, dx_1 \wedge \ldots \wedge dx_{i-1} \wedge dx_{i+1} \wedge \ldots \wedge dx_n,
$$

whose exterior derivative takes on the following form:

$$
\begin{aligned}
d\omega &= \sum_{i=1}^{n} (-1)^{i+1} \Big(da_i(x)\Big) \wedge dx_1 \wedge \ldots \wedge dx_{i-1} \wedge dx_{i+1} \wedge \ldots \wedge dx_n \\
&= \sum_{i,j=1}^{n} (-1)^{i+1} \frac{\partial a_i}{\partial x_j}(x)\, dx_j \wedge dx_1 \wedge \ldots \wedge dx_{i-1} \wedge dx_{i+1} \wedge \ldots \wedge dx_n \\
&= \sum_{i=1}^{n} (-1)^{i+1} \frac{\partial a_i}{\partial x_i}(x)\, dx_i \wedge dx_1 \wedge \ldots \wedge dx_{i-1} \wedge dx_{i+1} \wedge \ldots \wedge dx_n \\
&= \left(\sum_{i=1}^{n} \frac{\partial a_i}{\partial x_i}(x)\right) dx_1 \wedge \ldots \wedge dx_n \\
&= \Big(\operatorname{div} a(x)\Big)\, dx_1 \wedge \ldots \wedge dx_n.
\end{aligned}
$$

Definition 3.6. *The vector-field* $a(x) = \Big(a_1(x), \ldots, a_n(x)\Big) \in C^1(\mathcal{O}, \mathbb{R}^n)$ *on the open set* $\mathcal{O} \subset \mathbb{R}^n$ *possesses the* divergence

$$div\, a(x) := \sum_{i=1}^{n} \frac{\partial a_i}{\partial x_i}(x), \qquad x \in \mathcal{O}.$$

Example 3.7. We can integrate the n-form

$$d\omega = (\operatorname{div} a(x))\, dx_1 \wedge \ldots \wedge dx_n$$

over an n-dimensional rectangle. This differential form can also be integrated over a substantially larger class of domains in $\mathbb{R}^n$ - bounded by finitely many hypersurfaces - with the aid of the *Gaussian integral theorem*, one of the most important theorems in the higher-dimensional analysis.
At first, we integrate $d\omega$ over the following standard domain: For $r > 0$ we define the *semidisc*

$$H := \Big\{x = (x_1, \ldots, x_n) \in \mathbb{R}^n \mid x_1 \in (-r, 0),\ x_i \in (-r, +r),\ i = 2, \ldots, n\Big\}$$

with the upper bounding side

$$S := \Big\{x = (0, x_2, \ldots, x_n) \mid |x_i| < r,\ i = 2, \ldots, n\Big\}.$$

The exterior normal vector to the surface S is given by $e_1 = (1, 0, \ldots, 0) \in \mathbb{R}^n$ explicitly. Then we comprehend H and S as surfaces in $\mathbb{R}^n$ via the representations

$$H\,:\, X(t_1, \ldots, t_n) = (t_1, \ldots, t_n), \qquad (t_1, \ldots, t_n) \in H$$

and

$$S\,:\, Y(\widetilde{t}_1, \ldots, \widetilde{t}_{n-1}) = (0, \widetilde{t}_1, \ldots, \widetilde{t}_{n-1}), \qquad |\widetilde{t}_i| < r, \quad i = 1, \ldots, n-1,$$

respectively. With the assumption $\omega \in C_0^1(H \cup S)$, we obtain

$$\begin{aligned}
\int_H d\omega = \int_X d\omega &= \int_{-r}^{0} \int_{-r}^{+r} \cdots \int_{-r}^{+r} \left(\frac{\partial a_1}{\partial x_1} + \ldots + \frac{\partial a_n}{\partial x_n} \right) dx_1 \ldots dx_n \\
&= \int_{-r}^{+r} \cdots \int_{-r}^{+r} a_1(0, x_2, \ldots, x_n)\, dx_2 \ldots dx_n \;=\; \int_S \omega.
\end{aligned}$$

In the sequel, we shall investigate the behavior of differential forms with respect to transformations of the ambient space.

Definition 3.8. (Transformed differential form)
Let the symbol

$$\omega = \sum_{1 \le i_1 < \ldots < i_m \le n} a_{i_1 \ldots i_m}(x)\, dx_{i_1} \wedge \ldots \wedge dx_{i_m}$$

denote a continuous m-form in an open set $\mathcal{O} \subset \mathbb{R}^n$. *Furthermore, let* $T \subset \mathbb{R}^l$ *with* $l \in \mathbb{N}$ *describe an open set such that*

$$\begin{aligned} x = (x_1, \ldots, x_n) &= \Phi(y) \\ &= (\varphi_1(y_1, \ldots, y_l), \ldots, \varphi_n(y_1, \ldots, y_l)) \,:\, T \to \mathcal{O} \end{aligned}$$

defines a mapping of the class $C^1(T, \mathbb{R}^n)$. *With*

$$d\varphi_i = \sum_{j=1}^{l} \frac{\partial \varphi_i}{\partial y_j}(y)\, dy_j, \qquad i = 1, \ldots, n$$

and

$$\omega_\Phi := \sum_{1 \le i_1 < \ldots < i_m \le n} a_{i_1 \ldots i_m}\Big(\Phi(y)\Big)\, d\varphi_{i_1} \wedge \ldots \wedge d\varphi_{i_m},$$

we obtain the transformed m-form ω_Φ with respect to the mapping Φ.

Remarks:

1. When ω_1, ω_2 are two m-forms and $\alpha_1, \alpha_2 \in \mathbb{R}$ are given, we infer the identity
$$(\alpha_1 \omega_1 + \alpha_2 \omega_2)_\Phi = \alpha_1 (\omega_1)_\Phi + \alpha_2 (\omega_2)_\Phi.$$
2. When λ represents an l-form and ω an m-form, we have the rule
$$(\omega \wedge \lambda)_\Phi = \omega_\Phi \wedge \lambda_\Phi.$$

The following result is important for the evaluation of integrals for differential forms over surfaces.

Theorem 3.9. (Pull-back of differential forms)
Let ω *denote a continuous m-form in the open set* $\mathcal{O} \subset \mathbb{R}^n$. *On the open set* $T \subset \mathbb{R}^m$ *we define a surface* X *by the parametric representation*

$$x = \Phi(y) \,:\, T \longrightarrow \mathcal{O} \in C^1(T)$$

with $\Phi(T) \subset\subset \mathcal{O}$. *Finally, we define the surface*

$$Y(t) = (t_1, \ldots, t_m), \qquad t \in T$$

and note that

$$X(t) = \Phi \circ Y(t), \qquad t \in T.$$

Then the following identity holds true:

$$\int_X \omega = \int_Y \omega_\Phi.$$

Proof: We calculate

$$d\varphi_{i_1} \wedge \ldots \wedge d\varphi_{i_m} = \left(\sum_{j_1=1}^{m} \frac{\partial \varphi_{i_1}}{\partial y_{j_1}} \, dy_{j_1} \right) \wedge \ldots \wedge \left(\sum_{j_m=1}^{m} \frac{\partial \varphi_{i_m}}{\partial y_{j_m}} \, dy_{j_m} \right)$$

$$= \frac{\partial(\varphi_{i_1}, \ldots, \varphi_{i_m})}{\partial(y_1, \ldots, y_m)} \, dy_1 \wedge \ldots \wedge dy_m,$$

as well as

$$\omega_\Phi = \sum_{1 \le i_1 < \ldots < i_m \le n} a_{i_1 \ldots i_m}(\Phi(y)) \, \frac{\partial(\varphi_{i_1}, \ldots, \varphi_{i_m})}{\partial(y_1, \ldots, y_m)} \, dy_1 \wedge \ldots \wedge dy_m.$$

This implies

$$\int_Y \omega_\Phi = \int_T \sum_{1 \le i_1 < \ldots < i_m \le n} a_{i_1 \ldots i_m}(X(t)) \, \frac{\partial(x_{i_1}, \ldots, x_{i_m})}{\partial(t_1, \ldots, t_m)} \, dt_1 \ldots dt_m$$

$$= \int_X \omega,$$

and our theorem is proved. q.e.d.

Theorem 3.10. *Let ω denote an m-form in the open set $\mathcal{O} \subset \mathbb{R}^n$ of the regularity class $C^1(\mathcal{O})$. Furthermore, let the mapping*

$$x = \Phi(y) \, : \, T \longrightarrow \mathcal{O} \in C^2(T)$$

be given on the open set $T \subset \mathbb{R}^l$, where $l \in \mathbb{N}$ holds true. Then we have the calculus rule

$$d(\omega_\Phi) = (d\omega)_\Phi.$$

Proof: At first, an arbitrary function $\Psi(y) \in C^2(\mathcal{O})$ satisfies the identity

$$d^2\Psi = d(d\Psi) = d\left(\sum_{i=1}^{n} \Psi_{y_i} \, dy_i \right) = \sum_{i,j=1}^{n} \Psi_{y_i y_j} \, dy_j \wedge dy_i = 0.$$

Now we note that

$$\omega_\Phi = \sum_{1 \le i_1 < \ldots < i_m \le n} a_{i_1 \ldots i_m}\Big(\Phi(y)\Big) \, d\varphi_{i_1} \wedge \ldots \wedge d\varphi_{i_m},$$

and we arrive at

$$
\begin{aligned}
d\omega_\Phi &= \sum_{1\le i_1<\ldots<i_m\le n} da_{i_1\ldots i_m}\Big(\Phi(y)\Big) \wedge d\varphi_{i_1} \wedge \ldots \wedge d\varphi_{i_m} \\
&= \sum_{1\le i_1<\ldots<i_m\le n} \sum_{j=1}^{n} \sum_{k=1}^{l} \frac{\partial a_{i_1\ldots i_m}}{\partial x_j}\Big(\Phi(y)\Big)\frac{\partial \varphi_j}{\partial y_k}\, dy_k \wedge d\varphi_{i_1} \wedge \ldots \wedge d\varphi_{i_m} \\
&= \sum_{1\le i_1<\ldots<i_m\le n} \sum_{j=1}^{n} \frac{\partial a_{i_1\ldots i_m}}{\partial x_j}\Big(\Phi(y)\Big)\, d\varphi_j \wedge d\varphi_{i_1} \wedge \ldots \wedge d\varphi_{i_m},
\end{aligned}
$$

and consequently

$$
d\omega_\Phi = (d\omega)_\Phi .
$$

q.e.d.

Theorem 3.11. (Chain rule for differential forms)
Let ω denote a continuous m-form in an open set $\mathcal{O} \subset \mathbb{R}^n$. Furthermore, we consider the open sets $T' \subset \mathbb{R}^{l'}$ and $T'' \subset \mathbb{R}^{l''}$ - with $l', l'' \in \mathbb{N}$ - where the C^1-functions Φ, Ψ are defined due to

$$
\Psi : T'' \to T', \quad \Phi : T' \to \mathcal{O} \quad \text{with} \qquad z \stackrel{\Psi}{\longmapsto} y \stackrel{\Phi}{\longmapsto} x.
$$

Then the following identity holds true:

$$
(\omega_\Phi)_\Psi = \omega_{\Phi\circ\Psi}.
$$

Proof: We calculate

$$
\begin{aligned}
\omega_{\Phi\circ\Psi} &= \sum_{i_1,\ldots,i_m} a_{i_1\ldots i_m}\Big(\Phi\circ\Psi(z)\Big)\, d(\varphi_{i_1}\circ\Psi) \wedge \ldots \wedge d(\varphi_{i_m}\circ\Psi) \\
&= \sum_{\substack{i_1,\ldots,i_m \\ j_1,\ldots,j_m \\ k_1,\ldots,k_m}} a_{i_1\ldots i_m}\Big(\Phi\circ\Psi(z)\Big)\left(\frac{\partial\varphi_{i_1}}{\partial y_{j_1}}\frac{\partial\psi_{j_1}}{\partial z_{k_1}}\, dz_{k_1}\right) \wedge \ldots \wedge \left(\frac{\partial\varphi_{i_m}}{\partial y_{j_m}}\frac{\partial\psi_{j_m}}{\partial z_{k_m}}\, dz_{k_m}\right) \\
&= \sum_{\substack{i_1,\ldots,i_m \\ j_1,\ldots,j_m}} a_{i_1\ldots i_m}\Big(\Phi\circ\Psi(z)\Big)\left(\frac{\partial\varphi_{i_1}}{\partial y_{j_1}}\, d\psi_{j_1}\right) \wedge \ldots \wedge \left(\frac{\partial\varphi_{i_m}}{\partial y_{j_m}}\, d\psi_{j_m}\right) \\
&= \left(\sum_{i_1,\ldots,i_m} a_{i_1\ldots i_m}\Big(\Phi(y)\Big)\, d\varphi_{i_1} \wedge \ldots \wedge d\varphi_{i_m}\right)_{y=\Psi(z)},
\end{aligned}
$$

and consequently

$$
\omega_{\Phi\circ\Psi} = (\omega_\Phi)_\Psi .
$$

Here we perform our summation over the indices $i_1, \ldots, i_m \in \{1, \ldots, n\}$, $j_1, \ldots, j_m \in \{1, \ldots, l'\}$, and $k_1, \ldots, k_m \in \{1, \ldots, l''\}$. q.e.d.

4 The Stokes Integral Theorem for Manifolds

We choose $m \in \mathbb{N}$ and consider the m-dimensional plane

$$\mathbb{E}^m := \Big\{(0, y_1, \ldots, y_m) \in \mathbb{R}^{m+1} : (y_1, \ldots, y_m) \in \mathbb{R}^m\Big\}.$$

Parallel to the Example 3.7 from Section 3, we take the data $\eta \in \mathbb{R}^{m+1}$ and $r > 0$ in order to define the *semicube*

$$H_r(\eta) := \Big\{y \in \mathbb{R}^{m+1} : y_1 \in (\eta_1 - r, \eta_1),\ y_j \in (\eta_j - r, \eta_j + r) \text{ for } j = 2, \ldots, m+1\Big\}$$

with the lateral lengths $2r$. This object has the upper bounding side

$$S_r(\eta) := \Big\{y \in \mathbb{R}^{m+1} : y_1 = \eta_1, y_j \in (\eta_j - r, \eta_j + r) \text{ for } j = 2, \ldots, m+1\Big\}.$$

We comprehend $H_r(\eta)$ and $S_r(\eta)$ as surfaces in $\mathbb{R}^{m+1}$:

$$H_r(\eta) : Y(t_1, \ldots, t_{m+1}) = (\eta_1 + t_1, \ldots, \eta_{m+1} + t_{m+1})$$
$$\text{with} \quad -r < t_1 < 0, \quad |t_j| < r, \quad j = 2, \ldots, m+1$$

as well as

$$S_r(\eta) : Y(t_1, \ldots, t_m) := (\eta_1, \eta_2 + t_1, \ldots, \eta_{m+1} + t_m)$$
$$\text{with} \quad |t_j| < r, \quad j = 1, \ldots, m.$$

When $\eta \in \mathbb{E}^m$ and $r > 0$ are fixed, we define $H := H_r(\eta)$ and $S := S_r(\eta)$, respectively. With $n > m$ given, we denote by

$$\Phi = \Phi(y_1, \ldots, y_{m+1}) : \overline{H} \longrightarrow \mathbb{R}^n \in C^1(\overline{H}, \mathbb{R}^n)$$

a surface, which can be continued onto an open set containing $\overline{H}$ in $\mathbb{R}^{m+1}$. When we set

$$X(t_1, \ldots, t_{m+1}) := \Phi(t_1, \ldots, t_{m+1}), \quad (t_1, \ldots, t_{m+1}) \in \overline{H},$$

we obtain the following $(m+1)$-dimensional surface in $\mathbb{R}^n$, namely

$$\mathcal{F} := \Big\{X(t) \in \mathbb{R}^n : t \in H\Big\},$$

whose boundary contains the m-dimensional surface

$$\mathcal{S} := \Big\{X(t) \in \mathbb{R}^n : t \in S\Big\}.$$

Let the m-form be given on the set $\overline{\mathcal{F}} = \Phi(\overline{H})$ by the symbol

$$\omega = \sum_{i_1,\ldots,i_m=1}^{n} a_{i_1\ldots i_m}(x)\, dx_{i_1} \wedge \ldots \wedge dx_{i_m}, \quad x \in \overline{\mathcal{F}}$$

of the class $C_0^0(\mathcal{F} \cup \mathcal{S}) \cap C^1(\mathcal{F})$. Here the symbol $\omega \in C^1(\mathcal{F})$ means that we have an open set $\mathcal{O} \subset \mathbb{R}^n$ with $\mathcal{F} \subset \mathcal{O}$ satisfying $\omega \in C^1(\mathcal{O})$. Finally, let $d\omega$ be absolutely integrable over $\mathcal{F}$ in the following sense:

$$\int_{\mathcal{F}} |d\omega| := \int_H \Big| \sum_{i_1,\ldots,i_{m+1}=1}^{n} \frac{\partial a_{i_1\ldots i_m}}{\partial x_{i_{m+1}}}\Big(X(t)\Big) \frac{\partial(x_{i_1},\ldots,x_{i_{m+1}})}{\partial(t_1,\ldots,t_{m+1})} \Big| \, dt_1 \ldots dt_{m+1}$$
$$< +\infty.$$

Now we prove the basic

Proposition 4.1. (Local Stokes theorem)
Let the surface $\mathcal{F}$ with the boundary part $\mathcal{S}$ be given as above, and furthermore the symbol ω may denote an m-dimensional differential form of the class

$$C_0^0(\mathcal{F} \cup \mathcal{S}) \cap C^1(\mathcal{F})$$

satisfying

$$\int_{\mathcal{F}} |d\omega| < +\infty.$$

Then we have the identity

$$\int_{\mathcal{F}} d\omega = \int_{\mathcal{S}} \omega.$$

Proof:

1. At first, we prove this formula under the stronger assumptions $\Phi \in C^2(\overline{H})$ and $\omega \in C_0^1(\mathcal{F} \cup \mathcal{S})$. Utilizing Theorem 3.10 and Example 3.7 from Section 3, we infer the identity

$$\int_{\mathcal{F}} d\omega = \int_X d\omega = \int_H (d\omega)_\Phi = \int_H d(\omega_\Phi) = \int_S \omega_\Phi = \int_{\mathcal{S}} \omega.$$

2. When $\Phi \in C^1(\overline{H})$ and $\omega \in C^1(\mathcal{F}) \cap C_0^0(\mathcal{F} \cup \mathcal{S})$ hold true, we approximate Φ uniformly in H up to the first derivatives by the functions $\Phi^{(k)}(y) \in C^\infty$, due to the Weierstraß approximation theorem. Now we exhaust H by rectangles

$$H^{(l)} := H_{r-\frac{2}{l}}\Big(\eta_1 - \frac{1}{l}, \eta_2, \ldots, \eta_{m+1}\Big) \subset H$$

with the upper bounding sides

$$S^{(l)} := S_{r-\frac{2}{l}}\Big(\eta_1 - \frac{1}{l}, \eta_2, \ldots, \eta_{m+1}\Big).$$

The considerations in part 1.) reveal

$$\int\limits_{H^{(l)}} (d\omega)_{\Phi^{(k)}} = \int\limits_{S^{(l)}} \omega_{\Phi^{(k)}} \quad \text{for all} \quad k, l \geq N \in \mathbb{N}.$$

The transition to the limit $k \to \infty$ implies

$$\int\limits_{H^{(l)}} (d\omega)_{\Phi} = \int\limits_{S^{(l)}} \omega_{\Phi}.$$

On account of $\int_{\mathcal{F}} |d\omega| < +\infty$, the limit procedure $l \to \infty$ yields

$$\int\limits_{\mathcal{F}} d\omega = \int\limits_{H} (d\omega)_{\Phi} = \int\limits_{S} \omega_{\Phi} = \int\limits_{\mathcal{S}} \omega.$$

This is exactly the identity stated above. q.e.d.

Now we introduce the fundamental notion of a differentiable manifold.

Definition 4.2. *Let us fix the dimensions* $1 \leq m \leq n$ *as well as the set* $\mathcal{M} \subset \mathbb{R}^n$. *We name* $\mathcal{M}$ *an* m-dimensional C^k-manifold, *if each point* $\xi \in \mathcal{M}$ *possesses an element* $\eta \in \mathbb{R}^m$ *and open neighborhoods* $U \subset \mathbb{R}^n$ *of* $\xi \in U$ *and* $V \subset \mathbb{R}^m$ *of* $\eta \in V$ *as well as an embedded regular surface*

$$x = \Phi(y) \, : \, V \longrightarrow U \in C^k(V)$$

such that

$$\xi = \Phi(\eta) \quad \text{and} \quad \Phi(V) = \mathcal{M} \cap U$$

is correct; here we have chosen $k \in \mathbb{N}$ *adequately. We call* (Φ, V) *a* chart of the manifold. *All charts together*

$$\mathcal{A} := \Big\{ (\Phi_\iota, V_\iota) \, : \, \iota \in J \Big\}$$

constitute an atlas of the manifold. *When* $\Phi_j : V_j \to U_j \cap \mathcal{M}$ *with* $j = 1, 2$ *represent two charts of the atlas* $\mathcal{A}$ *such that*

$$W_{1,2} := \mathcal{M} \cap U_1 \cap U_2 \neq \emptyset$$

is correct, then we consider the parameter transformation $\Phi_{2,1} := \Phi_2^{-1} \circ \Phi_1$. *If the functional determinant satisfies* $J_{\Phi_{2,1}} > 0$ *on* $\Phi_1^{-1}(W_{1,2})$ *for such ar-*

bitrarily chosen charts from the atlas, the manifold is oriented *by the atlas.*

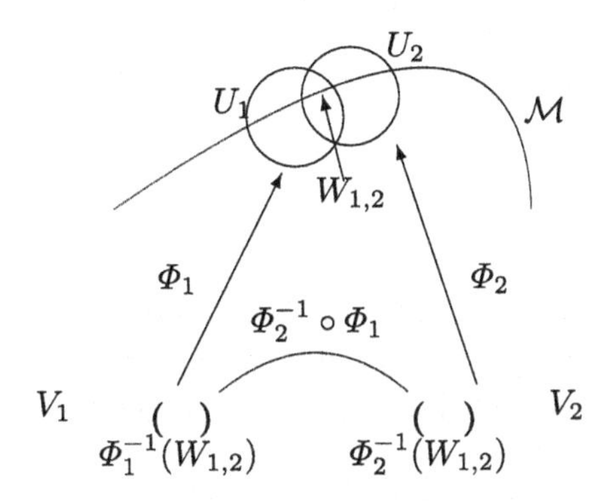

Definition 4.3. *Let $\mathcal{M}$ denote a bounded, $(m+1)$-dimensional, oriented C^1-manifold in $\mathbb{R}^n$ with $n > m$. We indicate the topological closure of the point set $\mathcal{M}$ by the symbol $\overline{\mathcal{M}}$ and the set of* boundary points *by the symbol $\dot{\mathcal{M}} := \overline{\mathcal{M}} \setminus \mathcal{M}$. We name $\xi \in \dot{\mathcal{M}}$ a* regular boundary point *of the manifold $\mathcal{M}$ if the following holds true:*

We have a semicube $H_r(\eta)$ in $\mathbb{R}^{m+1}$ with $\eta \in \mathbb{E}^m$ and $r > 0$, a regular embedded surface

$$\Phi(y) : \overline{H_r(\eta)} \to \mathbb{R}^n \in C^1(\overline{H_r(\eta)})$$

such that $\Phi|_{H_r(\eta)}$ belongs to the oriented atlas $\mathcal{A}$ of $\mathcal{M}$,

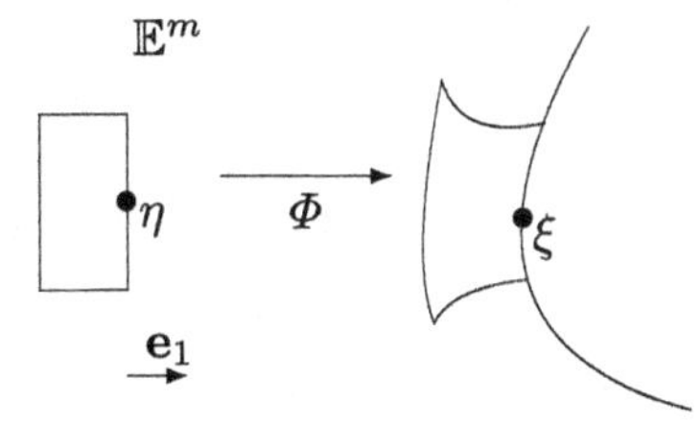

and an open neighborhood $U \subset \mathbb{R}^n$ of $\xi \in U$ with the following properties:

$$\Phi(\eta) = \xi, \quad \Phi\Big(S_r(\eta)\Big) = \dot{\mathcal{M}} \cap U, \quad \Phi\Big(H_r(\eta)\Big) = \mathcal{M} \cap U.$$

The set of regular boundary points *will be denoted by the symbol $\partial\mathcal{M}$.*

Definition 4.4. *For the bounded manifold $\mathcal{M}$ from Definition 4.3, we define the set of* singular boundary points *$\triangle\mathcal{M}$ according to*

$$\triangle\mathcal{M} := \dot{\mathcal{M}} \setminus \partial\mathcal{M}.$$

In the case $\triangle\mathcal{M} = \emptyset$, we obtain a compact manifold with regular boundary. *If the condition $\partial\mathcal{M} = \emptyset$ is fulfilled additionally, we speak of a* closed manifold.

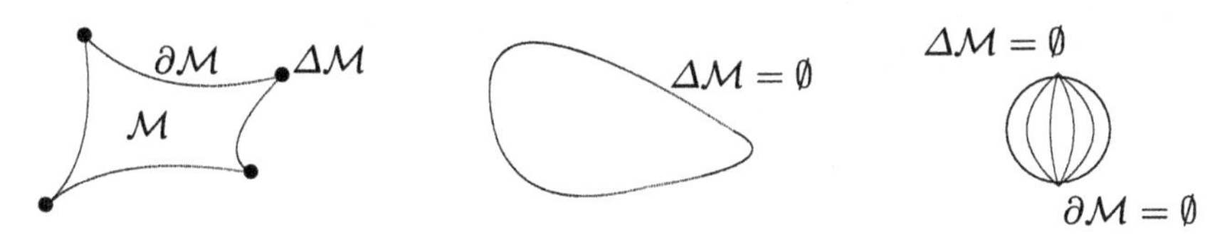

Proposition 4.5. (Induced orientation on $\partial\mathcal{M}$)
Let $\mathcal{M}$ and $\partial\mathcal{M}$ from Definition 4.3 with the charts $\Phi : \overline{H_r(\eta)} \to \mathbb{R}^n$ be given. Then the mappings

$$\left\{\Phi\big|_{S_r(\eta)} : \Phi\big|_{H_r(\eta)} \text{ belongs to the oriented atlas } \mathcal{A} \text{ of } \mathcal{M}\right\} =: \partial\mathcal{A}$$

constitute an oriented atlas of $\partial\mathcal{M}$. Consequently, $\partial\mathcal{M}$ represents an oriented C^1-manifold.

Proof: We consider $\Phi(\eta) = \xi = \widetilde{\Phi}(\widetilde{\eta})$. The vectors $\Phi_{y_2}(\eta), \ldots, \Phi_{y_{m+1}}(\eta)$ and $\widetilde{\Phi}_{y_2}(\widetilde{\eta}), \ldots, \widetilde{\Phi}_{y_{m+1}}(\widetilde{\eta})$ span the m-dimensional tangential space $T_{\partial\mathcal{M}}(\xi)$ to $\partial\mathcal{M}$ at the point ξ. When we add the vectors $\Phi_{y_1}(\eta)$ and $\widetilde{\Phi}_{y_1}(\widetilde{\eta})$, respectively, the tangential space $T_{\mathcal{M}}(\xi)$ to $\mathcal{M}$ is generated.
Now we construct an orthonormal system $N^1, \ldots, N^{n-m} \in \mathbb{R}^n$ which is orthogonal to $T_{\partial\mathcal{M}}(\xi)$. We choose the vector $N^1 \in T_{\mathcal{M}}(\xi)$, directed out of the surface at the point ξ, and obtain

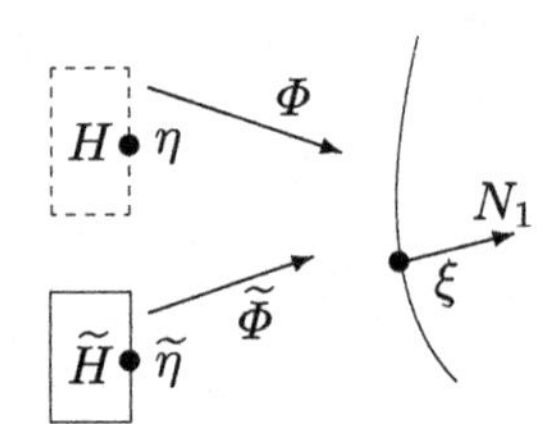

$$\Phi_{y_1}(\eta) \cdot N^1 > 0, \quad \widetilde{\Phi}_{y_1}(\widetilde{\eta}) \cdot N^1 > 0.$$

With the parameter $0 \le \tau \le 1$, we consider the matrices

$$M(\tau) := \begin{pmatrix} (1-\tau)\Phi_{y_1}(\eta) + \tau N^1 \\ \Phi_{y_2}(\eta) \\ \vdots \\ \Phi_{y_{m+1}}(\eta) \\ N^2 \\ \vdots \\ N^{n-m} \end{pmatrix}, \quad \widetilde{M}(\tau) := \begin{pmatrix} (1-\tau)\widetilde{\Phi}_{y_1}(\widetilde{\eta}) + \tau N^1 \\ \widetilde{\Phi}_{y_2}(\widetilde{\eta}) \\ \vdots \\ \widetilde{\Phi}_{y_{m+1}}(\widetilde{\eta}) \\ N^2 \\ \vdots \\ N^{n-m} \end{pmatrix}.$$

Furthermore, we define $\Psi := \Phi\big|_{S_r(\eta)}$ and $\widetilde{\Psi} := \widetilde{\Phi}\big|_{S_r(\widetilde{\eta})}$. Now the functions $\det M(\tau)$ and $\det \widetilde{M}(\tau)$ in [0,1] are continuous with $\det M(\tau) \neq 0$ and $\det \widetilde{M}(\tau) \neq 0$ for all $0 \le \tau \le 1$. Consequently, the following function is continuous in $[0,1]$, and we have

$$\det\left(\widetilde{M}(\tau)^{-1} \circ M(\tau)\right) \neq 0, \qquad 0 \le \tau \le 1.$$

By assumption we note that

$$\det\left(\widetilde{M}(0)^{-1}\circ M(0)\right)=\det\partial(\widetilde{\Phi}^{-1}\circ\Phi)\big|_\eta>0,$$

and a continuity argument implies

$$\det\partial(\widetilde{\Psi}^{-1}\circ\Psi)\big|_\eta=\det\left(\widetilde{M}(1)^{-1}\circ M(1)\right)>0.$$

Therefore, $\partial\mathcal{A}$ constitutes an oriented atlas of $\partial\mathcal{M}$. q.e.d.

We now intend to prove the Stokes integral theorem for manifolds $\mathcal{M}$ with the regular boundary $\partial\mathcal{M}$ and the singular boundary $\triangle\mathcal{M}$, namely the identity

$$\int\limits_{\mathcal{M}} d\omega=\int\limits_{\partial\mathcal{M}}\omega,$$

under weak assumptions. The transition from the local Stokes theorem to the global result is achieved by the *partition of unity.*

Let $\mathcal{M}$ denote an $(m+1)$-dimensional, bounded, oriented C^1-manifold in $\mathbb{R}^n$ with the regular boundary $\partial\mathcal{M}$. Furthermore, let the symbol

$$\lambda=\sum_{1\le i_1<\ldots<i_{m+1}\le n} b_{i_1\ldots i_{m+1}}(x)\,dx_{i_1}\wedge\ldots\wedge dx_{i_{m+1}},\qquad x\in\mathcal{M}$$

represent a continuous differential form on $\mathcal{M}$.

We shall investigate which conditions for λ allow us to define the *improper integral*

$$\int\limits_{\mathcal{M}}\lambda$$

of the differential form λ over the manifold $\mathcal{M}$.

1. At first, let the set

$$\operatorname{supp}\lambda:=\overline{\{x\in\mathcal{M}\,:\,\lambda(x)\neq 0\}}\subset\mathcal{M}\cup\partial\mathcal{M}$$

be compact. Then we have open sets $V_\iota\subset\mathbb{R}^{m+1}$ and $U_\iota\subset\mathbb{R}^n\setminus\triangle\mathcal{M}$ with $\iota\in J$ and, moreover, charts $\Phi_\iota\,:\,V_\iota\to U_\iota\cap\mathcal{M}$ such that the open sets $\{U_\iota\}_{\iota\in J}$ cover the compact set $\operatorname{supp}\lambda$. Now we choose a partition of unity in $\mathbb{R}^n$ subordinate to the sets $\{U_\iota\}$ and obtain

$$\alpha_k(x)\,:\,\mathcal{M}\longrightarrow[0,1]\in C^1\quad\text{with}\quad\operatorname{supp}\alpha_k\subset U_{\iota_k}\qquad\text{for}\quad k=1,\ldots,k_0$$

as well as

$$\sum_{k=1}^{k_0}\alpha_k(x)=1\qquad\text{for all}\quad x\in\operatorname{supp}\lambda.$$

We define

$$\int_{\mathcal{M}} \lambda := \sum_{k=1}^{k_0} \int_{\mathcal{M}} \alpha_k \lambda = \sum_{k=1}^{k_0} \int_{V_k} (\alpha_k \lambda)_{\Phi_k}, \tag{1}$$

if

$$\int_{\mathcal{M}} \alpha_k |\lambda| < +\infty \quad \text{for} \quad k = 1, \ldots, k_0$$

is correct.

We still have to show that the integral, given in equation (1), is independent of the covering for the support of λ and of the partition of unity used.

When $\widetilde{\Phi}_\iota : \widetilde{V}_\iota \to \widetilde{U}_\iota \cap \mathcal{M}$ with $\iota \in \widetilde{J}$ represents an alternative system of charts covering $\operatorname{supp} \lambda$, we choose again a partition of unity for $\operatorname{supp} \lambda$ subordinate to the system $\{\widetilde{U}_\iota\}_\iota$. We obtain

$$\widetilde{\alpha}_l : \mathcal{M} \to [0,1] \in C^1, \quad \operatorname{supp} \widetilde{\alpha}_l \subset \widetilde{U}_{\iota_l}, \qquad l = 1, \ldots, l_0$$

as well as

$$\sum_{l=1}^{l_0} \widetilde{\alpha}_l(x) = 1 \qquad \text{for all} \quad x \in \operatorname{supp} \lambda.$$

We note that $\operatorname{supp}(\alpha_k \widetilde{\alpha}_l) \subset U_k \cap U_l \cap \mathcal{M}$ holds true. Under the mapping $\Phi_k^{-1} \circ \widetilde{\Phi}_l$ for all indices $k = 1, \ldots, k_0$ and $l = 1, \ldots, l_0$ we transform the integrals

$$\int_{V_k} (\alpha_k \widetilde{\alpha}_l \lambda)_{\Phi_k} = \int_{\widetilde{V}_l} (\alpha_k \widetilde{\alpha}_l \lambda)_{\widetilde{\Phi}_l}. \tag{2}$$

The summation yields

$$\sum_{k=1}^{k_0} \int_{V_k} (\alpha_k \lambda)_{\Phi_k} = \sum_{k=1}^{k_0} \sum_{l=1}^{l_0} \int_{V_k} (\alpha_k \widetilde{\alpha}_l \lambda)_{\Phi_k}$$

$$= \sum_{k=1}^{k_0} \sum_{l=1}^{l_0} \int_{\widetilde{V}_l} (\alpha_k \widetilde{\alpha}_l \lambda)_{\widetilde{\Phi}_l} = \sum_{l=1}^{l_0} \int_{\widetilde{V}_l} (\widetilde{\alpha}_l \lambda)_{\widetilde{\Phi}_l}.$$

Consequently, the integral given in (1) is independent of the choice of charts and the partition of unity. Correspondingly, we define $\int_{\mathcal{M}} |\lambda|$ and $\int_{\partial \mathcal{M}} \lambda$.

2. The differential form $\lambda \in C^0(\mathcal{M})$ is *absolutely integrable over* $\mathcal{M}$, symbolically

$$\int_{\mathcal{M}} |\lambda| < +\infty,$$

if we have a constant $M \in [0, +\infty)$ such that the inequality

$$\int_{\mathcal{M}} |\beta\lambda| \leq M \quad \text{for all} \quad \beta \in C_0^0(\mathcal{M} \cup \partial\mathcal{M}, [0,1])$$

is correct. We say that the sequence of functions $\beta_k \in C_0^0(\mathcal{M} \cup \partial\mathcal{M}, [0,1])$ is *exhausting the manifold*, when each compact set $K \subset \mathcal{M} \cup \partial\mathcal{M}$ possesses an index $k_0 = k_0(K) \in \mathbb{N}$ such that

$$\beta_k(x) = 1 \qquad \text{for all} \quad x \in K, \quad k \geq k_0.$$

When $\int_{\mathcal{M}} |\lambda| < +\infty$ holds true, we show as in the theory of improper integrals that for each exhausting sequence of functions $\{\beta_k\}_{k=1,2,\ldots}$ the following expression

$$\lim_{k\to\infty} \int_{\mathcal{M}} \beta_k \lambda$$

exists and has the same value. We set

$$\int_{\mathcal{M}} \lambda := \lim_{k\to\infty} \int_{\mathcal{M}} \beta_k \lambda. \tag{3}$$

In this sense, we comprehend all improper integrals appearing in the sequel.

Definition 4.6. *The singular boundary $\triangle\mathcal{M}$ of the manifold $\mathcal{M}$ has* capacity zero *if we can find a function*

$$\chi \in C_0^1(\mathcal{M} \cup \partial\mathcal{M}, [0,1])$$

for each $\varepsilon > 0$ and each compact set $K \subset \mathcal{M} \cup \partial\mathcal{M}$ with the following properties:

1. *We have $\chi(x) = 1$ for all $x \in K$;*
2. *The following condition holds true:*

$$\int_{\mathcal{M}} \sqrt{\nabla(\chi,\chi)}\, d^{m+1}\sigma \leq \varepsilon.$$

Here $d^{m+1}\sigma$ denotes the $(m+1)$-dimensional surface element on $\mathcal{M}$, and we set

$$|\nabla(\chi)|^2\Big|_x = \nabla(\chi,\chi)\Big|_x := sup\Big\{|\nabla\chi\cdot\xi|^2 \,:\, \xi \in T_{\mathcal{M}}(x),\ |\xi| = 1\Big\}.$$

Now we arrive at our central result, namely

Theorem 4.7. (The Stokes integral theorem for manifolds)
Assumptions:

1. *Let $\mathcal{M}$ represent a bounded, oriented, $(m+1)$-dimensional C^1-manifold in $\mathbb{R}^n$ - where $n > m$ is correct - with the atlas $\mathcal{A}$. Via the induced atlas $\partial\mathcal{A}$, the regular boundary $\partial\mathcal{M}$ becomes a bounded, oriented, m-dimensional C^1-manifold. We assume that the regular boundary possesses finite surface area as follows:*
$$\int\limits_{\partial\mathcal{M}} d^m\sigma < +\infty.$$
Furthermore, the singular boundary $\Delta\mathcal{M}$ has capacity zero.
2. *Let the symbol*
$$\omega = \sum_{1\le i_1<\ldots<i_m\le n} a_{i_1\ldots i_m}(x)\,dx_{i_1}\wedge\ldots\wedge dx_{i_m}, \qquad x\in\overline{\mathcal{M}}$$
denote an m-dimensional differential form of the class $C^1(\mathcal{M})\cap C^0(\overline{\mathcal{M}})$, such that its exterior derivative $d\omega$ is absolutely integrable in the following sense:
$$\int\limits_{\mathcal{M}} |d\omega| < +\infty.$$

Statement: Then we have the identity
$$\int\limits_{\mathcal{M}} d\omega = \int\limits_{\partial\mathcal{M}} \omega.$$

Proof:

1. At first, let the condition $\omega \in C^1(\mathcal{M}) \cap C^0_0(\mathcal{M}\cup\partial\mathcal{M})$ be fulfilled. As above we choose a partition of unity $\{\alpha_k\}$ with $k = 1,\ldots,k_0$ on the set $\operatorname{supp}\omega \subset \mathcal{M}\cup\partial\mathcal{M}$ subordinate to the covering system of the charts. We utilize Proposition 4.1 and deduce
$$\int\limits_{\partial\mathcal{M}} \omega = \sum_{k=1}^{k_0}\int\limits_{\partial\mathcal{M}} \alpha_k\omega = \sum_{k=1}^{k_0}\int\limits_{\mathcal{M}} d(\alpha_k\omega) = \int\limits_{\mathcal{M}} d\omega.$$
2. Let the differential form ω be arbitrary now. Then we choose a sequence $\{\beta_k\}_{k=1,2,\ldots}$ of functions exhausting the manifold $\mathcal{M}$ with the property
$$\int\limits_{\mathcal{M}} \sqrt{\nabla(\beta_k,\beta_k)}\,d^{m+1}\sigma \to 0 \qquad \text{for} \quad k\to\infty.$$
According to part 1, we obtain the following identities for $k = 1,2,\ldots,$ namely

$$\int_{\partial\mathcal{M}} \beta_k \omega = \int_{\mathcal{M}} d(\beta_k \omega) = \int_{\mathcal{M}} \beta_k \, d\omega + \int_{\mathcal{M}} d\beta_k \wedge \omega. \tag{4}$$

At first, we see

$$\Big| \int_{\mathcal{M}} d\beta_k \wedge \omega \Big| \leq c \int_{\mathcal{M}} \sqrt{\nabla(\beta_k, \beta_k)} \, d^{m+1}\sigma \to 0 \quad \text{for} \quad k \to \infty.$$

Furthermore, we estimate

$$\int_{\partial\mathcal{M}} |\beta_k \omega| \leq \int_{\partial\mathcal{M}} |\omega| \leq c \int_{\partial\mathcal{M}} d^{m+1}\sigma < +\infty \quad \text{for} \quad k = 1, 2, \ldots$$

Therefore, we comprehend

$$\lim_{k \to \infty} \int_{\partial\mathcal{M}} \beta_k \omega =: \int_{\partial\mathcal{M}} \omega < +\infty.$$

On account of $\int_{\mathcal{M}} |d\omega| < +\infty$, we infer

$$\lim_{k \to \infty} \int_{\mathcal{M}} \beta_k \, d\omega =: \int_{\mathcal{M}} d\omega < +\infty.$$

The transition to the limit $k \to \infty$ in (4) reveals the identity

$$\int_{\partial\mathcal{M}} \omega = \int_{\mathcal{M}} d\omega,$$

which corresponds to the statement above. q.e.d.

5 The Integral Theorems of Gauß and Stokes

We endow the bounded open set $\Omega \subset \mathbb{R}^n$ with the chart $X(t) = t$, $t \in \Omega$ generating an atlas $\mathcal{A}$. In this way, we obtain a bounded oriented n-dimensional manifold $\mathcal{M} = \Omega$ in $\mathbb{R}^n$. When

$$f(x) = \Big(f_1(x), \ldots, f_n(x)\Big) \, : \, \Omega \longrightarrow \mathbb{R}^n \in C^1(\Omega, \mathbb{R}^n)$$

denotes an n-dimensional vector-field in $\mathbb{R}^n$ with its divergence

$$\operatorname{div} f(x) = \frac{\partial}{\partial x_1} f_1(x) + \ldots + \frac{\partial}{\partial x_n} f_n(x), \quad x \in \Omega,$$

we consider the $(n-1)$-form

$$\omega = \sum_{i=1}^{n} f_i(x)(-1)^{i+1}\, dx_1 \wedge \ldots \wedge dx_{i-1} \wedge dx_{i+1} \wedge \ldots \wedge dx_n.$$

The set of regular points $\partial\Omega$, endowed by the induced atlas $\partial\mathcal{A}$, becomes an $(n-1)$-dimensional bounded oriented manifold in $\mathbb{R}^n$. We show the identity

$$\int_{\partial\Omega} \omega = \int_{\partial\Omega} \Big(f(x)\cdot\xi(x)\Big)\, d^{n-1}\sigma$$

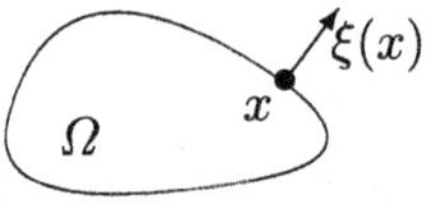

$\xi(x)$ x Ω

later, where $\xi(x)$ denotes the exterior normal to the domain Ω at the point x. When we take the relation

$$d\omega = \Big(\operatorname{div} f(x)\Big)\, dx_1 \wedge \ldots \wedge dx_n$$

into account, Theorem 4.7 from Section 4 reveals the fundamental *identity of Gauß:*

$$\int_{\Omega} \operatorname{div} f(x)\, d^n x = \int_{\partial\Omega} \Big(f(x)\cdot\xi(x)\Big)\, d^{n-1}\sigma. \tag{1}$$

With the aid of Theorem 4.7 from Section 4, we shall derive the identity (1) under very general conditions to Ω and f which are relevant for the applications in this textbook. Thus we shall obtain the *Gaussian integral theorem.*

Assumption (A):
Let $\Omega \subset \mathbb{R}^n$ denote a bounded open set, with the topological boundary $\dot{\Omega} = \overline{\Omega} \setminus \Omega$. For each boundary point $x \in \dot{\Omega}$, we can find a sequence of points

$$\Big\{x^{(p)}\Big\} \subset \mathbb{R}^n \setminus \overline{\Omega}, \qquad p = 1, 2, \ldots$$

satisfying $x^{(p)} \to x$ for $p \to \infty$; this means *each boundary point is attainable from outside.*

Assumption (B):
We choose $N \in \mathbb{N}$ bounded domains $T_i \subset \mathbb{R}^{n-1}$ with $i = 1, 2, \ldots, N$ as our parameter domains. Then we consider N regular hypersurfaces in $\mathbb{R}^n$ as follows:

$$\mathcal{F}_i : X^{(i)}(t) = \Big(x_1^{(i)}(t_1, \ldots, t_{n-1}), \ldots, x_n^{(i)}(t_1, \ldots, t_{n-1})\Big) : \overline{T}_i \to \mathbb{R}^n.$$

Here the mapping $X^{(i)}(t) \in C^1(T_i) \cap C^0(\overline{T}_i)$ is injective, and the rank of its functional matrix satisfies the condition $\operatorname{rg} \partial X^{(i)}(t) = n - 1$ for all points $t \in T_i$ and the indices $i = 1, \ldots, N$. Furthermore, their surface areas fulfill

$$A(\mathcal{F}_i) := \int_{T_i} d^{n-1}\sigma^{(i)}(t) < +\infty \quad \text{for} \quad i = 1, \ldots, N.$$

We define

$$F_i := X^{(i)}(T_i), \quad \overline{F}_i := X^{(i)}(\overline{T}_i), \quad \dot{F}_i := X^{(i)}(\dot{T}_i)$$

with $i = 1, \ldots, N$. Let the union of these finitely many hypersurfaces F_i constitute the boundary of Ω; more precisely

$$\dot{\Omega} = \overline{F}_1 \cup \ldots \cup \overline{F}_N.$$

Furthermore, we require the condition

$$\overline{F}_i \cap \overline{F}_j = \dot{F}_i \cap \dot{F}_j \quad \text{for all} \quad i, j \in \{1, \ldots, N\} \quad \text{with} \quad i \neq j.$$

Therefore, two different hypersurfaces possess common boundary points at most.

We need the following two auxiliary lemmas:

Proposition 5.1. *The point set $\Omega \subset \mathbb{R}^n$ may satisfy the assumptions (A) and (B). Furthermore, let $x^0 \in F_l$ denote an arbitrary point of the surface F_l with $l \in \{1, \ldots, N\}$. Then we find an index $k = k(x^0) \in \{1, \ldots, n\}$ as well as two positive numbers $\varrho = \varrho(x^0)$ and $\sigma = \sigma(x^0)$, such that the rectangle*

$$Q(x^0, \varrho, \sigma) := \Big\{x \in \mathbb{R}^n \, : \, |x_i - x_i^0| < \varrho, \; i = 1, \ldots, n \text{ with } i \neq k; \; |x_k - x_k^0| < \sigma\Big\}$$

is subject to the following conditions:

$$\dot{\Omega} \cap Q = \Big\{x \in \mathbb{R}^n \, : \, |x_i - x_i^0| < \varrho, \; i \neq k; \; x_k = \Phi(x_1, \ldots, x_{k-1}, x_{k+1}, \ldots, x_n)\Big\}.$$

Here Φ denotes a C^1-function on the domain of definition being given, such that $|\Phi - x_k^0| < \frac{1}{2}\sigma$ holds true. Furthermore, we have the alternative

$$\Omega \cap Q = \Big\{x \in \mathbb{R}^n \, : \, |x_i - x_i^0| < \varrho \quad \text{for } i \neq k, \\ |x_k - x_k^0| < \sigma, \; x_k < \Phi(x_1, \ldots, x_{k-1}, x_{k+1}, \ldots, x_n)\Big\}$$

or

$$\Omega \cap Q = \Big\{x \in \mathbb{R}^n \, : \, |x_i - x_i^0| < \varrho \quad \text{for } i \neq k, \\ |x_k - x_k^0| < \sigma, \; x_k > \Phi(x_1, \ldots, x_{k-1}, x_{k+1}, \ldots, x_n)\Big\}.$$

The adjacent diagram illustrates the statement of our proposition.

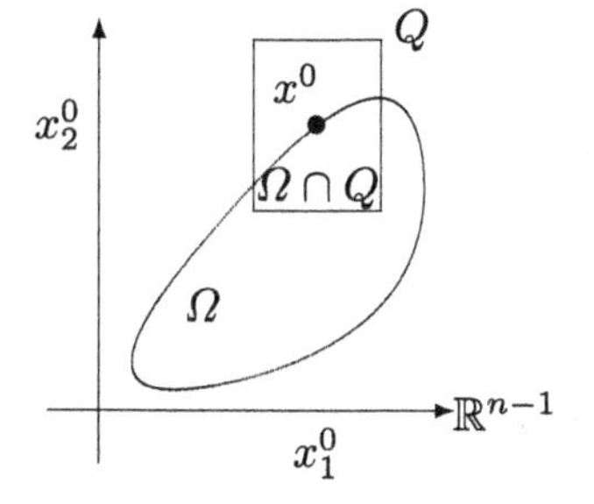

Proof:

1. With the open set $T \subset \mathbb{R}^{n-1}$, let us represent our surface $F = F_l$ by the mapping

$$X(t) = \Big(x_1(t_1, \ldots, t_{n-1}), \ldots, x_n(t_1, \ldots, t_{n-1})\Big) : T \longrightarrow \mathbb{R}^n.$$

On account of $\operatorname{rg} \partial X(t) = n - 1$ for all points $t \in T$, we find an index $k = k(x^0) \in \{1, \ldots, n\}$ with $x^0 = X(t^0)$, such that

$$\left.\frac{\partial(x_1, \ldots, x_{k-1}, x_{k+1}, \ldots, x_n)}{\partial(t_1, \ldots, t_{n-1})}\right|_{t=t^0} \neq 0$$

is correct. Now the theorem of the inverse mapping provides an open set $U \subset \mathbb{R}^{n-1}$ and a rectangle

$$\begin{aligned} R_\varrho := (x_1^0 - \varrho, x_1^0 + \varrho) \times \ldots \times (x_{k-1}^0 - \varrho, x_{k-1}^0 + \varrho) \\ \times (x_{k+1}^0 - \varrho, x_{k+1}^0 + \varrho) \times \ldots\ldots \times (x_n^0 - \varrho, x_n^0 + \varrho) \end{aligned}$$

with a sufficiently small quantity $\varrho = \varrho(x^0) > 0$, such that

$$f(t_1, \ldots, t_{n-1}) := \Big(x_1(t), \ldots, x_{k-1}(t), x_{k+1}(t), \ldots, x_n(t)\Big) : U \longrightarrow R_\varrho$$

constitutes a C^1-diffeomorphism. This means that f is bijective, f as well as f^{-1} are continuously differentiable, and we have the condition $J_f(t) \neq 0$ for all $t \in U$. We define

$$\overset{k}{\underset{\vee}{x}} := (x_1, \ldots, x_{k-1}, x_{k+1}, \ldots, x_n) \in R_\varrho \subset \mathbb{R}^{n-1}$$

and introduce the function

$$\Phi(\overset{k}{\underset{\vee}{x}}) := x_k\Big(f^{-1}(\overset{k}{\underset{\vee}{x}})\Big), \qquad \overset{k}{\underset{\vee}{x}} \in R_\varrho.$$

Then we observe

$$\Phi \in C^1(R_\varrho, \mathbb{R}), \quad X(U) = \Big\{(x_1, \ldots, x_n) : \overset{k}{\underset{\vee}{x}} \in R_\varrho,\ x_k = \Phi(\overset{k}{\underset{\vee}{x}})\Big\}.$$

Now we see

$$x^0 \in \dot{\Omega} \setminus \bigcup_{\substack{m=1 \\ m \neq l}}^{N} \overline{F}_m,$$

and consequently

$$\operatorname{dist}\,(x^0, \bigcup_{\substack{m=1 \\ m \neq l}}^{N} \overline{F}_m) > 0.$$

We choose the quantities $\varrho > 0$ and $\sigma > 0$ sufficiently small, such that

$$Q(x^0, \varrho, \sigma) \cap \dot{\Omega} = Q(x^0, \varrho, \sigma) \cap F_l \quad \text{as well as} \quad |\Phi(\overset{k}{\vee}{x}) - x_k^0| < \frac{1}{2}\sigma$$

holds true for all $\overset{k}{\vee}{x} \in R_\varrho$. We summarize our considerations and obtain

$$\dot{\Omega} \cap Q(x^0, \varrho, \sigma) = \Big\{x \in \mathbb{R}^n : \overset{k}{\vee}{x} \in R_\varrho,\ x_k = \Phi(\overset{k}{\vee}{x})\Big\}.$$

2. Now we define the point sets

$$P^+ := \Big\{x \in Q(x^0, \varrho, \sigma) \ :\ x_k > \Phi(\overset{k}{\vee}{x})\Big\},$$

$$P^0 := \Big\{x \in Q(x^0, \varrho, \sigma) \ :\ x_k = \Phi(\overset{k}{\vee}{x})\Big\},$$

$$P^- := \Big\{x \in Q(x^0, \varrho, \sigma) \ :\ x_k < \Phi(\overset{k}{\vee}{x})\Big\}.$$

These sets above decompose the set $Q(x^0, \varrho, \sigma)$ according to the prescription

$$Q(x^0, \varrho, \sigma) = P^- \cup P^0 \cup P^+. \tag{2}$$

From the first part of our proof we infer

$$\dot{\Omega} \cap Q(x^0, \varrho, \sigma) = P^0. \tag{3}$$

On account of $x^0 \in \dot{\Omega}$ and the assumption (A), we can find the two points $y \in \Omega \cap Q$ and $z \in (\mathbb{R}^n \setminus \overline{\Omega}) \cap Q$. We distinguish between two possible cases, namely the case 1: $y \in P^-$ and the case 2: $y \in P^+$.

Case 1. When we consider with $\widetilde{y} \in P^-$ an arbitrary further point, we find a continuous curve $\Gamma \subset P^-$ from y to $\widetilde{y}$, which does not intersect the surface P^0. Since $y \in \Omega$ holds true and the curve Γ does not intersect the set $\dot{\Omega}$ due to (3), we infer $\widetilde{y} \in \Omega$. We finally obtain the inclusion

$$P^- \subset \Omega \cap Q. \tag{4}$$

Now we arrive at $z \in P^+$. Each further point $\widetilde{z} \in P^+$ can be connected by a curve Γ in P^+ with the point z. Since this curve does not intersect $\dot{\Omega}$, the condition $z \in \mathbb{R}^n \setminus \overline{\Omega}$ implies $\widetilde{z} \in \mathbb{R}^n \setminus \overline{\Omega}$ as well. We conclude

$$P^+ \subset (\mathbb{R}^n \setminus \overline{\Omega}) \cap Q. \tag{5}$$

Furthermore, we observe

$$Q(x^0, \varrho, \sigma) = (\Omega \cap Q) \cup (\dot{\Omega} \cap Q) \cup \Big((\mathbb{R}^n \setminus \overline{\Omega}) \cap Q\Big). \tag{6}$$

We deduce $P^- = \Omega \cap Q$ and $P^+ = (\mathbb{R}^n \setminus \overline{\Omega}) \cap Q$ from the equations (2) to (6).

Case 2. In the same way as in the first case, we show $P^+ = \Omega \cap Q$ and $P^- = (\mathbb{R}^n \setminus \overline{\Omega}) \cap Q$.

q.e.d.

Remark: In the neighborhood of a regular boundary point

$$x^0 \in \bigcup_{i=1}^{N} F_i$$

we choose the function

$$\Psi(x) := \pm\Big(x_k - \Phi(x_1, \ldots, x_{k-1}, x_{k+1}, \ldots, x_n)\Big)$$

due to Proposition 5.1. Thus we can characterize the set Ω in this neighborhood by the inequality $\Psi(x) < 0$.

Proposition 5.2. *The set $\Omega \subset \mathbb{R}^n$ may satisfy the assumptions (A) and (B); let $x^0 \in F_l$ with $l \in \{1, \ldots, N\}$ denote a point of the surface F_l. Furthermore, we have an open set $U = U(x^0) \subset \mathbb{R}^n$ containing the point x^0 and a function $\Psi(x) \in C^1(U)$ with $|\nabla\Psi(x)| > 0$ for all points $x \in U$, such that*

$$\Omega \cap U = \{x \in U \,:\, \Psi(x) < 0\}.$$

Then the vector

$$\xi(x) := |\nabla\Psi(x)|^{-1}\nabla\Psi(x), \quad x \in \dot{\Omega} \cap U$$

has the following properties:

1. *We have $\xi\Big(X(t)\Big) \cdot X_{t_i}(t) = 0$ for $i = 1, \ldots, n-1$ near $t = t^0$;*
2. *The condition $|\xi| = 1$ on $\dot{\Omega} \cap U$ holds true;*
3. *For each point $x \in \dot{\Omega} \cap U$, we can find a number $\varrho_0(x) > 0$ such that*

$$x + \varrho\xi \in \begin{cases} \Omega & \text{for} \quad -\varrho_0 < \varrho < 0 \\ \mathbb{R}^n \setminus \overline{\Omega} & \text{for} \quad 0 < \varrho < +\varrho_0 \end{cases}.$$

The vector ξ is uniquely determined by these conditions.

Definition 5.3. *The function $\xi = \xi(x)$, defined in Proposition 5.2 for all points $x \in F_1 \cup \ldots \cup F_N$, is named the* exterior normal of $\dot{\Omega}$ at the point x.

Proof of Proposition 5.2: The uniqueness of ξ follows from the properties 1 to 3 above. Now we shall prove the properties given for the function ξ. At first, $\Psi = 0$ on $\dot{\Omega} \cap U$ holds true, and we infer

$$0 = \Psi\Big(x_1(t), \ldots, x_n(t)\Big), \quad t = (t_1, \ldots, t_{n-1}) \in V(t_1^0, \ldots, t_{n-1}^0) \subset \mathbb{R}^{n-1} \text{ open},$$

and consequently

$$0 = \sum_{i=1}^{n} \Psi_{x_i}\Big(X(t)\Big)\frac{\partial x_i}{\partial t_j}, \quad j = 1, \ldots, n-1.$$

This implies $\xi \cdot X_{t_j} = 0$ in V for $j = 1, \ldots, n-1$ and the property 1. Evidently, the condition $|\xi| = 1$ is valid on $\dot{\Omega} \cap U$. Therefore, it remains to show the property 3. When $0 < |\varrho| < \varrho_0$ holds true, we infer the inequality

$$\begin{aligned}\Psi(x + \varrho\xi) &= \Psi(x + \varrho\xi) - \Psi(x) = \varrho \sum_{i=1}^{n} \Psi_{x_i}(x + \kappa\varrho\xi)\xi_i \\ &= \varrho \frac{1}{|\nabla\Psi(x)|} \sum_{i=1}^{n} \Psi_{x_i}(x + \kappa\varrho\xi)\Psi_{x_i}(x) \begin{cases} < 0 \text{ if } -\varrho_0 < \varrho < 0 \\ > 0 \text{ if } 0 < \varrho < \varrho_0 \end{cases}\end{aligned}$$

for all points $x \in \dot{\Omega} \cap U$; with a number $\kappa = \kappa(\varrho) \in (0, 1)$. This implies

$$x + \varrho\xi \in \begin{cases} \Omega \text{ if } -\varrho_0 < \varrho < 0 \\ \mathbb{R}^n \setminus \overline{\Omega} \text{ if } 0 < \varrho < \varrho_0 \end{cases}.$$

q.e.d.

Remark: Let the surface patch $F = F_l$ bounding Ω be given by the parametric representation

$$X(t) = X(t_1, \ldots, t_{n-1}) : T \longrightarrow \mathbb{R}^n \quad \text{on the domain} \quad T \subset \mathbb{R}^{n-1}$$

with the normal

$$\begin{aligned}\nu(t) &= |X_{t_1} \wedge \ldots \wedge X_{t_{n-1}}|^{-1} X_{t_1} \wedge \ldots \wedge X_{t_{n-1}}(t) \\ &= \left[\sum_{j=1}^{n} \Big(D_j(t)\Big)^2\right]^{-\frac{1}{2}} (D_1(t), \ldots, D_n(t)), \qquad t \in T.\end{aligned}$$

With a fixed $\varepsilon \in \{\pm 1\}$, we observe

$$\xi\Big(X(t)\Big) = \varepsilon\nu(t) \qquad \text{for all} \quad t \in T.$$

Proof: At first, we see $\xi\Big(X(t)\Big) = \varepsilon(t)\nu(t)$, $t \in T$ with the orientation factor $\varepsilon(t) \in \{\pm 1\}$. Now the function

$$\varepsilon(t) = \xi\Big(X(t)\Big) \cdot \nu(t), \quad t \in T$$

is continuous on the domain T, and we obtain $\varepsilon(t) \equiv +1$ or $\varepsilon(t) \equiv -1$ on T.

q.e.d.

Definition 5.4. *The set $\Omega \subset \mathbb{R}^n$ may satisfy the assumptions (A) and (B). Then we define*

$$\partial\Omega := \bigcup_{j=1}^{N} F_j$$

as the regular boundary of Ω. *Furthermore, let* $g(x) : \partial\Omega \to \mathbb{R}$ *denote a continuous bounded function on* $\partial\Omega$. *We define the* surface integral of g over the regular boundary $\partial\Omega$ *by the expression*

$$\int_{\partial\Omega} g(x)\, d^{n-1}\sigma := \sum_{j=1}^{N} \int_{F_j} g(x)\, d^{n-1}\sigma_j.$$

Now we formulate the assumption for our vector-fields to be integrated.

Assumption (C):
The function $f(x) = (f_1(x), \ldots, f_n(x))$, $x \in \overline{\Omega}$ belongs to the regularity class $C^1(\Omega, \mathbb{R}^n) \cap C^0(\overline{\Omega}, \mathbb{R}^n)$, and we require

$$\int_{\Omega} |\operatorname{div} f(x)|\, dx < +\infty.$$

We present a condition on the singular boundary $\dot{F}_1 \cup \ldots \cup \dot{F}_N$, which guarantees the validity of the Gaussian identity (1):

Assumption (D):
The set $\dot{F}_1 \cup \ldots \cup \dot{F}_N$ has the $(n-1)$-dimensional Hausdorff content zero or equivalently represents an $(n-1)$-dimensional Hausdorff null-set. More precisely, for each quantity $\varepsilon > 0$ we have finitely many balls

$$K_j := \left\{x \in \mathbb{R}^n \, : \, |x - x^{(j)}| \le \varrho_j\right\} \quad \text{for} \quad j = 1, \ldots, J$$

with the centers $x^{(j)} \in \mathbb{R}^n$ and radii $\varrho_j > 0$, such that the following conditions hold true:

1. $\dot{F}_1 \cup \ldots \cup \dot{F}_N \subset \bigcup_{j=1}^{J} K_j$ (Covering property);

2. $\sum_{j=1}^{J} \varrho_j^{n-1} \le \varepsilon$ (Smallness of the total area).

Remark: The condition (D) is valid, if all surface patches F_l with $l = 1, \ldots, N$ fulfill the subsequent assumptions: When F_l is parametrized by the representation $X = X(t) : \overline{T}_l \to \overline{F}_l$, we require the following:

1. The set $\overline{T}_l$ constitutes a Jordan domain in $\mathbb{R}^{n-1}$, which means that T_l is compact and its boundary $\dot{T}_l$ represents a Jordan null-set in $\mathbb{R}^{n-1}$;

2. The mapping $X(t)$ satisfies a *Lipschitz condition* on $\overline{T}_l$, namely

$$|X(t') - X(t'')| \leq L|t' - t''| \qquad \text{for all} \quad t', t'' \in \overline{T}_l,$$

with the Lipschitz constant $L > 0$.

We now arrive at the central theorem of the n-dimensional integral-calculus.

Theorem 5.5. (Gaussian integral theorem)
Let $\Omega \subset \mathbb{R}^n$ denote a bounded open set satisfying the assumptions (A), (B), and (D). Furthermore, the vector-valued function $f(x)$ fulfills the assumption (C). Then we have the identity

$$\int_{\Omega} \operatorname{div} f(x)\, dx = \int_{\partial\Omega} f(x) \cdot \xi(x)\, d^{n-1}\sigma.$$

Proof: (E. Heinz)
We shall prove this statement by referring to Theorem 4.7 from Section 4.

1. We comprehend $\mathcal{M} = \Omega \subset \mathbb{R}^n$ as an n-dimensional manifold in $\mathbb{R}^n$ with the atlas $\mathcal{A} : X(t) = t$, $t \in \Omega$. For each point

$$x^0 \in \bigcup_{l=1}^{N} F_l \subset \dot{\Omega}$$

we now find a rectangle $Q(x^0, \varrho, \sigma)$ due to Proposition 5.1, such that

$$\Omega \cap Q = \Big\{ x \in \mathbb{R}^n \,:\, |x_i - x_i^0| < \varrho\, (i \neq k),$$
$$x_k \lesseqgtr \Phi(x_1, \ldots, x_{k-1}, x_{k+1}, \ldots, x_n),\ |x_k - x_k^0| < \sigma \Big\}.$$

On the semicube

$$H := \Big\{ t \in \mathbb{R}^n \,:\, t_1 \in (-\varrho, 0),\ |t_i| < \varrho,\ i = 2, \ldots, n \Big\}$$

with the upper bounding side

$$S := \Big\{ t \in \mathbb{R}^n \,:\, t_1 = 0,\ |t_i| < \varrho,\ i = 2, \ldots, n \Big\}$$

in the direction of e_1, we consider the transformation

$$Y(t) = \Big(x_1^0 + \varepsilon_2 t_2, \ldots, x_{k-1}^0 + \varepsilon_k t_k, \Phi(x_1^0 + \varepsilon_2 t_2, \ldots, x_{k-1}^0 + \varepsilon_k t_k,$$
$$x_{k+1}^0 + \varepsilon_{k+1} t_{k+1}, \ldots, x_n^0 + \varepsilon_n t_n) + \varepsilon_1 t_1, x_{k+1}^0 + \varepsilon_{k+1} t_{k+1}, \ldots, x_n^0 + \varepsilon_n t_n \Big)$$

where $\varepsilon_k \in \{\pm 1\}$ for $k = 1, \ldots, n$ holds true. Choosing the sign factors $\varepsilon_1, \ldots, \varepsilon_n$ suitably, we attain the conditions

$$Y(H) \subset \Omega \cap Q, \quad Y(S) = \dot{\Omega} \cap Q, \quad \text{and} \quad J_Y(0) = +1$$

for the functional determinant of Y. Therefore, the mapping Y is compatible with the chart X from above, and we endow $\partial\mathcal{M} = \partial\Omega$ with the induced atlas. On account of the condition $J_Y(0) > 0$, the normal $\nu(t)$ to a surface patch oriented by $\partial\Omega$ points in the direction of the exterior normal ξ to $\partial\Omega$.

We now consider the $(n-1)$-form

$$\omega = \sum_{i=1}^{n} (-1)^{i+1} f_i(x)\, dx_1 \wedge \ldots \wedge dx_{i-1} \wedge dx_{i+1} \wedge \ldots \wedge dx_n \in C^1(\mathcal{M}) \cap C^0(\overline{\mathcal{M}}).$$

From our considerations above we infer

$$\int_{\partial\Omega} \omega = \int_{\partial\Omega} f(x) \cdot \xi(x)\, d^{n-1}\sigma.$$

2. Due to the assumption (D), we have finitely many balls to each quantity $\varepsilon > 0$, namely

$$K_j := \left\{ x \in \mathbb{R}^n \,:\, |x - x^{(j)}| \le \varrho_j \right\} \quad \text{for} \quad j = 1, \ldots, J,$$

satisfying

$$\dot{F}_1 \cup \ldots \cup \dot{F}_N \subset \bigcup_{j=1}^{J} K_j \quad \text{and} \quad \sum_{j=1}^{J} \rho_j^{n-1} \le \epsilon.$$

Now we show that the capacity of the singular boundary vanishes. In this context we construct a function $\Psi(r) : [0, +\infty) \to [0, 1] \in C^1$ with

$$\Psi(r) = \begin{cases} 0, \; 0 \le r \le 2 \\ 1, \quad 3 \le r \end{cases} \quad \text{and} \quad M := \sup_{r \ge 0} |\Psi'(r)| < +\infty.$$

For the indices $j = 1, \ldots, J$ we consider the functions

$$\chi_j(x) := \Psi\left(|x - x^{(j)}| \,/\, \varrho_j \right), \qquad x \in \mathbb{R}^n,$$

satisfying $\chi_j \in C^1(\mathbb{R}^n)$ and

$$\chi_j(x) = \begin{cases} 1, \; |x - x^{(j)}| \ge 3\varrho_j \\ 0, \; |x - x^{(j)}| \le 2\varrho_j \end{cases}.$$

When E_n denotes the volume of the n-dimensional unit ball, we evaluate

$$\int_{\mathbb{R}^n} |\nabla \chi_j(x)|\, dx = \int_{2\varrho_j \le |x-x^{(j)}| \le 3\varrho_j} \left| \Psi'\Big(\frac{1}{\varrho_j}|x - x^{(j)}|\Big) \right| \frac{1}{\varrho_j}\, dx$$

$$\le \frac{M}{\varrho_j} E_n (3^n \varrho_j^n - 2^n \varrho_j^n)$$

$$= M E_n (3^n - 2^n) \varrho_j^{n-1}$$

for $j = 1, \ldots, J$. We obtain a function

$$\chi(x) := \chi_1(x) \cdot \ldots \cdot \chi_J(x) \in C_0^1\Big(\overline{\Omega} \setminus (\dot{F}_1 \cup \ldots \cup \dot{F}_N)\Big)$$

with

$$\int_{\Omega} |\nabla \chi(x)|\, dx \le \sum_{j=1}^{J} \int_{\mathbb{R}^n} |\nabla \chi_j(x)|\, dx$$

$$\le M E_n (3^n - 2^n) \sum_{j=1}^{J} \varrho_j^{n-1}$$

$$\le M E_n (3^n - 2^n) \varepsilon.$$

Therefore, the set $\dot{F}_1 \cup \ldots \cup \dot{F}_n \subset \dot{\Omega}$ has capacity zero.

3. The Stokes integral theorem for manifolds finally reveals

$$\int_{\partial\Omega} f(x) \cdot \xi(x)\, d^{n-1}\sigma = \int_{\partial\mathcal{M}} \omega = \int_{\mathcal{M}} d\omega = \int_{\Omega} \operatorname{div} f(x)\, dx.$$

This corresponds to the statement above. q.e.d.

We obtain immediately *Green's formula* from Theorem 5.5, which is fundamental for the *potential theory* presented in Chapter 5.

Theorem 5.6. (Green's formula)
Let $\Omega \subset \mathbb{R}^n$ denote an open bounded set in $\mathbb{R}^n$ satisfying the assumptions (A), (B), and (D). Furthermore, let the functions $f(x)$ and $g(x)$ belong to the class $C^1(\overline{\Omega}) \cap C^2(\Omega)$ subject to the integrability condition

$$\int_{\Omega} \Big(|\Delta f(x)| + |\Delta g(x)| \Big)\, dx < +\infty.$$

Here the symbol $\triangle$ denotes the Laplace operator due to

$$\triangle f(x) := \sum_{i=1}^{n} \frac{\partial^2 f}{\partial x_i \partial x_i}(x).$$

Then we have the identity

$$\int\limits_{\Omega} \Big(f\Delta g - g\Delta f\Big)\, dx = \int\limits_{\partial\Omega} \left(f\,\frac{\partial g}{\partial \xi} - g\,\frac{\partial f}{\partial \xi}\right)\, d^{n-1}\sigma$$

using the notations

$$\frac{\partial f}{\partial \xi} := \nabla f(x)\cdot \xi(x), \quad \frac{\partial g}{\partial \xi} := \nabla g(x)\cdot \xi(x), \qquad x\in\partial\Omega.$$

Proof: We apply the Gaussian integral theorem to the vector-field

$$h(x) := f(x)\nabla g(x) - g(x)\nabla f(x).$$

Now we deduce

$$\operatorname{div} h(x) = \nabla h(x) = f(x)\Delta g(x) - g(x)\Delta f(x),$$

and we obtain

$$\int\limits_{\Omega} \Big(f(x)\Delta g(x) - g(x)\Delta f(x)\Big)\, dx = \int\limits_{\partial\Omega} h(x)\cdot\xi(x)\, d^{n-1}\sigma$$

$$= \int\limits_{\partial\Omega} \left(f(x)\frac{\partial g}{\partial \xi}(x) - g(x)\frac{\partial f}{\partial \xi}(x)\right)\, d^{n-1}\sigma,$$

which implies the statement above. q.e.d.

We specialize the Stokes integral theorem for manifolds onto 2-dimensional surfaces in the Euclidean space $\mathbb{R}^3$. Since we even prove this theorem for surfaces with singular boundaries, we need the following result which is important to construct conformal mappings (in Chapter 4) and central within the theory of *Nonlinear Elliptic Systems* (in Chapter 12).

Theorem 5.7. (Oscillation lemma of R. Courant and H. Lebesgue) *Let*

$$B := \Big\{w = u + iv = (u,v)\in\mathbb{C}\cong\mathbb{R}^2 \,:\, |w| < 1\Big\}$$

denote the open unit disc and

$$X(u,v) = \Big(x_1(u,v),\ldots,x_n(u,v)\Big) \,:\, B\to\mathbb{R}^n \in C^1(B)$$

a vector-valued function with finite Dirichlet integral $D(X)$; more precisely

$$D(X) := \iint\limits_{B} \Big(|X_u(u,v)|^2 + |X_v(u,v)|^2\Big)\, du dv \le N < +\infty.$$

For each point $w_0 = u_0 + iv_0 \in \overline{B}$ and each quantity $\delta\in(0,1)$, we then find a number $\delta^*\in[\delta,\sqrt{\delta}]$, such that the estimate

$$L := \int\limits_{\substack{|w-w_0|=\delta^* \\ w\in B}} d\sigma(w) \ \leq \ 2\sqrt{\frac{\pi N}{\log\frac{1}{\delta}}}$$

is valid for the length L of the curve $X(w)$, $|w-w_0| = \delta^*$, $w \in B$.

For the proof of this theorem we add the elementary

Proposition 5.8. *Let the numbers $a < b$ be given and the function $f(x) : [a,b] \to \mathbb{R}$ be continuous. Then we have the estimate*

$$\int\limits_a^b |f(x)|\,dx \leq \sqrt{b-a}\sqrt{\int\limits_a^b |f(x)|^2\,dx}.$$

Proof: Let $\mathcal{Z} : a = x_0 < x_1 < \ldots < x_N = b$ represent an equidistant decomposition of the interval $[a,b]$ - with the partitioning points $x_j := a+j\frac{b-a}{N}$ for $j = 0,1,\ldots,N$. When $\xi_j \in [x_j, x_{j+1}]$ denote arbitrary intermediate points, the Cauchy-Schwarz inequality reveals

$$\sum_{j=0}^{N-1} |f(\xi_j)|(x_{j+1}-x_j) \leq \sqrt{\sum_{j=0}^{N-1} |f(\xi_j)|^2(x_{j+1}-x_j)}\sqrt{\sum_{j=0}^{N-1}(x_{j+1}-x_j)}$$

$$= \sqrt{b-a}\sqrt{\sum_{j=0}^{N-1} |f(\xi_j)|^2(x_{j+1}-x_j)}.$$

The transition to the limit $N \to \infty$ yields the inequality

$$\int\limits_a^b |f(x)|\,dx \leq \sqrt{b-a}\sqrt{\int_a^b |f(x)|^2\,dx},$$

which has been stated above. q.e.d.

Proof of Theorem 5.7: We introduce polar coordinates about the point $w_0 = u_0 + iv_0$ as follows:

$$u = u_0 + \varrho\cos\varphi, \quad v = v_0 + \varrho\sin\varphi, \qquad 0 \leq \varrho \leq \sqrt{\delta}, \quad \varphi_1(\varrho) \leq \varphi \leq \varphi_2(\varrho).$$

Furthermore, we define the function

$$\Psi(\varrho,\varphi) := X(u_0 + \varrho\cos\varphi, v_0 + \varrho\sin\varphi)$$

and calculate

$$\Psi_\varrho = X_u \cos\varphi + X_v \sin\varphi,$$

$$\Psi_\varphi = -X_u \varrho \sin\varphi + X_v \varrho \cos\varphi$$

as well as

$$|\Psi_\varrho|^2 + \frac{1}{\varrho^2}|\Psi_\varphi|^2 = |X_u|^2 + |X_v|^2.$$

Using the intermediate value theorem of the integral-calculus in combination with Proposition 5.8, we obtain

$$N \geq D(X) = \iint\limits_B (|X_u|^2 + |X_v|^2)\, du dv \geq \int\limits_\delta^{\sqrt{\delta}} \int\limits_{\varphi_1(\varrho)}^{\varphi_2(\varrho)} \left(|\Psi_\varrho|^2 + \frac{1}{\varrho^2}|\Psi_\varphi|^2\right) \varrho\, d\varrho d\varphi$$

$$\geq \int\limits_\delta^{\sqrt{\delta}} \frac{1}{\varrho} \left(\int\limits_{\varphi_1(\varrho)}^{\varphi_2(\varrho)} |\Psi_\varphi|^2\, d\varphi \right) d\varrho = \left(\int\limits_{\varphi_1(\delta^*)}^{\varphi_2(\delta^*)} |\Psi_\varphi(\delta^*,\varphi)|^2\, d\varphi \right) \int\limits_\delta^{\sqrt{\delta}} \frac{d\varrho}{\varrho}$$

$$\geq \frac{1}{2}\left(\log\frac{1}{\delta}\right) \frac{1}{\varphi_2(\delta^*) - \varphi_1(\delta^*)} \left(\int\limits_{\varphi_1(\delta^*)}^{\varphi_2(\delta^*)} |\Psi_\varphi(\delta^*,\varphi)|\, d\varphi \right)^2$$

$$\geq \frac{1}{4\pi}\log\left(\frac{1}{\delta}\right) \left(\int\limits_{\varphi_1(\delta^*)}^{\varphi_2(\delta^*)} |\Psi_\varphi(\delta^*,\varphi)|\, d\varphi \right)^2$$

for a number $\delta^* \in [\delta, \sqrt{\delta}]$. Finally, we infer the inequality

$$L = \int\limits_{\varphi_1(\delta^*)}^{\varphi_2(\delta^*)} |\Psi_\varphi(\delta^*,\varphi)|\, d\varphi \leq \sqrt{\frac{4\pi N}{\log\frac{1}{\delta}}} = 2\sqrt{\frac{\pi N}{\log\frac{1}{\delta}}}$$

and arrive at the statement above. q.e.d.

Remark: When we choose $w_0 \in B$ in Theorem 5.7, we have only to require the regularity $X \in C^1(B \setminus \{w_0\})$.

We are now prepared to prove the interesting

Theorem 5.9. (Classical Stokes integral theorem with singular boundary)

1. *On the boundary of the closed unit disc $\overline{B}$ we have given $k_0 \in \mathbb{N} \cup \{0\}$ points $w_k = exp(i\varphi_k)$ for $k = 1, \ldots, k_0$ with their associate angles $0 \leq \varphi_1 < \ldots < \varphi_{k_0} < 2\pi$. When we exempt the points w_k for $k = 1, \ldots, k_0$ from the sets $\overline{B}$ and ∂B, we obtain the sets $\overline{B}'$ and $\partial B'$, respectively.*

2. *Furthermore, let the injective mapping*

$$X(u,v) = \Big(x_1(u,v), x_2(u,v), x_3(u,v)\Big) \; : \; \overline{B} \longrightarrow \mathbb{R}^3 \in C^1(\overline{B}\,') \cap C^0(\overline{B})$$

with the property $X_u \wedge X_v \neq 0$ *for all* $(u,v) \in \overline{B}\,'$ *and finite Dirichlet integral* $D(X) < +\infty$ *be given. Let the surface be* conformally parametrized, *which means the conformality relations*

$$|X_u| = |X_v|, \quad X_u \cdot X_v = 0 \qquad \textit{for all} \quad (u,v) \in B$$

are satisfied. Denoting by

$$\overline{X}(\varphi) := X\Big(e^{i\varphi}\Big), \qquad 0 \leq \varphi \leq 2\pi$$

the restriction of X *onto* ∂B, *we obtain the line element*

$$d^1\sigma(\varphi) = |\overline{X}'(\varphi)|\, d\varphi, \qquad 0 \leq \varphi \leq 2\pi, \quad \varphi \notin \{\varphi_1, \ldots, \varphi_{k_0}\}.$$

We require finite length for the curve $\overline{X}(\varphi)$; *and more precisely*

$$L(\overline{X}) = \sum_{k=0}^{k_0-1} \int\limits_{\varphi_k}^{\varphi_{k+1}} d^1\sigma(\varphi) < +\infty,$$

where we defined $\varphi_0 := \varphi_{k_0} - 2\pi$.

3. *By the symbol*

$$\nu(u,v) := |X_u \wedge X_v|^{-1} X_u \wedge X_v\,, \qquad (u,v) \in \overline{B}\,'$$

we denote the unit normal vector and by

$$d^2\sigma(u,v) := |X_u \wedge X_v|\, dudv$$

the surface element of the surface $X(u,v)$. *The tangential vector to the boundary curve is abbreviated by*

$$T(\varphi) := \frac{\overline{X}\,'(\varphi)}{|\overline{X}\,'(\varphi)|}.$$

4. *Let* $\mathcal{O} \supset X(B) =: \mathcal{M}$ *constitute an open set in* $\mathbb{R}^3$, *and let the vector-field*

$$a(x) = \Big(a_1(x_1,x_2,x_3), a_2(x_1,x_2,x_3), a_3(x_1,x_2,x_3)\Big) \in C^1(\mathcal{O}) \cap C^0(\overline{\mathcal{M}})$$

be prescribed with the integrability property

$$\iint\limits_B |\operatorname{rot} a(X(u,v))|\, d^2\sigma(u,v) < +\infty.$$

Then we have the Stokes identity

$$\iint_B \left\{\operatorname{rot} a\Big(X(u,v)\Big)\cdot\nu(u,v)\right\} d^2\sigma(u,v) = \int_0^{2\pi} \left\{a\Big(\overline{X}(\varphi)\Big)\cdot T(\varphi)\right\} d^1\sigma(\varphi). \quad (7)$$

Remarks: Since the surface is *conformally parametrized*, our condition $D(X) < +\infty$ is equivalent to the finiteness of the surface area of X, on account of the relation

$$D(X) = 2\iint_B d^2\sigma(u,v) =: 2A(X).$$

The introduction of isothermal parameters in the large is treated in Section 8 of Chapter 12.

Proof of Theorem 5.9:

1. We intend to apply the Stokes integral theorem for manifolds: The set $\mathcal{M} := X(B)$ constitutes a bounded oriented 2-dimensional C^1-manifold in $\mathbb{R}^3$ with the chart $X(u,v) : B \to \mathcal{M}$. The regular boundary $\partial\mathcal{M} := X(\partial B')$ inherits its orientation by the mapping $\overline{X}(\varphi)$, $0 \le \varphi \le 2\pi$ and possesses finite length $L(\overline{X}) < +\infty$. At first, we show that the singular boundary $\Delta\mathcal{M} := X(\{w_1,\ldots,w_{k_0}\}) \subset \dot{\mathcal{M}} \subset \mathbb{R}^3$ has capacity zero.
2. When $w^* \in \partial B$ is a singular point of the surface, we introduce polar coordinates in a neighborhood of w^* as follows:

$$w = w^* + \varrho e^{i\varphi}, \qquad 0 < \varrho < \varrho^*, \quad \varphi_1(\varrho) < \varphi < \varphi_2(\varrho).$$

For the quantity $\eta > 0$ being given, the Courant-Lebesgue oscillation lemma provides a number $\delta \in (0,\rho^*)$ with the following property: Defining the function $Y(\varrho,\varphi) := X(w^* + \varrho e^{i\varphi})$, $0 < \rho < \rho^*$, $\varphi_1(\rho) < \varphi < \varphi_2(\rho)$, we have the inequality

$$\int_{\varphi_1(\delta^*)}^{\varphi_2(\delta^*)} |Y_\varphi(\delta^*,\varphi)|\,d\varphi \le 2\sqrt{\frac{\pi D(X)}{\log\frac{1}{\delta}}} \le \eta \quad (8)$$

for one number $\delta^* \in [\delta,\sqrt{\delta}]$ at least. Consequently, we find two numbers $0 < \varrho_1 < \delta^* < \varrho_2 < \varrho^*$ with the property

$$\int_{\varphi_1(\varrho)}^{\varphi_2(\varrho)} |Y_\varphi(\varrho,\varphi)|\,d\varphi \le 2\eta \qquad \text{for all} \quad \varrho \in [\varrho_1,\varrho_2].$$

Now we consider the weakly monotonic function

$$\Psi(\varrho) : [0,\varrho^*] \longrightarrow [0,1] \in C^1$$

with the properties

$$\Psi(\varrho) = \begin{cases} 0, \, 0 \leq \varrho \leq \varrho_1 \\ 1, \, \varrho_2 \leq \varrho \leq \varrho^* \end{cases} .$$

In a neighborhood of the surface $\mathcal{M}$, we now construct a function

$$\chi = \chi(x_1, x_2, x_3) \in C^1(\mathcal{M})$$

satisfying

$$\Psi(\varrho) = \chi \circ Y(\varrho, \varphi), \; 0 < \varrho < \varrho^*, \; \varphi_1(\varrho) < \varphi < \varphi_2(\varrho).$$

This implies

$$\Psi'(\varrho) = \nabla\chi\big|_{Y(\varrho,\varphi)} \cdot Y_\varrho(\varrho, \varphi) = |\boldsymbol{\nabla}\chi(Y(\varrho, \varphi))||Y_\varrho(\varrho, \varphi)|.$$

We conclude

$$\begin{aligned} &\iint\limits_{w \in B \cap B_{\varrho^*}(w^*)} |\boldsymbol{\nabla}\chi| \, d^2\sigma(u, v) \\ &\leq \int\limits_0^{\varrho^*} \left(\int\limits_{\varphi_1(\varrho)}^{\varphi_2(\varrho)} |\boldsymbol{\nabla}\chi(Y(\varrho, \varphi))||Y_\varrho||Y_\varphi| \, d\varphi \right) d\varrho \\ &= \int\limits_0^{\varrho^*} \Psi'(\varrho) \left(\int\limits_{\varphi_1(\varrho)}^{\varphi_2(\varrho)} |Y_\varphi(\varrho, \varphi)| \, d\varphi \right) d\varrho \\ &= \int\limits_{\varrho_1}^{\varrho_2} \Psi'(\varrho) \left(\int\limits_{\varphi_1(\varrho)}^{\varphi_2(\varrho)} |Y_\varphi(\varrho, \varphi)| \, d\varphi \right) d\varrho \leq 2\eta \int\limits_{\varrho_1}^{\varrho_2} \Psi'(\varrho) \, d\varrho = 2\eta \end{aligned}$$

for all $\eta > 0$. In this way, we see that the boundary point $X(w^*) \in \dot{\mathcal{M}}$ has capacity zero, and the finitely many boundary points $X(\{w_1, \ldots, w_{k_0}\})$ share this property.

3. Now we consider the Pfaffian form

$$\omega = a_1(x) \, dx_1 + a_2(x) \, dx_2 + a_3(x) \, dx_3 \in C^1(\mathcal{M}) \cap C^0(\overline{\mathcal{M}})$$

satisfying

$$\int\limits_{\mathcal{M}} |d\omega| \leq \iint\limits_B \left|\operatorname{rot} a\Big(X(u, v)\Big)\right| d^2\sigma(u, v) < +\infty.$$

Theorem 4.7 from Section 4 yields the identity

$$\iint\limits_B \Big\{\operatorname{rot} a\Big(X(u,v)\Big)\cdot\nu\Big\}\,d^2\sigma$$

$$= \int\limits_{\mathcal{M}} d\omega = \int\limits_{\partial\mathcal{M}} \omega = \int\limits_0^{2\pi} \Big\{a\Big(\overline{X}(\varphi)\Big)\cdot T(\varphi)\Big\}\,d^1\sigma(\varphi),$$

and our theorem is proved. q.e.d.

6 Curvilinear Integrals

We begin with the fundamental

Example 6.1. (Gravitational potentials)
Let the solid of the mass $M>0$ and another solid of the mass $m>0$ with $m \ll M$ be given (imagine the system *Sun - Earth*). Based on the *theory of gravitation* by I. Newton, the movement in the arising force-field can be described by the *Newtonian potential*

$$F(x)=\gamma\,\frac{mM}{r}, \qquad r=r(x)=\sqrt{x_1^2+x_2^2+x_3^2}, \quad x\in\mathbb{R}^3\setminus\{0\};$$

here $\gamma>0$ denotes the gravitational constant. We determine the work being performed during the movement from a given point P to another point Q in the Euclidean space by the formula $W=F(Q)-F(P)$. We can deduce the *force-field* by differentiation from the potential as follows:

$$\begin{aligned} f(x) &= \Big(f_1(x),f_2(x),f_3(x)\Big)=\nabla F(x) \\ &= -\gamma\,\frac{mM}{r^3}\,(x_1,x_2,x_3) = -\gamma\,\frac{mM}{r^3}\,x. \end{aligned}$$

Now we associate the Pfaffian form

$$\begin{aligned} \omega &= f_1(x)\,dx_1+f_2(x)\,dx_2+f_3(x)\,dx_3 \\ &= -\gamma\,\frac{mM}{r^3}(x_1\,dx_1+x_2\,dx_2+x_3\,dx_3). \end{aligned}$$

When

$$X(t)\,:\,[a,b]\longrightarrow\mathbb{R}^3\setminus\{0\}\in C^1([a,b])$$

denotes an arbitrary path satisfying $X(a)=P$ and $X(b)=Q$, we infer

$$\int_X \omega = \int_a^b \Big(F_{x_1} x_1'(t) + F_{x_2} x_2'(t) + F_{x_3} x_3'(t) \Big)\, dt$$

$$= \int_a^b \frac{d}{dt} \Big(F(X(t)) \Big)\, dt$$

$$= F\Big(X(b)\Big) - F\Big(X(a)\Big).$$

Consequently, this integral depends only on the end-points - and does not depend on the path chosen. Then we speak of a *conservative force-field*; movements along closed curves do not require energy.

We intend to present the theory of curvilinear integrals in the sequel.

Definition 6.2. *Let $\Omega \subset \mathbb{R}^n$ - with $n \geq 2$ - denote a domain and $P, Q \in \Omega$ two points. Then we define the* class $\mathcal{C}(\Omega, P, Q)$ of piecewise continuously differentiable paths (or synonymously, curves) in Ω from P to Q *as follows:*

$$\begin{aligned} \mathcal{C}(\Omega, P, Q) := \Big\{ X(t) \,:\, & [a, b] \longrightarrow \Omega \in C^0([a, b]) \;: \\ & -\infty < a < b < +\infty, \;\; X(a) = P, \;\; X(b) = Q; \\ & \textit{We have } a = t_0 < t_1 < \ldots < t_N = b \textit{ such that} \\ & X\big|_{[t_i, t_{i+1}]} \in C^1([t_i, t_{i+1}], \Omega) \textit{ for } i = 0, \ldots, N-1 \textit{ holds true} \Big\}. \end{aligned}$$

With the set

$$\mathcal{C}(\Omega) := \bigcup_{P \in \Omega} \mathcal{C}(\Omega, P, P),$$

we obtain the class of closed paths (or synonymously, closed curves) in Ω. *When $X(t) \equiv P$, $a \leq t \leq b$ holds true, we speak of a* point-curve.

Remark: In particular, the polygonal paths from P to Q are contained in $\mathcal{C}(\Omega, P, Q)$.

Definition 6.3. *Let*

$$\omega = \sum_{i=1}^{n} f_i(x)\, dx_i\,, \qquad x \in \Omega$$

denote a continuous Pfaffian form in the domain Ω and $X \in \mathcal{C}(\Omega, P, Q)$ a piecewise continuously differentiable path between the two points $P, Q \in \Omega$. Introducing

$$X^{(j)} := X\big|_{[t_j, t_{j+1}]} \in C^1([t_j, t_{j+1}]) \quad \textit{for} \quad j = 0, \ldots, N-1,$$

we define by

$$\int_X \omega := \sum_{j=0}^{N-1} \int_{X^{(j)}} \omega = \sum_{j=0}^{N-1} \int_{t_j}^{t_{j+1}} \sum_{i=1}^{n} f_i\Big(X(t)\Big)x_i'(t)\,dt$$

the curvilinear integral of ω over X.

Definition 6.4. *Let*

$$\omega = \sum_{i=1}^{n} f_i(x)\,dx_i, \qquad x \in \Omega$$

represent a continuous Pfaffian form in the domain $\Omega \subset \mathbb{R}^n$*. Then we call* $F(x) \in C^1(\Omega)$ *a* primitive *of* ω*, if the identity*

$$dF = \omega \qquad in \quad \Omega$$

or equivalently the equations

$$F_{x_i}(x) = f_i(x) \qquad for \quad x \in \Omega \quad and \quad i = 1, \ldots, n$$

hold true. When ω *possesses a primitive, we speak of an* exact Pfaffian form.

Theorem 6.5. (Curvilinear integrals)
Let $\Omega \subset \mathbb{R}^n$ *denote a domain and* ω *a continuous Pfaffian form in* Ω*. Then* ω *possesses a primitive* F *in* Ω *if and only if we have the identity* $\int_X \omega = 0$ *for each closed curve* $X \in \mathcal{C}(\Omega, P, P)$ *- with a point* $P \in \Omega$*. In the latter case, we obtain a primitive as follows: We take a fixed point* $P \in \Omega$ *and have the following representation for all arbitrary points* $Q \in \Omega$*, namely*

$$F(Q) := \gamma + \int_Y \omega \quad with \quad Y \in \mathcal{C}(\Omega, P, Q),$$

where $\gamma \in \mathbb{R}$ *is a constant.*

Proof:

1. When ω possesses a primitive F, we infer

$$\omega = \sum_{i=1}^{n} f_i(x)\,dx_i = \sum_{i=1}^{n} F_{x_i}(x)\,dx_i, \qquad x \in \Omega.$$

Let us consider $X \in \mathcal{C}(\Omega, P, P)$ with $P \in \Omega$ and

$$X^{(j)} := X\Big|_{[t_j, t_{j+1}]} \in C^1([t_j, t_{j+1}]) \quad \text{for} \quad j = 0, \ldots, N-1.$$

This implies

$$\int_X \omega = \sum_{j=0}^{N-1} \int_{X^{(j)}} \omega = \sum_{j=0}^{N-1} \int_{t_j}^{t_{j+1}} \left(\sum_{i=1}^{n} F_{x_i}\Big(X(t)\Big) x_i'(t)\,dt \right)$$

$$= \sum_{j=0}^{N-1} \int_{t_j}^{t_{j+1}} \frac{d}{dt} F\Big(X(t)\Big)\,dt = \sum_{j=0}^{N-1} \Big\{ F\Big(X(t_{j+1})\Big) - F\Big(X(t_j)\Big) \Big\}$$

$$= F\Big(X(t_N)\Big) - F\Big(X(t_0)\Big) = F(P) - F(P) = 0.$$

2. Now we start with the assumption

$$\int_X \omega = 0 \quad \text{for all curves} \quad X \in \mathcal{C}(\Omega, P, P) \quad \text{with} \quad P \in \Omega.$$

The point $P \in \Omega$ being fixed, we choose a path $X \in \mathcal{C}(\Omega, P, Q)$ for an arbitrary $Q \in \Omega$ and define $F(Q) := \int_X \omega$. Then we have to show the independence of this definition from the choice of the curve X: When $Y \in \mathcal{C}(\Omega, P, Q)$ represents another curve, we have to establish the identity

$$\int_X \omega = \int_Y \omega.$$

We associate the following closed curve to the curves $X : [a, b] \to \mathbb{R}^n$ and $Y : [c, d] \to \mathbb{R}^n$, namely

$$Z(t) := \begin{cases} X(t),\ t \in [a, b] \\ Y(b + d - t),\ t \in [b, b + d - c] \end{cases}.$$

Evidently, $Z \in \mathcal{C}(\Omega, P, P)$ holds true and

$$0 = \int_Z \omega = \int_X \omega - \int_Y \omega$$

follows, which implies

$$\int_X \omega = \int_Y \omega.$$

3. Finally, we have to deduce the formulas

$$F_{x_i}(Q) = f_i(Q) \quad \text{for} \quad i = 1, \ldots, n.$$

Here we proceed from Q to the point

$$Q_\varepsilon := Q + \varepsilon e_i, \quad e_i := (0, \ldots, \underbrace{1}_{i-th}, \ldots, 0)$$

along the path

$$Y(t) : [0, \varepsilon] \to \mathbb{R}^n, \quad Y(t) = Q + te_i$$

for a fixed index $i \in \{1, \ldots, n\}$. Now we evaluate

$$\begin{aligned} F(Q_\varepsilon) &= F(Q) + F(Q_\varepsilon) - F(Q) = F(Q) + \int_Y \omega \\ &= F(Q) + \int_0^\varepsilon \sum_{j=1}^n f_j\big(Y(t)\big) y_j'(t)\, dt \\ &= F(Q) + \int_0^\varepsilon f_i(Q + te_i)\, dt. \end{aligned}$$

Finally, we obtain

$$\frac{d}{dx_i} F\big|_Q = \frac{d}{d\varepsilon} F(Q_\varepsilon)\big|_{\varepsilon=0} = f_i(Q), \qquad i = 1, \ldots, n$$

proving the statement above. q.e.d

Let

$$\omega = \sum_{i=1}^n f_i(x)\, dx_i$$

represent an exact differential form of the class $C^1(\Omega)$ in a domain $\Omega \subset \mathbb{R}^n$. Then we have a function $F(x) : \Omega \longrightarrow \mathbb{R} \in C^2(\Omega)$ with the property

$$dF = \omega \quad \text{or equivalently} \quad f_i(x) = F_{x_i}(x).$$

Furthermore, we infer the identity

$$d\omega = d^2F = d\sum_{i=1}^n F_{x_i}\, dx_i = \sum_{i,j=1}^n F_{x_i x_j}\, dx_j \wedge dx_i = 0,$$

since the Hessian matrix $(F_{x_i x_j})_{i,j=1,\ldots,n}$ is symmetric.

Definition 6.6. *We name an m-form $\omega \in C^1(\Omega)$ in a domain $\Omega \subset \mathbb{R}^n$ as being* closed, *if the identity $d\omega = 0$ in Ω holds true.*

Remark: The Pfaffian form $\omega = \sum_{i=1}^n f_i(x)\, dx_i, \quad x \in \Omega$ is closed if and only if the matrix $\left(\frac{\partial f_i(x)}{\partial x_j}\right)$ is symmetric.

The considerations above show that an exact Pfaffian form is always closed. We shall now answer the question, which conditions guarantee that a closed Pfaffian form is necessarily exact - and consequently has a primitive.

Example 6.7. In the pointed plane $\mathbb{R}^2 \setminus \{(0,0)\}$, we consider the Pfaffian form

$$\omega = \frac{-y}{x^2+y^2}\,dx + \frac{x}{x^2+y^2}\,dy, \qquad x^2+y^2>0.$$

This 1-form is closed, since we have

$$\frac{\partial}{\partial y}\left(\frac{-y}{x^2+y^2}\right) = \frac{-(x^2+y^2)-(-y)2y}{(x^2+y^2)^2} = \frac{-x^2+y^2}{(x^2+y^2)^2}$$

as well as

$$\frac{\partial}{\partial x}\left(\frac{x}{x^2+y^2}\right) = \frac{x^2+y^2-x(2x)}{(x^2+y^2)^2} = \frac{y^2-x^2}{(x^2+y^2)^2},$$

and consequently

$$d\omega = \frac{\partial}{\partial y}\left(\frac{-y}{x^2+y^2}\right) dy \wedge dx + \frac{\partial}{\partial x}\left(\frac{x}{x^2+y^2}\right) dx \wedge dy = 0.$$

We observe the closed curve

$$X(t) := (\cos t, \sin t), \qquad 0 \le t \le 2\pi$$

and evaluate

$$\int\limits_X \omega = \int\limits_0^{2\pi} \Big(-\sin t(-\sin t) + \cos t \cos t\Big)\,dt = 2\pi.$$

According to Theorem 6.5, a primitive to ω in $\mathbb{R}^2 \setminus \{0,0\}$ does not exist - and the differential form is not exact there.

The nonvanishing of this curvilinear integral is caused by the fact that the curve X in $\mathbb{R}^2 \setminus \{(0,0)\}$ cannot be contracted to a point-curve.

Definition 6.8. *Let $\Omega \subset \mathbb{R}^n$ denote a domain. Two* closed curves

$$X(t) : [a,b] \longrightarrow \Omega \quad \textit{and} \quad Y(t) : [a,b] \longrightarrow \Omega, \qquad X,Y \in \mathcal{C}(\Omega)$$

are named homotopic in Ω, *if we have a mapping*

$$Z(t,s) : [a,b] \times [0,1] \longrightarrow \Omega \in C^0([a,b]\times[0,1],\mathbb{R}^n)$$

with the properties

$$Z(a,s) = Z(b,s) \qquad \textit{for all} \quad s \in [0,1]$$

as well as

$$Z(t,0) = X(t), \quad Z(t,1) = Y(t) \qquad \textit{for all} \quad t \in [a,b].$$

Now we establish the profound

Theorem 6.9. (Curvilinear integrals)
Let $\Omega \subset \mathbb{R}^n$ constitute a domain, where the two closed curves $X, Y \in \mathcal{C}(\Omega)$ are homotopic to each other. Finally, let

$$\omega = \sum_{i=1}^{n} f_i(x)\, dx_i, \qquad x \in \Omega$$

represent a closed Pfaffian form of the class $C^1(\Omega)$. Then we have the identity

$$\int_X \omega = \int_Y \omega.$$

For our proof we need the following

Proposition 6.10. (Smoothing of a closed curve)
Let

$$X(t) \,:\, [a,b] \longrightarrow \mathbb{R}^n \in \mathcal{C}(\Omega)$$

represent a closed curve, which is continued periodically via

$$X\Big(t + k(b-a)\Big) = X(t), \qquad t \in \mathbb{R}, \quad k \in \mathbb{Z}$$

onto the entire real line $\mathbb{R}$ with the period $(b-a)$. Furthermore, let the function

$$\chi(t) \in C_0^\infty((-1,+1),[0,\infty))$$

give us a mollifier with the properties

$$\chi(-t) = \chi(t) \qquad \textit{for all} \quad \in (-1,1)$$

and

$$\int_{-1}^{+1} \chi(t)\, dt = 1.$$

When we define

$$\chi_{t,\varepsilon}(\tau) := \frac{1}{\varepsilon}\chi\left(\frac{\tau - t}{\varepsilon}\right), \qquad \tau \in \mathbb{R},$$

we obtain the smoothed function

$$X^\varepsilon(t) := \int_{-\infty}^{+\infty} X(\tau)\chi_{t,\varepsilon}(\tau)\, d\tau = \int_{-\infty}^{+\infty} X(\tau)\frac{1}{\varepsilon}\chi\left(\frac{\tau - t}{\varepsilon}\right)\, d\tau,$$

which has the period $(b-a)$ again. Then we observe

$$\lim_{\varepsilon\to 0+} X^\varepsilon(t) = X(t) \qquad \textit{uniformly on} \quad [a,b].$$

Furthermore, the function $X^\varepsilon(t)$ belongs to the class $C^\infty(\mathbb{R})$, and we obtain the estimate

$$\left|\frac{d}{dt}X^\varepsilon(t)\right| \le C \qquad \textit{for all} \quad t\in[a,b], \quad 0<\varepsilon<\varepsilon_0,$$

with a constant $C>0$ and a sufficiently small ε_0. For all compact subsets

$$T\subset (t_0,t_1)\cup(t_1,t_2)\cup\ldots\cup(t_{N-1},t_N)\subset(a,b)$$

we infer

$$\frac{d}{dt}X^\varepsilon(t)\longrightarrow X'(t) \qquad \textit{for} \quad \varepsilon\to 0+ \quad \textit{uniformly in} \quad T.$$

Proof: We show parallel to Proposition 1.2 in Section 1 that

$$X^\varepsilon(t)\longrightarrow X(t) \quad \text{for all} \quad t\in[a,b] \quad \text{uniformly, where} \quad \varepsilon\to 0+ \quad \text{holds true.}$$

Since X is piecewise differentiable and continuous, a partial integration yields

$$\begin{aligned}\frac{d}{dt}X^\varepsilon(t) &= \int_{-\infty}^{+\infty} X(\tau)\frac{d}{dt}\chi_{t,\varepsilon}(\tau)\,d\tau = \int_{-\infty}^{+\infty} X(\tau)\left(-\frac{d}{d\tau}\chi_{t,\varepsilon}(\tau)\right)\,d\tau\\ &= \int_{-\infty}^{+\infty} X'(\tau)\chi_{t,\varepsilon}(\tau)\,d\tau.\end{aligned}$$

Therefore, we obtain

$$\left|\frac{d}{dt}X^\varepsilon(t)\right| \le \int_{-\infty}^{+\infty}|X'(\tau)|\chi_{t,\varepsilon}(\tau)\,d\tau \le C\int_{-\infty}^{+\infty}\chi_{t,\varepsilon}(\tau)\,d\tau = C \qquad \text{for all} \quad t\in\mathbb{R},$$

using the estimate $|X'(\tau)|\le C$ on $\mathbb{R}$. Finally, we show - parallel to Proposition 1.2 in Section 1 again - the relation

$$\lim_{\varepsilon\to 0+}\frac{d}{dt}X^\epsilon(t) = X'(t) \qquad \text{uniformly in} \quad T\subset(t_0,t_1)\cup\ldots\cup(t_{N-1},t_N),$$

which had to be proved. q.e.d.

Proof of Theorem 6.9:

1. Let $X,Y\in\mathcal{C}(\Omega)$ represent two homotopic closed curves. Then we have a continuous function

$$Z(t,s)\,:\,[a,b]\times[0,1]\longrightarrow\Omega\in C^0([a,b]\times[0,1],\mathbb{R}^n)$$

with the properties

$$Z(a,s) = Z(b,s) \qquad \text{for all} \quad s \in [0,1]$$

and

$$Z(t,0) = X(t), \quad Z(t,1) = Y(t) \qquad \text{for all} \quad t \in [a,b].$$

We continue Z onto the rectangle $[a,b] \times [-2,3]$ to the function

$$\Phi(t,s) := \begin{cases} X(t),\ (t,s) \in [a,b] \times [-2,0] \\ Z(t,s),\ (t,s) \in [a,b] \times [0,1] \\ Y(t),\ (t,s) \in [a,b] \times [1,3] \end{cases}.$$

Via the prescription

$$\Phi\Big(t + k(b-a), s\Big) = \Phi(t,s) \quad \text{for} \quad t \in \mathbb{R}, \quad s \in [-2,3] \quad \text{and} \quad k \in \mathbb{Z},$$

we extend the function onto the stripe $\mathbb{R} \times [-2,3]$ to a continuous function, which is periodic in the first variable with the period $(b-a)$.

2. On the rectangle $Q := [a,b] \times [-1,2]$ we consider the function

$$\Phi^\varepsilon(u,v) := \int\limits_{-\infty}^{+\infty}\int\limits_{-\infty}^{+\infty} \Phi(\xi,\eta)\chi_{u,\varepsilon}(\xi)\chi_{v,\varepsilon}(\eta)\,d\xi d\eta \qquad \text{for all} \quad 0 < \varepsilon < 1.$$

Now the regularity $\Phi^\varepsilon \in C^\infty(Q)$ is fulfilled, and we have the limit relation

$$\Phi^\varepsilon(u,v) \longrightarrow \Phi(u,v) \quad \text{for} \quad \varepsilon \to 0 \quad \text{uniformly in} \quad [a,b] \times [-1,2].$$

This implies the property $\Phi^\varepsilon(Q) \subset \Omega$, $0 < \varepsilon < \varepsilon_0$ and the periodicity

$$\Phi^\varepsilon\Big(u + k(b-a), v\Big) = \Phi^\varepsilon(u,v) \quad \text{for all} \quad (u,v) \in \mathbb{R} \times [-1,2], \quad k \in \mathbb{Z}.$$

For all parameters $a \leq u \leq b$ we have

$$\begin{aligned} \Phi^\varepsilon(u,-1) &= \int\limits_{-\infty}^{+\infty}\int\limits_{-\infty}^{+\infty} \Phi(\xi,\eta)\chi_{u,\epsilon}(\xi)\chi_{-1,\varepsilon}(\eta)\,d\xi d\eta \\ &= \int\limits_{-\infty}^{+\infty}\int\limits_{-\infty}^{+\infty} X(\xi)\chi_{u,\varepsilon}(\xi)\chi_{-1,\varepsilon}(\eta)\,d\xi d\eta \\ &= \int\limits_{-\infty}^{+\infty} X(\xi)\chi_{u,\epsilon}(\xi)\,d\xi \;=\; X^\epsilon(u) \end{aligned}$$

and additionally

$$\Phi^\varepsilon(u,2) = Y^\varepsilon(u).$$

3. By the Stokes integral theorem on the rectangle Q, we obtain the following identity for all $0 < \varepsilon < \varepsilon_0$, namely

$$\int_{X^\varepsilon} \omega - \int_{Y^\varepsilon} \omega = \oint_{\partial Q} \omega_{\Phi^\varepsilon} = \int_Q d(\omega_{\Phi^\varepsilon}) = \int_Q (d\omega)_{\Phi^\varepsilon} = 0.$$

We observe $\varepsilon \to 0+$, and Proposition 6.10 yields

$$0 = \lim_{\varepsilon \to 0+} \left(\int_{X^\varepsilon} \omega - \int_{Y^\varepsilon} \omega \right) = \int_X \omega - \int_Y \omega$$

and therefore our statement above. q.e.d

Definition 6.11. *Let the domain $\Omega \subset \mathbb{R}^n$ as well as the points $P, Q \in \Omega$ be given. We name two curves*

$$X(t), Y(t) \,:\, [a,b] \longrightarrow \Omega \in \mathcal{C}(\Omega, P, Q)$$

as being homotopic in Ω with the fixed start-point P and end-point Q, *if we have a continuous mapping*

$$Z(t,s) \,:\, [a,b] \times [0,1] \longrightarrow \Omega$$

with the following properties:

$$Z(a,s) = P, \quad Z(b,s) = Q \qquad \text{for all} \quad s \in [0,1]$$

as well as

$$Z(t,0) = X(t), \quad Z(t,1) = Y(t) \qquad \text{for all} \quad t \in [a,b].$$

We deduce immediately the following result from Theorem 6.9.

Theorem 6.12. (Monodromy)
Let $\Omega \subset \mathbb{R}^n$ denote a domain and $P, Q \in \Omega$ two arbitrary points. Furthermore, let the two curves $X(t), Y(t) \in \mathcal{C}(\Omega, P, Q)$ be homotopic to each other with fixed start- and end-point. Finally, let

$$\omega = \sum_{i=1}^n f_i(x)\, dx_i, \qquad x \in \Omega$$

represent a closed Pfaffian form of the class $C^1(\Omega)$. Then we have the identity

$$\int_X \omega = \int_Y \omega.$$

Proof: We consider the following homotopy of closed curves in Ω, namely

$$\Phi(t,s)\,:\,[a,2b-a]\times[0,1]\longrightarrow\Omega$$

with

$$\Phi(t,s)=\begin{cases} X(t),\, a\le t\le b \\ Z(2b-t,s),\, b\le t\le 2b-a \end{cases}.$$

Now we note that

$$\Phi(t,0)=\begin{cases} X(t),\, a\le t\le b \\ X(2b-t),\, b\le t\le 2b-a \end{cases}.$$

Here the curve X is run through from P to Q and then backwards from Q to P. Therefore, we infer

$$\int\limits_{\Phi(\cdot,0)}\omega=0.$$

Furthermore, we deduce

$$\Phi(t,1)=\begin{cases} X(t),\, a\le t\le b \\ Y(2b-t),\, b\le t\le 2b-a \end{cases}.$$

Here the curve X is run through from P to Q at first, and the curve Y is run through from Q to P afterwards. Finally, Theorem 6.9 reveals the identity

$$0=\int\limits_{\Phi(\cdot,0)}\omega=\int\limits_{\Phi(\cdot,1)}\omega=\int\limits_X\omega-\int\limits_Y\omega.$$

q.e.d.

The study of curvilinear integrals becomes very simple in the following domains.

Definition 6.13. *A domain $\Omega\subset\mathbb{R}^n$ is named* simply connected, *if each closed curve $X(t)\in\mathcal{C}(\Omega)$ is homotopic to a point-curve in Ω. This means geometrically that each closed curve is contractible to one point.*

Theorem 6.14. (Curvilinear integrals in simply connected domains) *Let $\Omega\subset\mathbb{R}^n$ constitute a simply connected domain and*

$$\omega=\sum_{i=1}^n f_i(x)\,dx_i,\qquad x\in\Omega$$

a Pfaffian form of the class $C^1(\Omega)$. Then the following statements are equivalent:

1. *The Pfaffian form ω is exact, and therefore possesses a primitive F.*
2. *For all curves $X \in \mathcal{C}(\Omega, P, P)$ - with a point $P \in \Omega$ - we have the identity $\int\limits_X \omega = 0$.*
3. *The Pfaffian form ω is closed, which means*

$$d\omega = 0 \quad in \quad \Omega$$

or equivalently that the matrix $\left(\frac{\partial f_i}{\partial x_j}(x)\right)_{i,j=1,\dots,n}$ is symmetric for all points $x \in \Omega$.

Proof: From the first theorem on curvilinear integrals we infer the equivalence '1. $\Leftrightarrow$ 2.'. The statement '1. $\Rightarrow$ 3.' is revealed by the considerations preceding Definition 6.6. We only have to show the direction '3. $\Rightarrow$ 2.': Here we choose an arbitrary closed curve $X(t) \in \mathcal{C}(\Omega, P, P)$, which is homotopic to the closed curve $Y(t) \equiv P, \quad a \le t \le b$, due to the assumption on the domain Ω. The application of Theorem 6.9 yields

$$\int\limits_X \omega = \int\limits_Y \omega = \int\limits_a^b \sum_{i=1}^n f_i\Big(Y(t)\Big) y_i'(t)\, dt = 0,$$

which implies our theorem. q.e.d.

Remark: In the Euclidean space $\mathbb{R}^3$, our condition 3 from Theorem 6.14 implies that the vector-field $f(x) = \Big(f_1(x), f_2(x), f_3(x)\Big), \quad x \in \Omega$ is irrotational, which means

$$\operatorname{rot} f(x) = 0 \qquad \text{in} \quad \Omega.$$

In simply connected domains $\Omega \subset \mathbb{R}^3$, Theorem 6.14 guarantees the existence of a primitive $F : \Omega \to \mathbb{R} \in C^2(\Omega)$ with the property $\nabla F(x) = f(x), \quad x \in \Omega$.

7 The Lemma of Poincaré

The theory of curvilinear integrals was transferred to the higher-dimensional situation of surface-integrals especially by de Rham (compare G. de Rham: *Varietés differentiables*, Hermann, Paris 1955). In this context we refer the reader to Paragraph 20 in the textbook by H. Holmann and H. Rummler: *Alternierende Differentialformen*, BI-Wissenschaftsverlag, 2.Auflage, 1981.

We shall construct primitives for arbitrary m-forms, which correspond to vector-potentials - however, in 'contractible domains' only. Here we do not need the Stokes integral theorem!

Definition 7.1. *A continuous m-form with $1 \le m \le n$ in an open set $\Omega \subset \mathbb{R}^n$ with $n \in \mathbb{N}$, namely*

$$\omega = \sum_{1 \le i_1 < \ldots < i_m \le n} a_{i_1 \ldots i_m}(x)\, dx_{i_1} \wedge \ldots \wedge dx_{i_m}, \qquad x \in \Omega,$$

is named exact *if we have an* $(m-1)$*-form*

$$\lambda = \sum_{1 \le i_1 < \ldots < i_{m-1} \le n} b_{i_1 \ldots i_{m-1}}(x)\, dx_{i_1} \wedge \ldots \wedge dx_{i_{m-1}}, \qquad x \in \Omega$$

of the class $C^1(\Omega)$ *with the property*

$$d\lambda = \omega \qquad \text{in} \quad \Omega.$$

We begin with the easy

Theorem 7.2. *An exact differential form* $\omega \in C^1(\Omega)$ *is closed.*

Proof: We calculate

$$\begin{aligned} d\omega = d(d\lambda) &= d \sum_{1 \le i_1 < \ldots < i_{m-1} \le n} db_{i_1 \ldots i_{m-1}}(x) \wedge dx_{i_1} \wedge \ldots \wedge dx_{i_{m-1}} \\ &= \sum_{1 \le i_1 < \ldots < i_{m-1} \le n} \Big(d\, db_{i_1 \ldots i_{m-1}}(x) \Big) \wedge dx_{i_1} \wedge \ldots \wedge dx_{i_{m-1}} = 0, \end{aligned}$$

which implies the statement above. q.e.d.

We now provide a condition on the domain Ω, which guarantees that a closed differential form is necessarily exact.

Definition 7.3. *Let* $\Omega \subset \mathbb{R}^n$ *denote a domain with the associate cylinder*

$$\widehat{\Omega} := \Omega \times [0,1] \subset \mathbb{R}^{n+1}.$$

Furthermore, we have a point $x_0 \in \Omega$ *and a mapping*

$$F = F(x,t) = \Big(f_1(x_1, \ldots, x_n, t), \ldots, f_n(x_1, \ldots, x_n, t) \Big) : \widehat{\Omega} \longrightarrow \Omega$$

of the class $C^2(\widehat{\Omega}, \mathbb{R}^n)$ *as follows:*

$$F(x,0) = x_0\,, \quad F(x,1) = x \qquad \text{for all} \quad x \in \Omega.$$

Then we name the domain Ω contractible *(onto the point* x_0*).*

Remarks:

1. Let the domain Ω be *star-shaped* with respect to the point $x_0 \in \Omega$, which means

$$(tx + (1-t)x_0) \in \Omega \qquad \text{for all} \quad t \in [0,1], \quad x \in \Omega.$$

Then Ω is contractible with the contraction-mapping

$$F(x,t) := tx + (1-t)x_0\,, \qquad x \in \Omega, \quad t \in [0,1].$$

2. Each contractible domain $\Omega \subset \mathbb{R}^n$ is simply connected as well. When $X(s)$, $0 \le s \le 1$ with $X(0) = X(1)$ represents a closed curve in Ω, it is contractible onto the point x_0 via
$$Y(s,t) := F\Big(X(s), t\Big), \qquad 0 \le s \le 1, \quad 0 \le t \le 1.$$
In a contractible domain, we can perform the contraction of an arbitrary curve $X(s)$ by the joint mapping F. Therefore, the contraction is independent from the choice of the curve X.
3. The following chain of implications for domains in $\mathbb{R}^n$ holds true:
$$\begin{aligned} \text{convex} &\Longrightarrow \text{star-shaped} \\ &\Longrightarrow \text{contractible} \\ &\Longrightarrow \text{simply connected.} \end{aligned}$$

On the cylinder $\widehat{\Omega}$ we consider the l-form
$$\gamma(x,t) := \sum_{1 \le i_1 < \ldots < i_l \le n} c_{i_1 \ldots i_l}(x,t)\, dx_{i_1} \wedge \ldots \wedge dx_{i_l}$$
of the class $C^1(\widehat{\Omega})$. We use the abbreviation $\frac{d}{dt} := \dot{}$ for the time-derivative and define
$$\dot{\gamma}(x,t) := \sum_{1 \le i_1 < \ldots < i_l \le n} \dot{c}_{i_1 \ldots i_l}(x,t)\, dx_{i_1} \wedge \ldots \wedge dx_{i_l}.$$
Furthermore, we set
$$\int_0^1 \gamma(x,t)\, dt := \sum_{1 \le i_1 < \ldots < i_l \le n} \left(\int_0^1 c_{i_1 \ldots i_l}(x,t)\, dt \right) dx_{i_1} \wedge \ldots \wedge dx_{i_l}.$$
The fundamental theorem of the differential- and integral-calculus reveals
$$\int_0^1 \dot{\gamma}(x,t)\, dt = \gamma(x,1) - \gamma(x,0). \tag{1}$$
The function $g(x,t) : \widehat{\Omega} \to \mathbb{R} \in C^1(\widehat{\Omega})$ being given, we determine its exterior derivative
$$dg = \sum_{k=1}^n \frac{\partial g}{\partial x_k}\, dx_k + \dot{g}(x,t)\, dt =: d_x g + \dot{g}\, dt.$$
Consequently, we obtain
$$d\gamma = d_x \gamma + dt \wedge \dot{\gamma}$$
abbreviating

$$d_x\gamma := \sum_{1\le i_1<\ldots<i_l\le n} \Big(d_x c_{i_1\ldots i_l}(x,t)\Big) \wedge dx_{i_1} \wedge \ldots \wedge dx_{i_l}.$$

Finally, we deduce the identity

$$d\left(\int_0^1 \gamma(x,t)\,dt\right) = \int_0^1 \Big(d_x\gamma(x,t)\Big)\,dt. \tag{2}$$

Therefore, we calculate

$$\begin{aligned}
&d\left(\int_0^1 \gamma(x,t)\,dt\right)\\
&= \sum_{1\le i_1<\ldots<i_l\le n} \sum_{i=1}^n \frac{\partial}{\partial x_i}\left(\int_0^1 c_{i_1\ldots i_l}(x,t)\,dt\right) dx_i \wedge dx_{i_1} \wedge \ldots \wedge dx_{i_l}\\
&= \sum_{1\le i_1<\ldots<i_l\le n} \sum_{i=1}^n \left(\int_0^1 \frac{\partial}{\partial x_i} c_{i_1\ldots i_l}(x,t)\,dt\right) dx_i \wedge dx_{i_1} \wedge \ldots \wedge dx_{i_l}\\
&= \int_0^1 \left\{\sum_{1\le i_1<\ldots<i_l\le n} \left(\sum_{i=1}^n \frac{\partial}{\partial x_i} c_{i_1\ldots i_l}(x,t)\,dx_i\right) \wedge dx_{i_1} \wedge \ldots \wedge dx_{i_l}\right\} dt\\
&= \int_0^1 \Big(d_x\gamma(x,t)\Big)\,dt.
\end{aligned}$$

We are now prepared to prove the central result of this section.

Theorem 7.4. (Lemma of Poincaré)
Let $\Omega \subset \mathbb{R}^n$ denote a contractible domain, and choose a dimension $1 \le m \le n$. Then each closed m-form ω in Ω is exact.

Proof (A. Weil):

1. Since Ω is contractible, we have a mapping

$$F = F(x,t) : \widehat{\Omega} \longrightarrow \Omega \in C^2(\widehat{\Omega})$$

satisfying

$$F(x,0) = x_0, \quad F(x,1) = x \qquad \text{for all} \quad x \in \Omega.$$

On the set $\widehat{\Omega} = \Omega \times [0,1]$, we consider the transformed differential form

$$\begin{aligned}\widehat{\omega}(x,t) &:= \omega \circ F(x,t)\\ &= \sum_{1\le i_1<\ldots<i_m\le n} a_{i_1\ldots i_m}(F(x,t))\, df_{i_1}\wedge\ldots\wedge df_{i_m}\\ &= \sum_{1\le i_1<\ldots<i_m\le n} a_{i_1\ldots i_m}(F(x,t))\, d_x f_{i_1}\wedge\ldots\wedge d_x f_{i_m} + dt\wedge\omega_2(x,t)\\ &= \omega_1 + dt\wedge\omega_2.\end{aligned}$$

Here we used the identities

$$df_{i_k} = d_x f_{i_k} + \dot{f}_{i_k}\, dt \qquad \text{for} \qquad k = 1,\ldots,m.$$

The differential forms $\omega_1(x,t)$ and $\omega_2(x,t)$ are independent of dt and have the degrees m and $(m-1)$, respectively. Furthermore, we note that

$$\omega_1(x,0) = 0 \quad \text{and} \quad \omega_1(x,1) = \omega(x).$$

2. We evaluate

$$\begin{aligned} 0 &= (d\omega)\circ F = d(\omega\circ F) = d\widehat{\omega}\\ &= d\omega_1 + d(dt\wedge\omega_2) = d_x\omega_1 + dt\wedge\dot{\omega}_1 - dt\wedge d\omega_2\\ &= d_x\omega_1 + dt\wedge\dot{\omega}_1 - dt\wedge(d_x\omega_2 + dt\wedge\dot{\omega}_2)\\ &= d_x\omega_1 + dt\wedge(\dot{\omega}_1 - d_x\omega_2).\end{aligned}$$

This implies

$$\dot{\omega}_1 = d_x\omega_2. \tag{3}$$

3. Now we define the $(m-1)$-form

$$\lambda := \int_0^1 \omega_2(x,t)\, dt.$$

With the aid of the identities (1), (2), and (3) we calculate

$$d\lambda = \int_0^1 \Big(d_x\omega_2(x,t)\Big)\, dt = \int_0^1 \dot{\omega}_1(x,t)\, dt = \omega_1(x,1) - \omega_1(x,0) = \omega(x),$$

which completes the proof. q.e.d.

Example 7.5. In a star-shaped domain $\Omega\subset\mathbb{R}^3$, let the source-free vector-field

$$b(x) = \Big(b_1(x), b_2(x), b_3(x)\Big) \,:\, \Omega \longrightarrow \mathbb{R}^3 \in C^1(\Omega,\mathbb{R}^3)$$

with

$$\operatorname{div} b(x) = 0$$

be given. Then its associate 2-form

$$\omega = b_1(x)\, dx_2 \wedge dx_3 + b_2(x)\, dx_3 \wedge dx_1 + b_3(x)\, dx_1 \wedge dx_2$$

is closed. Theorem 7.4 gives us a Pfaffian form

$$\lambda = a_1(x)\, dx_1 + a_2(x)\, dx_2 + a_3(x)\, dx_3 \in C^2(\Omega)$$

satisfying $d\lambda = \omega$. The calculations in Section 3 imply the following identity for the vector-field $a(x) = (a_1(x), a_2(x), a_3(x))$, namely

$$\operatorname{rot} a(x) = b(x) \qquad \text{for all} \qquad x \in \Omega.$$

Therefore, we have constructed a vector-potential $a(x)$ for the source-free vector-field $b(x)$.

8 Co-derivatives and the Laplace-Beltrami Operator

In this section we introduce an inner product for differential forms. We consider the space

$$\mathbb{R}^n := \Big\{\overline{x} = (\overline{x}_1, \ldots, \overline{x}_n) \,:\, \overline{x}_i \in \mathbb{R},\ i = 1, \ldots, n\Big\}$$

with the subset $\Theta \subset \mathbb{R}^n$. Furthermore, we have given two continuous m-forms on Θ, namely

$$\overline{\alpha} := \sum_{1 \le i_1 < \ldots < i_m \le n} \overline{a}_{i_1 \ldots i_m}(\overline{x})\, d\overline{x}_{i_1} \wedge \ldots \wedge d\overline{x}_{i_m}, \qquad \overline{x} \in \Theta,$$

as well as

$$\overline{\beta} := \sum_{1 \le i_1 < \ldots < i_m \le n} \overline{b}_{i_1 \ldots i_m}(\overline{x})\, d\overline{x}_{i_1} \wedge \ldots \wedge d\overline{x}_{i_m}, \qquad \overline{x} \in \Theta.$$

We define an *inner product* between the m-forms $\overline{\alpha}$ and $\overline{\beta}$ as follows:

$$(\overline{\alpha}, \overline{\beta})_m := \sum_{1 \le i_1 < \ldots < i_m \le n} \overline{a}_{i_1 \ldots i_m}(\overline{x})\, \overline{b}_{i_1 \ldots i_m}(\overline{x}), \qquad m = 0, 1, \ldots, n. \tag{1}$$

Consequently, the inner product attributes a 0-form to a pair of m-forms. It represents a symmetric bilinear form on the vector space of m-forms.

Now we consider the parameter transformation

$$\overline{x} = \Phi(x) = \Big(\Phi_1(x_1, \ldots, x_n), \ldots, \Phi_n(x_1, \ldots, x_n)\Big) \,:\, \Omega \longrightarrow \Theta \in C^2(\Omega)$$

on the open set $\Omega \subset \mathbb{R}^n$. The mapping Φ satisfies

$$J_\Phi(x) = \det\Big(\partial\Phi(x)\Big) \neq 0 \qquad \text{for all} \quad x \in \Omega. \tag{2}$$

We set

$$g(x) := \Big(J_\Phi(x)\Big)^2 = \det\Big(\partial\Phi(x)^t \circ \partial\Phi(x)\Big), \qquad x \in \Omega.$$

The volume form

$$\omega = \sqrt{g(x)}\, dx_1 \wedge \ldots \wedge dx_n, \qquad x \in \Omega \tag{3}$$

is associated with the transformation $\overline{x} = \Phi(x)$ in a natural way. The m-forms $\overline{\alpha}$ and $\overline{\beta}$ are transformed into the m-forms

$$\begin{aligned}\alpha := \overline{\alpha}_\Phi &= \sum_{1 \leq i_1 < \ldots < i_m \leq n} \overline{a}_{i_1 \ldots i_m}\Big(\Phi(x)\Big)\, d\Phi_{i_1}(x) \wedge \ldots \wedge d\Phi_{i_m}(x) \\ &=: \sum_{1 \leq i_1 < \ldots < i_m \leq n} a_{i_1 \ldots i_m}(x)\, dx_{i_1} \wedge \ldots \wedge dx_{i_m}\end{aligned}$$

and

$$\begin{aligned}\beta := \overline{\beta}_\Phi &= \sum_{1 \leq i_1 < \ldots < i_m \leq n} \overline{b}_{i_1 \ldots i_m}\Big(\Phi(x)\Big)\, d\Phi_{i_1}(x) \wedge \ldots \wedge d\Phi_{i_m}(x) \\ &=: \sum_{1 \leq i_1 < \ldots < i_m \leq n} b_{i_1 \ldots i_m}(x)\, dx_{i_1} \wedge \ldots \wedge dx_{i_m},\end{aligned}$$

respectively. We shall define an inner product $(\alpha, \beta)_m$ between the transformed m-forms α and β such that it is parameter-invariant:

$$(\alpha, \beta)_m(x) = (\overline{\alpha}, \overline{\beta})_m\Big(\Phi(x)\Big), \qquad x \in \Omega. \tag{4}$$

We shall explicitly represent this inner product for differential forms of the orders $0, 1, n-1, n$ in the sequel.

1. Let $m = 0$ hold true. We consider the 0-forms

$$\overline{\alpha} = \overline{a}(\overline{x}), \quad \overline{\beta} = \overline{b}(\overline{x}).$$

Then we see

$$\alpha = \overline{\alpha}_\Phi = \overline{a}\Big(\Phi(x)\Big), \quad \beta = \overline{\beta}_\Phi = \overline{b}\Big(\Phi(x)\Big).$$

Setting

$$(\alpha, \beta)_0(x) := a(x) b(x),$$

we obtain

$$\begin{aligned}(\alpha, \beta)_0(x) = a(x) b(x) &= \overline{a}\Big(\Phi(x)\Big)\, \overline{b}\Big(\Phi(x)\Big) \\ &= (\overline{\alpha}, \overline{\beta})_0\Big(\Phi(x)\Big), \qquad x \in \Omega.\end{aligned}$$

2. Let $m = n$ hold true. We consider the n-forms

$$\overline{\alpha} = \overline{a}(\overline{x})\, d\overline{x}_1 \wedge \ldots \wedge d\overline{x}_n, \quad \overline{\beta} = \overline{b}(\overline{x})\, d\overline{x}_1 \wedge \ldots \wedge d\overline{x}_n.$$

We calculate

$$\begin{aligned}\alpha = \overline{\alpha}_\Phi &= \overline{a}\Big(\Phi(x)\Big)\, d\Phi_1 \wedge \ldots \wedge d\Phi_n \\ &= \overline{a}\Big(\Phi(x)\Big) \left(\sum_{i_1=1}^n \frac{\partial \Phi_1}{\partial x_{i_1}}\, dx_{i_1}\right) \wedge \ldots \wedge \left(\sum_{i_n=1}^n \frac{\partial \Phi_n}{\partial x_{i_n}}\, dx_{i_n}\right) \\ &= \overline{a}\Big(\Phi(x)\Big) J_\Phi(x)\, dx_1 \wedge \ldots \wedge dx_n.\end{aligned}$$

Therefore, we have

$$a(x) = \overline{a}\Big(\Phi(x)\Big) J_\Phi(x), \quad b(x) = \overline{b}\Big(\Phi(x)\Big) J_\Phi(x), \qquad x \in \Omega.$$

Now we set

$$(\alpha, \beta)_n(x) := \frac{1}{g(x)}\, a(x) b(x), \qquad x \in \Omega,$$

observe $g(x) = \Big(J_\Phi(x)\Big)^2$, and infer

$$\begin{aligned}(\alpha, \beta)_n(x) &= \frac{1}{\Big(J_\Phi(x)\Big)^2}\, \overline{a}\Big(\Phi(x)\Big)\, J_\Phi(x)\, \overline{b}\Big(\Phi(x)\Big)\, J_\Phi(x) \\ &= \overline{a}\Big(\Phi(x)\Big)\, \overline{b}\Big(\Phi(x)\Big) = (\overline{\alpha}, \overline{\beta})_n\Big(\Phi(x)\Big).\end{aligned}$$

3. Let $m = 1$ hold true. We consider the Pfaffian forms

$$\overline{\alpha} = \sum_{i=1}^n \overline{a}_i(\overline{x})\, d\overline{x}_i, \quad \overline{\beta} = \sum_{i=1}^n \overline{b}_i(\overline{x})\, d\overline{x}_i$$

and calculate

$$\begin{aligned}\alpha = \overline{\alpha}_\Phi &= \sum_{i=1}^n \overline{a}_i\Big(\Phi(x)\Big)\, d\Phi_i \\ &= \sum_{i=1}^n \overline{a}_i\Big(\Phi(x)\Big) \left(\sum_{j=1}^n \frac{\partial \Phi_i}{\partial x_j}\, dx_j\right) \\ &= \sum_{j=1}^n \left(\sum_{i=1}^n \overline{a}_i\Big(\Phi(x)\Big) \frac{\partial \Phi_i}{\partial x_j}\right) dx_j.\end{aligned}$$

Thus we obtain

$$\alpha = \overline{\alpha}_\Phi = \sum_{j=1}^{n} a_j(x)\,dx_j \quad \text{with} \quad a_j(x) = \sum_{i=1}^{n} \overline{a}_i\Big(\Phi(x)\Big)\frac{\partial \Phi_i}{\partial x_j},$$

$$\beta = \overline{\beta}_\Phi = \sum_{j=1}^{n} b_j(x)\,dx_j \quad \text{with} \quad b_j(x) = \sum_{i=1}^{n} \overline{b}_i\Big(\Phi(x)\Big)\frac{\partial \Phi_i}{\partial x_j},$$

where $j = 1, \ldots, n$ is valid. We introduce the following abbreviation for the *functional matrix*

$$F(x) := \left(\frac{\partial \Phi_i}{\partial x_j}(x)\right)_{i,j=1,\ldots,n}, \quad x \in \Omega.$$

The vectors

$$a(x) = \Big(a_1(x), \ldots, a_n(x)\Big), \quad \overline{a}(x) = \Big(\overline{a}_1(\overline{x}), \ldots, \overline{a}_n(\overline{x})\Big)$$

and

$$b(x) = \Big(b_1(x), \ldots, b_n(x)\Big), \quad \overline{b}(x) = \Big(\overline{b}_1(\overline{x}), \ldots, \overline{b}_n(\overline{x})\Big)$$

are subject to the transformation laws

$$a(x) = \overline{a}\Big(\Phi(x)\Big) \circ F(x), \quad b(x) = \overline{b}\Big(\Phi(x)\Big) \circ F(x),$$

and

$$a(x) \circ F^{-1}(x) = \overline{a}\Big(\Phi(x)\Big), \quad b(x) \circ F^{-1}(x) = \overline{b}\Big(\Phi(x)\Big),$$

respectively. We define the *transformation matrix*

$$G(x) = \Big(g_{ij}(x)\Big)_{i,j=1,\ldots,n} := F(x)^t \circ F(x)$$

with the inverse matrix

$$G^{-1}(x) = \Big(g^{ij}(x)\Big)_{i,j=1,\ldots,n} = F^{-1}(x) \circ \Big(F^{-1}(x)\Big)^t.$$

Evidently, we have

$$\sum_{j=1}^{n} g^{ij}(x) g_{jk}(x) = \delta_k^i, \qquad i,k = 1, \ldots, n$$

and

$$g(x) = \Big(J_\Phi(x)\Big)^2 = \det G(x).$$

Now we define

$$(\alpha, \beta)_1(x) := \sum_{i,j=1}^{n} g^{ij}(x) a_i(x) b_j(x).$$

Then we infer

$$(\alpha, \beta)_1(x) = a(x) \circ G^{-1}(x) \circ \Big(b(x)\Big)^t$$

$$= \overline{a}\Big(\Phi(x)\Big) \circ F(x) \circ F^{-1}(x) \circ \Big(F^{-1}(x)\Big)^t \circ \Big(F(x)\Big)^t \circ \Big(\overline{b}(\Phi(x))\Big)^t$$

$$= \overline{a}\Big(\Phi(x)\Big) \circ \Big(\overline{b}(\Phi(x))\Big)^t$$

$$= (\overline{\alpha}, \overline{\beta})_1\Big(\Phi(x)\Big).$$

4. Let $m = n - 1$ hold true. We define the $(n-1)$-forms

$$\overline{\theta}_i := (-1)^{i-1} \, d\overline{x}_1 \wedge \ldots \wedge d\overline{x}_{i-1} \wedge d\overline{x}_{i+1} \wedge \ldots \wedge d\overline{x}_n$$

for $1 \le i \le n$ and consider the $(n-1)$-forms

$$\overline{\alpha} = \sum_{i=1}^{n} \overline{a}_i(\overline{x}) \overline{\theta}_i, \quad \overline{\beta} = \sum_{i=1}^{n} \overline{b}_i(\overline{x}) \overline{\theta}_i.$$

We use the symbol $\check{}$ to indicate that we omit this factor. Defining

$$\theta_j := (-1)^{j-1} \, dx_1 \wedge \ldots \wedge dx_{j-1} \wedge dx_{j+1} \wedge \ldots \wedge dx_n$$

for $j = 1, \ldots, n$, we calculate

$$\alpha = \overline{\alpha}_\Phi = \sum_{i=1}^{n} \overline{a}_i\Big(\Phi(x)\Big)(-1)^{i-1} \, d\Phi_1 \wedge \ldots \wedge d\Phi_{i-1} \wedge d\Phi_{i+1} \wedge \ldots \wedge d\Phi_n$$

$$= \sum_{i=1}^{n} \overline{a}_i\Big(\Phi(x)\Big)(-1)^{i-1} \left(\sum_{j_1=1}^{n} \frac{\partial \Phi_1}{\partial x_{j_1}} \, dx_{j_1} \right) \wedge \ldots \wedge \left(\sum_{j_{i-1}=1}^{n} \frac{\partial \Phi_{i-1}}{\partial x_{j_{i-1}}} \, dx_{j_{i-1}} \right)$$

$$\wedge \left(\sum_{j_{i+1}=1}^{n} \frac{\partial \Phi_{i+1}}{\partial x_{j_{i+1}}} \, dx_{j_{i+1}} \right) \wedge \ldots \wedge \left(\sum_{j_n=1}^{n} \frac{\partial \Phi_n}{\partial x_{j_n}} \, dx_{j_n} \right)$$

$$= \sum_{i=1}^{n} \overline{a}_i\Big(\Phi(x)\Big)(-1)^{i-1} \sum_{j=1}^{n} \frac{\partial(\Phi_1, \ldots, \check{\Phi}_i, \ldots, \Phi_n)}{\partial(x_1, \ldots, \check{x}_j, \ldots, x_n)} \cdot$$

$$\cdot \, dx_1 \wedge \ldots \wedge d\check{x}_j \wedge \ldots \wedge dx_n$$

$$= \sum_{j=1}^{n} \left(\sum_{i=1}^{n} \overline{a}_i\Big(\Phi(x)\Big)(-1)^{i+j} \frac{\partial(\Phi_1, \ldots, \check{\Phi}_i, \ldots, \Phi_n)}{\partial(x_1, \ldots, \check{x}_j, \ldots, x_n)} \right) \theta_j \;\; =: \; \sum_{j=1}^{n} a_j(x) \theta_j.$$

Correspondingly, we define $b_j(x)$ for $j = 1, \ldots, n$. The matrix of adjoints for $F(x)$, namely

$$E(x) := \left((-1)^{i+j} \frac{\partial(\Phi_1, \ldots, \check{\Phi}_i, \ldots, \Phi_n)}{\partial(x_1, \ldots, \check{x}_j, \ldots, x_n)} \right)_{i,j=1,\ldots,n},$$

satisfies the identity

$$\left(F(x)^t\right)^{-1} = \left(\left(\frac{\partial \Phi_j}{\partial x_i}(x) \right)_{i,j=1,\ldots,n} \right)^{-1} = \frac{1}{J_\Phi(x)} E(x),$$

and equivalently

$$E(x) = J_\Phi(x) \left(F(x)^t\right)^{-1}. \tag{5}$$

When

$$\overline{\alpha}_\Phi = \alpha = \sum_{j=1}^{n} a_j(x)\theta_j, \quad \overline{\beta}_\Phi = \beta = \sum_{j=1}^{n} b_j(x)\theta_j$$

denote the transformed $(n-1)$-forms, their coefficient vectors

$$a(x) = \left(a_1(x), \ldots, a_n(x)\right), \quad \overline{a}(x) = \left(\overline{a}_1(\overline{x}), \ldots, \overline{a}_n(\overline{x})\right)$$

and

$$b(x) = \left(b_1(x), \ldots, b_n(x)\right), \quad \overline{b}(x) = \left(\overline{b}_1(\overline{x}), \ldots, \overline{b}_n(\overline{x})\right)$$

are subject to the transformation laws

$$a(x) = \overline{a}\left(\Phi(x)\right) \circ E(x) = J_\Phi(x)\overline{a}\left(\Phi(x)\right) \circ \left(F(x)^t\right)^{-1},$$

$$b(x) = \overline{b}\left(\Phi(x)\right) \circ E(x) = J_\Phi(x)\overline{b}\left(\Phi(x)\right) \circ \left(F(x)^t\right)^{-1}.$$

Now we define as the inner product

$$(\alpha, \beta)_{n-1}(x) := \frac{1}{g(x)} \sum_{i,j=1}^{n} g_{ij}(x) a_i(x) b_j(x).$$

Finally, we infer

$$(\alpha, \beta)_{n-1}(x) = \frac{1}{\left(J_\Phi(x)\right)^2} a(x) \circ G(x) \circ \left(b(x)\right)^t$$

$$= \overline{a}\left(\Phi(x)\right) \circ \left(F(x)^t\right)^{-1} \circ F(x)^t \circ F(x) \circ \left(F(x)\right)^{-1} \circ \left(\overline{b}(\Phi(x))\right)^t$$

$$= \overline{a}\left(\Phi(x)\right) \circ \left(\overline{b}(\Phi(x))\right)^t = (\overline{\alpha}, \overline{\beta})_{n-1}\left(\Phi(x)\right).$$

Now we introduce another operation in the set of differential forms.

Definition 8.1. *When $k \in K := \{0, 1, n-1, n\}$ holds true, we attribute to each k-form α its dual $(n-k)$-form $*\alpha$ as follows:*

1. *Let $k = 0$ and $\alpha = a(x)$ be given. Then we define*

$$*\alpha := a(x)\omega,$$

where

$$\omega = \sqrt{g(x)}\, dx_1 \wedge \ldots \wedge dx_n$$

denotes the volume form (compare (3)).

2. *Let $k = 1$ and*

$$\alpha = \sum_{i=1}^{n} a_i(x)\, dx_i$$

be given. Then we define

$$*\alpha := \sqrt{g(x)} \sum_{i=1}^{n} \left(\sum_{j=1}^{n} g^{ij}(x) a_j(x) \right) \theta_i.$$

3. *Let $k = n-1$ and*

$$\alpha = \sum_{i=1}^{n} a_i(x)\theta_i$$

be given. Then we define

$$*\alpha := \frac{(-1)^{n-1}}{\sqrt{g(x)}} \sum_{i=1}^{n} \left(\sum_{j=1}^{n} g_{ij}(x) a_j(x) \right) dx_i.$$

4. *Let $k = n$ and $\alpha = a(x)\omega$ be given. Then we define*

$$*\alpha := a(x).$$

We collect some *properties of the $*$-operator.*

1. The $*$-operator represents a linear operator from the vector space of k-forms into the vector space of $(n-k)$-forms. It gives us an *involution*, which means

$$* * \alpha = (-1)^{k(n-k)}\alpha$$

for all k-forms α with $k \in K$.

2. The k-form α and the $(n-k)$-form β fulfill the identity

$$(\alpha, *\beta)_k = (*\alpha, \beta)_{n-k}(-1)^{k(n-k)}, \qquad k \in K.$$

We prove this statement for all $k \in K$:

a) Let $k = 0$, $\alpha = a(x)$, $\beta = b(x)\omega$, $*\beta = b(x)$, $*\alpha = a(x)\omega$ be given. Then we obtain

$$(\alpha, *\beta)_0 = a(x)b(x) = a(x)b(x)(\omega, \omega)_n = (a(x)\omega, b(x)\omega)_n = (*\alpha, \beta)_n.$$

b) Let $k = n$, α an n-form, β a 0-form be given. We calculate with the aid of property 1 and (a) as follows:

$$(\alpha, *\beta)_n = (*(*\alpha), *\beta)_n = (*\alpha, *(*\beta))_0 = (*\alpha, \beta)_0.$$

c) Let $k = 1$ be given. We consider the forms

$$\alpha = \sum_{i=1}^{n} a_i(x)\, dx_i, \quad \beta = \sum_{i=1}^{n} b_i(x)\theta_i.$$

Then we obtain

$$\begin{aligned}
(\alpha, *\beta)_1 &= \frac{(-1)^{n-1}}{\sqrt{g(x)}} \sum_{i,j=1}^{n} g^{ij}(x)a_i(x) \left(\sum_{k=1}^{n} g_{jk}(x)b_k(x) \right) \\
&= \frac{(-1)^{n-1}}{\sqrt{g(x)}} \sum_{i,j=1}^{n} a_i(x) \left(\sum_{k=1}^{n} g^{ij}(x)g_{jk}(x)b_k(x) \right) \\
&= \frac{(-1)^{n-1}}{\sqrt{g(x)}} \sum_{i=1}^{n} a_i(x) \left(\sum_{k=1}^{n} \delta^i_k b_k(x) \right) \\
&= \frac{(-1)^{n-1}}{\sqrt{g(x)}} \sum_{i=1}^{n} a_i(x)b_i(x),
\end{aligned}$$

as well as

$$\begin{aligned}
(*\alpha, \beta)_{n-1} &= \frac{\sqrt{g(x)}}{g(x)} \sum_{i,j=1}^{n} g_{ij}(x) \left(\sum_{k=1}^{n} g^{ik}(x)a_k(x) \right) b_j(x) \\
&= \frac{1}{\sqrt{g(x)}} \sum_{i,j=1}^{n} b_j(x) \left(\sum_{k=1}^{n} g_{ij}(x)g^{ik}(x)a_k(x) \right) \\
&= \frac{1}{\sqrt{g(x)}} \sum_{j,k=1}^{n} b_j(x) \left(\delta^k_j a_k(x) \right) \\
&= \frac{1}{\sqrt{g(x)}} \sum_{i=1}^{n} a_i(x)b_i(x).
\end{aligned}$$

This implies $(\alpha, *\beta)_1 = (-1)^{n-1}(*\alpha, \beta)_{n-1}$.

d) The case $k = n - 1$ remains. With the aid of property 1 and (c), we deduce for the $(n-1)$-form α and the 1-form β as follows:

$$\begin{aligned}(\alpha, *\beta)_{n-1} &= (-1)^{n-1}(*(*\alpha), *\beta)_{n-1} \\ &= (*\alpha, *(*\beta))_1 = (-1)^{n-1}(*\alpha, \beta)_1.\end{aligned}$$

3. Taking the two k-forms α and β with $k \in K$, we infer

$$\begin{aligned}(*\alpha, *\beta)_{n-k} &= (-1)^{k(n-k)}(*(*\alpha), \beta)_k \\ &= \left((-1)^{k(n-k)}\right)^2 (\alpha, \beta)_k = (\alpha, \beta)_k.\end{aligned}$$

Consequently, the $*$-operator represents an *isometry.*

4. Two k-forms α and β satisfy the identity

$$\alpha \wedge (*\beta) = (-1)^{k(n-k)}(*\alpha) \wedge \beta = (\alpha, \beta)_k \omega, \qquad k \in K.$$

For the proof, we show the relation

$$\alpha \wedge (*\beta) = (\alpha, \beta)_k \omega. \tag{6}$$

Then the $(n-k)$-form $*\alpha$ and the k-form β satisfy

$$(-1)^{k(n-k)}(*\alpha) \wedge \beta = \beta \wedge (*\alpha) = (\beta, \alpha)_k \omega = (\alpha, \beta)_k \omega = \alpha \wedge (*\beta).$$

a) Let $k = 0$, $\alpha = a(x)$, $\beta = b(x)$, $*\beta = b(x)\omega$ be given. Then we see

$$\alpha \wedge (*\beta) = a(x)b(x)\omega = (\alpha, \beta)_0 \omega.$$

b) Let $k = 1$ as well as

$$\alpha = \sum_{i=1}^{n} a_i(x)\, dx_i, \quad \beta = \sum_{i=1}^{n} b_i(x)\, dx_i$$

and

$$*\beta = \sqrt{g(x)} \sum_{i=1}^{n} \left(\sum_{j=1}^{n} g^{ij}(x) b_j(x) \right) \theta_i$$

be given. Now we evaluate

$$\alpha \wedge (*\beta) = \sqrt{g(x)} \left(\sum_{i,j=1}^{n} g^{ij}(x) a_i(x) b_j(x) \right) dx_1 \wedge \ldots \wedge dx_n = (\alpha, \beta)_1 \omega.$$

c) For $k = n - 1$ and

$$\alpha = \sum_{i=1}^{n} a_i(x)\theta_i, \quad \beta = \sum_{i=1}^{n} b_i(x)\theta_i$$

as well as

$$*\beta = \frac{(-1)^{n-1}}{\sqrt{g(x)}} \sum_{i=1}^{n} \left(\sum_{j=1}^{n} g_{ij}(x) b_j(x) \right) dx_i,$$

we infer

$$\alpha \wedge (*\beta) = \left(\sum_{i=1}^{n} a_i(x)\theta_i \right) \wedge \left(\frac{(-1)^{n-1}}{\sqrt{g(x)}} \sum_{i=1}^{n} \left(\sum_{j=1}^{n} g_{ij}(x) b_j(x) \right) dx_i \right)$$

$$= \left(\frac{1}{\sqrt{g(x)}} \sum_{i,j=1}^{n} g_{ij}(x) a_i(x) b_j(x) \right) dx_1 \wedge \ldots \wedge dx_n$$

$$= (\alpha, \beta)_{n-1} \sqrt{g(x)}\, dx_1 \wedge \ldots \wedge dx_n = (\alpha, \beta)_{n-1} \omega.$$

d) Finally, let $k = n$, $\alpha = a(x)\omega$, and $\beta = b(x)\omega$ be given. This implies

$$\alpha \wedge (*\beta) = a(x)\omega b(x) = a(x)b(x)\omega = (\alpha, \beta)_n \omega.$$

5. Let

$$\alpha = \sum_{i=1}^{n} a_i(x) dx_i$$

denote a Pfaffian form and

$$x = \Phi(\overline{x}) = \Big(\Phi_1(\overline{x}_1, \ldots, \overline{x}_n), \ldots, \Phi_n(\overline{x}_1, \ldots, \overline{x}_n) \Big)$$

a parameter transformation. Then we observe $(*\alpha)_\Phi = *(\alpha_\Phi)$.

We use the invariance of the inner product as well as the property 4: For an arbitrary 1-form

$$\beta = \sum_{i=1}^{n} b_i(x)\, dx_i$$

with the transformed 1-form β_Φ, we infer the identity

$$\beta_\Phi \wedge *(\alpha_\Phi) = (\beta_\Phi, \alpha_\Phi)_1 \omega_\Phi = \{(\beta, \alpha)_1\}_\Phi \omega_\Phi$$

$$= \{(\beta, \alpha)_1 \omega\}_\Phi = \{\beta \wedge (*\alpha)\}_\Phi = \beta_\Phi \wedge (*\alpha)_\Phi.$$

Then we obtain

$$\beta_\Phi \wedge (*(\alpha_\Phi) - (*\alpha)_\Phi) = 0 \qquad \text{for all} \quad \beta,$$

and consequently

$$*(\alpha_\Phi) = (*\alpha)_\Phi.$$

Definition 8.2. *Given a 1-form*

$$\alpha = \sum_{i=1}^{n} a_i(x)\, dx_i\,, \qquad x \in \Omega$$

of the class $C^1(\Omega)$, we define the co-derivative *$\delta\alpha$ due to*

$$\delta\alpha := *d * \alpha.$$

Remark: Now δ represents a parameter-invariant differential operator of first order - and attributes a 0-form to each 1-form. We determine the operator δ in arbitrary coordinates. Let us consider

$$\alpha = \sum_{i=1}^{n} a_i(x)\, dx_i, \quad *\alpha = \sqrt{g(x)} \sum_{i=1}^{n} \left(\sum_{j=1}^{n} g^{ij}(x) a_j(x) \right) \theta_i.$$

Then we evaluate

$$d * \alpha = \sum_{i=1}^{n} \frac{\partial}{\partial x_i} \left(\sqrt{g(x)} \sum_{j=1}^{n} g^{ij}(x) a_j(x) \right) dx_1 \wedge \ldots \wedge dx_n$$

$$= \frac{1}{\sqrt{g(x)}} \sum_{i=1}^{n} \frac{\partial}{\partial x_i} \left(\sqrt{g(x)} \sum_{j=1}^{n} g^{ij}(x) a_j(x) \right) \omega.$$

The application of the $*$-operator on $d * \alpha$ yields

$$\delta\alpha = *d * \alpha = \frac{1}{\sqrt{g(x)}} \sum_{i=1}^{n} \frac{\partial}{\partial x_i} \left(\sqrt{g(x)} \sum_{j=1}^{n} g^{ij}(x) a_j(x) \right). \tag{7}$$

Theorem 8.3. (Partial integration in arbitrary parameters)
Let $\Omega \subset \mathbb{R}^n$ denote a domain satisfying the assumptions (A), (B), and (D) for the Gaussian integral theorem. The parameter transformation

$$\overline{x} = \Phi(x) \,:\, \Omega \longrightarrow \Theta \in C^1(\overline{\Omega})$$

may be bijective and subject to the condition

$$J_\Phi(x) \geq \eta > 0 \qquad \text{for all points} \quad x \in \overline{\Omega}.$$

Furthermore, let a 1-form

$$\alpha = \sum_{i=1}^{n} a_i(x)\, dx_i, \qquad x \in \overline{\Omega}$$

and a 0-form $\beta = b(x)$, $x \in \overline{\Omega}$ *of the class* $C^1(\overline{\Omega})$ *be given. Then we have the identity*

$$\int\limits_{\Omega} (\alpha, d\beta)_1 \omega + \int\limits_{\Omega} (\delta\alpha, \beta)_0 \omega = \int\limits_{\partial\Omega} (*\alpha) \wedge \beta.$$

Here the boundary $\partial\Omega$ *is endowed with the induced canonical orientation of* $\mathbb{R}^n$.

Proof: The assumptions on the parameter transformation Φ guarantee that all functions appearing belong to the regularity class $C^1(\overline{\Omega})$. We apply the Stokes integral theorem and obtain - with the aid of (6) - our statement as follows:

$$\begin{aligned}
\int\limits_{\Omega} (\alpha, d\beta)_1 \omega &= \int\limits_{\Omega} \alpha \wedge (*d\beta) = (-1)^{n-1} \int\limits_{\Omega} (*\alpha) \wedge d\beta \\
&= \int\limits_{\Omega} d\Big((*\alpha) \wedge \beta\Big) - \int\limits_{\Omega} (d * \alpha) \wedge \beta \\
&= \int\limits_{\partial\Omega} (*\alpha) \wedge \beta - \int\limits_{\Omega} (d * \alpha) \wedge (* * \beta) \\
&= \int\limits_{\partial\Omega} (*\alpha) \wedge \beta - \int\limits_{\Omega} (d * \alpha, *\beta)_n \omega \\
&= \int\limits_{\partial\Omega} (*\alpha) \wedge \beta - \int\limits_{\Omega} (*d * \alpha, \beta)_0 \omega \\
&= \int\limits_{\partial\Omega} (*\alpha) \wedge \beta - \int\limits_{\Omega} (\delta\alpha, \beta)_0 \omega.
\end{aligned}$$

q.e.d.

Corollary: When we require zero-boundary-values in Theorem 8.3 for the function β, or more precisely $\beta \in C_0^1(\Omega)$, we deduce the identity

$$\int\limits_{\Omega} (\alpha, d\beta)_1 \omega + \int\limits_{\Omega} (\delta\alpha, \beta)_0 \omega = 0.$$

Therefore, we name δ the *adjoint derivative* to the exterior derivative d.

Definition 8.4. *The two functions* $\psi(x)$ *and* $\chi(x)$ *of the class* $C^1(\Omega)$ *with their associate differentials*

$$d\psi = \sum_{i=1}^{n} \psi_{x_i} \, dx_i, \quad d\chi = \sum_{i=1}^{n} \chi_{x_i} \, dx_i$$

being given, we define the Beltrami operator of first order *via*

$$\nabla(\psi,\chi) := (d\psi, d\chi)_1(x) = \sum_{i,j=1}^{n} g^{ij}(x)\psi_{x_i}(x)\chi_{x_j}(x).$$

Remark: Evidently, the property

$$\nabla(\psi,\chi)(x) = \nabla(\overline{\psi},\overline{\chi})\Big(\Phi(x)\Big)$$

holds true, where we note that

$$\overline{\psi}\Big(\Phi(x)\Big) = \psi(x), \quad \overline{\chi}\Big(\Phi(x)\Big) = \chi(x).$$

Consequently, ∇ represents a parameter-invariant differential operator of first order.

Definition 8.5. *We define the* Laplace-Beltrami operator

$$\Delta\psi(x) := \delta d\psi(x), \qquad x \in \Omega$$

for functions $\psi(x) \in C^2(\Omega)$.

Remark: Since the operators d and δ are parameter-invariant, the operator Δ is parameter-invariant as well:

$$\Delta\psi(x) = \Delta\overline{\psi}\Big(\Phi(x)\Big), \qquad x \in \Omega.$$

Using (7), we now describe Δ in coordinates:

$$\begin{aligned}\Delta\psi = \delta d\psi &= \delta\left(\sum_{j=1}^{n}\psi_{x_j}\,dx_j\right)\\ &= \frac{1}{\sqrt{g(x)}}\sum_{i=1}^{n}\frac{\partial}{\partial x_i}\left(\sqrt{g(x)}\sum_{j=1}^{n}g^{ij}(x)\psi_{x_j}\right).\end{aligned} \tag{8}$$

Theorem 8.6. *Let $\Omega \subset \mathbb{R}^n$ denote a domain satisfying the assumptions (A), (B), and (D) of the Gaussian integral theorem. Furthermore, the parameter transformation*

$$\overline{x} = \Phi(x) : \overline{\Omega} \longrightarrow \overline{\Theta}$$

belongs to the class $C^2(\overline{\Omega})$ and is bijective subject to the condition

$$J_\Phi(x) \ge \eta > 0 \qquad \text{for all points} \quad x \in \overline{\Omega}.$$

Finally, let the functions $\psi(x) \in C^2(\overline{\Omega})$ as well as $\chi(x) \in C^1(\overline{\Omega})$ be given. Then we have the identity

$$\int_\Omega \nabla(\psi,\chi)\omega + \int_\Omega (\Delta\psi,\chi)_0\omega = \int_{\partial\Omega}(*d\psi)\chi.$$

Proof: We apply Theorem 8.3 and insert

$$\alpha = d\psi \in C^1(\overline{\Omega}), \quad \beta = \chi(x) \in C^1(\overline{\Omega}).$$

At first, we obtain

$$\int_{\Omega} (d\psi, d\chi)_1 \omega + \int_{\Omega} (\delta d\psi, \beta)_0 \omega = \int_{\partial\Omega} (*d\psi)\chi.$$

Using the Definitions 8.4 and 8.5, we infer the identity

$$\int_{\Omega} \boldsymbol{\nabla}(\psi, \chi)\omega + \int_{\Omega} (\boldsymbol{\Delta}\psi, \chi)_0 \omega = \int_{\partial\Omega} (*d\psi)\chi$$

stated above. q.e.d.

Remark:

1. We evaluate the Laplace operator in cylindrical coordinates,

$$x = r\cos\varphi, \quad y = r\sin\varphi, \quad z = h,$$

where $0 < r < +\infty$, $0 \le \varphi < 2\pi$, $-\infty < h < +\infty$ hold true. Therefore, we consider the case $n = 3$ and choose

$$x_1 = r, \quad x_2 = \varphi, \quad x_3 = h.$$

The fundamental tensor appears in the following form:

$$(g_{ij}) = \begin{pmatrix} 1 & 0 & 0 \\ 0 & r^2 & 0 \\ 0 & 0 & 1 \end{pmatrix}, \qquad (g^{ij}) = \begin{pmatrix} 1 & 0 & 0 \\ 0 & \frac{1}{r^2} & 0 \\ 0 & 0 & 1 \end{pmatrix}.$$

This implies

$$g(x) = \det(g_{ij}) = r^2.$$

In our calculations we have to respect only those elements on the principal diagonal. With the aid of (7), we then obtain

$$\begin{aligned}
\boldsymbol{\Delta} &= \frac{1}{r}\left\{\frac{\partial}{\partial r}\left(r\frac{\partial}{\partial r}\right) + \frac{\partial}{\partial\varphi}\left(\frac{1}{r}\frac{\partial}{\partial\varphi}\right) + \frac{\partial}{\partial h}\left(r\frac{\partial}{\partial h}\right)\right\} \\
&= \frac{1}{r}\left(\frac{\partial}{\partial r} + r\frac{\partial^2}{\partial r^2} + \frac{1}{r}\frac{\partial^2}{\partial\varphi^2} + r\frac{\partial^2}{\partial h^2}\right) \\
&= \frac{\partial^2}{\partial r^2} + \frac{1}{r}\frac{\partial}{\partial r} + \frac{1}{r^2}\frac{\partial^2}{\partial\varphi^2} + \frac{\partial^2}{\partial h^2}.
\end{aligned}$$

For plane polar coordinates we set $z \equiv 0$, and the expression above is reduced to

$$\boldsymbol{\Delta} = \frac{\partial^2}{\partial r^2} + \frac{1}{r}\frac{\partial}{\partial r} + \frac{1}{r^2}\frac{\partial^2}{\partial \varphi^2}.$$

Defining

$$\boldsymbol{\Lambda} := \frac{\partial^2}{\partial \varphi^2}$$

for the angular expression, we rewrite $\boldsymbol{\Delta}$ into the form

$$\boldsymbol{\Delta} = \frac{\partial^2}{\partial r^2} + \frac{1}{r}\frac{\partial}{\partial r} + \frac{1}{r^2}\boldsymbol{\Lambda}.$$

(compare the Laplace operator in spherical coordinates).

2. We introduce spherical coordinates

$$x = r\cos\varphi\sin\theta, \quad y = r\sin\varphi\sin\theta, \quad z = r\cos\theta$$

with $0 < r < +\infty$, $0 \le \varphi < 2\pi$, and $0 < \theta < \pi$. Calculations parallel to Remark 1 yield

$$\begin{aligned}
\boldsymbol{\Delta} &= \frac{1}{r^2}\left\{\frac{\partial}{\partial r}\left(r^2\frac{\partial}{\partial r}\right) + \frac{1}{\sin\theta}\frac{\partial}{\partial\theta}\left(\sin\theta\frac{\partial}{\partial\theta}\right) + \frac{1}{\sin^2\theta}\frac{\partial^2}{\partial\varphi^2}\right\} \\
&= \frac{\partial^2}{\partial r^2} + \frac{2}{r}\frac{\partial}{\partial r} + \frac{1}{r^2}\left\{\frac{1}{\sin\theta}\frac{\partial}{\partial\theta}\left(\sin\theta\frac{\partial}{\partial\theta}\right) + \frac{1}{\sin^2\theta}\frac{\partial^2}{\partial\varphi^2}\right\} \\
&=: \frac{\partial^2}{\partial r^2} + \frac{2}{r}\frac{\partial}{\partial r} + \frac{1}{r^2}\boldsymbol{\Lambda}.
\end{aligned}$$

Here the operator $\boldsymbol{\Lambda}$ does not depend on r again. However, it is only dependent on the angles φ, θ.

When we investigate *spherical harmonic functions* in Chapter 5, we need the Laplace operator for spherical coordinates in n dimensions. Now we treat this general case.

Let the unit sphere in $\mathbb{R}^n$, namely

$$\Sigma = \Big\{\xi = (\xi_1, \ldots, \xi_n) \in \mathbb{R}^n \, : \, |\xi| = 1\Big\},$$

by parametrized by

$$\xi = \xi(t) = \Big(\xi_1(t_1, \ldots, t_{n-1}), \ldots, \xi_n(t_1, \ldots, t_{n-1})\Big)^t : T \longrightarrow \Sigma \in C^2(T),$$

with the open set $T \subset \mathbb{R}^{n-1}$. Via the mapping

$$X(r,t) := r\xi(t_1, \ldots, t_{n-1}), \qquad r \in (0, +\infty), \quad t \in T,$$

we obtain polar coordinates in $\mathbb{R}^n$. Furthermore, the functional matrix appears in the form

$$\partial X(r,t) = (X_r, X_{t_1}, \dots, X_{t_{n-1}}) = (\xi, r\xi_{t_1}, \dots, r\xi_{t_{n-1}}).$$

We determine the metric tensor as follows:

$$G(r,t) = \Big(g_{ij}(r,t)\Big)_{i,j} = \begin{pmatrix} 1 & 0 & \cdots & 0 \\ 0 & r^2 h_{11} & \cdots & r^2 h_{1,n-1} \\ \vdots & \cdots & & \vdots \\ 0 & r^2 h_{n-1,1} & \cdots & r^2 h_{n-1,n-1} \end{pmatrix} = \begin{pmatrix} 1 & 0 \cdots & 0 \\ 0 & & \\ \vdots & r^2 H(t) & \\ 0 & & \end{pmatrix},$$

where we abbreviate

$$H(t) = \Big(h_{ij}(t)\Big)_{i,j=1,\dots,n-1} := \Big(\xi_{t_i}(t) \cdot \xi_{t_j}(t)\Big)_{i,j=1,\dots,n-1}.$$

Using the convention

$$H^{-1}(t) = \Big(h^{ij}(t)\Big)_{i,j=1,\dots,n-1}, \quad G^{-1}(r,t) = \Big(g^{ij}(r,t)\Big)_{i,j=1,\dots,n},$$

we infer

$$G^{-1}(r,t) = \Big(g^{ij}(r,t)\Big)_{i,j} = \begin{pmatrix} 1 & 0 \cdots & 0 \\ 0 & & \\ \vdots & \dfrac{H^{-1}(t)}{r^2} & \\ 0 & & \end{pmatrix} = \begin{pmatrix} 1 & 0 & \cdots & 0 \\ 0 & \frac{h^{11}}{r^2} & \cdots & \frac{h^{1,n-1}}{r^2} \\ \vdots & \cdots & & \vdots \\ 0 & \frac{h^{n-1,1}}{r^2} & \cdots & \frac{h^{n-1,n-1}}{r^2} \end{pmatrix}.$$

Furthermore, we define

$$g(r,t) := \det G(r,t), \quad h(t) := \det H(t)$$

and obtain

$$g(r,t) = r^{2(n-1)} h(t).$$

When $u = u(r,t)$ and $v = v(r,t)$ are two functions, we determine the Beltrami differential operator of first order due to

$$\begin{aligned} \boldsymbol{\nabla}(u,v) &= \sum_{i,j=1}^{n} g^{ij}(x) u_{x_i} v_{x_j} \\ &= \frac{\partial u}{\partial r}\frac{\partial v}{\partial r} + \frac{1}{r^2}\sum_{i,j=1}^{n-1} h^{ij}(t) \frac{\partial u}{\partial t_i}\frac{\partial v}{\partial t_j}. \end{aligned}$$

We express the *invariant Beltrami operator of first order on the sphere* Σ via

$$\boldsymbol{\Gamma}(u,v) := \sum_{i,j=1}^{n-1} h^{ij}(t) \frac{\partial u}{\partial t_i}\frac{\partial v}{\partial t_j}$$

and deduce

$$\boldsymbol{\nabla}(u,v) = \frac{\partial u}{\partial r}\frac{\partial v}{\partial r} + \frac{1}{r^2}\boldsymbol{\Gamma}(u,v) \qquad \text{for all} \quad u = u(r,t), \quad v = v(r,t). \tag{9}$$

Now we represent the Laplace-Beltrami operator in spherical coordinates: We take the function

$$u = u(r,t) = u(r,t_1,\ldots,t_{n-1}),$$

utilize the identity $\sqrt{g(r,t)} = r^{n-1}\sqrt{h(t)}$ as well as formula (8), and obtain

$$\begin{aligned}
\boldsymbol{\Delta} u &= \frac{1}{\sqrt{g(r,t)}}\operatorname{div}_{(r,t)}\left\{\sqrt{g(r,t)}\; G^{-1}(r,t)\circ\begin{pmatrix} u_r \\ u_{t_1} \\ \vdots \\ u_{t_{n-1}}\end{pmatrix}\right\} \\
&= \frac{1}{\sqrt{g(r,t)}}\frac{\partial}{\partial r}\left(\sqrt{g(r,t)}\,\frac{\partial u}{\partial r}\right) \\
&\quad + \frac{1}{\sqrt{g(r,t)}}\operatorname{div}_t\left\{r^{n-1}\sqrt{h(t)}\,\frac{1}{r^2}H^{-1}(t)\circ\begin{pmatrix} u_{t_1} \\ \vdots \\ u_{t_{n-1}}\end{pmatrix}\right\} \\
&= \frac{\partial^2 u}{\partial r^2} + \frac{n-1}{r}\frac{\partial u}{\partial r} + \frac{1}{r^2}\frac{1}{\sqrt{h(t)}}\operatorname{div}_t\left\{\sqrt{h(t)}\; H^{-1}(t)\circ\begin{pmatrix} u_{t_1} \\ \vdots \\ u_{t_{n-1}}\end{pmatrix}\right\}.
\end{aligned}$$

Defining the *Laplace-Beltrami operator on the sphere* Σ by

$$\boldsymbol{\Lambda} u := \frac{1}{\sqrt{h(t)}}\sum_{i=1}^{n-1}\frac{\partial}{\partial t_i}\left(\sqrt{h(t)}\sum_{j=1}^{n-1} h^{ij}(t)\frac{\partial u}{\partial t_j}\right), \qquad t \in T,$$

we obtain the following identity

$$\boldsymbol{\Delta} u = \frac{\partial^2 u}{\partial r^2} + \frac{n-1}{r}\frac{\partial u}{\partial r} + \frac{1}{r^2}\boldsymbol{\Lambda} u \qquad \text{for all} \quad u = u(r,t) \in C^2((0,+\infty)\times T). \tag{10}$$

We still show the symmetry of the Laplace-Beltrami operator on the sphere for later use.

Theorem 8.7. *Taking the functions* $f, g \in C^2(\Sigma)$, *we have the relation*

$$\int\limits_{\Sigma} f(\xi)\Big(\boldsymbol{\Lambda} g(\xi)\Big)\,d\sigma(\xi) = -\int\limits_{\Sigma}\boldsymbol{\Gamma}(f,g)\,d\sigma(\xi) = \int\limits_{\Sigma}\Big(\boldsymbol{\Lambda} f(\xi)\Big)g(\xi)\,d\sigma(\xi).$$

Here $d\sigma$ *denotes the surface element on* Σ.

Proof: Let $0 < \varepsilon < 1$ be given, and we consider the domain

$$\Omega_\varepsilon := \Big\{ x \in \mathbb{R}^n \, : \, 1 - \varepsilon < |x| < 1 + \varepsilon \Big\}.$$

Furthermore, we have

$$u(r,\xi) := f(\xi), \quad v(r,\xi) := g(\xi), \qquad r \in (1-\varepsilon, 1+\varepsilon), \quad \xi \in \Sigma.$$

Theorem 8.6 yields

$$\int\limits_{\Omega_\varepsilon} \boldsymbol{\nabla}(u,v)\,\omega + \int\limits_{\Omega_\varepsilon} (\boldsymbol{\Delta} u, v)_0\,\omega = \int\limits_{\partial\Omega_\varepsilon} (*du)v = \int\limits_{\partial\Omega_\varepsilon} v\,\frac{\partial u}{\partial \nu}\,d\sigma,$$

where ν denotes the exterior normal to $\partial\Omega_\varepsilon$. These parameter-invariant integrals are evaluated in (r,ξ)-coordinates: Via the identities (9) as well as (10) and noting that

$$\frac{\partial u}{\partial \nu} = \pm \frac{\partial u}{\partial r} \equiv 0 \quad \text{on} \quad \partial\Omega_\varepsilon,$$

we arrive at the relation

$$0 = \int\limits_{1-\varepsilon}^{1+\varepsilon} \left(\int\limits_{\Sigma} \frac{1}{r^2}\, \boldsymbol{\Gamma}(f,g)\, d\sigma(\xi)\, r^{n-1} \right) dr + \int\limits_{1-\varepsilon}^{1+\varepsilon} \left(\int\limits_{\Sigma} \frac{1}{r^2}\, \boldsymbol{\Lambda}(f)\, g\, d\sigma(\xi)\, r^{n-1} \right) dr$$

$$= \left(\int\limits_{1-\epsilon}^{1+\epsilon} r^{n-3}\, dr \right) \int\limits_{\Sigma} \Big(\boldsymbol{\Gamma}(f,g) + \boldsymbol{\Lambda}(f)\, g \Big) d\sigma(\xi).$$

This implies

$$\int_\Sigma \Big(\boldsymbol{\Lambda} f(\xi) \Big) g(\xi)\, d\sigma(\xi) = - \int_\Sigma \boldsymbol{\Gamma}(f,g)\, d\sigma(\xi).$$

Correspondingly, we deduce the second identity stated above. q.e.d.

9 Some Historical Notices to Chapter 1

The theory of partial differential equations in the classical sense is treated within the framework of the continuously differentiable functions. The profound integral theorem of Gauß constitutes the center for the classical investigations of partial differential equations. This might explain the title *Princeps Mathematicorum* attributed to him. His tomb in Göttingen and the monument for him, together with the physicist W. Weber, express the great respect, which is given to C.F. Gauß.

Our treatment within the framework of differential forms, created by E. Cartan (1869–1961), simplifies the various integral theorems and classifies them geometrically. Though differential forms are systematically used, with great success, in differential geometry, analysts mostly refrain from their application in the theory of partial differential equations. We owe the introduction of invariant differential operators to E. Beltrami (1835–1900) – the first representative of a great differential-geometric tradition in Italy.

Figure 1.1 PORTRAIT OF CARL FRIEDRICH GAUSS (1777–1855)
Lithography by Siegried Detlef Bendixen published in Schumacher's *Astronomische Nachrichten* in 1828; taken from the inner titel-page of the biography by *Horst Michling: Carl Friedrich Gauß – Aus dem Leben des Princeps Mathematicorum*, Verlag Göttinger Tageblatt, Göttingen (1976).

Chapter 2

Foundations of Functional Analysis

We start with the Riemannian integral - and their Riemann integrable functions - and construct a considerably larger class of integrable functions via an extension procedure. Then we obtain Lebesgue's integral, which is distinguished by general convergence theorems for *pointwise* convergent sequences of functions. This extension procedure - from the Riemannian integral to Lebesgue's integral - will be provided by the Daniell integral. The measure theory for Lebesgue measurable sets will appear in this context as the theory of integration for characteristic functions. We shall present classical results from the theory of measure and integration in this chapter, e.g. the theorems of Egorov and Lusin.

Then we treat the Lebesgue spaces L^p with the exponents $1 \leq p \leq +\infty$ as classical Banach spaces. We investigate orthogonal systems of functions in the Hilbert space L^2. With ideas of J. von Neumann we determine the dual spaces $(L^p)^* = L^q$ and show the weak compactness of the Lebesgue spaces.

1 Daniell's Integral with Examples

Our point of departure is the following

Definition 1.1. *We consider an arbitrary set X, and by $M = M(X)$ we denote a space of functions $f : X \to \mathbb{R}$ which have the following properties:*

- *M is a linear space, which means*

$$\text{for all } f, g \in M \text{ and all } \alpha, \beta \in \mathbb{R} \text{ we have } \quad \alpha f + \beta g \in M. \tag{1}$$

- *M is closed with respect to the modulus operation, which means*

$$\text{for all } f \in M \text{ we have } \quad |f| \in M. \tag{2}$$

Furthermore, the symbol $I : M \to \mathbb{R}$ denotes a functional on M satisfying the following conditions:

F. Sauvigny, *Partial Differential Equations 1*, Universitext,
DOI 10.1007/978-1-4471-2981-3_2, © Springer-Verlag London 2012

- *I is linear, which means*

$$\text{for all } f, g \in M \text{ and all } \alpha, \beta \in \mathbb{R} \text{ we have} \quad I(\alpha f + \beta g) = \alpha I(f) + \beta I(g). \tag{3}$$

- *I is nonnegative, which says*

$$\text{for all } f \in M \text{ with } f \geq 0 \text{ we have} \quad I(f) \geq 0. \tag{4}$$

Here the relation $f \geq 0$ *indicates that* $f(x) \geq 0$ *for all* $x \in X$ *is correct.*

- *I is continuous with respect to monotone convergence in* M*, which means*

$$\begin{array}{ll} \text{for each sequence} \{f_n\}_{n=1,2,\ldots} \subset M & \text{with} \quad f_n \downarrow 0 \\ \text{we have} \quad \lim_{n\to\infty} I(f_n) = I(0) = 0. & \end{array} \tag{5}$$

Here we comprehend by $f_n \downarrow 0$ *that the sequence* $\{f_n(x)\}_{n=1,2,\ldots} \subset \mathbb{R}$ *is weakly monotonically decreasing for all* $x \in X$ *and* $\lim\limits_{n\to\infty} f_n(x) = 0$ *holds true.*

Then this functional I is named Daniell's integral *defined on* M.

Remarks:

1. From the linearity (1) and the lattice property (2) we infer

$$\max(f, g) = \frac{1}{2}\Big(f + g + |f - g|\Big) \in M$$

as well as

$$\min(f, g) = \frac{1}{2}\Big(f + g - |f - g|\Big) \in M$$

for two elements $f, g \in M$. In particular, with each element $f \in M$ we have

$$f^+(x) := \max\Big(f(x), 0\Big) = \frac{1}{2}\Big(f(x) + |f(x)|\Big) \in M$$

as well as

$$f^-(x) := \max\Big(-f(x), 0\Big) = (-f)^+(x) \in M.$$

We address f^+ as the *positive part of* f and f^- as the *negative part of* f. The definitions of f^+ and f^- imply the identities

$$f = f^+ - f^- \quad \text{and} \quad |f| = f^+ + f^- = f^+ + (-f)^+.$$

Consequently, the lattice condition (2) is equivalent to

$$f \in M \quad \Longrightarrow \quad f^+ \in M. \tag{2'}$$

More generally, we see that finitely many functions $f_1, \ldots, f_m \in M$ with $m \in \mathbb{N}$ imply the inclusion

$$\max(f_1, \ldots, f_m) \in M \quad \text{and} \quad \min(f_1, \ldots, f_m) \in M.$$

2. The condition (4) is equivalent to the monotonicity of the integral, namely
$$I(f) \geq I(g) \quad \text{for all} \quad f, g \in M \text{ with } f \geq g. \tag{4'}$$
3. The condition (5) is equivalent to the following property:

   All sequences $\{f_n\}_{n=1,2,\ldots} \subset M$ with $f_n \uparrow f$ and $f, g \in M$ with $g \leq f$ fulfill (5′)
   $$I(g) \leq \lim_{n\to\infty} I(f_n).$$

   Proof: At first, we show the direction '(5′) ⇒ (5)'. Let the sequence of functions $\{f_n\}_{n=1,2,\ldots} \subset M$ with $f_n \downarrow 0$ be given. Then we infer $(-f_n) \uparrow 0$. We set $f(x) \equiv 0 \equiv g(x)$. The linearity of I implies $I(g) = 0$ immediately. The combination of (5′) and (4) reveals the relation
   $$0 = I(g) \leq \lim_{n\to\infty} I(-f_n) = -\lim_{n\to\infty} \underbrace{I(f_n)}_{\geq 0} \leq 0.$$
   This yields $\lim\limits_{n\to\infty} I(f_n) = I(0) = 0$.

   Now we show the implication '(5) ⇒ (5′)'.
   The sequence $\{f_n\}_{n=1,2,\ldots}$ may satisfy $f_n \uparrow f$ with an element $f \in M$, which immediately implies $(f - f_n) \downarrow 0$. From (5) we infer $0 = \lim_{n\to\infty} I(f - f_n)$, and the linearity of I yields
   $$0 = I(f) - \lim_{n\to\infty} I(f_n).$$
   With $g \leq f$ and (4′) we obtain
   $$\lim_{n\to\infty} I(f_n) = I(f) \geq I(g),$$
   and the proof is complete. q.e.d.

Now we provide examples of Daniell integrals, where we need the following

Theorem 1.2. (U. Dini)
Let the continuous functions $f_1, f_2, \ldots$ and $f \in C^0(K, \mathbb{R})$ be defined on the compact set $K \subset \mathbb{R}^n$. We have the relation $f_l \uparrow f$, which means that the sequence $\{f_l(x)\} \subset \mathbb{R}$ is weakly monotonically increasing for all $x \in K$ and furthermore
$$\lim_{l\to\infty} f_l(x) = f(x).$$
Then the sequence $\{f_l\}_{l=1,2,\ldots}$ converges uniformly on the set K towards the function f.

Remark: The transition to functions $g_l := f - f_l$ implies that the statement above is equivalent to the following:

A sequence of functions $\{g_l\}_{l=1,2,\ldots} \subset C^0(K, \mathbb{R})$ with $g_l \downarrow 0$ has necessarily the

property that $\{g_l\}_{l=1,2,\ldots}$ converges uniformly on K towards 0.

Proof of Theorem 1.2: Let $\{g_l\}_{l=1,2,\ldots} \subset C^0(K,\mathbb{R})$ denote a sequence satisfying $g_l \downarrow 0$. We have to show that

$$\sup_{x\in K} |g_l(x)| \longrightarrow 0$$

is correct. If this property was not valid, then we could find indices $\{l_i\}$ with $l_i < l_{i+1}$ and points $\xi_i \in K$ such that

$$g_{l_i}(\xi_i) \geq \varepsilon > 0 \qquad \text{for all} \quad i \in \mathbb{N}$$

hold true with a fixed quantity $\varepsilon > 0$. According to the Weierstraß compactness theorem, we can assume - without loss of generality - that the relation $\xi_i \to \xi$ for $i \to \infty$ is valid, with the limit point $\xi \in K$. For the fixed index l_*, we now choose an index $i_* = i(l_*) \in \mathbb{N}$ such that $l_i \geq l_*$holds true for all $i \geq i_*$. Now the monotonicity of the sequence of functions $\{g_l\}$ implies

$$g_{l_*}(\xi_i) \geq g_{l_i}(\xi_i) \geq \varepsilon \qquad \text{for all} \quad i \geq i_*.$$

Since the function g_{l_*} is assumed to be continuous, we infer

$$g_{l_*}(\xi) = \lim_{i\to\infty} g_{l_*}(\xi_i) \geq \varepsilon \qquad \text{for all} \quad l_* \in \mathbb{N}.$$

Therefore, $\{g_l(\xi)\}$ does not constitute a null-sequence, which gives an obvious contradiction to the assumption.

q.e.d.

Main example 1: Let us consider $X = \Omega$ with the open set $\Omega \subset \mathbb{R}^n$ and the linear space

$$M_1 = M_1(X) := \left\{ f(x) \in C^0(\Omega,\mathbb{R}) : \int_\Omega |f(x)|\,dx < +\infty \right\}.$$

Here the symbol

$$\int_\Omega |f(x)|\,dx$$

means the improper Riemannian integral over the open set Ω. Then our space M_1 satisfies the conditions (1) and (2). Now we choose the functional

$$I_1(f) := \int_\Omega f(x)\,dx, \qquad f \in M_1,$$

where the improper Riemannian integral over Ω appears again on the right-hand side. Because the Riemannian integral is linear and nonnegative, the conditions (3) and (4) are evident. We still have to establish the continuity of our functional with respect to monotone convergence, namely (5). Let us consider with $\{f_n\}_{n=1,2,\ldots} \subset M_1$ a sequence of functions satisfying $f_n \downarrow 0$. If $K \subset \Omega$ denotes a compact subset, Dini's theorem tells us that $\{f_n\}$ converges uniformly on K towards 0. When we observe the properties $0 \leq f_n(x) \leq f_1(x)$ for all $n \in \mathbb{N}$ and $x \in \Omega$ as well as $\int\limits_\Omega |f_1(x)|\,dx < +\infty$, the fundamental convergence theorem for improper Riemannian integrals implies

$$\lim_{n\to\infty} I_1(f_n) = \lim_{n\to\infty} \int\limits_\Omega f_n(x)\,dx = \int\limits_\Omega \underbrace{\Big(\lim_{n\to\infty} f_n(x)\Big)}_{=0}\,dx = 0.$$

Therefore, I_1 represents a Daniell integral on the space M_1.

Remark: The set M_1 does not contain *all* functions whose improper Riemannian integral exists. The concept of Daniell's integral additionally necessitates the function space being closed with respect to the modulus operation, namely the lattice property (2). For instance, the integral

$$\int\limits_1^\infty \frac{\sin x}{x^\alpha}\,dx \qquad \text{for all powers} \quad \alpha \in (0,1)$$

does not converge absolutely, although it exists as an improper Riemannian integral.

Main example 2: As we described in Section 4 of Chapter 1, let $\mathcal{M} \subset \mathbb{R}^n$ denote a bounded m-dimensional manifold of the class C^1 with the regular boundary $\partial\mathcal{M}$. Then we can cover $\overline{\mathcal{M}}$ by finitely many charts, and we define the Riemannian integral over $\mathcal{M}$ via *partition of unity*, namely

$$I_2(f) := \int\limits_{\overline{\mathcal{M}}} f(x)\,d^m\sigma(x), \qquad f \in M_2$$

for all functions of the class

$$M_2 := \Big\{f(x) : \overline{\mathcal{M}} \to \mathbb{R} \;:\; f \text{ is continuous on } \overline{\mathcal{M}}\Big\}.$$

Here the symbol $d^m\sigma$ means the m-dimensional surface element on $\mathcal{M}$. This integral I_2 gives us a further interesting Daniell integral: The linear space M_2 is closed with respect to the modulus operation. The properties (1) and (2) are consequently fulfilled. The existence of the integral above follows from the continuity - and therefore the boundedness - of f on the compact manifold $\overline{\mathcal{M}}$. The linearity and the positive-semidefinite character of I_2 are evident. The continuity of I_2 with respect to monotone convergence follows from Dini's theorem again.

2 Extension of Daniell's Integral to Lebesgue's Integral

In our main examples from Section 1, we already have an integral which allows, at least, to integrate the continuous functions with compact support. Now we consider an arbitrary Daniell integral $I : M \to \mathbb{R}$ due to Definition 1.1 in Section 1. We intend to extend this integral onto the larger linear space

$$L(X) \supset M(X),$$

in order to study convergence properties of the created integral on the space $L(X)$. This extension procedure is essentially based on the monotonicity property (4) and the associate continuity property (5) of this integral.

Developing our theory of integration simultaneously for *characteristic functions*

$$\chi_A(x) := \begin{cases} 1, \; x \in A \\ 0, \; x \in X \setminus A \end{cases}$$

of the subsets $A \subset X$, we obtain a measure theory which depends on our Daniell integral I for the subsets of X.

The extension procedure presented here was initiated by Carathéodory, later Daniell considered these particular functionals I, and Stone established the connection to measure theory. The consideration of minimal surfaces gave H. Lebesgue the impetus to study thoroughly the concept of surface area.

We prepare our considerations and introduce the function

$$\Phi(t) := \begin{cases} 0, \; t \le 0 \\ t, \; t \ge 0 \end{cases}$$

which is continuous and weakly monotonically increasing. Furthermore, we define

$$f^+(x) := \Phi(f(x)) = \max(f(x), 0), \quad x \in X$$

and study the following properties of the prescription $f \mapsto f^+$:

i.) $f(x) \le f^+(x)$ for all $x \in X$;

ii.) $f_1(x) \le f_2(x) \quad \Longrightarrow \quad f_1^+(x) \le f_2^+(x)$ for all $x \in X$;

iii.) $f_n(x) \to f(x) \quad \Longrightarrow \quad f_n^+(x) \to f^+(x)$ for all $x \in X$;

iv.) $f_n(x) \downarrow f(x) \quad \Longrightarrow \quad f_n^+(x) \downarrow f^+(x)$ for all $x \in X$;

v.) $f_n(x) \uparrow f(x) \quad \Longrightarrow \quad f_n^+(x) \uparrow f^+(x)$ for all $x \in X$.

Proposition 2.1. *Let $\{g_n\} \subset M$ and $\{g'_n\} \subset M$, $n = 1, 2, \ldots$ denote two sequences satisfying $g_n(x) \uparrow g(x)$ and $g'_n(x) \uparrow g'(x)$ defined on X. Here $g, g' : X \longrightarrow \mathbb{R} \cup \{+\infty\}$ represent two functions with the property $g'(x) \geq g(x)$. Then we infer the inequality*

$$\lim_{n\to\infty} I(g'_n) \geq \lim_{n\to\infty} I(g_n).$$

Proof: Since $\{I(g_n)\}_{n=1,2,\ldots}$ and $\{I(g'_n)\}_{n=1,2,\ldots}$ represent monotonically non-decreasing sequences, their limits exist for $n \to \infty$ in $\mathbb{R} \cup \{+\infty\}$. In the case $\lim\limits_{n\to\infty} I(g'_n) = +\infty$, the inequality above evidently holds true. Therefore, we can assume $\lim\limits_{n\to\infty} I(g'_n) < +\infty$ without loss of generality. With the index m being fixed, we observe

$$(g_m - g'_n)^+ \downarrow (g_m - g')^+ = 0 \qquad \text{for} \quad n \to \infty.$$

Then we invoke the properties of Daniell's integral I as follows:

$$\begin{aligned} I(g_m) - \lim_{n\to\infty} I(g'_n) &= \lim_{n\to\infty} \Big(I(g_m) - I(g'_n) \Big) = \lim_{n\to\infty} I(g_m - g'_n) \\ &\leq \lim_{n\to\infty} I\Big((g_m - g'_n)^+ \Big) = 0. \end{aligned}$$

Now we see

$$I(g_m) \leq \lim_{n\to\infty} I(g'_n) \qquad \text{for all} \quad m \in \mathbb{N},$$

and we arrive at the relation

$$\lim_{m\to\infty} I(g_m) \leq \lim_{n\to\infty} I(g'_n).$$

q.e.d.

When we assume $g = g'$ on X in Proposition 2.1, we obtain equality for the two limits above. This justifies the following

Definition 2.2. *Let the symbol $V(X)$ denote the set of all functions $f : X \to \mathbb{R} \cup \{+\infty\}$, which can be approximated weakly monotonically increasing from $M(X)$ as follows: Each such element f possesses a sequence $\{f_n\}_{n=1,2,\ldots}$ in $M(X)$ with the property*

$$f_n(x) \uparrow f(x) \qquad \textit{for} \quad n \to \infty \quad \textit{and for all} \quad x \in X.$$

For the element $f \in V$, we then define

$$I(f) := \lim_{n\to\infty} I(f_n),$$

and we observe $I(f) \in \mathbb{R} \cup \{+\infty\}$.

Definition 2.3. *We set*

$$-V := \Big\{ f : X \to \mathbb{R} \cup \{-\infty\} \;:\; -f \in V \Big\}$$

and define

$$I(f) := -I(-f) \in \mathbb{R} \cup \{-\infty\} \qquad \textit{for all} \quad f \in -V.$$

Remarks:

1. The set $-V$ represents the set of all functions f which can be approximated weakly monotonically decreasing from M as follows: There exists a sequence $\{f_n\}_{n=1,2,\ldots} \subset M$ satisfying $f_n \downarrow f$. Then we obtain
$$I(f) = \lim_{n\to\infty} I(f_n).$$
2. If $f \in V \cap (-V)$ holds true, we find sequences $\{f_n'\}_{n=1,2,\ldots}$ and $\{f_n''\}_{n=1,2,\ldots}$ in M which fulfill the approximative relations $f_n' \uparrow f$ and $f_n'' \downarrow f$, respectively. Now we see $f_n'' - f_n' \downarrow 0$, and the property (5) implies
$$0 = \lim_{n\to\infty} I(f_n'' - f_n') = \lim_{n\to\infty} I(f_n'') - \lim_{n\to\infty} I(f_n')$$
as well as
$$\lim_{n\to\infty} I(f_n'') = \lim_{n\to\infty} I(f_n').$$
Consequently, the functional I is uniquely defined on the set $V \cup (-V) \supset V \cap (-V) \supset M$.
3. The set V contains the element $f(x) \equiv +\infty$ as the monotonically increasing limit of $f_n(x) = n$; however, it does not contain the element $g(x) \equiv -\infty$. Therefore, the set V does not represent a linear space.

   According to Proposition 2.1, the functional I is monotonic on V as follows: Each two elements $f, g \in V$ with $f \le g$ fulfill $I(f) \le I(g)$. Furthermore, the linear combination $\alpha f + \beta g$ of two elements $f, g \in V$ with nonnegative scalars $\alpha \ge 0$ and $\beta \ge 0$ belongs to V as well, and we have
$$I(\alpha f + \beta g) = \alpha I(f) + \beta I(g).$$

Proposition 2.4. *The function $f : X \to [0, +\infty]$ satisfies the equivalence*

$$f \in V \quad \Longleftrightarrow \quad f(x) = \sum_{n=1}^{\infty} \varphi_n(x),$$

where $\varphi_n \in M(X)$ and $\varphi_n \ge 0$ for all $n \in \mathbb{N}$ hold true.

Proof: The direction '$\Longleftarrow$' is evident from the definition of the space V: The element f is constructed monotonically by the functions $\varphi_n \in M$, and this implies the conclusion.

Now we show the opposite direction '$\Longrightarrow$' as follows: Taking $f \in V$, we find a sequence $\{f_n\}_{n=1,2,\ldots} \subset M$ such that $f_n \uparrow f$, and we infer $f_n^+ \uparrow f^+ = f$. When we define

$$f_0(x) \equiv 0 \quad \text{and} \quad \varphi_n(x) := f_n^+(x) - f_{n-1}^+(x),$$

we observe

$$f_k^+(x) = \sum_{n=1}^{k} \varphi_n(x) \uparrow f(x)$$

and consequently

$$\sum_{n=1}^{\infty} \varphi_n(x) = f(x).$$

Obviously, the functions fulfill $\varphi_n(x) \in M$ and $\varphi_n(x) \geq 0$ for all $n \in \mathbb{N}$.

q.e.d.

Proposition 2.5. *Let the elements $f_i \in V$ with $f_i \geq 0$ for $i = 1, 2, \ldots$ be given. Then the function*

$$f(x) := \sum_{i=1}^{\infty} f_i(x)$$

belongs to the set V, and we have

$$I(f) = \sum_{i=1}^{\infty} I(f_i).$$

Proof: The double sequence $c_{ij} \in \mathbb{R}$ with $c_{ij} \geq 0$ satisfies the following equation:

$$\sum_{i,j=1}^{\infty} c_{ij} = \sum_{i=1}^{\infty} \left(\sum_{j=1}^{\infty} c_{ij} \right) = \lim_{n \to \infty} \sum_{i,j=1}^{n} c_{ij}. \tag{1}$$

This equation holds true for convergent as well as for definitely divergent double series. On account of $f_i \in V$, we have functions $\varphi_{ij} \in M$ satisfying $\varphi_{ij} \geq 0$ such that

$$f_i(x) = \sum_{j=1}^{\infty} \varphi_{ij}(x) \quad \text{for all} \quad x \in X \quad \text{and all} \quad i \in \mathbb{N}$$

is correct. From Definition 2.2 we infer

$$I(f_i) = \lim_{n\to\infty} I\left(\sum_{j=1}^{n}\varphi_{ij}\right) = \lim_{n\to\infty}\left\{\sum_{j=1}^{n} I(\varphi_{ij})\right\} = \sum_{j=1}^{\infty} I(\varphi_{ij}).$$

Furthermore, we have the following representation for all $x \in X$:

$$f(x) = \sum_{i=1}^{\infty} f_i(x) = \sum_{i=1}^{\infty}\left(\sum_{j=1}^{\infty}\varphi_{ij}(x)\right) = \sum_{i,j=1}^{\infty}\varphi_{ij}(x) = \lim_{n\to\infty}\left(\sum_{i,j=1}^{n}\varphi_{ij}(x)\right).$$

Consequently, $f \in V$ holds true and Definition 2.2 yields

$$I(f) = \lim_{n\to\infty} I\left(\sum_{i,j=1}^{n}\varphi_{ij}\right) = \lim_{n\to\infty}\sum_{i,j=1}^{n} I(\varphi_{ij})$$

$$= \sum_{i,j=1}^{\infty} I(\varphi_{ij}) = \sum_{i=1}^{\infty}\left(\sum_{j=1}^{\infty} I(\varphi_{ij})\right) = \sum_{i=1}^{\infty} I(f_i).$$

q.e.d.

Definition 2.6. *We consider an arbitrary function* $f : X \to \overline{\mathbb{R}} = \mathbb{R} \cup \{\pm\infty\}$ *and define*

$$I^+(f) := \inf\Big\{I(h) \,:\, h \in V,\, h \geq f\Big\}, \quad I^-(f) := \sup\Big\{I(g) \,:\, g \in -V,\, g \leq f\Big\}.$$

We name $I^+(f)$ *the* upper *and* $I^-(f)$ *the* lower Daniell integral *of* f.

Proposition 2.7. *Let* $f : X \to \overline{\mathbb{R}}$ *denote an arbitrary function and* (g, h) *a pair of functions satisfying* $g \in -V$ *and* $h \in V$ *as well as* $g(x) \leq f(x) \leq h(x)$ *for all* $x \in X$. *Then we infer*

$$I(g) \leq I^-(f) \leq I^+(f) \leq I(h).$$

Proof: Definition 2.6 implies $I(h) \geq I^+(f)$ and $I(g) \leq I^-(f)$. Furthermore, we find sequences $\{g_n\}_{n=1,2,\ldots} \subset -V$ and $\{h_n\}_{n=1,2,\ldots} \subset V$ satisfying

$$g_n \leq f \leq h_n \quad \text{for all} \quad n \in \mathbb{N},$$

such that

$$\lim_{n\to\infty} I(g_n) = I^-(f) \quad \text{and} \quad \lim_{n\to\infty} I(h_n) = I^+(f)$$

holds true. On account of $0 \leq h_n + (-g_n) \in V$ for arbitrary $n \in \mathbb{N}$, we see

$$0 \leq I\Big(h_n + (-g_n)\Big) = I(h_n) + I(-g_n)$$

and consequently

$$I(g_n) \le I(h_n)$$

and finally

$$I^-(f) = \lim_{n\to\infty} I(g_n) \le \lim_{n\to\infty} I(h_n) = I^+(f).$$

q.e.d.

In the sequel, we consider functions with values in the extended real number system $\overline{\mathbb{R}} = \mathbb{R} \cup \{-\infty\} \cup \{+\infty\}$. Within the set $\overline{\mathbb{R}}$ we need the following calculus rules:

- *Addition:*

$$\begin{aligned} a + (+\infty) = \quad (+\infty) + a \quad &= +\infty \text{ for all } a \in \mathbb{R} \cup \{+\infty\} \\ a + (-\infty) = \quad (-\infty) + a \quad &= -\infty \text{ for all } a \in \mathbb{R} \cup \{-\infty\} \\ (-\infty) + (+\infty) = (+\infty) + (-\infty) = \quad & 0 \end{aligned}$$

- *Multiplication:*

$$\left.\begin{aligned} a\,(+\infty) &= (+\infty)\,a = +\infty \\ a\,(-\infty) &= (-\infty)\,a = -\infty \end{aligned}\right\} \quad \text{for all } 0 < a \le +\infty$$

$$\begin{aligned} 0\,(+\infty) &= (+\infty)\,0 = +\infty \\ 0\,(-\infty) &= (-\infty)\,0 = -\infty \end{aligned}$$

$$\left.\begin{aligned} a\,(+\infty) &= (+\infty)\,a = -\infty \\ a\,(-\infty) &= (-\infty)\,a = +\infty \end{aligned}\right\} \quad \text{for all } -\infty \le a < 0$$

- *Subtraction:* For $a, b \in \overline{\mathbb{R}}$ we define

$$a - b := a + (-b),$$

where we set

$$-(+\infty) = -\infty \quad \text{and} \quad -(-\infty) = +\infty.$$

- *Ordering:* We have

$$-\infty \le a \le +\infty \qquad \text{for all} \quad a \in \overline{\mathbb{R}}.$$

Remark: Algebraically the set $\overline{\mathbb{R}}$ does not constitute a field, because the addition is not associative; consider for instance:

$$(-\infty) + \Big((+\infty) + (+\infty)\Big) = (-\infty) + (+\infty) = 0,$$

$$\Big((-\infty) + (+\infty)\Big) + (+\infty) = 0 + (+\infty) = +\infty.$$

With these calculus operations in $\overline{\mathbb{R}}$, we can uniquely define the functions $f+g$, $f-g$, cf for two functions $f : X \to \overline{\mathbb{R}}$ and $g : X \to \overline{\mathbb{R}}$ and arbitrary scalars $c \in \mathbb{R}$. Furthermore, we have the inequality $f \le g$ if and only if $g - f \ge 0$ is correct.

Definition 2.8. *The function $f : X \to \overline{\mathbb{R}}$ belongs to the class $L = L(X) = L(X, I)$ if and only if*

$$-\infty < I^-(f) = I^+(f) < +\infty$$

holds true. Then we define

$$I(f) := I^-(f) = I^+(f),$$

and we say that f is Lebesgue integrable *with respect to I.*

Remark: In our main example 1 from Section 1, we consider the open subset $\Omega \subset \mathbb{R}^n$ and obtain the class $L(X) =: L(\Omega)$ of *Lebesgue integrable functions in Ω*. In our main example 2, we get the class of *Lebesgue integrable functions on the manifold $\mathcal{M}$* with $L(X) =: L(\mathcal{M})$.

Proposition 2.9. *The function $f : X \to \overline{\mathbb{R}}$ belongs to the class $L(X)$ if and only if each quantity $\varepsilon > 0$ admits two functions $g \in -V$ and $h \in V$ satisfying*

$$g(x) \le f(x) \le h(x), \quad x \in X \quad \text{and} \quad I(h) - I(g) < \varepsilon.$$

In particular, $I(g)$ and $I(h)$ are finite.

Proof:

'$\Longrightarrow$' We consider $f \in L(X)$ and note that $I^-(f) = I^+(f) \in \mathbb{R}$. According to Definition 2.6, we find functions $g \in -V$ and $h \in V$ with $g \le f \le h$ and $I(h) - I(g) < \varepsilon$.

'$\Longleftarrow$' For each quantity $\varepsilon > 0$, we have functions $g \in -V$ and $h \in V$ with $g \le f \le h$ and $I(h) - I(g) < \varepsilon$. On account of $I(h) \in (-\infty, +\infty]$ and $I(g) \in [-\infty, +\infty)$, we infer $I(h), I(g) \in \mathbb{R}$. Now Proposition 2.7 implies the estimate

$$0 \le I^+(f) - I^-(f) \le I(h) - I(g) < \varepsilon$$

for arbitrary $\varepsilon > 0$. Consequently, $I^+(f) = I^-(f) \in \mathbb{R}$ holds true and finally $f \in L(X)$. q.e.d.

Theorem 2.10. (Calculus rules for Lebesgue integrable functions) *The set $L(X)$ of Lebesgue integrable functions has the following properties:*

a) The statement

$$f \in L(X) \qquad \textit{for each} \quad f \in V(X) \quad \textit{with} \quad I(f) < +\infty$$

is correct, and the integrals from Definition 2.2 and Definition 2.8 coincide. Consequently, the functional $I : M(X) \to \mathbb{R}$ *has been extended onto* $L(X) \supset M(X)$. *Furthermore, we have*

$$I(f) \geq 0 \qquad \textit{for all} \quad f \in L(X) \quad \textit{with} \quad f \geq 0.$$

b) The space $L(X)$ *is linear, which means*

$$c_1 f_1 + c_2 f_2 \in L(X) \qquad \textit{for all} \quad f_1, f_2 \in L(X) \quad \textit{and} \quad c_1, c_2 \in \mathbb{R}.$$

Furthermore, $I : L(X) \to \mathbb{R}$ *represents a linear functional. Therefore, we have the calculus rule*

$$I(c_1 f_1 + c_2 f_2) = c_1 I(f_1) + c_2 I(f_2) \quad \textit{for all} \quad f_1, f_2 \in L(X), \quad c_1, c_2 \in \mathbb{R}.$$

c) When $f \in L(X)$ *is given, then* $|f| \in L(X)$ *holds true and the estimate* $\big|I(f)\big| \leq I\big(|f|\big)$ *is valid.*

Proof:

a) Consider $f \in V(X)$ with $I(f) < +\infty$. Then we find a sequence

$$\{f_n\}_{n=1,2,\ldots} \subset M(X)$$

such that $f_n \uparrow f$ holds true. When we define $g_n := f_n$ and $h_n := f$ for all $n \in \mathbb{N}$, we infer $g_n \leq f \leq h_n$ with $g_n \in -V$ and $h_n \in V$, and we observe $I(h_n) - I(g_n) = I(f) - I(f_n) \to 0$. Proposition 2.9 tells us that $f \in L(X)$, and Definition 2.8 implies

$$-\infty < I(f) := I^+(f) = I^-(f) = \lim_{n \to \infty} I(f_n) < +\infty.$$

We consider $0 \leq f \in L(X)$, and we infer from $0 \in -V$ the statement $0 \leq I^-(f) = I(f)$.

b) At first, we show: If $f \in L(X)$ is chosen, we have $-f \in L(X)$ as well as $I(-f) = -I(f)$.

With $f \in L(X)$ given, each quantity $\varepsilon > 0$ admits a pair of functions $g \in -V$ and $h \in V$ satisfying $g \leq f \leq h$ as well as $I(h) - I(g) < \varepsilon$. This implies $-h \leq -f \leq -g$ with $-h \in -V$ and $-g \in V$. We note that $I(-g) = -I(g)$ and $I(-h) = -I(h)$ hold true, and we obtain

$$I(-g) - I(-h) = -I(g) + I(h) < \varepsilon \quad \text{for all} \quad \varepsilon > 0.$$

Finally, we arrive at $-f \in L(X)$ and $I(-f) = -I(f)$.

Now we show: With $f \in L(X)$ and $c > 0$, we have $cf \in L(X)$ and $I(cf) = cI(f)$.

Therefore, we consider $f \in L(X), c > 0$, and each $\varepsilon > 0$ admits functions $g \in -V$ and $h \in V$ with $g \leq f \leq h$ as well as $I(h) - I(g) < \varepsilon$. This implies $cg \leq cf \leq ch$, $cg \in -V$, $ch \in V$ and finally

$$I(ch) - I(cg) = c\Big(I(h) - I(g)\Big) < c\varepsilon.$$

We have thus proved $cf \in L(X)$ and $I(cf) = cI(f)$.

Finally, we deduce the calculus rule: From $f_1, f_2 \in L(X)$ we infer $f_1 + f_2 \in L(X)$ and $I(f_1 + f_2) = I(f_1) + I(f_2)$.

The elements $f_1, f_2 \in L(X)$ being given, we find to each $\varepsilon > 0$ the functions $g_1, g_2 \in -V$ and $h_1, h_2 \in V$ satisfying $g_i \leq f_i \leq h_i$ and $I(h_i) - I(g_i) < \varepsilon$ for $i = 1, 2$. This immediately implies $h_1 + h_2 \in V$, $g_1 + g_2 \in -V$, $g_1 + g_2 \leq f_1 + f_2 \leq h_1 + h_2$ and $I(h_1 + h_2) - I(g_1 + g_2) < 2\varepsilon$. We conclude $f_1 + f_2 \in L(X)$ and obtain the calculus rule $I(f_1 + f_2) = I(f_1) + I(f_2)$.

Therefore, $I : L(X) \to \mathbb{R}$ represents a linear functional on the linear space $L(X)$ of Lebesgue integrable functions.

c) With $f \in L(X)$, we find functions $g \in -V$ and $h \in V$ satisfying $g \leq f \leq h$ and $I(h) - I(g) < \varepsilon$ to each $\varepsilon > 0$, and we see $g^+ \leq f^+ \leq h^+$. Furthermore, we have sequences $g_n \downarrow g$ and $h_n \uparrow h$ in $M(X)$, which give us the approximations $g_n^+ \downarrow g^+$ and $h_n^+ \uparrow h^+$, respectively. Therefore, $h^+ \in V$ and $g^+ \in -V$ holds true as well as $h^+ - g^+ \in V$. From $h \geq g$ we infer $h^+ - g^+ \leq h - g$ and see

$$\begin{aligned} I(h^+) - I(g^+) = I(h^+) + I(-g^+) &= I(h^+ - g^+) \\ \leq I(h - g) &= I(h) - I(g) < \varepsilon. \end{aligned}$$

Consequently, the statements $f^+ \in L(X)$ and $|f| = f^+ + (-f)^+ \in L(X)$ are established. With $f \in L(X)$, the elements $-f$ and $|f|$ belong to $L(X)$ as well, and the inequalities $f \leq |f|$, $-f \leq |f|$ imply $I(f) \leq I(|f|)$, $-I(f) = I(-f) \leq I(|f|)$ and finally $|I(f)| \leq I(|f|)$.

q.e.d.

Now we deduce convergence theorems for Lebesgue's integral: Fundamental is the following

Proposition 2.11. *Let the sequence* $\{f_k\}_{k=1,2,\ldots} \subset L(X)$ *with* $f_k \geq 0,\ k \in \mathbb{N}$ *and* $\sum\limits_{k=1}^{\infty} I(f_k) < +\infty$ *be given. Then the property*

$$f(x) := \sum_{k=1}^{\infty} f_k(x) \in L(X)$$

is fulfilled, and we have

$$I(f) = \sum_{k=1}^{\infty} I(f_k).$$

Proof: Given the quantity $\varepsilon > 0$, we find functions $g_k \in -V$ and $h_k \in V$ with $0 \le g_k \le f_k \le h_k$ and $I(h_k) - I(g_k) < \varepsilon\, 2^{-k}$ for all $k \in \mathbb{N}$, on account of $f_k \in L(X)$. Therefore, we have the inequalities

$$I(g_k) > I(h_k) - \frac{\varepsilon}{2^k} \ge I(f_k) - \frac{\varepsilon}{2^k} \quad \text{and} \quad I(h_k) < I(g_k) + \frac{\varepsilon}{2^k} \le I(f_k) + \frac{\varepsilon}{2^k}.$$

Now we choose n so large that $\sum\limits_{k=n+1}^{\infty} I(f_k) \le \varepsilon$ is correct. When we set

$$g := \sum_{k=1}^{n} g_k, \qquad h := \sum_{k=1}^{\infty} h_k,$$

we observe $g \in -V$ and $h \in V$, due to Proposition 2.5, as well as $g \le f \le h$. Furthermore, we see

$$I(g) = \sum_{k=1}^{n} I(g_k) > \sum_{k=1}^{n} \Big(I(f_k) - \frac{\varepsilon}{2^k} \Big) \ge \sum_{k=1}^{\infty} I(f_k) - 2\varepsilon$$

and

$$I(h) = \sum_{k=1}^{\infty} I(h_k) < \sum_{k=1}^{\infty} \Big(I(f_k) + \frac{\varepsilon}{2^k} \Big) = \sum_{k=1}^{\infty} I(f_k) + \varepsilon.$$

Consequently, we obtain $I(h) - I(g) < 3\varepsilon$ and additionally $f \in L(X)$. Finally, our estimates yield the identity

$$I(f) = \sum_{k=1}^{\infty} I(f_k).$$

q.e.d.

Theorem 2.12. (B.Levi's theorem on monotone convergence) *Let $\{f_n\}_{n=1,2,\ldots} \subset L(X)$ denote a sequence satisfying*

$$f_n(x) \ne \pm\infty \qquad \textit{for all} \quad x \in X \quad \textit{and all} \quad n \in \mathbb{N}.$$

Furthermore, let the conditions

$$f_n(x) \uparrow f(x), \quad x \in X, \quad \textit{and} \quad I(f_n) \le C, \quad n \in \mathbb{N}$$

be valid, with a constant $C \in \mathbb{R}$. Then we have $f \in L(X)$ and

$$\lim_{n\to\infty} I(f_n) = I(f).$$

Proof: On account of $f_k(x) \in \mathbb{R}$, the addition is associative there. Setting

$$\varphi_k(x) := (f_k(x) - f_{k-1}(x)) \in L(X), \qquad k = 2, 3, \ldots,$$

we infer $\varphi_k \geq 0$ as well as

$$\sum_{k=2}^{n} \varphi_k(x) = f_n(x) - f_1(x), \qquad x \in X.$$

Now we observe

$$C - I(f_1) \geq I(f_n) - I(f_1) = \sum_{k=2}^{n} I(\varphi_k) \qquad \text{for all} \quad n \geq 2.$$

Proposition 2.11 implies

$$f - f_1 = \sum_{k=2}^{\infty} \varphi_k \in L(X)$$

and furthermore

$$\lim_{n\to\infty} I(f_n) - I(f_1) = \sum_{k=2}^{\infty} I(\varphi_k) = I\left(\sum_{k=2}^{\infty} \varphi_k\right) = I(f - f_1) = I(f) - I(f_1).$$

Therefore, we obtain $f \in L(X)$ and the following limit relation:

$$\lim_{n\to\infty} I(f_n) = I(f).$$

q.e.d.

Remark: The restrictive assumption $f_n(x) \neq \pm\infty$ will be eliminated in the next section.

Theorem 2.13. (Fatou's convergence theorem)
Let $\{f_n\}_{n=1,2,\ldots} \subset L(X)$ denote a sequence of functions such that

$$0 \leq f_n(x) < +\infty \qquad \textit{for all} \quad x \in X \quad \textit{and all} \quad n \in \mathbb{N}$$

holds true. Furthermore, we assume

$$\liminf_{n\to\infty} I(f_n) < +\infty.$$

Then the function $g(x) := \liminf_{n\to\infty} f_n(x)$ belongs to the space $L(X)$, and we observe the lower semicontinuity

$$I(g) \leq \liminf_{n\to\infty} I(f_n).$$

Proof: We note that

$$g(x) = \liminf_{n\to\infty} f_n(x) = \lim_{n\to\infty}\Big(\inf_{m\geq n} f_m(x)\Big) = \lim_{n\to\infty}\Big(\lim_{k\to\infty} g_{n,k}(x)\Big)$$

holds true with

$$g_{n,k}(x) := \min\Big(f_n(x), f_{n+1}(x), \ldots, f_{n+k}(x)\Big) \in L(X).$$

When we define

$$g_n(x) := \inf_{m\geq n} f_m(x),$$

we infer the relations $g_{n,k} \downarrow g_n$ and $-g_{n,k} \uparrow -g_n$ for $k \to \infty$. Furthermore, we obtain $I(-g_{n,k}) \leq 0$ due to $f_n(x) \geq 0$. From Theorem 2.12 we infer $-g_n \in L(X)$ and consequently $g_n \in L(X)$ for all $n \in \mathbb{N}$.

Furthermore, we see $g_n(x) \leq f_m(x)$, $x \in X$ for all $m \geq n$. Therefore, the inequality

$$I(g_n) \leq \inf_{m\geq n} I(f_m) \leq \lim_{n\to\infty}\Big(\inf_{m\geq n} I(f_m)\Big) = \liminf_{n\to\infty} I(f_n) < +\infty$$

is correct for all $n \in \mathbb{N}$. We utilize $g_n \uparrow g$ as well as Theorem 2.12, and we obtain $g \in L(X)$ and, moreover,

$$I(g) = \lim_{n\to\infty} I(g_n) \leq \liminf_{n\to\infty} I(f_n).$$

q.e.d.

Theorem 2.14. *Let $\{f_n\}_{n=1,2,\ldots} \subset L(X)$ denote a sequence with*

$$|f_n(x)| \leq F(x) < +\infty, \qquad n \in \mathbb{N}, \quad x \in X,$$

where $F(x) \in L(X)$ is correct. Furthermore, let us define

$$g(x) := \liminf_{n\to\infty} f_n(x) \quad \textit{and} \quad h(x) := \limsup_{n\to\infty} f_n(x).$$

Then the elements g and h belong to $L(X)$, and we have the inequalities

$$I(g) \leq \liminf_{n\to\infty} I(f_n), \quad I(h) \geq \limsup_{n\to\infty} I(f_n).$$

Proof: We apply Theorem 2.13 on both sequences $\{F + f_n\}$ and $\{F - f_n\}$ of nonnegative finite-valued functions from $L(X)$. We observe the inequality

$$I(F \pm f_n) \leq I(F + F) \leq 2I(F) < +\infty \qquad \text{for all} \quad n \in \mathbb{N}.$$

Thus we obtain

$$L(X) \ni \liminf_{n\to\infty}(F + f_n) = F + \liminf_{n\to\infty} f_n = F + g$$

as well as $g \in L(X)$. Now Theorem 2.13 yields

$$I(F) + I(g) = I(F+g) \le \liminf_{n\to\infty} I(F+f_n) = I(F) + \liminf_{n\to\infty} I(f_n)$$

and

$$I(g) \le \liminf_{n\to\infty} I(f_n).$$

In the same way we deduce

$$L(X) \ni \liminf_{n\to\infty}(F - f_n) = F - \limsup_{n\to\infty} f_n = F - h$$

and consequently $h \in L(X)$. This implies

$$I(F) - I(h) = I(F-h) \le \liminf_{n\to\infty} I(F-f_n) = I(F) - \limsup_{n\to\infty} I(f_n)$$

and finally

$$I(h) \ge \limsup_{n\to\infty} I(f_n).$$

q.e.d.

Theorem 2.15. (H.Lebesgue's theorem on dominated convergence) *Let* $\{f_n\}_{n=1,2,\dots} \subset L(X)$ *denote a sequence with*

$$f_n(x) \to f(x) \qquad \text{for} \quad n \to \infty, \quad x \in X.$$

Furthermore, we assume

$$|f_n(x)| \le F(x) < +\infty, \qquad n \in \mathbb{N}, \quad x \in X$$

where $F \in L(X)$ *is valid. Then we infer* $f \in L(X)$ *as well as*

$$\lim_{n\to\infty} I(f_n) = I(f).$$

Proof: The limit relation

$$\lim_{n\to\infty} f_n(x) = f(x), \qquad x \in X$$

implies

$$\liminf_{n\to\infty} f_n(x) = f(x) = \limsup_{n\to\infty} f_n(x).$$

According to Theorem 2.14, we have $f \in L(X)$ and

$$\limsup_{n\to\infty} I(f_n) \le I(f) \le \liminf_{n\to\infty} I(f_n).$$

Therefore, the subsequent limit exists

$$\lim_{n\to\infty} I(f_n),$$

and we deduce

$$I(f) = \lim_{n\to\infty} I(f_n).$$

q.e.d.

3 Measurable Sets

Beginning with this section, we have to require the following

Additional assumptions for the sets X and $M(X)$:

- We assume $X \subset \mathbb{R}^n$ with the dimension $n \in \mathbb{N}$. Then X becomes a topological space as follows: A subset $A \subset X$ is open (closed) if and only if we have an open (closed) subset $\widehat{A} \subset \mathbb{R}^n$ such that $A = X \cap \widehat{A}$ holds true.
- Furthermore, we assume that the inclusion $C_b^0(X, \mathbb{R}) \subset M(X) \subset C^0(X, \mathbb{R})$ is fulfilled. Here $C_b^0(X, \mathbb{R})$ describes the set of bounded continuous functions. This is valid for our main example 2. In our main example 1, this is fulfilled as well if the open set $\Omega \subset \mathbb{R}^n$ is subject to the following condition:

$$\int_{\Omega} 1\,dx < +\infty.$$

We see immediately that the function $f_0 \equiv 1$, $x \in X$ then belongs to the class $M(X)$.

Now we specialize our theory of integration from Section 2 to characteristic functions and obtain a measure theory. For an arbitrary set $A \subset X$ we define its *characteristic function* by

$$\chi_A(x) := \begin{cases} 1, & x \in A \\ 0, & x \in X \setminus A \end{cases}.$$

Definition 3.1. *A subset $A \subset X$ is called* finitely measurable *(or alternatively* integrable*) if its characteristic function satisfies $\chi_A \in L(X)$. We name*

$$\mu(A) := I(\chi_A)$$

the measure of the set A with respect to the integral I. *The* set of all finitely measurable sets in X *is denoted by $\mathcal{S}(X)$.*

From the additional assumptions above, namely $f_0 \equiv 1 \in M(X)$, we infer $\chi_X \in M(X) \subset L(X)$ and consequently $X \in \mathcal{S}(X)$. Therefore, we speak equivalently of *finitely measurable* and *measurable* sets.

Proposition 3.2. **(σ-Additivity)**
Let $\{A_i\}_{i=1,2,\ldots} \subset \mathcal{S}(X)$ denote a sequence of mutually disjoint sets. Then the set

$$A := \bigcup_{i=1}^{\infty} A_i$$

belongs to $\mathcal{S}(X)$ as well, and we have

$$\mu(A) = \sum_{i=1}^{\infty} \mu(A_i).$$

Proof: We consider the sequence of functions

$$f_k := \sum_{l=1}^{k} \chi_{A_l} \uparrow \chi_A \le \chi_X \in L(X)$$

and note that $f_k \in L(X)$ for all $k \in \mathbb{N}$ holds true. Now Lebesgue's convergence theorem yields $\chi_A \in L(X)$ and consequently $A \in \mathcal{S}(X)$. Finally, we evaluate

$$\begin{aligned}\mu(A) = I(\chi_A) &= \lim_{k\to\infty} I(f_k) = \lim_{k\to\infty} I(\chi_{A_1} + \ldots + \chi_{A_k}) \\ &= \lim_{k\to\infty} \Big(\mu(A_1) + \ldots + \mu(A_k)\Big) = \sum_{l=1}^{\infty} \mu(A_l).\end{aligned}$$

q.e.d.

We show that with $A, B \in \mathcal{S}(X)$ their intersection $A \cap B$ belongs to $\mathcal{S}(X)$ as well. On account of $\chi_{A\cap B} = \chi_A \chi_B$, we have to verify that with $\chi_A, \chi_B \in L(X)$ their product satisfies $\chi_A\chi_B \in L(X)$ as well. In general, the product of two functions in $L(X)$ need not lie in $L(X)$ as demonstrated by the following

Example 3.3. With $X = (0,1)$, we define the space

$$M(X) = \left\{ f : (0,1) \to \mathbb{R} \in C^0\Big((0,1), \mathbb{R}\Big) \; : \; \int_0^1 |f(x)|\, dx < +\infty \right\}$$

and the improper Riemannian integral $I(f) := \int_0^1 f(x)\, dx$. Then we observe

$$f(x) := \frac{1}{\sqrt{x}} \in L(X); \quad \text{however,} \quad f^2(x) := \frac{1}{x} \notin L(X).$$

Now we establish the following

Theorem 3.4. (Continuous combination of bounded L-functions) *Let $f_k(x) \in L(X)$ for $k = 1, \ldots, \kappa$ denote finitely many bounded functions, such that the estimate*

$$|f_k(x)| \le c \qquad \textit{for all points} \quad x \in X \quad \textit{and all indices} \quad k \in \{1, \ldots, \kappa\}$$

is valid, with a constant $c \in (0, +\infty)$. Furthermore, let the function $\Phi = \Phi(y_1, \ldots, y_\kappa) : \mathbb{R}^\kappa \to \mathbb{R} \in C^0(\mathbb{R}^\kappa, \mathbb{R})$ be given. Then the composition

$$g(x) := \Phi\Big(f_1(x), \ldots, f_\kappa(x)\Big), \qquad x \in X$$

belongs to the class $L(X)$ and is bounded.

Proof:

1. With $f : X \to \mathbb{R} \in L(X)$ let us consider a bounded function. At first, we show that its square satisfies $f^2 \in L(X)$. We observe

$$f^2(x) = \{f(x) - \lambda\}^2 + 2\lambda f(x) - \lambda^2$$

and infer

$$f^2(x) \geq 2\lambda f(x) - \lambda^2 \qquad \text{for all} \quad \lambda \in \mathbb{R},$$

where equality is attained only for $\lambda = f(x)$. Therefore, we can rewrite the square-function as follows:

$$f^2(x) = \sup_{\lambda \in \mathbb{R}} \Big(2\lambda f(x) - \lambda^2\Big).$$

Since the function $\lambda \mapsto (2\lambda f(x) - \lambda^2)$ is continuous with respect to λ for each fixed $x \in X$, it is sufficient to evaluate this supremum only over the set of rational numbers. Furthermore, we have $\mathbb{Q} = \{\lambda_l\}_{l=1,2,\dots}$ and see

$$f^2(x) = \sup_{l \in \mathbb{N}} \Big(2\lambda_l f(x) - \lambda_l^2\Big) = \lim_{m \to \infty} \left(\max_{1 \leq l \leq m} \Big(2\lambda_l f(x) - \lambda_l^2\Big)\right).$$

With the aid of

$$\varphi_m(x) := \max_{1 \leq l \leq m} \Big(2\lambda_l f(x) - \lambda_l^2\Big)$$

we obtain

$$f^2(x) = \lim_{m \to \infty} \varphi_m(x) = \lim_{m \to \infty} \varphi_m^+(x),$$

where the last equality is inferred from the positivity of $f^2(x)$. Since $f \in L(X)$ holds true, the linearity and the closedness with respect to the maximum operation of $L(X)$ imply: The elements φ_m and consequently φ_m^+ belong to the space $L(X)$. Furthermore, for all points $x \in X$ and all $m \in \mathbb{N}$ we have the estimate

$$0 \leq \varphi_m^+(x) \leq f^2(x) \leq c$$

with a constant $c \in (0, +\infty)$. From the property $f_0(x) \equiv 1 \in L(X)$ we infer $f_c(x) \equiv c \in L(X)$, and the functions φ_m^+ have an integrable dominating function. Now Lebesgue's convergence theorem yields

$$f^2(x) = \lim_{m \to \infty} \varphi_m^+(x) \in L(X).$$

2. When $f, g \in L(X)$ represent bounded functions, its product $f \cdot g$ is bounded as well. On account of part 1 of our proof and the identity

$$fg = \frac{1}{4}(f+g)^2 - \frac{1}{4}(f-g)^2,$$

we deduce $fg \in L(X)$.

3. On the rectangle

$$Q := \Big\{ y = (y_1, \ldots, y_\kappa) \in \mathbb{R}^\kappa \; : \; |y_k| \le c, \; k = 1, \ldots, \kappa \Big\}$$

we can approximate the continuous function Φ uniformly by polynomials

$$\Phi_l = \Phi_l(y_1, \ldots, y_\kappa), \qquad l = 1, 2, \ldots .$$

From part 2 we infer that the functions

$$g_l(x) := \Phi_l\Big(f_1(x), \ldots, f_\kappa(x)\Big), \qquad x \in X$$

are bounded and belong to the class $L(X)$. We have the estimate

$$|g_l(x)| \le C \qquad \text{for all} \quad x \in X \quad \text{and all} \quad l \in \mathbb{N}$$

with a fixed constant $C \in (0, +\infty)$. Since the function satisfies $\varphi(x) \equiv C \in L(X)$, Lebesgue's convergence theorem yields

$$g(x) = \Phi\Big(f_1(x), \ldots, f_\kappa(x)\Big) = \lim_{l \to \infty} g_l(x) \in L(X).$$

q.e.d.

Corollary from Theorem 3.4: If $f(x) \in L(X)$ represents a bounded function, its power $|f|^p$ belongs to the class $L(X)$ for all exponents $p > 0$.

Proposition 3.5. *With the sets $A, B \in \mathcal{S}(X)$ the following sets $A \cap B$, $A \cup B$, $A \setminus B$, $A^c := X \setminus A$ belong to $\mathcal{S}(X)$ as well.*

Proof: Let us take $A, B \in \mathcal{S}(X)$, and the associate characteristic functions χ_A, χ_B are bounded and belong to the class $L(X)$. Via Proposition 3.4, we deduce

$$\chi_{A \cap B} = \chi_A \chi_B \in L(X) \quad \text{and consequently} \quad A \cap B \in \mathcal{S}(X).$$

Now we see $A \cup B \in \mathcal{S}(X)$ due to $\chi_{A \cup B} = \chi_A + \chi_B - \chi_{A \cap B} \in L(X)$. Furthermore, we observe

$$\chi_{A \setminus B} = \chi_{A \setminus (A \cap B)} = \chi_A - \chi_{A \cap B} \in L(X) \quad \text{and consequently} \quad A \setminus B \in L(X).$$

On account of $X \in \mathcal{S}(X)$, we finally infer $A^c = (X \setminus A) \in \mathcal{S}(X)$. q.e.d.

Proposition 3.6. (σ-Subadditivity)
Let $\{A_i\}_{i=1,2,\ldots} \subset \mathcal{S}(X)$ denote a sequence of sets. Then their denumerable union

$$A := \bigcup_{i=1}^{\infty} A_i$$

belongs to $\mathcal{S}(X)$ as well, and we have the following estimate:

$$\mu(A) \le \sum_{i=1}^{\infty} \mu(A_i) \; \in [0, +\infty].$$

Proof: We make the transition from the sequence $\{A_i\}_{i=1,2,\ldots}$ to the sequence $\{B_i\}_{i=1,2,\ldots}$ of mutually disjoint sets:

$$B_1 := A_1,\ B_2 := A_2 \setminus B_1, \ldots,\ B_k := A_k \setminus (B_1 \cup \cdots \cup B_{k-1}), \ldots$$

Now Proposition 3.5 yields $\{B_i\}_{i=1,2,\ldots} \subset \mathcal{S}(X)$. Furthermore, we note that $B_i \subset A_i$ holds true for all $i \in \mathbb{N}$ and, moreover, $A = \bigcup\limits_{i=1}^{\infty} B_i$. Then Proposition 3.2 implies $A \in \mathcal{S}(X)$ as well as $\mu(A) = \sum_{i=1}^{\infty} \mu(B_i) \leq \sum_{i=1}^{\infty} \mu(A_i)$.
q.e.d.

Definition 3.7. *A system $\mathcal{A}$ of subsets of a set X is called* σ-algebra *if we have the following properties :*

1. $X \in \mathcal{A}$.
2. *With $B \in \mathcal{A}$, its complement satisfies $B^c = (X \setminus B) \in \mathcal{A}$ as well.*
3. *For each sequence of sets $\{B_i\}_{i=1,2,\ldots}$ in $\mathcal{A}$, their denumerable union $\bigcup\limits_{i=1}^{\infty} B_i$ belongs to $\mathcal{A}$ as well.*

Remark: We infer $\emptyset \in \mathcal{A}$ immediately from these conditions. Furthermore, with the sets $\{B_i\}_{i=1,2,\ldots} \subset \mathcal{A}$ their denumerable intersection satisfies $\bigcap\limits_{i=1}^{\infty} B_i \in \mathcal{A}$ as well.

Definition 3.8. *We name the function $\mu : \mathcal{A} \to [0, +\infty]$ on a σ-algebra $\mathcal{A}$ a* measure *if the following conditions are fulfilled:*

1. $\mu(\emptyset) = 0$.
2. $\mu\left(\bigcup\limits_{i=1}^{\infty} B_i\right) = \sum\limits_{i=1}^{\infty} \mu(B_i)$ *for all mutually disjoint sets $\{B_i\}_{i=1,2,\ldots} \subset \mathcal{A}$.*

We call this measure finite *if $\mu(X) < +\infty$ holds true.*

Remark: Property 2 is called the σ-additivity of the measure. If we only have finite additivity - that means $\mu\left(\bigcup_{i=1}^{N} B_i\right) = \sum_{i=1}^{N} \mu(B_i)$ for all mutually disjoint sets $\{B_i\}_{i=1,2,\ldots,N} \subset \mathcal{A}$ - we speak of a *content*.

From our Propositions 3.2 to 3.6, we immediately infer

Theorem 3.9. *The set $\mathcal{S}(X)$ of the finitely measurable subsets of X constitutes a σ-algebra. The prescription*

$$\mu(A) := I(\chi_A), \qquad A \in \mathcal{S}(X)$$

defines a finite measure on the σ-algebra $\mathcal{S}(X)$.

Remark: Carathéodory developed axiomatically the measure theory, on which the integration theory can be based. We have presented the inverse approach here. The axiomatic measure theory begins with Definitions 3.7 and 3.8 above.

Definition 3.10. *A set $A \subset X$ is named* null-set *if $A \in \mathcal{S}(X)$ and $\mu(A) = 0$ hold true.*

Remark: The measure μ from Definition 3.1 has the property that each subset of a null-set is a null-set again. For $B \subset A$ and $A \in \mathcal{S}(X)$ with $\mu(A) = 0$ we namely deduce

$$0 = I^+(\chi_A) \geq I^+(\chi_B) \geq I^-(\chi_B) \geq 0,$$

and consequently

$$I^+(\chi_B) = I^-(\chi_B) = 0.$$

Therefore, we obtain $\chi_B \in L(X)$ and finally $B \in \mathcal{S}(X)$ with $\mu(B) = 0$.

Proposition 3.6 immediately implies

Theorem 3.11. *The denumerable union of null-sets is a null-set again.*

Now we show the following

Theorem 3.12. *Each open and each closed set $A \subset X$ belongs to $\mathcal{S}(X)$.*

Proof:

1. At first, let the set A be closed in X and bounded in $\mathbb{R}^n \supset X$. Then we have a compact set $\widehat{A}$ in $\mathbb{R}^n$ satisfying $A = \widehat{A} \cap X$. For the set $\widehat{A}$ we construct - with the aid of Tietze's extension theorem - a sequence of functions $f_l : \mathbb{R}^n \to \mathbb{R} \in C_0^0(\mathbb{R}^n)$ such that

$$f_l(x) = \begin{cases} 1, & x \in \widehat{A} \\ 0, & x \in \mathbb{R}^n \text{ with dist}\,(x, \widehat{A}) \geq \dfrac{1}{l} \\ \in [0,1], \text{ elsewhere} & \end{cases}$$

holds true for $l = 1, 2, \ldots$. We observe $f_l(x) \to \chi_{\widehat{A}}(x)$, set $g_l = f_l\big|_X$, and obtain

$$g_l \in C_b^0(X) \subset M(X) \subset L(X)$$

as well as

$$0 \leq g_l(x) \leq 1 \quad \text{and} \quad g_l(x) \to \chi_A(x), \qquad x \in X.$$

On account of $f_0(x) \equiv 1 \in M(X)$, we can apply Lebesgue's convergence theorem and see

$$\chi_A(x) = \lim_{l \to \infty} g_l(x) \ \in L(X).$$

Therefore, $A \in \mathcal{S}(X)$ is satisfied.

2. For an arbitrary closed set $A \subset X$ we consider the sequence

$$A_l := A \cap \Big\{x \in \mathbb{R}^n \,:\, |x| \le l\Big\}.$$

Due to part 1 of our proof, the sets A_l belong to the system $\mathcal{S}(X)$ and consequently $A = \bigcup\limits_{l=1}^{\infty} A_l$ as well. Finally, the open sets belong to $\mathcal{S}(X)$ as complements of closed sets. q.e.d.

Proposition 3.13. *Let us consider $f \in V(X)$. Then the level set*

$$\mathcal{O}(f,a) := \Big\{x \in X \,:\, f(x) > a\Big\} \subset X$$

is open for all $a \in \mathbb{R}$.

Proof: We note that $f \in V(X)$ holds true and find a sequence

$$\{f_n\}_{n=1,2,\ldots} \subset M(X) \subset C^0(X,\mathbb{R})$$

satisfying $f_n \uparrow f$ on X. Let us consider a point $\xi \in \mathcal{O}(f,a)$ which means $f(\xi) > a$. Then we have an index $n_0 \in \mathbb{N}$ with $f_{n_0}(\xi) > a$. Since the function $f_{n_0} : X \to \mathbb{R}$ is continuous, there exists an open neighborhood $U \subset X$ of ξ such that $f_{n_0}(x) > a$ for all $x \in U$ holds true. Due to $f_{n_0} \le f$ on X, we infer $f(x) > a$ for all $x \in U$, which implies $U \subset \mathcal{O}(f,a)$. Consequently, the level set $\mathcal{O}(f,a)$ is open. q.e.d.

The following criterion illustrates the connection between open and measurable sets.

Theorem 3.14. *A set $B \subset X$ belongs to the system $\mathcal{S}(X)$ if and only if the following condition is valid: For all $\delta > 0$ we can find a closed set $A \subset X$ and an open set $O \subset X$, such that the properties $A \subset B \subset O$ and $\mu(O \setminus A) < \delta$ hold true.*

Proof:

'$\Longrightarrow$' When we take $B \in \mathcal{S}(X)$, we infer $\chi_B \in L(X)$ and Proposition 2.9 in Section 2 gives us a function $f \in V(X)$ satisfying $0 \le \chi_B \le f$ and $I(f) - \mu(B) < \varepsilon$ for all $\varepsilon > 0$. According to Proposition 3.13, the level sets

$$\mathcal{O}_\varepsilon := \{x \in X \mid f(x) > 1 - \varepsilon\} \supset B$$

with $\varepsilon > 0$ are open in X. Now we deduce

$$\chi_B \le \chi_{\mathcal{O}_\varepsilon} = \frac{1}{1-\varepsilon}(1-\varepsilon)\chi_{\mathcal{O}_\varepsilon} \le \frac{1}{1-\varepsilon} f \qquad \text{in } X,$$

and we see

$$\mu(\mathcal{O}_\varepsilon) - \mu(B) = I(\chi_{\mathcal{O}_\varepsilon}) - \mu(B) \le \frac{1}{1-\varepsilon} I(f) - \mu(B)$$

$$= \frac{1}{1-\varepsilon}\Big(I(f) - \mu(B)\Big) + \frac{\varepsilon}{1-\varepsilon}\mu(B) < \frac{\varepsilon}{1-\varepsilon}\Big(1 + \mu(B)\Big)$$

for all $\varepsilon > 0$. For the quantity $\delta > 0$ being given, we now choose a sufficiently small $\varepsilon > 0$ such that the set $O := \mathcal{O}_\varepsilon \supset B$ satisfies the estimate

$$\mu(O) - \mu(B) < \frac{\delta}{2}.$$

Furthermore, we attribute to each measurable set $B^c = X \setminus B$ an open set $\widetilde{O} = A^c$ such that $A^c = \widetilde{O} \supset B^c$ and $\mu(\widetilde{O} \cap B) < \frac{\delta}{2}$ hold true. Therefore, the closed set $A \subset X$ fulfills the inclusion $A \subset B \subset O$ and the estimate

$$\mu(O \setminus A) = \mu(O) - \mu(A) = \Big(\mu(O) - \mu(B)\Big) + \Big(\mu(B) - \mu(A)\Big)$$

$$< \frac{\delta}{2} + \mu(B \setminus A) = \frac{\delta}{2} + \mu(B \cap \widetilde{O}) < \delta.$$

'$\Longleftarrow$' The quantity $\delta > 0$ being given, we find an open set $O \supset B$ and a closed set $A \subset B$ - they are measurable due to Proposition 3.13 - such that the estimate $I(\chi_O - \chi_A) < \delta$ is fulfilled. Since $\chi_A, \chi_O \in L(X)$ is fulfilled, Proposition 2.9 in Section 2 provides functions $g \in -V(X)$ and $h \in V(X)$ satisfying

$$g \le \chi_A \le \chi_B \le \chi_O \le h \quad \text{in} \quad X \quad \text{and} \quad I(h-g) < 3\delta.$$

Using Proposition 2.9 in Section 2 again, we deduce $\chi_B \in L(X)$ and consequently $B \in \mathcal{S}(X)$. q.e.d.

In the sequel, we shall intensively study the null-sets. These appear as sets of exemption for Lebesgue integrable functions and can be neglected in the Lebesgue integration. We start our investigations with the following

Proposition 3.15. *A set $N \subset X$ is a null-set if and only if we have a function $h \in V(X)$ satisfying $h(x) \ge 0$ for all $x \in X$, $h(x) = +\infty$ for all $x \in N$, and $I(h) < +\infty$.*

Proof:

'$\Longrightarrow$' Let $N \subset X$ denote a null-set. Then $\chi_N \in L(X)$ and $I(\chi_N) = 0$ hold true. For each index $k \in \mathbb{N}$ we obtain a function $h_k \in V(X)$ satisfying $0 \le \chi_N \le h_k$ in X and $I(h_k) \le 2^{-k}$, due to Proposition 2.9 in Section 2. According to Proposition 2.5 in Section 2, the element

$$h(x) := \sum_{k=1}^{\infty} h_k(x)$$

belongs to the space $V(X)$ and fulfills

$$I(h) = \sum_{k=1}^{\infty} I(h_k) \leq 1.$$

On the other hand, the estimates $h_k(x) \geq 1$ in N for all $k \in \mathbb{N}$ imply that the relation $h(x) = +\infty$ for all $x \in N$ is correct. We note that $h_k(x) \geq 0$ in X holds true, and we deduce $h(x) \geq 0$ for all $x \in X$.

'$\Longleftarrow$' Let the conditions $h \in V(X)$, $h(x) \geq 0$ for all $x \in X$, $h(x) = +\infty$ for all $x \in N$, and $I(h) < +\infty$ be fulfilled. When we define

$$h_\varepsilon(x) := \frac{\varepsilon}{1 + I(h)}\, h(x),$$

we immediately deduce $h_\varepsilon \in V(X)$, $h_\varepsilon(x) \geq 0$ for all $x \in X$, and $I(h_\varepsilon) < \varepsilon$ for all $\varepsilon > 0$. On account of $h(x) = +\infty$ for all $x \in N$, we infer

$$0 \leq \chi_N(x) \leq h_\varepsilon(x) \qquad \text{in} \quad X \quad \text{for all} \quad \varepsilon > 0.$$

Proposition 2.9 in Section 2 yields $I(\chi_N) = 0$, which means that N is a null-set. q.e.d.

Definition 3.16. *A property holds true* almost everywhere *in X (symbolically: a.e.), if there exists a null-set $N \subset X$ such that this property is valid for all points $x \in X \setminus N$.*

Theorem 3.17. (a.e.-Finiteness of L-functions)
Let the function $f \in L(X)$ be given. Then the set

$$N := \Big\{x \in X \,:\, |f(x)| = +\infty\Big\}$$

constitutes a null-set.

Proof: With $f \in L(X)$ being given, we obtain $|f| \in L(X)$ and find a function $h \in V(X)$ satisfying $0 \leq |f(x)| \leq h(x)$ in X as well as $I(h) < +\infty$. Furthermore, $h(x) = +\infty$ in N holds true and Proposition 3.15 tells us that N represents a null-set.

q.e.d.

Theorem 3.18. *Let the function $f \in L(X)$ be given such that $I(|f|) = 0$ is correct. Then the set*

$$N := \Big\{x \in X \,:\, f(x) \neq 0\Big\}$$

constitutes a null-set.

Proof: With $f \in L(X)$ being given, we infer $|f| \in L(X)$. Setting

$$f_k(x) := |f(x)|, \qquad k \in \mathbb{N},$$

we observe

$$\sum_{k=1}^{\infty} I(f_k) = 0.$$

According to Proposition 2.11 in Section 2, the function

$$g(x) := \sum_{k=1}^{\infty} f_k(x)$$

is Lebesgue integrable as well. Now we see $N = \{x \in X \, : \, g(x) = +\infty\}$, and Theorem 3.17 implies that N is a null-set. q.e.d.

Now we want to show that an L-function can be arbitrarily modified on a null-set, without the value of the integral being changed! In this way we can confine ourselves to consider *finite-valued functions* $f \in L(X)$, which are functions f with $f(x) \in \mathbb{R}$ for all $x \in X$, more precisely. A bounded function is finite-valued; however, a finite-valued function is not necessarily bounded. In this context, we mention the function $f(x) = \frac{1}{x}$, $x \in (0,1)$.

Proposition 3.19. *Let $N \subset X$ denote a null-set. Furthermore, the function $f : X \to \overline{\mathbb{R}}$ may satisfy $f(x) = 0$ for all $x \in X \setminus N$. Then we infer $f \in L(X)$ as well as $I(f) = 0$.*

Proof: Due to Proposition 3.15, we find a function $h \in V(X)$ satisfying $h(x) \geq 0$ for all $x \in X$, $h(x) = +\infty$ for all $x \in N$, and $I(h) < +\infty$. For all numbers $\varepsilon > 0$, we see $\varepsilon h \in V$ and $-\varepsilon h \in -V$ as well as

$$-\varepsilon h(x) \leq f(x) \leq \varepsilon h(x) \qquad \text{for all} \quad x \in X.$$

Furthermore, the identity

$$I(\varepsilon h) - I(-\varepsilon h) = 2\varepsilon I(h) \qquad \text{for all} \quad \varepsilon > 0$$

is correct. We infer $f \in L(X)$ and, moreover, $I(f) = 0$ from Proposition 2.9 in Section 2.

q.e.d.

Theorem 3.20. *Consider the function $f \in L(X)$ and the null-set $N \subset X$. Furthermore, let the function $\widetilde{f} : X \to \overline{\mathbb{R}}$ with the property $\widetilde{f}(x) = f(x)$ for all $x \in X \setminus N$ be given. Then we infer $\widetilde{f} \in L(X)$ as well as $I(|f - \widetilde{f}|) = 0$, and consequently $I(f) = I(\widetilde{f})$.*

Proof: Since $f \in L(X)$ holds true, the following set

$$N_1 := \left\{x \in X \,:\, |f(x)| = +\infty\right\}$$

constitutes a null-set, due to Theorem 3.17. Now we find a function $\varphi(x) : X \to \overline{\mathbb{R}}$ such that

$$\widetilde{f}(x) = f(x) + \varphi(x) \qquad \text{for all} \quad x \in X.$$

Evidently, we have the identity $\varphi(x) = 0$ outside the null-set $N \cup N_1$. Proposition 3.19 yields $\varphi \in L(X)$ and $I(\varphi) = 0$. Consequently, $\widetilde{f} \in L(X)$ is correct and we see

$$I(\widetilde{f}) = I(f + \varphi) = I(f) + I(\varphi) = I(f).$$

When we apply these arguments on the function

$$\psi(x) := |f(x) - \widetilde{f}(x)|, \qquad x \in X,$$

Proposition 3.19 shows us $\psi \in L(X)$ and finally

$$0 = I(\psi) = I(|f - \widetilde{f}|).$$

q.e.d.

Remark: When a function $\widetilde{f}$ coincides a.e. with an L-function f, then $\widetilde{f} \in L(X)$ holds true and their integrals are identical.

We are now prepared to provide general convergence theorems of the Lebesgue integration theory.

Theorem 3.21. (General convergence theorem of B.Levi)
Let $\{f_k\}_{k=1,2,\ldots} \subset L(X)$ denote a sequence of functions satisfying $f_k \uparrow f$ a.e. in X. Furthermore, let $I(f_k) \le c$ for all $k \in \mathbb{N}$ be valid - with the constant $c \in \mathbb{R}$. Then we infer $f \in L(X)$ and

$$\lim_{k\to\infty} I(f_k) = I(f).$$

Proof: We consider the null-sets

$$N_k := \left\{x \in X \,:\, |f_k(x)| = +\infty\right\} \qquad \text{for} \quad k \in \mathbb{N}$$

as well as

$$N_0 := \left\{x \in X \,:\, f_k(x) \uparrow f(x) \text{ is not valid}\right\}.$$

We define the null-set

$$N := \bigcup_{k=0}^{\infty} N_k,$$

and modify f, f_k on N to 0. Then we obtain the functions $\widetilde{f_k} \in L(X)$ with

$$I(\widetilde{f_k}) = I(f_k) \leq c \qquad \text{for all} \quad k \in \mathbb{N}$$

and $\widetilde{f}$ with $\widetilde{f_k} \uparrow \widetilde{f}$. According to Theorem 2.12 from Section 2, we deduce $\widetilde{f} \in L(X)$ as well as

$$\lim_{k\to\infty} I(\widetilde{f_k}) = I(\widetilde{f}).$$

Now Theorem 3.20 yields $f \in L(X)$ and

$$I(f) = I(\widetilde{f}) = \lim_{k\to\infty} I(\widetilde{f_k}) = \lim_{k\to\infty} I(f_k).$$

q.e.d.

Modifying the functions to 0 on the relevant null-sets as above, we easily prove the following Theorems 3.22 and 3.23 with the aid of Theorem 2.13 and 2.15 from Section 2, respectively.

Theorem 3.22. (General convergence theorem of Fatou)
Let $\{f_k\}_{k=1,2,\ldots} \subset L(X)$ denote a sequence of functions with $f_k(x) \geq 0$ a.e. in X for all $k \in \mathbb{N}$, and we assume

$$\liminf_{k\to\infty} I(f_k) < +\infty.$$

Then the function

$$g(x) := \liminf_{k\to\infty} f_k(x)$$

belongs to the class $L(X)$ as well, and we have lower semicontinuity as follows:

$$I(g) \leq \liminf_{k\to\infty} I(f_k).$$

Theorem 3.23. (General convergence theorem of Lebesgue)
Let $\{f_k\}_{k=1,2,\ldots} \subset L(X)$ denote a sequence with $f_k \to f$ a.e. on X and $|f_k(x)| \leq F(x)$ a.e. in X for all $k \in \mathbb{N}$, where $F \in L(X)$ holds true. Then we infer $f \in L(X)$ and the identity

$$\lim_{k\to\infty} I(f_k) = I(f).$$

We conclude this section with the following

Theorem 3.24. *Lebesgue's integral $I : L(X) \to \mathbb{R}$ constitutes a Daniell integral.*

Proof: We invoke Theorem 2.10 in Section 2 and obtain the following: The space $L(X)$ is linear and closed with respect to the modulus operation. Furthermore, $L(X)$ satisfies the properties (1) and (2) in Section 1. The Lebesgue integral I is nonnegative, linear, and closed with respect to monotone convergence - due to Theorem 3.21. Therefore, the functional I fulfills the conditions (3)–(5) in Section 1. Consequently, Lebesgue's integral $I : L(X) \to \mathbb{R}$ represents a Daniell integral as described in Definition 1.1 from Section 1.

q.e.d.

4 Measurable Functions

Fundamental is the following

Definition 4.1. *The function* $f : X \to \overline{\mathbb{R}}$ *is named* measurable *if the* level set - above the level a -

$$\mathcal{O}(f,a) := \Big\{ x \in X \,:\, f(x) > a \Big\}$$

is measurable for all $a \in \mathbb{R}$.

Remark: Each continuous function $f : X \to \mathbb{R} \in C^0(X,\mathbb{R})$ is measurable. Then $\mathcal{O}(f,a) \subset X$ is an open set for all $a \in \mathbb{R}$, which is measurable due to Section 3, Theorem 3.12. Furthermore, Proposition 3.13 in Section 3 shows us that each function $f \in V(X)$ is measurable as well.

Proposition 4.2. *Let* $f : X \to \overline{\mathbb{R}}$ *denote a measurable function. Furthermore, let us consider the numbers* $a, b \in \overline{\mathbb{R}}$ *with* $a \leq b$ *and the interval* $I = [a,b]$; *for* $a < b$ *we consider the intervals* $I = (a,b]$, $I = [a,b)$, $I = (a,b)$ *as well. Then the following sets*

$$A := \Big\{ x \in X \,:\, f(x) \in I \Big\}$$

are measurable.

Proof: Definition 4.1 implies that the level sets

$$\mathcal{O}_1(f,c) := \mathcal{O}(f,c) = \Big\{ x \in X \,:\, f(x) > c \Big\}$$

are measurable for all $c \in \mathbb{R}$. For a given $c \in \mathbb{R}$, we now choose a sequence $\{c_n\}_{n=1,2,\ldots}$ satisfying $c_n \uparrow c$, and we obtain again a measurable set via

$$\mathcal{O}_2(f,c) := \Big\{ x \in X \,:\, f(x) \geq c \Big\} = \bigcap_{n=1}^{\infty} \Big\{ x \in X \,:\, f(x) > c_n \Big\}.$$

The measurable sets $\mathcal{S}(X)$ namely constitute a σ-algebra due to Section 3, Definition 3.7 and Theorem 3.9. Furthermore, we have the relations

$$\mathcal{O}_2(f,+\infty) = \bigcap_{n=1}^{\infty} \mathcal{O}_2(f,n), \quad \mathcal{O}_1(f,-\infty) = \bigcup_{n=1}^{\infty} \mathcal{O}_1(f,-n),$$

and these sets are measurable as well. The transition to their complements shows that

$$\mathcal{O}_3(f,c) := \Big\{ x \in X \,:\, f(x) \leq c \Big\} \quad \text{and} \quad \mathcal{O}_4(f,c) := \Big\{ x \in X \,:\, f(x) < c \Big\}$$

are measurable for all $c \in \overline{\mathbb{R}}$. Here

$$A := \Big\{ x \in X \,:\, f(x) \in I \Big\}$$

can be generated by an intersection of the sets $\mathcal{O}_1$–$\mathcal{O}_4$, when we replace c by a or b, respectively. This proves the measurability of the sets A. q.e.d.

For $a, b \in \overline{\mathbb{R}}$ with $a < b$, we define the function

$$\phi_{a,b}(t) := \begin{cases} a\,, & -\infty \leq t \leq a \\ t\,, & a \leq t \leq b \\ b\,, & b \leq t \leq +\infty \end{cases}$$

as a *cut-off function*. Given the function $f : X \to \overline{\mathbb{R}}$, we set

$$f_{a,b}(x) := \phi_{a,b}(f(x)) := \begin{cases} a\,, & -\infty \leq f(x) \leq a \\ f(x)\,, & a \leq f(x) \leq b \\ b\,, & b \leq f(x) \leq +\infty \end{cases} .$$

Evidently, we have the estimate

$$|\, f_{a,b}(x)| \leq \max\,(|\, a|, |\, b|) < +\infty \qquad \text{for all} \quad x \in X, \quad a, b \in \mathbb{R}.$$

Furthermore, we note that

$$f^+(x) = f_{0,+\infty}(x) \quad \text{and} \quad f^-(x) = f_{-\infty,0}(x), \qquad x \in X.$$

Theorem 4.3. *A function $f : X \to \overline{\mathbb{R}}$ is measurable if and only if the function $f_{a,b}$ belongs to $L(X)$ for all $a, b \in \mathbb{R}$ with $a < b$.*

Proof:

'$\Longrightarrow$' Let $f : X \to \overline{\mathbb{R}}$ be measurable and $-\infty < a < b < +\infty$ hold true. We define the intervals

$$I_0 := [-\infty, a); \; I_k := \Big[a + (k-1)\frac{b-a}{m}\,,\; a + k\frac{b-a}{m} \Big); \; I_{m+1} := [b, +\infty]$$

with $k = 1, \ldots, m$ for arbitrary $m \in \mathbb{N}$. Furthermore, we choose the intermediate values

$$\eta_l = a + (l-1)\frac{b-a}{m}\,, \qquad l = 0, \ldots, m+1.$$

We infer from Proposition 4.2 that the sets

$$A_l := \Big\{ x \in X \,:\, f(x) \in I_l \Big\}$$

are measurable. The function

$$f_m := \sum_{l=0}^{m+1} \eta_l \, \chi_{A_l}$$

is Lebesgue integrable, and we observe

$$|\, f_m(x)| \leq \max\left(|\, 2a - b|, |\, b|\right) \quad \text{for all} \quad x \in X \quad \text{and all} \quad m \in \mathbb{N}.$$

Since constant functions are integrable, Lebesgue's convergence theorem yields

$$f_{a,b}(x) = \lim_{m \to \infty} f_m(x) \, \in L(X).$$

'$\Longleftarrow$' We have to show that the set $\mathcal{O}(f, \tilde{a})$ is measurable for all $\tilde{a} \in \mathbb{R}$. Here we prove: The set $\{x \in X \, : \, f(x) \geq b\}$ is measurable for all $b \in \mathbb{R}$. Then we obtain the measurability of

$$\mathcal{O}(f, \tilde{a}) = \bigcup_{l=1}^{\infty} \left\{ x \in X \mid f(x) \geq \tilde{a} + \frac{1}{l} \right\}$$

via Proposition 3.6 from Section 3. Choosing $b \in \mathbb{R}$ arbitrarily, we take $a = b - 1$ and consider the function

$$g(x) := f_{a,b}(x) - a \, \in L(X).$$

Evidently, $g : X \to [\,0, 1\,]$ holds true and, moreover,

$$g(x) = 1 \quad \Longleftrightarrow \quad f(x) \geq b.$$

The corollary from Theorem 3.4 in Section 3 yields $g^l(x) \in L(X)$ for all $l \in \mathbb{N}$. Now Lebesgue's convergence theorem implies

$$\chi(x) := \lim_{l \to \infty} g^l(x) = \begin{cases} 1\,, & x \in X \text{ with } f(x) \geq b \\ 0\,, & x \in X \text{ with } f(x) < b \end{cases} \quad \in L(X),$$

and consequently $\{x \in X \, : \, f(x) \geq b\}$ is measurable for all $b \in \mathbb{R}$. q.e.d.

Corollary: Each function $f \in L(X)$ is measurable.

Proof: We take $f \in L(X)$, and see that $N := \{x \in X \, : \, |f(x)| = +\infty\}$ is a null-set. Then we define

$$\widetilde{f}(x) := \begin{cases} f(x)\,, & x \in X \setminus N \\ 0\,, & x \in N \end{cases} \quad \in L(X).$$

According to Definition 4.1, the function f is measurable if and only if $\widetilde{f}$ is measurable. We now apply the criterion of Theorem 4.3 on $\widetilde{f}$. When $-\infty < a < b < +\infty$ is arbitrary, we immediately infer

$$\widetilde{f}_{-\infty,b}(x) = \min\left(\widetilde{f}(x), b\right) = \frac{1}{2}\left(\widetilde{f}(x) + b\right) - \frac{1}{2}\,|\,\widetilde{f}(x) - b| \quad \in L(X),$$

because $\widetilde{f} \in L(X)$. Analogously, we deduce $g_{a,+\infty} \in L(X)$ for $g \in L(X)$. Taking the following relation

$$\widetilde{f}_{a,b} = \big(\widetilde{f}_{-\infty,b}\big)_{a,+\infty}$$

into account, we infer $\widetilde{f}_{a,b} \in L(X)$. q.e.d.

In the next theorem there will appear an adequate notion of convergence for measurable functions.

Theorem 4.4. (a.e.-Convergence)
Let $\{f_k\}_{k=1,2,\ldots}$ denote a sequence of measurable functions with the property $f_k(x) \to f(x)$ a.e. in X. Then f is measurable.

Proof: Let us take $a, b \in \mathbb{R}$ with $a < b$. Then the functions $(f_k)_{a,b}$ belong to $L(X)$ for all $k \in \mathbb{N}$, and we have

$$|(f_k)_{a,b}(x)| \le \max(|a|,|b|) \quad \text{and} \quad (f_k)_{a,b} \to f_{a,b} \quad \text{a.e. in} \quad X.$$

The general convergence theorem of Lebesgue yields $f_{a,b} \in L(X)$. Due to Theorem 4.3, the function f is measurable.

q.e.d.

Theorem 4.5. (Combination of measurable functions)
We have the following statements:

a) Linear Combination: *When f, g are measurable and $\alpha, \beta \in \mathbb{R}$ are chosen, the four functions $\alpha f + \beta g$, $\max(f,g)$, $\min(f,g)$, $|f|$ are measurable as well.*

b) Nonlinear Combination: *Let the $\kappa \in \mathbb{N}$ finite-valued measurable functions $f_1, \ldots, f_\kappa$ be given, and furthermore the continuous function $\phi = \phi(y_1, \ldots, y_\kappa) \in C^0(\mathbb{R}^\kappa, \mathbb{R})$. Then the composed function*

$$g(x) := \phi\Big(f_1(x), \ldots, f_\kappa(x)\Big), \quad x \in X$$

is measurable.

Proof:

a) According to Theorem 4.3, we have $f_{-p,p}, g_{-p,p} \in L(X)$ for all $p \in \mathbb{R}$. When we note that $f = \lim\limits_{p\to\infty} f_{-p,p}$ holds true, Theorem 4.4 combined with the linearity of the space $L(X)$ imply that the function

$$\alpha f + \beta g = \lim_{p\to+\infty} (\alpha f_{-p,p} + \beta g_{-p,p})$$

is measurable for all $\alpha, \beta \in \mathbb{R}$. In the same way, we see the measurability of the functions

$$\max(f,g) = \lim_{p\to+\infty} \max(f_{-p,p}, g_{-p,p})$$

and

$$\min(f,g) = \lim_{p\to+\infty} \min(f_{-p,p}, g_{-p,p}),$$

as well as $|f|$ - due to $|f| = \max(f,-f)$.

b) The functions $(f_k)_{-p,p} \in L(X)$ are bounded for all $p > 0$ and $k = 1,\ldots,\kappa$. According to Theorem 3.4 in Section 3 and Theorem 4.3 in Section 4, the function $\phi\Big((f_1)_{-p,p}(x),\ldots,(f_\kappa)_{-p,p}(x)\Big)$ belongs to the class $L(X)$. Furthermore, we have the limit relation

$$g(x) = \lim_{p\to+\infty} \phi\Big((f_1)_{-p,p}(x),\ldots,(f_\kappa)_{-p,p}(x)\Big)$$

for all $x \in X$, and Theorem 4.4 finally yields the measurablity of g.q.e.d.

Now we define improper Lebesgue integrals.

Definition 4.6. *We set for a nonnegative measurable function f the integral*

$$I(f) := \lim_{N\to+\infty} I(f_{0,N}) \;\in [0,+\infty].$$

Theorem 4.7. *A measurable function f belongs to the class $L(X)$ if and only if the following limit*

$$\lim_{\substack{a\to-\infty\\ b\to+\infty}} I(f_{a,b}) \;\in \mathbb{R}$$

exists. In this case we have the identity

$$I(f) = \lim_{\substack{a\to-\infty\\ b\to+\infty}} I(f_{a,b}) = I(f^+) - I(f^-).$$

Therefore, a measurable function f belongs to $L(X)$ if and only if $I(f^+) < +\infty$ as well as $I(f^-) < +\infty$ are valid.

Proof: On account of $f_{a,b} = (f^+)_{0,b} - (f^-)_{0,-a}$ for all $-\infty < a < 0 < b < +\infty$ we see

$$\lim_{\substack{a\to-\infty\\ b\to+\infty}} I(f_{a,b}) \text{ exists in } \mathbb{R} \quad\Longleftrightarrow\quad \lim_{N\to+\infty} I\Big((f^\pm)_{0,N}\Big) \text{ exist in } \mathbb{R}.$$

Consequently, it suffices to show:

$$f \in L(X) \quad\Longleftrightarrow\quad \lim_{N\to+\infty} I\Big((f^\pm)_{0,N}\Big) \text{ exist in } \mathbb{R}.$$

'$\Longrightarrow$' : Let us take $f \in L(X)$. Then we infer $f^\pm \in L(X)$, and B.Levi's theorem on monotone convergence yields

$$\lim_{N\to+\infty} I\Big((f^{\pm})_{0,N}\Big) = I(f^{\pm}) \in \mathbb{R}.$$

'$\Longleftarrow$' : If

$$\lim_{N\to+\infty} I\Big((f^{\pm})_{0,N}\Big)$$

in $\mathbb{R}$ exist, the theorem of B.Levi implies $f^{\pm} \in L(X)$, and together with the identity $f = f^+ - f^-$ the property $f \in L(X)$ is deduced.

q.e.d.

Theorem 4.8. *Let $f : X \to \overline{\mathbb{R}}$ denote a measurable function satisfying*

$$|f(x)| \leq F(x), \quad x \in X,$$

with a dominating function $F \in L(X)$. Then we have

$$f \in L(X) \quad \text{and} \quad I(|f|) \leq I(F).$$

Proof: According to Theorem 4.5, the functions f^+ and f^- are measurable, and we see $0 \leq f^{\pm} \leq F$. Consequently, the estimates $0 \leq (f^{\pm})_{0,N} \leq F$ and $(f^{\pm})_{0,N} \in L(X)$ are correct. Furthermore, we have

$$I\Big((f^{\pm})_{0,N}\Big) \leq I(F) < +\infty \qquad \text{for all} \quad N > 0.$$

B.Levi's theorem now yields $I(f^{\pm}) < +\infty$ and $f^{\pm} \in L(X)$, which implies $f \in L(X)$. On account of the monotonicity of Lebesgue's integral, the estimate $I(|f|) \leq I(F)$ follows from the inequality $|f(x)| \leq F(x)$.

q.e.d.

Theorem 4.9. *Let $\{f_l\}_{l=1,2,\ldots}$ denote a sequence of nonnegative measurable functions satisfying $f_l(x) \uparrow f(x)$, $x \in X$. Then the function f is measurable, and we have*

$$I(f) = \lim_{l\to\infty} I(f_l).$$

Proof: From Theorem 4.4 we infer the measurability of f. According to Definition 4.6, two measurable functions $0 \leq g \leq h$ satisfy the inequality $I(g) \leq I(h)$. Therefore, $\{I(f_l)\}_{l=1,2,\ldots} \in [0,+\infty]$ represents a monotonically nondecreasing sequence, such that $I(f) \geq I(f_l)$ for all $l \in \mathbb{N}$ holds true. We distinguish between the following two cases:

a) Let us consider

$$\lim_{l\to\infty} I(f_l) \leq c < +\infty.$$

Then we have $I(f_l) \leq c$, which implies $f_l \in L(X)$ due to Theorem 4.7. B.Levi's theorem now yields $f \in L(X)$ and

$$I(f) = \lim_{l\to\infty} I(f_l).$$

b) Let us consider
$$\lim_{l\to\infty} I(f_l) = +\infty.$$
Then we note that $I(f) \geq I(f_l)$ for all $l \in \mathbb{N}$ holds true, and we obtain immediately
$$I(f) = +\infty = \lim_{l\to\infty} I(f_l).$$
q.e.d.

Definition 4.10. *We name a function* $g : X \to \overline{\mathbb{R}}$ simple *if there exist finitely many mutually disjoint sets* $A_1, \ldots, A_{n^*} \in \mathcal{S}(X)$ *and numbers* $\eta_1, \ldots, \eta_{n^*} \in \mathbb{R}$ *with* $n^* \in \mathbb{N}$, *such that the following representation holds true in* X:
$$g = \sum_{k=1}^{n^*} \eta_k \, \chi_{A_k}.$$

Remark: Evidently, we then have $g \in L(X)$ and
$$I(g) = \sum_{k=1}^{n^*} \eta_k \, \mu(A_k).$$

Let us take an arbitrary decomposition $\mathcal{Z} : -\infty < y_0 < y_1 < \ldots < y_{n^*} < +\infty$ in the real line $\mathbb{R}$, with the intervals $I_k := [y_{k-1}, y_k)$ for $k = 1, \ldots, n^*$. Furthermore, we consider an arbitrary measurable function $f : X \to \overline{\mathbb{R}}$ and select arbitrary intermediate values $\eta_k \in I_k$ for $k = 1, \ldots, n^*$. Now we attribute the following simple function to the data $f, \mathcal{Z}$ and η, namely
$$f^{(\mathcal{Z},\eta)} := \sum_{k=1}^{n^*} \eta_k \, \chi_{A_k}$$
with $A_k := \{x \in X : f(x) \in I_k\}$ for $k = 1, \ldots, n^*$. Then we observe
$$I\Big(f^{(\mathcal{Z},\eta)}\Big) = \sum_{k=1}^{n^*} \eta_k \, \mu(A_k).$$

We denote by a *canonical sequence of decompositions* such a sequence of decompositions, whose start- and end-points tend towards $-\infty$ and $+\infty$, respectively, and whose maximal interval-lengths tend to 0.

Theorem 4.11. *When we consider* $f : X \to \overline{\mathbb{R}} \in L(X)$, *each canonical sequence of decompositions* $\{\mathcal{Z}^{(p)}\}_{p=1,2,\ldots}$ *in* $\mathbb{R}$ *and each choice of intermediate values* $\{\eta^{(p)}\}_{p=1,2,\ldots}$ *gives us the asymptotic identity*
$$I(f) = \lim_{p\to\infty} I\Big(f^{(\mathcal{Z}^{(p)},\eta^{(p)})}\Big) = \lim_{p\to\infty} \sum_{k=1}^{n^{(p)}} \eta_k^{(p)} \mu(A_k^{(p)}).$$

Remark: Therefore, Lebesgue's integral can be approximated by the Lebesgue sums as above, and the notation

$$I(f) = \int\limits_X f(x)\, d\mu(x)$$

is justified. However, the Riemannian intermediate sums can be evaluated numerically much better than the Lebesgue sums.

Proof of Theorem 4.11: Let us consider the function $f \in L(X)$, a decomposition $\mathcal{Z}$ with its *fineness* $\delta(\mathcal{Z}) = \max\{(y_k - y_{k-1}) \ : \ k = 1, \ldots, n^*\}$, and arbitrary intermediate values $\{\eta_k\}_{k=1,\ldots,n^*}$. Then we infer the estimate

$$|f^{(\mathcal{Z},\eta)}(x)| \leq \delta(\mathcal{Z}) + |f(x)| \quad \text{for all} \quad x \in X.$$

When $\{\mathcal{Z}^{(p)}\}_{p=1,2,\ldots}$ describes a canonical sequence of decompositions and $\{\eta^{(p)}\}_{p=1,2,\ldots}$ denote arbitrary intermediate values, we observe the limit relation

$$f^{(\mathcal{Z}^{(p)},\eta^{(p)})}(x) \to f(x) \quad \text{a.e. for} \quad p \to \infty,$$

which is valid for all $x \in X$ with $|f(x)| \neq +\infty$. Now Lebesgue's convergence theorem yields

$$I(f) = \lim_{p\to\infty} I\Big(f^{(\mathcal{Z}^{(p)},\eta^{(p)})}\Big) = \lim_{p\to\infty} \sum_{k=1}^{n^{(p)}} \eta_k^{(p)} \mu(A_k^{(p)}).$$

q.e.d.

Now we shall present a selection theorem related to a.e.-convergence.

Theorem 4.12. (Lebesgue's selection theorem)
Let $\{f_k\}_{k=1,2,\ldots}$ denote a sequence in $L(X)$ satisfying

$$\lim_{k,l\to\infty} I(|\, f_k - f_l|) = 0.$$

Then a null-set $N \subset X$ as well as a monotonically increasing subsequence $\{k_m\}_{m=1,2,\ldots}$ exist, such that the sequence of functions $\{f_{k_m}(x)\}_{m=1,2,\ldots}$ converges for all points $x \in X \setminus N$ and their limit fulfills

$$\lim_{m\to\infty} f_{k_m}(x) =: f(x) \in L(X).$$

Therefore, we can select an a.e. convergent subsequence from a Cauchy sequence with respect to the integral I.

Proof: On the null-set

$$N_1 := \bigcup_{k=1}^{\infty} \Big\{x \in X \ : \ |f_k(x)| = +\infty\Big\}$$

we modify the functions f_k and obtain

$$\widetilde{f}_k(x) := \begin{cases} f_k(x)\,, & x \in X \setminus N_1 \\ 0\,, & x \in N_1 \end{cases}.$$

Without loss of generality, we can assume the functions $\{f_k\}_{k=1,2,\ldots}$ to be finite-valued. On account of

$$\lim_{p,l\to\infty} I(|\, f_p - f_l|) = 0,$$

we find a subsequence $k_1 < k_2 < \cdots$ with the property

$$I(|\, f_p - f_l|) \le \frac{1}{2^m} \quad \text{for all} \quad p, l \ge k_m, \quad m = 1, 2, \ldots.$$

In particular, we infer the following estimates:

$$I(|\, f_{k_{m+1}} - f_{k_m}|) \le \frac{1}{2^m}, \qquad m = 1, 2, \ldots$$

and

$$\sum_{m=1}^{\infty} I(|\, f_{k_{m+1}} - f_{k_m}|) \le 1.$$

B.Levi's theorem tells us that the function

$$g(x) := \sum_{m=1}^{\infty} |\, f_{k_{m+1}}(x) - f_{k_m}(x)|, \qquad x \in X$$

belongs to $L(X)$, and $N_2 := \{x \in X \setminus N_1 \;:\; |g(x)| = +\infty\}$ represents a null-set. Therefore, the series

$$\sum_{m=1}^{\infty} |\, f_{k_{m+1}}(x) - f_{k_m}(x)| \quad \text{for all} \quad x \in X \setminus N \quad \text{with} \quad N := N_1 \cup N_2$$

converges, as well as the series

$$\sum_{m=1}^{\infty} \Big(f_{k_{m+1}}(x) - f_{k_m}(x) \Big).$$

Consequently, the limit

$$\lim_{m\to\infty} \Big(f_{k_m}(x) - f_{k_1}(x) \Big) =: f(x) - f_{k_1}(x)$$

exists for all points $x \in X \setminus N$, and the sequence $\{f_{k_m}\}_{m=1,2,\ldots}$ converges on $X \setminus N$ towards f. We note that $g \in L(X)$ and $|\, f_{k_m}(x) - f_{k_1}(x)| \le |\, g(x)|$ are valid, and Lebesgue's convergence theorem is applicable. Finally, we infer $f \in L(X)$ and the relation

$$I(f) = \lim_{m\to\infty} I(f_{k_m}).$$

q.e.d.

Proposition 4.13. (Approximation in the integral)
Let the function $f \in L(X)$ be given. To each quantity $\varepsilon > 0$, we then find a function $f_\varepsilon \in M(X)$ satisfying

$$I(|f - f_\varepsilon|) < \varepsilon.$$

Proof: Since $f \in L(X)$ holds true, Proposition 2.9 from Section 2 provides two functions $g \in -V$ and $h \in V$ such that

$$g(x) \le f(x) \le h(x), \quad x \in X, \quad \text{and} \quad I(h) - I(g) < \frac{\varepsilon}{2}.$$

Recalling the definition of the space $V(X)$, we find a function $h'(x) \in M(X)$ satisfying

$$h'(x) \le h(x), \quad x \in X, \quad \text{and} \quad I(h) - I(h') < \frac{\varepsilon}{2}.$$

This implies

$$|f - h'| \le |f - h| + |h - h'| \le (h - g) + (h - h'),$$

and the monotonicity and linearity of the integral yield

$$I(|f - h'|) \le (I(h) - I(g)) + (I(h) - I(h')) < \frac{\varepsilon}{2} + \frac{\varepsilon}{2} = \varepsilon.$$

With $f_\varepsilon := h'$ we obtain the desired function. q.e.d.

Theorem 4.14. (a.e.-Approximation)
Let f denote a measurable function satisfying $|f(x)| \le c$, $x \in X$ with the constant $c \in (0, +\infty)$. Then we have a sequence $\{f_k\}_{k=1,2,\ldots} \subset M(X)$ satisfying $|f_k(x)| \le c$, $x \in X$ for all $k \in \mathbb{N}$, such that $f_k(x) \to f(x)$ a.e. in X holds true.

Proof: Since f is measurable and dominated by the constant function $c \in L(X)$, we infer $f \in L(X)$ from Theorem 4.8. Now Proposition 4.13 allows us to find a sequence $\{g_k(x)\}_{k=1,2,\ldots} \subset M(X)$ satisfying $I(|f - g_k|) \to 0$ for $k \to \infty$. We set

$$h_k(x) := (g_k)_{-c,c}(x)$$

and observe $h_k \in M(X)$ as well as $|h_k(x)| \le c$ for all $x \in X$ and all $k \in \mathbb{N}$. We note that

$$|h_k - f| = |(g_k)_{-c,c} - f_{-c,c}| = |(g_k - f)_{-c,c}| \le |g_k - f|$$

is correct and see

$$\lim_{k\to\infty} I(|h_k - f|) \le \lim_{k\to\infty} I(|g_k - f|) = 0.$$

On account of the relation

$$I(|h_k - h_l|) \le I(|h_k - f|) + I(|f - h_l|) \longrightarrow 0 \qquad \text{for} \quad k, l \to \infty,$$

Lebesgue's selection theorem yields a null-set $N_1 \subset X$ and a monotonically increasing subsequence $\{k_m\}_{m=1,2,\dots}$ such that the following limit exists:

$$h(x) := \lim_{m\to\infty} h_{k_m}(x) \quad \text{for all} \quad x \in X \setminus N_1.$$

We extend h onto the null-set by the prescription $h(x) := 0$ for all $x \in N_1$. Now we conclude

$$\lim_{m\to\infty} |h_{k_m}(x) - f(x)| = |h(x) - f(x)| \quad \text{in} \quad X \setminus N_1.$$

The theorem of Fatou yields

$$I(|h - f|) \le \lim_{m\to\infty} I(|h_{k_m} - f|) = 0.$$

Consequently, we find a null-set $N_2 \subset X$ such that

$$f(x) = h(x) \qquad \text{for all} \quad x \in X \setminus N_2$$

holds true. When we define $N := N_1 \cup N_2$ and $f_m(x) := h_{k_m}(x)$, we obviously infer $f_m(x) \in M(X)$, $|f_m(x)| \le c$ for all $x \in X$ and all $m \in \mathbb{N}$, and the following limit relation:

$$\lim_{m\to\infty} f_m(x) = \lim_{m\to\infty} h_{k_m} \overset{x\notin N_1}{=} h(x) \overset{x\notin N_2}{=} f(x) \qquad \text{for all} \quad x \in X \setminus N.$$

Consequently, we obtain $f_m(x) \to f(x)$ for all $x \in X \setminus N$. q.e.d.

Uniform convergence and a.e.-convergence are connected by the following result.

Theorem 4.15. (Egorov)
Let the measurable set $B \subset X$ as well as the measurable a.e.-finite-valued functions $f : B \to \overline{\mathbb{R}}$ and $f_k : B \to \overline{\mathbb{R}}$ for all $k \in \mathbb{N}$ be given, with the convergence property $f_k(x) \to f(x)$ a.e. in B. To each quantity $\delta > 0$, we then find a closed set $A \subset B$ satisfying $\mu(B \setminus A) < \delta$ such that the limit relation, $f_k(x) \to f(x)$ uniformly on A, holds true.

Proof: We consider the null-set

$$\begin{aligned} N &:= \Big\{x \in B : f_k(x) \to f(x) \text{ is not satisfied}\Big\} \\ &= \left\{ x \in B : \begin{array}{l} \text{To } m \in \mathbb{N} \text{ and for all } l \in \mathbb{N} \text{ exists} \\ \text{an index } k \ge l \text{ with } |f_k(x) - f(x)| > \dfrac{1}{m} \end{array} \right\} \\ &= \bigcup_{m=1}^{\infty} \bigcap_{l=1}^{\infty} \bigcup_{k \ge l} \left\{ x \in B : |f_k(x) - f(x)| > \frac{1}{m} \right\} = \bigcup_{m=1}^{\infty} B_m, \end{aligned}$$

where

$$B_m := \bigcap_{l=1}^{\infty} \bigcup_{k \geq l} \left\{ x \in B \,:\, |f_k(x) - f(x)| > \frac{1}{m} \right\}$$

has been defined. We observe $B_m \subset N$ and consequently $\mu(B_m) = 0$ for all $m \in \mathbb{N}$. We note that

$$B_{m,l} := \bigcup_{k \geq l} \left\{ x \in B \,:\, |f_k(x) - f(x)| > \frac{1}{m} \right\}$$

holds true and infer $B_{m,l} \supset B_{m,l+1}$ for all $m, l \in \mathbb{N}$. From the relation

$$B_m = \bigcap_{l=1}^{\infty} B_{m,l}$$

we then obtain

$$0 = \mu(B_m) = \lim_{l \to \infty} \mu(B_{m,l}).$$

Consequently, to each index $m \in \mathbb{N}$ we find an index $l_m \in \mathbb{N}$ with $l_m < l_{m+1}$ such that

$$\mu\left(\bigcup_{k \geq l_m} \left\{ x \in B \,:\, |f_k(x) - f(x)| > \frac{1}{m} \right\} \right) = \mu(B_{m,l_m}) < \frac{\delta}{2^{m+1}}$$

holds true. We define

$$\widehat{B}_m := B_{m,l_m} \quad \text{and} \quad \widehat{B} := \bigcup_{m=1}^{\infty} \widehat{B}_m.$$

Evidently, the set $\widehat{B}$ is measurable and the estimate

$$\mu(\widehat{B}) \leq \sum_{m=1}^{\infty} \mu(\widehat{B}_m) \leq \frac{\delta}{2}$$

is fulfilled. When we still define $\widehat{A} := B \setminus \widehat{B}$, we comprehend

$$\widehat{A} = B \cap \left(\bigcup_{m=1}^{\infty} \widehat{B}_m \right)^c = B \cap \left(\bigcap_{m=1}^{\infty} \widehat{B}_m^c \right)$$

$$= \bigcap_{m=1}^{\infty} \left\{ x \in B \,:\, |f_k(x) - f(x)| \leq \frac{1}{m} \text{ for all } k \geq l_m \right\}.$$

For all points $x \in \widehat{A}$, we find an index $l_m \in \mathbb{N}$ to a given $m \in \mathbb{N}$ such that

$$|f_k(x) - f(x)| \leq \frac{1}{m} \quad \text{for all} \quad k \geq l_m$$

holds true. Consequently, the sequence $\{f_k|_{\widehat{A}}\}_{k=1,2,\ldots}$ converges uniformly towards $f|_{\widehat{A}}$. According to Theorem 3.14 in Section 3, we now choose a closed set $A \subset \widehat{A}$ with

$$\mu(\widehat{A} \setminus A) < \frac{\delta}{2}.$$

We note that $A \subset \widehat{A}$ holds true, and the sequence of functions $\{f_k|_A\}_{k=1,2,\ldots}$ converges uniformly towards $f|_A$. When we additionally observe $B \setminus \widehat{A} = \widehat{B}$, we finally see

$$\mu(B \setminus A) = \mu(B \setminus \widehat{A}) + \mu(\widehat{A} \setminus A) < \frac{\delta}{2} + \frac{\delta}{2} = \delta.$$

q.e.d.

The interrelation between measurable and continuous functions is revealed by the following result.

Theorem 4.16. (Lusin)
Let $f : B \to \mathbb{R}$ denote a measurable function on the measurable set $B \subset X$. To each quantity $\delta > 0$, we then find a closed set $A \subset X$ with the property $\mu(B \setminus A) < \delta$ such that the restriction $f|_A : A \to \mathbb{R}$ is continuous.

Proof: For $j = 1, 2, \ldots$ we consider the truncated functions

$$f_j(x) := \begin{cases} -j\,, & f(x) \in [-\infty, -j] \\ f(x)\,, & f(x) \in [-j, +j] \\ +j\,, & f(x) \in [+j, +\infty] \end{cases}.$$

All functions $f_j : B \to \mathbb{R}$ are measurable, and we infer

$$|f_j(x)| \le j \qquad \text{for all} \quad x \in B.$$

We utilize Theorem 4.14 and the property $M(X) \subset C^0(X)$: For each index $j \in \mathbb{N}$, there exists a sequence of continuous functions $f_{j,k} : B \to \mathbb{R}$ satisfying

$$\lim_{k\to\infty} f_{j,k}(x) = f_j(x) \qquad \text{a.e. in} \quad B.$$

Via Egorov's theorem, we find a closed set $A_j \subset B$ to each $j = 1, 2, \ldots$ satisfying

$$\mu(B \setminus A_j) < \frac{\delta}{2^{j+1}},$$

such that the sequence of functions $\{f_{j,k}|_{A_j}\}_{k=1,2,\ldots}$ converges uniformly towards the function $f_j|_{A_j}$. The Weierstraß convergence theorem reveals continuity of the functions $f_j|_{A_j}$ for all $j \in \mathbb{N}$. The set

$$\widehat{A} := \bigcap_{j=1}^{\infty} A_j \subset B$$

is closed, and we arrive at the estimate

$$\mu(B \setminus \widehat{A}) \le \sum_{j=1}^{\infty} \mu(B \setminus A_j) < \sum_{j=1}^{\infty} \frac{\delta}{2^{j+1}} = \frac{\delta}{2}.$$

Now the functions $f_j : \widehat{A} \to \mathbb{R}$ are continuous for all $j \in \mathbb{N}$, and we recall

$$f(x) = \lim_{j \to \infty} f_j(x) \qquad \text{in} \quad \widehat{A}.$$

Egorov's theorem supplies a closed set $A \subset \widehat{A}$ with

$$\mu(\widehat{A} \setminus A) < \frac{\delta}{2},$$

such that f_j converges uniformly on A towards f. Consequently, the function $f|_A$ is continuous, and we estimate as follows:

$$\mu(B \setminus A) = \mu(B \setminus \widehat{A}) + \mu(\widehat{A} \setminus A) < \frac{\delta}{2} + \frac{\delta}{2} = \delta.$$

q.e.d.

Remark: We have learned the *Three principles of Littlewood* in Lebesgue's theory of measure and integration. J.E.LITTLEWOOD: "There are three principles roughly expressible in the following terms: Every measurable set is nearly a finite union of intervals; every measurable function is nearly continuous; every a.e. convergent sequence of measurable functions is nearly uniformly convergent."

5 Riemann's and Lebesgue's Integral on Rectangles

With $d \in (0, +\infty)$ being given, we consider the rectangle

$$Q := \Big\{ x = (x_1, \ldots, x_n) \in \mathbb{R}^n \, : \, |x_j| \le d \, , \; j = 1, \ldots, n \Big\}, \quad \text{where} \quad n \in \mathbb{N}.$$

In our main example from Section 1, we choose $X = \Omega := \overset{\circ}{Q}$ and extend the improper Riemannian integral

$$I \, : \, M(X) \longrightarrow \mathbb{R}, \qquad \text{with} \quad f \mapsto I(f) := \int_{\Omega} f(x) \, dx$$

from the space

$$M(X) := \left\{ f \in C^0(\Omega) \, : \, \int_{\Omega} |f(x)| \, dx < +\infty \right\}$$

onto the space $L(X) \supset M(X)$ and obtain Lebesgue's integral $I : L(X) \to \mathbb{R}$.

Theorem 5.1. *For the set $E \subset \Omega$ being given, the following statements are equivalent:*

(1) E is a null-set.
(2) To each quantity $\varepsilon > 0$, we find with $\{Q_k\}_{k=1,2,\dots} \subset \Omega$ denumerably many rectangles satisfying $E \subset \bigcup_{k=1}^{\infty} Q_k$ *and* $\sum_{k=1}^{\infty} |Q_k| < \varepsilon$.

Proof:

(1)$\Longrightarrow$(2): Since E represents a null-set, Proposition 3.15 from Section 3 provides a function $h \in V(X)$ with $h \geq 0$ on X, $h = +\infty$ on E, and $I(h) < +\infty$. With the constant $c \in [1, +\infty)$ chosen arbitrarily, we consider the open - and consequently measurable - set

$$E_c := \Big\{x \in \Omega \,:\, h(x) > c\Big\} \supset E.$$

Then we observe

$$\mu(E_c) = I(\chi_{E_c}) = \frac{1}{c}\, I(c\,\chi_{E_c}) \leq \frac{1}{c}\, I(h) < \varepsilon$$

for $c > \frac{I(h)}{\varepsilon}$. The open set E_c can be represented as a denumerable union of closed rectangles Q_k which intersect, at most, in boundary points. Therefore, we deduce

$$E \subset E_c = \bigcup_{k=1}^{\infty} Q_k.$$

We note that the boundary points of a rectangle constitute a null-set and see

$$\sum_{k=1}^{\infty} |Q_k| = \mu(E_c) < \varepsilon.$$

(2)$\Longrightarrow$(1): For each index $k \in \mathbb{N}$ we find a function $h_k \in C_0^0(\Omega)$ satisfying

$$h_k(x) = \begin{cases} 1 & , \; x \in Q_k \\ \in [0,1] \, , & x \in \mathbb{R}^n \setminus Q_k \end{cases} \qquad \text{and} \quad I(h_k) \leq 2|Q_k|.$$

The sequence $\{g_l(x)\}_{l=1,2,\dots}$, defined by $g_l(x) := \sum_{k=1}^{l} h_k(x)$, converges monotonically and belongs to $M(X)$. This implies

$$h(x) := \sum_{k=1}^{\infty} h_k(x) \in V(X).$$

Furthermore, we have $\chi_E(x) \leq h(x)$, $x \in \mathbb{R}^n$ and estimate

$$0 \leq I^-(\chi_E) \leq I^+(\chi_E) \leq I(h) = \sum_{k=1}^{\infty} I(h_k) \leq 2 \sum_{k=1}^{\infty} |Q_k| < 2\varepsilon$$

for all $\varepsilon > 0$. Therefore, E is a null-set. q.e.d.

Riemann's and Lebesgue's integral are compared as follows:

Theorem 5.2. *A bounded function $f : \Omega \to \mathbb{R}$ is Riemann integrable if and only if the set K, containing all points of discontinuities, constitutes a null-set. In this case the function f belongs to the class $L(\Omega)$, and we have the identity*

$$I(f) = \int\limits_{\Omega} f(x)\,dx = \int_{Q} f(x)\,dx;$$

this means that Riemann's integral of f coincides with Lebesgue's integral. Here we have to extend f to 0 onto the whole space $\mathbb{R}^n$.

Proof: We consider the functions

$$m^+(x) := \lim_{\varepsilon\to 0+} \sup_{|y-x|<\varepsilon} f(y) \quad \text{and} \quad m^-(x) := \lim_{\varepsilon\to 0+} \inf_{|y-x|<\varepsilon} f(y)\,, \; x \in \mathbb{R}^n.$$

We have the identity $m^+(x) = m^-(x)$ if and only if f is continuous at the point x. Let

$$\mathcal{Z} \, : \, Q = \bigcup_{k=1}^{N} Q_k$$

denote a canonical decomposition of Q into N closed rectangles Q_k. We define the simple functions

$$m_k^+ := \sup_{Q_k} f(y), \quad m_k^- := \inf_{Q_k} f(y) \quad \text{and} \quad f_{\mathcal{Z}}^{\pm}(x) := \sum_{k=1}^{N} m_k^{\pm} \chi_{Q_k}(x) \in L(X).$$

We observe the identity

$$I(f_{\mathcal{Z}}^{\pm}) = \sum_{k=1}^{N} m_k^{\pm} |\,Q_k|.$$

Therefore, Lebesgue's integral of the functions $f_{\mathcal{Z}}^{\pm}$ coincides with the Riemannian upper and lower sums, respectively, of the function f - associated with the decomposition $\mathcal{Z}$. When we denote by

$$\partial\mathcal{Z} := \bigcup_{k=1}^{N} \partial Q_k$$

the set of the boundary points for the decomposition $\mathcal{Z}$, then $\partial\mathcal{Z}$ constitutes a null-set in $\mathbb{R}^n$. Now we observe an arbitrary canonical sequence of decompositions $\{\mathcal{Z}_p\}_{p=1,2,\ldots}$ for the rectangle Q, such that its fineness tends to 0. We obtain the limit relation

$$\lim_{p\to\infty} f_{\mathcal{Z}_p}^{\pm}(x) = m^{\pm}(x) \quad \text{for all} \quad x \in \Omega \setminus N,$$

where

$$N = \bigcup_{p=1}^{\infty} \partial \mathcal{Z}_p \subset Q$$

is a null-set. Now we select an adequate canonical sequence of decompositions such that

$$\underline{\int_Q} f(x)\, dx = \lim_{p\to\infty} I(f^-_{\mathcal{Z}_p}) \quad \text{and} \quad \overline{\int_Q} f(x)\, dx = \lim_{p\to\infty} I(f^+_{\mathcal{Z}_p}).$$

Lebesgue's convergence theorem implies

$$\underline{\int_Q} f(x)\, dx = I(m^-) \quad \text{and} \quad \overline{\int_Q} f(x)\, dx = I(m^+).$$

Now we note that the function $f : \Omega \to \mathbb{R}$ is Riemann integrable if and only if

$$I(m^+) = \overline{\int_Q} f(x)\, dx = \underline{\int_Q} f(x)\, dx = I(m^-) \quad \text{or equivalently} \quad I(m^+ - m^-) = 0$$

holds true. Due to $m^+ \geq m^-$, this is exactly the case if $m^+ = m^-$ a.e. in Q holds true, or equivalently if f is continuous a.e. on Q.

q.e.d.

We intend to prove Fubini's theorem interchanging the order of integration for Lebesgue integrable functions. Here we consider two open bounded rectangles $Q \subset \mathbb{R}^p$ and $R \subset \mathbb{R}^q$ and begin with the following

Proposition 5.3. *Let $f = f(x,y) : Q \times R \to \overline{\mathbb{R}} \in V(Q \times R)$ be given. Then the function $f(x,y)$, $y \in R$ belongs to the class $V(R)$ for each $x \in Q$, and the function*

$$\varphi(x) := \int_R f(x,y)\, dy$$

belongs to the class $V(Q)$. Furthermore, we have

$$\iint_{Q\times R} f(x,y)\, dxdy = \int_Q \varphi(x)\, dx.$$

Proof: Since $f \in V(Q \times R)$ holds true, we find a sequence $\{f_n(x,y)\}_{n=1,2,\ldots} \subset C^0_0(Q \times R)$ satisfying $f_n(x,y) \uparrow f(x,y)$. For each $x \in Q$, the functions $f_n(x,y)$, $\quad y \in R$ belong to the class $C^0_0(R)$ and consequently $f(x,y)$ to $V(R)$. When we define

$$\varphi_n(x) := \int_R f_n(x,y)\, dy, \qquad x \in Q,$$

we infer $\varphi_n \in C_0^0(Q)$ and $\varphi_n(x) \uparrow \varphi(x)$ in Q. This implies

$$\iint\limits_{Q\times R} f(x,y)\,dxdy := \lim_{n\to\infty} \iint\limits_{Q\times R} f_n(x,y)\,dxdy = \lim_{n\to\infty} \int\limits_Q \varphi_n(x)\,dx = \int\limits_Q \varphi(x)\,dx.$$

q.e.d.

Proposition 5.4. *Let N denote a null-set in $Q \times R$ and define*

$$N_x := \Big\{ y \in R \,:\, (x,y) \in N \Big\}.$$

Then we have a null-set $E \subset Q$, such that N_x constitutes a null-set in R for all points $x \in Q \setminus E$.

Proof: Since N is a null-set, we find a function $h(x,y) \in V(Q \times R)$ with $h \geq 0$ on $Q \times R$ and $h(x,y) = +\infty$ for all $(x,y) \in N$, such that the property

$$+\infty > \iint\limits_{Q\times R} h(x,y)\,dxdy = \int\limits_Q \varphi(x)\,dx \qquad \text{with} \quad \varphi(x) := \int\limits_R h(x,y)\,dy \geq 0$$

holds true - due to Proposition 5.3. We note that $\varphi \in V(Q)$ and

$$\int\limits_Q \varphi(x)\,dx < +\infty$$

is satisfied and deduce $\varphi \in L(Q)$. Furthermore, we find a null-set $E \subset Q$ with $\varphi(x) < +\infty$ for all $x \in Q \setminus E$. On account of $h = +\infty$ on N, the set N_x is a null-set for all $x \in Q \setminus E$.

q.e.d.

Theorem 5.5. (Fubini) *Let $f(x,y) : Q \times R \to [0, +\infty]$ represent a measurable function. Then we have a null-set $E \subset Q$, such that the function $f(x,y)$, $y \in R$ is measurable for all points $x \in Q \setminus E$. When we define*

$$\varphi(x) := \begin{cases} \displaystyle\int\limits_R f(x,y)\,dy\,, & x \in Q \setminus E \\ 0\,, & x \in E \end{cases},$$

the function φ is nonnegative and measurable. Furthermore, we have Fubini's identity

$$\iint\limits_{Q\times R} f(x,y)\,dxdy = \int\limits_Q \varphi(x)\,dx.$$

Proof: For $n = 1, 2, \ldots$ we consider the functions

$$f_n(x,y) := \begin{cases} f(x,y), & \text{if } f(x,y) \in [0,n] \\ n, & \text{otherwise} \end{cases}$$

with $f_n \in L(Q \times R)$. Applying Theorem 4.14 from Section 4, we find for each number $n \in \mathbb{N}$ a null-set $N_n \subset Q \times R$ and a sequence of functions

$$f_{n,m}(x,y) \in C_0^0(Q \times R) \quad \text{with} \quad |f_{n,m}| \leq n \quad \text{on} \quad Q \times R,$$

such that

$$\lim_{m \to \infty} f_{n,m}(x,y) = f_n(x,y) \qquad \text{for all} \quad (x,y) \in (Q \times R) \setminus N_n.$$

Each fixed number $n \in \mathbb{N}$ admits a null-set $E_n \subset Q$, such that

$$\{y \in R : (x,y) \in N_n\} \subset R$$

represents a null-set for all points $x \in Q \setminus E_n$. Now Lebesgue's convergence theorem yields

$$\iint\limits_{Q \times R} f_n(x,y)\,dxdy$$

$$= \lim_{m \to \infty} \iint\limits_{Q \times R} f_{n,m}(x,y)\,dxdy = \lim_{m \to \infty} \int\limits_Q \left(\int\limits_R f_{n,m}(x,y)\,dy \right) dx$$

$$= \lim_{m \to \infty} \int\limits_{Q \setminus E_n} \left(\int\limits_R f_{n,m}(x,y)\,dy \right) dx = \int\limits_{Q \setminus E_n} \left(\int\limits_R \underbrace{f_n(x,y)}_{\in L(R)}\,dy \right) dx.$$

In addition,

$$E := \bigcup_{n=1}^{\infty} E_n \subset Q$$

constitutes a null-set, and we see

$$\iint\limits_{Q \times R} f_n(x,y)\,dxdy = \int\limits_{Q \setminus E} \left(\int\limits_R f_n(x,y)\,dy \right) dx.$$

Finally, Theorem 4.9 from Section 4 yields

$$\iint\limits_{Q \times R} f(x,y)\,dxdy = \lim_{n \to \infty} \left(\iint\limits_{Q \times R} f_n(x,y)\,dxdy \right)$$

$$= \lim_{n \to \infty} \int\limits_{Q \setminus E} \left(\int\limits_R f_n(x,y)\,dy \right) dx = \int\limits_{Q \setminus E} \left(\int\limits_R f(x,y)\,dy \right) dx = \int\limits_Q \varphi(x)\,dx.$$

q.e.d.

6 Banach and Hilbert Spaces

We owe the basic concepts for linear spaces, which appear in the next sections, to the mathematicians D. Hilbert and S. Banach. Here we can equally consider real and complex vector spaces.

Definition 6.1. *Let $\mathcal{M}$ denote a real (or complex) linear space, which means*

$$f, g \in \mathcal{M}, \ \alpha, \beta \in \mathbb{R} \ (or \mathbb{C}) \quad \Longrightarrow \quad \alpha f + \beta g \in \mathcal{M}.$$

Then we name $\mathcal{M}$ a normed real (or complex) linear space *and equivalently a* normed vector space *if we have a function*

$$\| \cdot \| : \mathcal{M} \longrightarrow [0, +\infty)$$

with the following properties:

(N1) $\|f\| = 0 \iff f = 0$;
(N2) Triangle inequality: $\|f + g\| \le \|f\| + \|g\|$ *for all* $f, g \in \mathcal{M}$;
(N3) Homogeneity: $\|\lambda f\| = |\lambda| \|f\|$ *for all* $f \in \mathcal{M}$, $\lambda \in \mathbb{R} \ (or \mathbb{C})$.

The function $\| \cdot \|$ is called the norm on $\mathcal{M}$.

Remark: From the axioms (N1), (N2), and (N3) we immediately infer the inequality

$$\|f - g\| \ge \Big| \|f\| - \|g\| \Big| \quad \text{for all} \quad f, g \in \mathcal{M},$$

because we have

$$\|f\| - \|g\| = \|f - g + g\| - \|g\| \le \|f - g\| + \|g\| - \|g\| = \|f - g\|,$$

which yields our statement by interchanging f and g.

Definition 6.2. *The normed vector space $\mathcal{M}$ is named* complete, *if each Cauchy sequence in $\mathcal{M}$ converges. This means, to each sequence $\{f_n\} \subset \mathcal{M}$ satisfying* $\lim_{k,l \to \infty} \|f_k - f_l\| = 0$ *we find an element $f \in \mathcal{M}$ with* $\lim_{k \to \infty} \|f - f_k\| = 0$.

Definition 6.3. *A complete normed vector space is named a* Banach space.

Example 6.4. Choosing the compact set $K \subset \mathbb{R}^n$, we endow the space $\mathcal{B} := C^0(K, \mathbb{R})$ with the norm

$$\|f\| := \sup_{x \in K} |f(x)| = \max_{x \in K} |f(x)|, \qquad f \in \mathcal{B},$$

and thus obtain a Banach space. This norm generates the uniform convergence - a concept already introduced by Weierstraß.

Definition 6.5. *A complex linear space $\mathcal{H}'$ is named* pre-Hilbert-space *if an* inner product *is defined in $\mathcal{H}'$; more precisely, we have a function*

$$(\cdot,\cdot) : \mathcal{H}' \times \mathcal{H}' \longrightarrow \mathbb{C}$$

with the following properties:

(H1) $(f+g,h) = (f,h) + (g,h)$ *for all* $f,g,h \in \mathcal{H}'$;
(H2) $(f,\lambda g) = \lambda(f,g)$ *for all* $f,g \in \mathcal{H}',\ \lambda \in \mathbb{C}$;
(H3) Hermitian character: $(f,g) = \overline{(g,f)}$ *for all* $f,g \in \mathcal{H}'$;
(H4) Positive-definite character: $(f,f) > 0$, *if* $f \neq 0$.

Remarks:

1. We infer the following calculus rule from the axioms (H1) - (H4) immediately:
   (H5) For all $f,g,h \in \mathcal{H}'$ we have
   $$(f,g+h) = \overline{(g+h,f)} = \overline{(g,f)} + \overline{(h,f)} = (f,g) + (f,h).$$
   (H6) Furthermore, the relation
   $$(\lambda f,g) = \overline{\lambda}(f,g) \qquad \text{for all} \quad f,g \in \mathcal{H}', \quad \lambda \in \mathbb{C}$$
   is satisfied.
   Therefore, the inner product is antilinear in its first and linear in its second argument.
2. In a *real linear space* $\mathcal{H}'$, an inner product is characterized by the properties (H1) - (H4) as well, where (H3) then reduces to the symmetry condition
   $$(f,g) = (g,f) \qquad \text{for all} \quad f,g \in \mathcal{H}'.$$

Example 6.6. Let us consider the numbers $-\infty < a < b < +\infty$ and the space $\mathcal{H}' := C^0([a,b],\mathbb{C})$ of continuous functions. Via the inner product

$$(f,g) := \int_a^b \overline{f(x)}g(x)\,dx,$$

the set $\mathcal{H}'$ becomes a pre-Hilbert-space.

Theorem 6.7. *Let $\mathcal{H}'$ represent a pre-Hilbert-space. With the aid of the norm*

$$\|f\| := \sqrt{(f,f)},$$

the set $\mathcal{H}'$ becomes a normed vector space.

Proof:

1. At first, we show that the following inequality is valid in $\mathcal{H}'$, namely

$$|(g, f)| = |(f, g)| \leq \|f\|\|g\| \quad \text{for all} \quad f, g \in \mathcal{H}'.$$

With $f, g \in \mathcal{H}'$, we associate a quadratic form in $\lambda, \mu \in \mathbb{C}$ as follows:

$$\begin{aligned} 0 \leq Q(\lambda, \mu) &:= (\lambda f - \mu g, \lambda f - \mu g) \\ &= |\lambda|^2 (f, f) - \lambda\overline{\mu}(g, f) - \overline{\lambda}\mu(f, g) + |\mu|^2 (g, g). \end{aligned}$$

When $(g, f) = (f, g) = 0$ - in particular $f = 0$ or $g = 0$ - holds true, this inequality is evident. In the other case, we choose

$$\lambda = 1, \quad \overline{\mu} = \frac{\|f\|^2}{(g, f)}.$$

The nonnegative character of Q - easily seen from the property $(H4)$ - implies the inequality

$$0 \leq -\|f\|^2 + \frac{\|f\|^4 \|g\|^2}{|(f, g)|^2}$$

and finally by rearrangement

$$|(f, g)| \leq \|f\| \, \|g\| \quad \text{for all} \quad f, g \in \mathcal{H}'.$$

2. Now we show that $\|f\| := \sqrt{(f, f)}$ satisfies the norm conditions (N1) - (N3). We infer for all elements $f, g \in \mathcal{H}'$ and $\lambda \in \mathbb{C}$ the following properties:
   i.) $\|f\| \geq 0$, and (H4) tells us that $\|f\| = 0$ is fulfilled if and only if $f = 0$ is correct;
   ii.) $\|\lambda f\| = \sqrt{(\lambda f, \lambda f)} = \sqrt{\lambda\overline{\lambda}(f, f)} = |\lambda| \, \|f\|$;
   iii.)
$$\begin{aligned} \|f + g\|^2 &= (f + g, f + g) = (f, f) + 2\text{Re}(f, g) + (g, g) \\ &\leq \|f\|^2 + 2|(f, g)| + \|g\|^2 \\ &\leq \|f\|^2 + 2\|f\| \, \|g\| + \|g\|^2 \\ &= (\|f\| + \|g\|)^2, \end{aligned}$$
   and consequently
$$\|f + g\| \leq \|f\| + \|g\|.$$

Therefore, $\| \cdot \|$ gives us a norm on $\mathcal{H}'$. q.e.d.

Definition 6.8. *A pre-Hilbert-space $\mathcal{H}$ is named* Hilbert space, *if $\mathcal{H}$ endowed with the norm*

$$\|f\| := \sqrt{(f, f)}, \qquad f \in \mathcal{H}$$

is complete and consequently a Banach space.

Remarks:

1. We prove that the inner product (f,g) is continuous in $\mathcal{H}$. Here we note the following estimate for the elements $f, g, f_n, g_n \in \mathcal{H}$:
$$\begin{aligned}|(f_n, g_n) - (f,g)| &= |(f_n, g_n) - (f_n, g) + (f_n, g) - (f,g)| \\ &\leq |(f_n, g_n) - (f_n, g)| + |(f_n, g) - (f,g)| \\ &\leq |(f_n, g_n - g)| + |(f_n - f, g)| \\ &\leq \|f_n\| \, \|g_n - g\| + \|f_n - f\| \, \|g\|.\end{aligned}$$
Therefore, when the limit relations $f_n \to f$ and $g_n \to g$ for $n \to \infty$ in $\mathcal{H}$ hold true, we infer
$$\lim_{n\to\infty} (f_n, g_n) = (f,g).$$
We observe that the completeness of the space $\mathcal{H}$ is not needed for the proof of the continuity of the inner product.
2. The pre-Hilbert-space from Example 6.6 is not complete and consequently does not represent a Hilbert space.
3. In Section 3 from Chapter 8, we shall embed - parallel to the transition from rational numbers to real numbers - each pre-Hilbert-space $\mathcal{H}'$ into a Hilbert space $\mathcal{H}$. This means $\mathcal{H}' \subset \mathcal{H}$ and $\mathcal{H}'$ is dense in $\mathcal{H}$.
4. Hilbert spaces represent particular Banach spaces. The existence of an inner product in $\mathcal{H}$ allows us to introduce the notion of *orthogonality*: Two elements $f, g \in \mathcal{H}$ are named *orthogonal to each other* if $(f,g) = 0$ holds true.

Let $\mathcal{M} \subset \mathcal{H}$ denote an arbitrary linear subspace. We define the *orthogonal space* to $\mathcal{M}$ via
$$\mathcal{M}^\perp := \Big\{ g \in \mathcal{H} \, : \, (g,f) = 0 \text{ for all } f \in \mathcal{M} \Big\}.$$

We see immediately that $M^\perp$ is a linear subspace of $\mathcal{H}$, and the continuity of the inner product justifies the following

Remark: For an arbitrary linear subspace $\mathcal{M} \subset \mathcal{H}$, its associate orthogonal space $\mathcal{M}^\perp$ is closed. More precisely, each sequence
$$\{f_n\} \subset \mathcal{M}^\perp \quad \text{in} \quad \mathcal{M}^\perp \quad \text{satisfying} \quad f_n \to f \quad \text{for} \quad n \to \infty$$
fulfills $f \in \mathcal{M}^\perp$.

Proof: Since $\{f_n\} \subset \mathcal{M}^\perp$ holds true, we infer $(f_n, g) = 0$ for all $n \in \mathbb{N}$ and $g \in \mathcal{M}$. This implies
$$0 = \lim_{n\to\infty} (f_n, g) = (f,g) \quad \text{for all} \quad g \in \mathcal{M}.$$

q.e.d.

Fundamentally important is the following

Theorem 6.9. (Orthogonal projection)
Let $\mathcal{M} \subset \mathcal{H}$ denote a closed linear subspace of the Hilbert space $\mathcal{H}$. Then each element $f \in \mathcal{H}$ possesses the following representation:

$$f = g + h \qquad \textit{with} \quad g \in \mathcal{M} \quad \textit{and} \quad h \in \mathcal{M}^{\perp}.$$

Here the elements g and h are uniquely determined.

This theorem says that the Hilbert space $\mathcal{H}$ can be decomposed into two orthogonal subspaces $\mathcal{M}$ and $\mathcal{M}^{\perp}$ such that $\mathcal{H} = \mathcal{M} \oplus \mathcal{M}^{\perp}$ holds true.
Proof:

1. At first, we show the uniqueness. Let us consider an element $f \in \mathcal{H}$ with

$$f = g_1 + h_1 = g_2 + h_2, \qquad g_j \in \mathcal{M}, \quad h_j \in \mathcal{M}^{\perp}.$$

Then we deduce

$$0 = f - f = (g_1 - g_2) + (h_1 - h_2).$$

The uniqueness follows from the identity

$$\begin{aligned} 0 &= \|(g_1 - g_2) + (h_1 - h_2)\|^2 \\ &= ((g_1 - g_2) + (h_1 - h_2), (g_1 - g_2) + (h_1 - h_2)) \\ &= \|g_1 - g_2\|^2 + \|h_1 - h_2\|^2. \end{aligned}$$

2. Now we have to establish the existence of the desired representation. The element $f \in \mathcal{H}$ being given, we solve the subsequent variational problem: Find an element $g \in \mathcal{M}$ such that

$$\|f - g\| = \inf_{\widetilde{g} \in \mathcal{M}} \|f - \widetilde{g}\| =: d$$

holds true. We choose a sequence $\{g_k\} \subset \mathcal{M}$ with the property

$$\lim_{k \to \infty} \|f - g_k\| = d.$$

Then we prove that this sequence converges towards an element $g \in \mathcal{M}$. Here we utilize the *parallelogram identity*

$$\left\|\frac{\varphi + \psi}{2}\right\|^2 + \left\|\frac{\varphi - \psi}{2}\right\|^2 = \frac{1}{2}\left(\|\varphi\|^2 + \|\psi\|^2\right) \qquad \text{for all} \quad \varphi, \psi \in \mathcal{H},$$

which we easily check by evaluating the inner products on both sides. Now we apply this identity to the elements

$$\varphi = f - g_k, \quad \psi = f - g_l, \qquad k, l \in \mathbb{N}$$

and obtain

$$\left\| f - \frac{g_k + g_l}{2} \right\|^2 + \left\| \frac{g_k - g_l}{2} \right\|^2 = \frac{1}{2} \left(\|f - g_k\|^2 + \|f - g_l\|^2 \right).$$

Rearrangement of these equations implies

$$0 \leq \left\| \frac{g_k - g_l}{2} \right\|^2 = \frac{1}{2} \left(\|f - g_k\|^2 + \|f - g_l\|^2 \right) - \left\| f - \frac{g_k + g_l}{2} \right\|^2$$

$$\leq \frac{1}{2} \left(\|f - g_k\|^2 + \|f - g_l\|^2 \right) - d^2.$$

The passage to the limit $k, l \to \infty$ reveals that $\{g_k\}$ represents a Cauchy sequence. Since the linear subspace $\mathcal{M}$ is closed, we infer the existence of the limit $g \in \mathcal{M}$ for the sequence $\{g_k\}$.
Finally, we prove $h = (f - g) \in \mathcal{M}^\perp$ and obtain the desired representation $f = g + (f - g) = g + h$.
When $\varphi \in \mathcal{M}$ is chosen arbitrarily as well as the number $\varepsilon \in (-\varepsilon_0, \varepsilon_0)$, we infer the inequality

$$\|(f - g) + \varepsilon\varphi\|^2 \geq d^2 = \|f - g\|^2.$$

We note that

$$\|f - g\|^2 + 2\varepsilon \operatorname{Re}(f - g, \varphi) + \varepsilon^2 \|\varphi\|^2 \geq \|f - g\|^2,$$

and deduce

$$2\varepsilon \operatorname{Re}(f - g, \varphi) + \varepsilon^2 \|\varphi\|^2 \geq 0$$

for all $\varphi \in \mathcal{M}$ and all $\varepsilon \in (-\varepsilon_0, \varepsilon_0)$. Therefore, the identity

$$\operatorname{Re}(f - g, \varphi) = 0 \qquad \text{for all} \quad \varphi \in \mathcal{M}$$

must be valid. When we replace φ by $i\varphi$, we obtain $(f - g, \varphi) = 0$. Since the element φ has been chosen arbitrarily within $\mathcal{M}$, the property

$$(f - g) \in \mathcal{M}^\perp$$

is shown. q.e.d.

The subsequent concepts on the continuity of linear operators in infinite-dimensional vector spaces have been created by S. Banach.

Definition 6.10. *Let $\{\mathcal{M}_1, \|\cdot\|_1\}$ and $\{\mathcal{M}_2, \|\cdot\|_2\}$ denote two normed linear spaces and $A : \mathcal{M}_1 \to \mathcal{M}_2$ a linear mapping. Then A is called* continuous *at the point $f \in \mathcal{M}_1$, if we can find a quantity $\delta = \delta(\varepsilon, f) > 0$ for all $\varepsilon > 0$ such that*

$$g \in \mathcal{M}_1, \ \|g - f\|_1 < \delta \quad \Longrightarrow \quad \|A(g) - A(f)\|_2 < \varepsilon.$$

Theorem 6.11. *Consider the linear functional $A : \mathcal{M} \to \mathbb{C}$ on the linear normed space $\mathcal{M}$, which means*

$$A(\alpha f + \beta g) = \alpha A(f) + \beta A(g) \quad \textit{for all} \quad f, g \in \mathcal{M}, \quad \alpha, \beta \in \mathbb{C}.$$

Then the following statements are equivalent:

(i) A is continuous at all points $f \in \mathcal{M}$;
(ii) A is continuous at one point $f \in \mathcal{M}$;
(iii) A is bounded in the following sense: There exists a constant $\alpha \in [0, +\infty)$ such that

$$|A(f)| \leq \alpha \|f\| \qquad \textit{for all} \quad f \in \mathcal{M}$$

holds true.

Proof:

$(i) \Rightarrow (iii)$: Let A be continuous in $\mathcal{M}$, then this holds true at the origin $0 \in \mathcal{M}$ in particular. For $\varepsilon = 1$ we find a quantity $\delta(\varepsilon) > 0$ such that $\|f\| \leq \delta$ implies $|A(f)| \leq 1$. We obtain

$$|A(f)| \leq \frac{1}{\delta}\|f\| \qquad \text{for all} \quad f \in \mathcal{M}.$$

$(iii) \Rightarrow (ii)$: We immediately infer the continuity of A at the origin 0 from the boundedness of A.

$(ii) \Rightarrow (i)$: Let A be be continuous at one point $f_0 \in \mathcal{M}$. For a number $\varepsilon > 0$ being given, we find a quantity $\delta > 0$ satisfying

$$\varphi \in \mathcal{M}, \ \|\varphi\| \leq \delta \quad \Longrightarrow \quad |A(f_0 + \varphi) - A(f_0)| \leq \varepsilon.$$

The linearity of our functional A gives us the following estimate for all $f \in \mathcal{M}$:

$$\varphi \in \mathcal{M}, \ \|\varphi\| \leq \delta \quad \Longrightarrow \quad |A(f + \varphi) - A(f)| \leq \varepsilon.$$

Therefore, A is continuous for all $f \in \mathcal{M}$. q.e.d.

Remark: This theorem remains true for linear mappings $A : \mathcal{M}_1 \to \mathcal{M}_2$ between the normed vector spaces $\{\mathcal{M}_1, \|\cdot\|_1\}$ and $\{\mathcal{M}_2, \|\cdot\|_2\}$. Here we mean by the notion *'A is bounded'* that we can find a number $\alpha \in [0, +\infty)$ such that

$$\|A(f)\|_2 \leq \alpha \|f\|_1 \quad \text{for all} \quad f \in \mathcal{M}_1$$

holds true.

Definition 6.12. *When we consider a bounded linear functional $A : \mathcal{M} \to \mathbb{C}$ on the normed linear space $\mathcal{M}$, we introduce the* norm of the functional A *as follows:*

$$\|A\| := \sup_{f \in \mathcal{M}, \ \|f\| \leq 1} |A(f)|.$$

Definition 6.13. *By the symbol*

$$\mathcal{M}^* := \Big\{ A : \mathcal{M} \to \mathbb{C} \; : \; A \text{ is bounded on } \mathcal{M} \Big\},$$

we denote the dual space *of the normed linear space* $\mathcal{M}$.

Remarks:

1. We easily show that $\mathcal{M}^*$, endowed with the norm from Definition 6.12, constitutes a Banach space.
2. Let $\mathcal{H}$ denote a Hilbert space. Then its dual space $\mathcal{H}^*$ is isomorphic to $\mathcal{H}$, as we shall show now.

Theorem 6.14. (Representation theorem of Fréchet and Riesz)
Each bounded linear functional $A : \mathcal{H} \to \mathbb{C}$*, defined on a Hilbert space* $\mathcal{H}$*, can be represented in the form*

$$A(f) = (g, f) \qquad \text{for all} \quad f \in \mathcal{H},$$

with a generating element $g \in \mathcal{H}$ *which is uniquely determined.*

Proof:

1. At first, we show the uniqueness. Let $f \in \mathcal{H}$ and $g_1, g_2 \in \mathcal{H}$ denote two generating elements. Then we see

   $$A(f) = (g_1, f) = (g_2, f) \qquad \text{for all} \quad f \in \mathcal{H}.$$

   We subtract these equations and obtain

   $$(g_1, f) - (g_2, f) = (g_1 - g_2, f) = 0 \qquad \text{for all} \quad f \in \mathcal{H}.$$

   When we choose $f = g_1 - g_2$, we infer $g_1 = g_2$ on account of

   $$0 = (g_1 - g_2, g_1 - g_2) = \|g_1 - g_2\|^2.$$

2. In order to prove the existence of g, we consider

   $$\mathcal{M} := \Big\{ f \in \mathcal{H} \; : \; A(f) = 0 \Big\} \subset \mathcal{H}$$

   representing a closed linear subspace of $\mathcal{H}$.
   i.) Let $\mathcal{M} = \mathcal{H}$ be satisfied. Then we set $g = 0 \in \mathcal{H}$ and obtain the identity

   $$A(f) = (g, f) = 0 \qquad \text{for all} \quad f \in \mathcal{H}.$$

ii.) Let $\mathcal{M}\subsetneqq\mathcal{H}$ be satisfied. We invoke the theorem of the orthogonal projection and see $\mathcal{H}=\mathcal{M}\oplus\mathcal{M}^{\perp}$ with $\{0\}\neq\mathcal{M}^{\perp}$. Consequently, there exists an element $h\in\mathcal{M}^{\perp}$ with $h\neq 0$. We now determine a number $\alpha\in\mathbb{C}$, such that the identity $A(h)=(g,h)$ for $g=\alpha h$ is correct. This is equivalent to

$$A(h)=(g,h)=(\alpha h,h)=\overline{\alpha}\,(h,h)=\overline{\alpha}\,\|h\|^2$$

and

$$g=\frac{\overline{A(h)}}{\|h\|^2}\,h.$$

Now the identity $A(f)=(g,f)$ is valid for all $f\in\mathcal{M}$ and for $f=h$. When $f\in\mathcal{H}$ is arbitrary, we define $c:=\frac{A(f)}{A(h)}$. With $\widetilde{f}:=f-ch$, we obtain

$$A(\widetilde{f})=A(f)-cA(h)=A(f)-\frac{A(f)}{A(h)}\,A(h)=0$$

and consequently $\widetilde{f}\in\mathcal{M}$. Therefore, we have the representation

$$f=\widetilde{f}+ch \quad \text{for} \quad f\in\mathcal{H}, \quad \text{where} \quad \widetilde{f}\in\mathcal{M} \quad \text{and} \quad ch\in\mathcal{M}^{\perp}.$$

This implies

$$A(f)=A(\widetilde{f})+cA(h)=(g,\widetilde{f})+c(g,h)=(g,\widetilde{f}+ch)=(g,f)$$

for all $f\in\mathcal{H}$. q.e.d.

Definition 6.15. *We name a Banach space* separable *if a sequence* $\{f_k\}\subset\mathcal{B}$ *exists, which lies densely in* $\mathcal{B}$. *More precisely, we find an index* $k\in\mathbb{N}$ *to each element* $f\in\mathcal{B}$ *and every* $\varepsilon>0$ *such that* $\|f-f_k\|<\varepsilon$ *holds true.*

Definition 6.16. *In a pre-Hilbert-space* $\mathcal{H}'$, *we name the denumerably infinite many elements* $\{\varphi_1,\varphi_2,\ldots\}\subset\mathcal{H}'$ orthonormal *if*

$$(\varphi_i,\varphi_j)=\delta_{ij} \quad \textit{for all} \quad i,j\in\mathbb{N}$$

is valid.

Remark: When we have the system of denumerably many linearly independent elements in $\mathcal{H}'$, we can apply the *orthonormalization procedure of E. Schmidt* in order to transfer this into an orthonormal system.

Here we start with the linearly independent elements $\{f_1,\ldots,f_N\}\subset\mathcal{H}'$ of the pre-Hilbert-space $\mathcal{H}'$. Then we define inductively

$$\varphi_1:=\frac{1}{\|f_1\|}\,f_1, \quad \varphi_2:=\frac{f_2-(\varphi_1,f_2)\varphi_1}{\|f_2-(\varphi_1,f_2)\varphi_1\|},\ldots\varphi_N:=\frac{f_N-\sum\limits_{j=1}^{N-1}(\varphi_j,f_N)\varphi_j}{\left\|f_N-\sum\limits_{j=1}^{N-1}(\varphi_j,f_N)\varphi_j\right\|}.$$

The vector spaces spanned by $\{f_1, \ldots, f_N\}$ and $\{\varphi_1, \ldots, \varphi_N\}$ coincide, and we note that

$$(\varphi_i, \varphi_j) = \delta_{ij} \quad \text{for} \quad i, j = 1, \ldots, N.$$

Proposition 6.17. *Let $\{\varphi_k\}$ with $k = 1, \ldots, N$ represent a system of orthonormal elements in the pre-Hilbert-space $\mathcal{H}'$ and assume $f \in \mathcal{H}'$. Then we have the identity*

$$\Big\| f - \sum_{k=1}^{N} c_k \varphi_k \Big\|^2 = \Big\| f - \sum_{k=1}^{N} (\varphi_k, f) \varphi_k \Big\|^2 + \sum_{k=1}^{N} |c_k - (\varphi_k, f)|^2$$

for all numbers $c_1, \ldots, c_N \in \mathbb{C}$.

Proof: At first, we define

$$g := f - \sum_{k=1}^{N} (\varphi_k, f) \varphi_k, \quad h := \sum_{k=1}^{N} \Big((\varphi_k, f) - c_k \Big) \varphi_k.$$

Then we deduce the equation

$$f - \sum_{k=1}^{N} c_k \varphi_k = f - \sum_{k=1}^{N} (\varphi_k, f) \varphi_k + \sum_{k=1}^{N} \Big((\varphi_k, f) - c_k \Big) \varphi_k = g + h.$$

Now we evaluate

$$\begin{aligned} (g, h) &= \left(f - \sum_{k=1}^{N} (\varphi_k, f) \varphi_k \, , \sum_{l=1}^{N} \Big((\varphi_l, f) - c_l \Big) \varphi_l \right) \\ &= \sum_{l=1}^{N} \Big((\varphi_l, f) - c_l \Big) \overline{(\varphi_l, f)} - \sum_{k,l=1}^{N} \overline{(\varphi_k, f)} \Big((\varphi_l, f) - c_l \Big) (\varphi_k, \varphi_l). \end{aligned}$$

We note that $(\varphi_k, \varphi_l) = \delta_{kl}$ and obtain $(g, h) = 0$. This implies

$$\begin{aligned} \Big\| f - \sum_{k=1}^{N} c_k \varphi_k \Big\|^2 &= (g + h, g + h) = \|g\|^2 + \|h\|^2 \\ &= \Big\| f - \sum_{k=1}^{N} (\varphi_k, f) \varphi_k \Big\|^2 + \sum_{k,l=1}^{N} \overline{\Big((\varphi_k, f) - c_k \Big)} \Big((\varphi_l, f) - c_l \Big) (\varphi_k, \varphi_l) \\ &= \Big\| f - \sum_{k=1}^{N} (\varphi_k, f) \varphi_k \Big\|^2 + \sum_{k=1}^{N} |(\varphi_k, f) - c_k|^2. \end{aligned}$$

q.e.d.

Corollary: For all numbers $c_1, \ldots, c_N \in \mathbb{C}$ we have

$$\left\| f - \sum_{k=1}^{N} c_k \varphi_k \right\|^2 \geq \left\| f - \sum_{k=1}^{N} (\varphi_k, f) \varphi_k \right\|^2,$$

and equality is attained only if $c_k = (\varphi_k, f)$ for $k = 1, \ldots, N$ holds true. We name these numbers c_k the *Fourier coefficients* of f (with respect to the system (φ_k)).

When we set $c_1 = \ldots = c_N = 0$, we obtain

Proposition 6.18. *The following relation*

$$\left\| f - \sum_{k=1}^{N} (\varphi_k, f) \varphi_k \right\|^2 = \| f \|^2 - \sum_{k=1}^{N} |(\varphi_k, f)|^2 \geq 0$$

holds true.

From the last proposition we immediately infer

Theorem 6.19. *Let $\{\varphi_k\}$, $k = 1, 2, \ldots$ represent an orthonormal system in the pre-Hilbert-space $\mathcal{H}'$. For all elements $f \in \mathcal{H}'$,* Bessel's inequality

$$\sum_{k=1}^{\infty} |(\varphi_k, f)|^2 \leq \| f \|^2$$

holds true. An element $f \in \mathcal{H}'$ satisfies the equation

$$\sum_{k=1}^{\infty} |(\varphi_k, f)|^2 = \| f \|^2$$

if and only if the limit relation

$$\lim_{N \to \infty} \left\| f - \sum_{k=1}^{N} (\varphi_k, f) \varphi_k \right\| = 0$$

is valid.

Remark: The last statement means that $f \in \mathcal{H}'$ can be represented by its *Fourier series*

$$\sum_{k=1}^{\infty} (\varphi_k, f) \varphi_k$$

with respect to the Hilbert-space-norm $\| \cdot \|$.

Definition 6.20. *We say that an orthonormal system $\{\varphi_k\}$ is* complete *- we abbreviate this as c.o.n.s - if each element $f \in \mathcal{H}'$ of the pre-Hilbert-space $\mathcal{H}'$ satisfies the* completeness relation

$$\| f \|^2 = \sum_{k=1}^{\infty} |(\varphi_k, f)|^2.$$

Remarks:

1. In Section 4 and Section 5 of Chapter 5, we shall present explicit c.o.n.s. with the classical Fourier series and the spherical harmonic functions. More profound results are contained in Chapter 8 about Linear Operators in Hilbert Spaces.
2. With the aid of E. Schmidt's orthonormalization procedure, we can construct a complete orthonormal system in each separable Hilbert space.
3. When we have a complete orthonormal system $\{\varphi_k\} \subset \mathcal{H}'$ with $k = 1, 2, \dots$ in the pre-Hilbert-space $\mathcal{H}'$, the representation via the Fourier series
$$f = \sum_{k=1}^{\infty} (\varphi_k, f)\varphi_k$$
holds true with respect to convergence in the Hilbert-space-norm. The interesting question remains open, whether a Fourier series converges pointwise or even uniformly (see e.g. H. Heuser: *Analysis II.* B. G. Teubner-Verlag, Stuttgart, 1992).

7 The Lebesgue Spaces $L^p(X)$

Now we continue our considerations from Section 1 to Section 4. We assume $n \in \mathbb{N}$ as usual, and we consider subsets $X \subset \mathbb{R}^n$ which we endow with the *relative topology* of the Euclidean space $\mathbb{R}^n$ as follows:

$$A \subset X \text{ is} \left\{ \begin{array}{c} \text{open} \\ \text{closed} \end{array} \right\}$$
$$\Longleftrightarrow \quad \text{There exists } B \subset \mathbb{R}^n \left\{ \begin{array}{c} \text{open} \\ \text{closed} \end{array} \right\} \text{with } A = B \cap X.$$

By the symbol $M(X)$ we denote a linear space of continuous functions $f : X \to \overline{\mathbb{R}} = \mathbb{R} \cup \{\pm\infty\}$ with the following properties:

(M1) *Linearity:* With $f, g \in M(X)$ and $\alpha, \beta \in \mathbb{R}$ we have $\alpha f + \beta g \in M(X)$.
(M2) *Lattice property:* From $f \in M(X)$ we infer $|f| \in M(X)$.
(M3) *Global property:* The function $f(x) \equiv 1$, $x \in X$ belongs to $M(X)$.

We name a linear functional $I : M \to \mathbb{R}$, which is defined on $M = M(X)$, *Daniell's integral* if the following properties are valid:

(D1) *Linearity:* $I(\alpha f + \beta g) = \alpha I(f) + \beta I(g)$ for all $f, g \in M$ and $\alpha, \beta \in \mathbb{R}$;
(D2) *Nonnegativity:* $I(f) \geq 0$ for all $f \in M$ with $f \geq 0$;
(D3) *Monotone continuity:* For all $\{f_k\} \subset M(X)$ with $f_k(x) \downarrow 0$ $(k \to \infty)$ on X we infer $I(f_k) \to 0$ $(k \to \infty)$.

Example 7.1. Let $X = \Omega \subset \mathbb{R}^n$ denote an open bounded set, and we define the linear space

$$M = M(X) := \left\{ f : X \to \mathbb{R} \in C^0(X) : \int_{\Omega} |f(x)|\,dx < +\infty \right\}.$$

We utilize the improper Riemannian integral on the set X, namely

$$I(f) := \int_{\Omega} f(x)\,dx, \qquad f \in M$$

as our linear functional.

Example 7.2. On the sphere $X = S^{n-1} := \left\{ x \in \mathbb{R}^n \ : \ |x| = 1 \right\}$, we consider the linear space of all continuous functions $M(X) = C^0(S^{n-1})$, and we introduce the Daniell integral

$$I(f) := \int_{S^{n-1}} f(x)\,d\sigma^{n-1}(x), \qquad f \in M.$$

In Section 2 we have extended the functional I from $M(X)$ onto the space $L(X)$ of the *Lebesgue integrable functions.* In Section 3 we investigated sets which are Lebesgue measurable, more precisely those sets A whose characteristic functions χ_A are Lebesgue integrable.

Definition 7.3. *Let the exponent satisfy* $1 \le p < +\infty$. *We name a measurable function* $f : X \to \overline{\mathbb{R}}$ p-times integrable *if* $|f|^p \in L(X)$ *is correct. In this case we write* $f \in L^p(X)$. *With*

$$\|f\|_p := \|f\|_{L^p(X)} := \left(\int_X |f(x)|^p\,d\mu(x) \right)^{\frac{1}{p}} = \left(I(|f|^p) \right)^{\frac{1}{p}}$$

we obtain the L^p-norm *of the function* $f \in L^p(X)$; *here the symbol* μ *denotes the* Lebesgue measure *on* X.

Remark: Evidently, we have the identity $L^1(X) = L(X)$.

The central tool, when dealing with Lebesgue spaces, is provided by the subsequent result.

Theorem 7.4. (Hölder's inequality)
Let the exponents $p, q \in (1, +\infty)$ *be conjugate, which means* $p^{-1} + q^{-1} = 1$ *holds true. Furthermore, we assume* $f \in L^p(X)$ *and* $g \in L^q(X)$ *being given. Then we infer the property* $fg \in L^1(X)$ *and the inequality*

$$\|fg\|_{L^1(X)} \le \|f\|_{L^p(X)} \|g\|_{L^q(X)}.$$

Proof: We have to investigate only the case $\|f\|_p > 0$ and $\|g\|_q > 0$. Alternatively, we had $\|f\|_p = 0$, and consequently $f = 0$ a.e. as well as $f \cdot g = 0$ a.e. would hold. Analogously, we treat the case $\|g\|_q = 0$. Then we apply *Young's inequality*

$$ab \leq \frac{a^p}{p} + \frac{b^q}{q}$$

to the functions

$$\varphi(x) = \frac{1}{\|f\|_p}|f(x)|, \quad \psi(x) = \frac{1}{\|g\|_q}|g(x)|, \qquad x \in X,$$

and we obtain

$$\frac{1}{\|f\|_p\|g\|_q}\,|f(x)g(x)| = \varphi(x)\psi(x) \leq \frac{1}{p}\,\frac{|f(x)|^p}{\|f\|_p^p} + \frac{1}{q}\,\frac{|g(x)|^q}{\|g\|_q^q}$$

for all points $x \in X$. Theorem 4.8 from Section 4 implies $fg \in L(X) = L^1(X)$. Now integration yields the inequality

$$\frac{1}{\|f\|_p\|g\|_q} I(|fg|) \leq \frac{1}{p}\,\frac{1}{\|f\|_p^p}\,I(|f|^p) + \frac{1}{q}\,\frac{1}{\|g\|_q^q}\,I(|g|^q) = 1,$$

and finally

$$I(|fg|) \leq \|f\|_p\|g\|_q.$$

q.e.d.

Theorem 7.5. (Minkowski's inequality)
With the exponent $p \in [1, +\infty)$, let us consider the functions $f, g \in L^p(X)$. Then we infer $f + g \in L^p(X)$ and we have

$$\|f + g\|_{L^p(X)} \leq \|f\|_{L^p(X)} + \|g\|_{L^p(X)}.$$

Proof: The case $p = 1$ can be easily derived by application of the triangle inequality on the integrand $|f + g|$. Therefore, we assume $p, q \in (1, +\infty)$ with $p^{-1} + q^{-1} = 1$. At first, convexity arguments yield

$$|f(x) + g(x)|^p \leq 2^{p-1}\left(|f(x)|^p + |g(x)|^p\right)$$

and consequently $f + g \in L^p$ or equivalently $I(|f + g|^p) < +\infty$. Now we have

$$\begin{aligned} |f(x) + g(x)|^p &= |f(x) + g(x)|^{p-1}|f(x) + g(x)| \\ &\leq |f(x) + g(x)|^{p-1}|f(x)| + |f(x) + g(x)|^{p-1}|g(x)| \\ &= |f(x) + g(x)|^{\frac{p}{q}}|f(x)| + |f(x) + g(x)|^{\frac{p}{q}}|g(x)|. \end{aligned}$$

The factors of the summands on the right-hand side are L^q- and L^p-functions, respectively. Therefore, we obtain

$$I(|f+g|^p) \le I(|f+g|^p)^{\frac{1}{q}}(\|f\|_p + \|g\|_p).$$

Finally, we see

$$(I(|f+g|^p)^{\frac{1}{p}} \le \|f\|_p + \|g\|_p$$

and the desired inequality

$$\|f+g\|_p \le \|f\|_p + \|g\|_p.$$

q.e.d.

Remark: Minkowski's inequality represents the triangle inequality for the $\|\cdot\|_p$-norm in the space L^p.

The following result guarantees the completeness of L^p-spaces, which means: Each Cauchy sequence converges towards a function in the respective space.

Theorem 7.6. (Fischer, Riesz)
Let us consider the exponent $p \in [1, +\infty)$ and a sequence $\{f_k\}_{k=1,2,\ldots} \subset L^p(X)$ satisfying

$$\lim_{k,l\to\infty} \|f_k - f_l\|_{L^p(X)} = 0.$$

Then we have a function $f \in L^p(X)$ with the property

$$\lim_{k\to\infty} \|f_k - f\|_{L^p(X)} = 0.$$

Proof: With the aid of Hölder's inequality we show the identity

$$\lim_{k,l\to\infty} I(|f_k - f_l|) = 0.$$

Here we estimate in the case $p > 1$ as follows:

$$I(|f_k - f_l|) = I(|f_k - f_l| \cdot 1) \le \|f_k - f_l\|_p \|1\|_q \longrightarrow 0.$$

The Lebesgue selection theorem gives us a subsequence $k_1 < k_2 < k_3 < \ldots$ and a null-set $N \subset X$, such that

$$\lim_{m\to\infty} f_{k_m}(x) = f(x), \qquad x \in X \setminus N$$

holds true. We observe that the function f is measurable. Now we choose $l \ge N(\varepsilon)$ and $k_m \ge N(\varepsilon)$, where $\|f_k - f_l\|_p \le \varepsilon$ for all $k, l \ge N(\varepsilon)$ is valid, and we infer

$$I(|f_{k_m} - f_l|^p) = \|f_{k_m} - f_l\|^p_{L^p(X)} \le \varepsilon^p.$$

For $m \to \infty$, Fatou's theorem implies the inequality

$$I(|f - f_l|^p) \le \varepsilon^p \qquad \text{for all} \quad l \ge N(\varepsilon)$$

and consequently

$$\|f - f_l\|_{L^p(X)} \le \varepsilon \qquad \text{for all} \quad l \ge N(\varepsilon).$$

Since $L^p(X)$ is linear and f_l as well as $(f - f_l)$ belong to this space, we infer $f \in L^p(X)$. Furthermore, we observe

$$\lim_{l\to\infty} \|f - f_l\|_p = 0.$$

q.e.d.

Definition 7.7. *A measurable function $f : X \to \overline{\mathbb{R}}$ belongs to the class $L^\infty(X)$ if we have a null-set $N \subset X$ and a constant $c \in [0, +\infty)$ with the property*

$$|f(x)| \le c \qquad \textit{for all} \quad x \in X \setminus N.$$

We name

$$\begin{aligned}\|f\|_\infty = \|f\|_{L^\infty(X)} &= \operatorname*{ess\,sup}_{x\in X} |f(x)| \\ &= \inf\left\{c \ge 0 : \begin{array}{l}\textit{There exists a null-set } N \subset X \\ \textit{with } |f(x)| \le c \textit{ for all } x \in X \setminus N\end{array}\right\}\end{aligned}$$

the L^∞-norm *or equivalently the* essential supremum *of the function f.*

Remark: Evidently, we have the inclusion

$$L^\infty(X) \subset \bigcap_{p\in[1,+\infty)} L^p(X).$$

Theorem 7.8. *A function $f \in \bigcap\limits_{p\ge 1} L^p(X)$ belongs to the class $L^\infty(X)$, if the condition*

$$\limsup_{p\to\infty} \|f\|_{L^p(X)} < +\infty$$

is correct. In this case we have

$$\|f\|_{L^\infty(X)} = \lim_{p\to\infty} \|f\|_{L^p(X)} < +\infty,$$

where the limit on the right-hand side exists.

Proof: Let $f \in \bigcap\limits_{p\ge 1} L^p(X)$ hold true. When we assume $f \in L^\infty(X)$, we infer $0 \le \|f\|_\infty < +\infty$ as well as

$$|f|^p = |f|^q |f|^{p-q} \le |f|^q \|f\|_\infty^{p-q} \quad \text{a.e. on} \quad X.$$

Therefore, we obtain

$$\|f\|_p \le \|f\|_\infty^{1-\frac{q}{p}} \|f\|_q^{\frac{q}{p}}$$

and finally

$$\limsup_{p\to\infty} \|f\|_p \le \|f\|_\infty < +\infty. \tag{1}$$

In order to show the inverse direction, we consider the set

$$A_a := \Big\{x \in X \,:\, |f(x)| > a\Big\}$$

for an arbitrary number $a < \|f\|_\infty$. Therefore, A_a does not constitute a null-set. We obtain the estimate

$$\begin{aligned} +\infty > \limsup_{p\to\infty} \|f\|_p \;&\ge\; \liminf_{p\to\infty} \|f\|_p \\ = \liminf_{p\to\infty} \Big(I(|f|^p)\Big)^{\frac{1}{p}} \;&\ge\; a \liminf_{p\to\infty} \Big(\mu(A_a)\Big)^{\frac{1}{p}} \;=\; a. \end{aligned}$$

Now we infer

$$+\infty > \liminf_{p\to\infty} \|f\|_p \ge \|f\|_\infty \tag{2}$$

and consequently $f \in L^\infty(X)$. These inequalities immediately imply the existence of

$$\lim_{p\to\infty} \|f\|_p = \|f\|_\infty.$$

q.e.d.

Corollary: Hölder's inequality remains valid for the case $p = 1$ and $q = \infty$. Furthermore, Minkowski's inequality holds true in the case $p = \infty$ as well.

Definition 7.9. *Let $1 \le p \le +\infty$ be satisfied. Then we introduce an equivalence relation on the space $L^p(X)$ as follows:*

$$f \sim g \quad \Longleftrightarrow \quad f(x) = g(x) \;\; a.e. \text{ in } X.$$

By the symbol $[f]$ we denote the equivalence class belonging to the element $f \in L^p(X)$. We name

$$\mathcal{L}^p(X) := \Big\{[f] \,:\, f \in L^p(X)\Big\}$$

the Lebesgue space of order $1 \le p \le +\infty$.

We summarize our considerations to the subsequent

Theorem 7.10. *For each fixed p with $1 \le p \le +\infty$, the Lebesgue space $\mathcal{L}^p(X)$ constitutes a real Banach space with the given L^p-norm. Furthermore, we have the inclusion*

$$\mathcal{L}^r(X) \supset \mathcal{L}^s(X)$$

for all $1 \le r < s \le +\infty$. Moreover, the estimate

$$\|f\|_{\mathcal{L}^r(X)} \le C(r,s)\|f\|_{\mathcal{L}^s(X)} \qquad \text{for all} \quad f \in \mathcal{L}^s(X)$$

holds true with a constant $C(r,s) \in [0,+\infty)$. This means, the mapping for the embedding

$$\Phi : \mathcal{L}^s(X) \longrightarrow \mathcal{L}^r(X), \quad f \mapsto \Phi(f) = f$$

is continuous. Therefore, a sequence converging in the space $\mathcal{L}^s(X)$ is convergent in the space $\mathcal{L}^r(X)$ as well.

Proof:

1. At first, we show that $\mathcal{L}^p(X)$ constitute normed spaces. Let us consider $[f] \in \mathcal{L}^p(X)$: We have $\|[f]\|_p = 0$ if and only if $\|f\|_p = 0$ and consequently $f = 0$ a.e. in X is fulfilled. This implies $[f] = 0$ and gives us the norm property (N1). Minkowski's inequality from Theorem 7.5 ascertains the norm property (N2), where Theorem 7.8 provides the triangle inequality in the space $L^\infty(X)$. The norm property (N3), namely the homogeneity, is obvious.
2. The Fischer-Riesz theorem implies completeness of the spaces $\mathcal{L}^p$ for $1 \le p < +\infty$. Therefore, only completeness of the space $\mathcal{L}^\infty$ has to be shown. Here we consider a Cauchy sequence $\{f_k\} \subset L^\infty$ satisfying

$$\|f_k - f_l\|_\infty \to 0 \qquad \text{for} \quad k,l \to \infty.$$

We infer the inequality $\|f_k\|_\infty \le c$ for all $k \in \mathbb{N}$, with a constant $c \in (0,+\infty)$. Then we find a null-set $N_0 \subset X$ with $|f_k(x)| \le c$ for all points $x \in X \setminus N_0$ and all indices $k \in \mathbb{N}$. Furthermore, we have null-sets $N_{k,l}$ with

$$|f_k(x) - f_l(x)| \le \|f_k - f_l\|_\infty \quad \text{for} \quad x \in X \setminus N_{k,l}.$$

We define

$$N := N_0 \cup \bigcup_{k,l} N_{k,l}$$

and observe

$$\lim_{k,l\to\infty} \sup_{x \in X\setminus N} |f_k(x) - f_l(x)| = 0.$$

When we introduce the function

$$f(x) := \begin{cases} \lim\limits_{k\to\infty} f_k(x) \,, & x \in X \setminus N \\ \qquad 0 & , \; x \in N \end{cases} \in L^\infty(X)$$

we infer

$$\lim_{k\to\infty} \sup_{x \in X\setminus N} |f_k(x) - f(x)| = 0$$

and finally

$$\lim_{k\to\infty} \|f_k - f\|_{L^\infty(X)} = 0.$$

3. Let us assume $1 \le r < s \le +\infty$. The function $f \in L^s(X)$ satisfies

$$\|f\|_r = \Big(I(|f|^r \cdot 1)\Big)^{\frac{1}{r}} \le \Big\{\Big(I(|f|^s)\Big)^{\frac{r}{s}} \Big(\mu(X)\Big)^{\frac{s-r}{s}}\Big\}^{\frac{1}{r}} = \Big(\mu(X)\Big)^{\frac{s-r}{rs}} \|f\|_s$$

for all elements $f \in L^s(X)$. q.e.d.

Definition 7.11. *Let $\mathcal{B}_1$ and $\mathcal{B}_2$ denote two Banach spaces with $\mathcal{B}_1 \subset \mathcal{B}_2$. Then we say $\mathcal{B}_1$* is continuously embedded into *$\mathcal{B}_2$ if the mapping*

$$I_1 : \mathcal{B}_1 \longrightarrow \mathcal{B}_2, \quad f \mapsto I_1(f) = f$$

is continuous. This means, the inequality

$$\|f\|_{\mathcal{B}_2} \le c\,\|f\|_{\mathcal{B}_1} \quad \textit{for all} \quad f \in \mathcal{B}_1$$

holds true with a constant $c \in [0, +\infty)$. Then we use the notation $\mathcal{B}_1 \hookrightarrow \mathcal{B}_2$.

Remarks:

1. The transition to equivalence classes will be made tacitly - such that we can identify $\mathcal{L}^p(X)$ and $L^p(X)$.
2. We have the embedding $\mathcal{L}^s(X) \hookrightarrow \mathcal{L}^r(X)$ for all $1 \le r \le s \le +\infty$.
3. On the space $C^0(X)$, we obtain with

$$\|f\|_0 := \sup_{x \in X} |f(x)|, \qquad f \in C^0(X)$$

the *supremum-norm* which induces uniform convergence. With the L^p-norms $\|\cdot\|_p$ for $1 \le p \le +\infty$, we have constructed a family of norms which constitute a continuum beginning with the weakest norm, namely the L^1-norm, and ending with the strongest norm, namely the L^∞-norm or the C^0-norm, respectively. Exactly in the centrum for $p = 2$, we find the Hilbert space $\mathcal{H} = L^2(X)$.

Example 7.12. Let the space

$$\mathcal{H} = L^2(X, \mathbb{C}) := \Big\{f = g + ih \;:\; g, h \in L^2(X, \mathbb{R})\Big\}$$

be endowed with the inner product

$$(f_1, f_2)_{\mathcal{H}} := I(\overline{f_1} f_2) \quad \text{for} \quad f_j = g_j + ih_j \in \mathcal{H} \quad \text{and} \quad j = 1, 2.$$

Here we define $I(f) = I(g + ih) := I(g) + i\,I(h)$. Then $\mathcal{H}$ represents a Hilbert space.

In the sequel, we use the space of functions

$$M^\infty(X) := \Big\{f \in M(X) \;:\; \sup_{x \in X} |f(x)| < +\infty\Big\} = M(X) \cap L^\infty(X).$$

Theorem 7.13. (Approximation of L^p-functions)
Given the exponent $p \in [1, +\infty)$, the space $M^\infty(X)$ lies densely in $L^p(X)$, which means: For each function $f \in L^p(X)$ and each $\varepsilon > 0$, we have a function $f_\varepsilon \in M^\infty(X)$ satisfying

$$\|f - f_\varepsilon\|_{L^p(X)} < \varepsilon.$$

Proof: Let $\varepsilon > 0$ be given. We choose $K > 0$ and consider the truncated function

$$f_{-K,+K}(x) := \begin{cases} f(x), \; x \in X \text{ with } |f(x)| \leq K \\ -K, \;\; x \in X \text{ with } f(x) \leq -K \\ +K, \;\; x \in X \text{ with } f(x) \geq +K \end{cases}$$

subject to the inequality

$$|f(x) - f_{-K,+K}(x)|^p \leq |f(x)|^p.$$

Furthermore, we have

$$\lim_{K \to \infty} |f(x) - f_{-K,+K}(x)|^p = 0$$

almost everywhere in X. Lebesgue's convergence theorem implies

$$\lim_{K \to \infty} I(|f - f_{-K,+K}|^p) = 0,$$

and we find a number $K = K(\varepsilon) > 0$ with

$$\|f(x) - f_{-K,+K}(x)\|_p \leq \frac{\varepsilon}{2}.$$

According to Theorem 4.14 in Section 4, the function $f_{-K,+K}$ possesses a sequence $\{\varphi_k\}_{k=1,2,\ldots} \subset M(X)$ with $|\varphi_k(x)| \leq K$ satisfying

$$\varphi_k(x) \longrightarrow f_{-K,+K}(x) \qquad \text{a.e. in } X.$$

The Lebesgue convergence theorem yields

$$\|f_{-K,+K} - \varphi_k\|_p^p = I(|f_{-K,+K} - \varphi_k|^p) \longrightarrow 0$$

for $k \to \infty$. Consequently, we find an index $k = k(\varepsilon)$ with

$$\|f_{-K,+K} - \varphi_k\|_p \leq \frac{\varepsilon}{2}.$$

The function $f_\varepsilon := \varphi_{k(\varepsilon)} \in M(X)$, which is uniformly bounded by $K(\varepsilon)$ on X, satisfies

$$\|f - f_\varepsilon\|_p \leq \|f - f_{-K,+K}\|_p + \|f_{-K,+K} - \varphi_{k(\varepsilon)}\|_p \leq \frac{\varepsilon}{2} + \frac{\varepsilon}{2} = \varepsilon.$$

q.e.d.

Theorem 7.14. (Separability of L^p-spaces)
Let $X \subset \mathbb{R}^n$ be an open bounded set and $p \in [1,+\infty)$ the exponent given. Then the Banach space $L^p(X)$ is separable: More precisely, there exists a sequence of functions $\{\varphi_k(x)\}_{k=1,2,\ldots} \subset C_0^\infty(X) \subset L^p(X)$ which lies densely in $L^p(X)$.

Proof: Let us consider the set

$$\mathcal{R} := \left\{ g(x) = \sum_{i_1,\ldots,i_n=0}^{N} a_{i_1\ldots i_n} x_1^{i_1}\ldots x_n^{i_n} \;:\; a_{i_1\ldots i_n} \in \mathbb{Q},\; N \in \mathbb{N}\cup\{0\} \right\}$$

of polynomials in $\mathbb{R}^n$ with rational coefficients. Furthermore, let

$$\chi_j(x) : X \longrightarrow \mathbb{R} \in C_0^\infty(X), \qquad j = 1,2,\ldots$$

denote an exhausting sequence for the set X, which means

$$\chi_j(x) \le \chi_{j+1}(x), \quad \lim_{j\to\infty} \chi_j(x) = 1 \quad \text{for all} \quad x \in X.$$

Now we show that the denumerable set

$$\mathcal{D}(X) := \Big\{ h(x) = \chi_j(x) g(x) \;:\; j \in \mathbb{N},\; g \in \mathcal{R} \Big\}$$

lies densely in $L^p(X)$. Here we take the function $f \in L^p(X)$ and the quantity $\varepsilon > 0$ arbitrarily. Then we find a function $g \in M^\infty(X)$ with $\|f-g\|_p \le \varepsilon$. Now we infer

$$\begin{aligned}\|g-\chi_j g\|_p^p &= \int_X |g(x)-\chi_j(x)g(x)|^p \, d\mu(x)\\ &= \int_X \Big(1-\chi_j(x)\Big)^p |g(x)|^p \, d\mu(x) \longrightarrow 0,\end{aligned}$$

and consequently we find an index $j \in \mathbb{N}$ satisfying $\|g-\chi_j g\|_p \le \varepsilon$. Now the function $\chi_j g$ has compact support in X. Via the Weierstraß approximation theorem, there exists a polynomial $h(x) \in \mathcal{R}$ such that $\sup_{x\in X} \chi_j |g-h| \le \delta(\varepsilon)$ is correct - with a quantity $\delta(\varepsilon) > 0$ given. Consequently, we find a polynomial $h(x) \in \mathcal{R}$ with the property

$$\|\chi_j g - \chi_j h\|_p \le \varepsilon.$$

This implies

$$\|f-\chi_j h\|_p \le \|f-g\|_p + \|g-\chi_j g\|_p + \|\chi_j g - \chi_j h\|_p \le 3\varepsilon.$$

Consequently, $\mathcal{D}(X)$ lies densely in $L^p(X)$. q.e.d.

8 Bounded Linear Functionals on $L^p(X)$ and Weak Convergence

We begin with

Theorem 8.1. (Extension of linear functionals)
Take $p \in [1, +\infty)$ and let $A : M^\infty(X) \to \mathbb{R}$ denote a linear functional with the following property: We have a constant $\alpha \in [0, +\infty)$ such that

$$|A(f)| \leq \alpha \|f\|_{L^p(X)} \qquad \text{for all} \quad f \in M^\infty(X)$$

holds true. Then there exists exactly one bounded linear functional $\widehat{A}$: $L^p(X) \to \mathbb{R}$ satisfying

$$\|\widehat{A}\| \leq \alpha \quad \text{and} \quad \widehat{A}(f) = A(f) \quad \text{for all} \quad f \in M^\infty(X).$$

Consequently, the functional $\widehat{A}$ can be uniquely continued from $M^\infty(X)$ onto $L^p(X)$.

Proof: The linear functional A is bounded on $\{M^\infty(X), \|\cdot\|_{L^p(X)}\}$ and therefore continuous. According to Theorem 7.13 from Section 7, each element $f \in L^p(X)$ possesses a sequence $\{f_k\}_{k=1,2,\ldots} \subset M^\infty(X)$ satisfying

$$\|f_k - f\|_{L^p(X)} \to 0 \quad \text{for} \quad k \to \infty.$$

Now we define

$$\widehat{A}(f) := \lim_{k\to\infty} A(f_k).$$

We immediately verify that $\widehat{A}$ has been defined independently of the sequence $\{f_k\}_{k=1,2,\ldots}$ chosen, and that the mapping $\widehat{A} : L^p(X) \to \mathbb{R}$ is linear. Furthermore, we have

$$\|\widehat{A}\| = \sup_{f\in L^p,\ \|f\|_p\leq 1} |\widehat{A}(f)| = \sup_{f\in M^\infty,\ \|f\|_p\leq 1} |A(f)| \leq \alpha.$$

When we consider with $\widehat{A}$ and $\widehat{B}$ two extensions of A onto $L^p(X)$, we infer $\widehat{A} = \widehat{B}$ on $M^\infty(X)$. Since the functionals $\widehat{A}$ and $\widehat{B}$ are continuous, and $M^\infty(X)$ lies densely in $L^p(X)$, we obtain the identity $\widehat{A} = \widehat{B}$ on $L^p(X)$.

q.e.d.

Now we consider *multiplication functionals* A_g as follows:

Theorem 8.2. *Let us choose the exponent $1 \leq p \leq +\infty$ and with $q \in [1, +\infty]$ its conjugate exponent satisfying*

$$\frac{1}{p} + \frac{1}{q} = 1.$$

For each function $g \in L^q(X)$ being given, the symbol $A_g : L^p(X) \to \mathbb{R}$ with

$$A_g(f) := I(fg), \qquad f \in L^p(X)$$

represents a bounded linear functional such that $\|A_g\| = \|g\|_q$ holds true.

Proof: Obviously, $A_g : L^p(X) \to \mathbb{R}$ constitutes a linear functional. Hölder's inequality yields the estimate

$$|A_g(f)| = |I(fg)| \leq I(|f||g|) \leq \|f\|_p \|g\|_q \quad \text{for all} \quad f \in L^p(X),$$

and we see

$$\|A_g\| \leq \|g\|_q.$$

In the case $1 < p < +\infty$, we choose the function

$$f(x) = |g(x)|^{\frac{q}{p}} \operatorname{sign} g(x)$$

and calculate

$$A_g(f) = I(fg) = I\left(|g|^{\frac{q}{p}+1}\right) = I(|g|^q)$$

$$= \|g\|_q^q = \|g\|_q \|g\|_q^{\frac{q}{p}} = \|g\|_q \Big(I(|f|^p)\Big)^{\frac{1}{p}} = \|g\|_q \|f\|_p.$$

This implies

$$\frac{A_g(f)}{\|f\|_p} = \|g\|_q \qquad \text{and therefore} \quad \|A_g\| \geq \|g\|_q \tag{1}$$

and consequently $\|A_g\| = \|g\|_q$ for all $1 < p < +\infty$. In the case $p = +\infty$, we choose

$$f(x) = \operatorname{sign} g(x)$$

and we obtain

$$A_g(f) = I(g \operatorname{sign} g) = I(|g|) = \|g\|_1 \, \|f\|_\infty.$$

This implies

$$\frac{A_g(f)}{\|f\|_\infty} = \|g\|_1 \qquad \text{and therefore} \quad \|A_g\| = \|g\|_1.$$

In the case $p = 1$, we choose the following function to the element $g \in L^q(X) = L^\infty(X)$ and for all quantities $\varepsilon > 0$, namely

$$f_\varepsilon(x) := \begin{cases} 1, & x \in X \text{ with } \; g(x) \geq \|g\|_\infty - \varepsilon \\ 0, & x \in X \text{ with } |g(x)| < \|g\|_\infty - \varepsilon \\ -1, & x \in X \text{ with } g(x) \leq -\|g\|_\infty + \varepsilon \end{cases}.$$

Therefore, we have

$$A_g(f_\varepsilon) = I(gf_\varepsilon) \geq (\|g\|_\infty - \varepsilon)\|f_\varepsilon\|_1 \qquad \text{for all} \quad \varepsilon > 0,$$

which reveals

$$\frac{A_g(f_\varepsilon)}{\|f_\varepsilon\|_1} \geq \|g\|_\infty - \varepsilon.$$

Consequently, $\|A_g\| \geq \|g\|_\infty - \varepsilon$ is correct and finally $\|A_g\| = \|g\|_\infty$. q.e.d.

We want to show that each bounded linear functional on $L^p(X)$ with $1 \leq p < \infty$ can be represented as a multiplication functional A_g via a generating element $g \in L^q(X)$, where $p^{-1} + q^{-1} = 1$ holds true.

Theorem 8.3. (Regularity in $L^p(X)$)
Let us consider $1 \leq p < +\infty$ and $g \in L^1(X)$. Furthermore, we have a constant $\alpha \in [0, +\infty)$ such that

$$|A_g(f)| = |I(fg)| \leq \alpha \|f\|_p \qquad \text{for all} \quad f \in M^\infty(X) \tag{2}$$

holds true. Then we infer the property $g \in L^q(X)$ and the estimate $\|g\|_q \leq \alpha$.

Proof:

1. At first, we deduce the following inequality from (2), namely

$$|I(fg)| \leq \alpha \|f\|_p \qquad \text{for all} \quad f \text{ measurable and bounded.} \tag{3}$$

According to Theorem 4.14 from Section 4, the bounded measurable function $f : X \to \mathbb{R}$ possesses a sequence of functions $\{f_k\}_{k=1,2,\ldots} \subset M^\infty(X)$ with

$$f_k(x) \to f(x) \qquad \text{a.e. in} \quad X$$

and

$$\sup_X |f_k(x)| \leq \sup_X |f(x)| =: c \in [0, +\infty).$$

Now Lebesgue's convergence theorem yields

$$|I(fg)| = \lim_{k \to \infty} |I(f_k g)| \leq \lim_{k \to \infty} \alpha \|f_k\|_p = \alpha \|f\|_p.$$

2. Let us assume $1 < p < +\infty$, at first. Then we consider the functions

$$g_k(x) := \begin{cases} g(x) \ , & x \in X \text{ with } |g(x)| \leq k \\ 0 \ , & x \in X \text{ with } |g(x)| > k \end{cases}.$$

Now the functions

$$f_k(x) = |g_k(x)|^{\frac{q}{p}} \operatorname{sign} g_k(x), \qquad x \in X,$$

are measurable and bounded. Consequently, we are allowed to insert $f_k(x)$ into (3) and obtain

$$I(f_k g) = I\Big(|g_k|^{\frac{q}{p}+1}\Big) = I(|g_k|^q) = \|g_k\|_q^q \ .$$

Then (3) implies

$$I(f_k g) \le \alpha \|f_k\|_p = \alpha (I(|g_k|^q))^{\frac{1}{p}} = \alpha \|g_k\|_q^{\frac{q}{p}}.$$

For $k = 1, 2, \ldots$ we have the estimate

$$\alpha \ge \|g_k\|_q^{q-\frac{q}{p}} = \|g_k\|_q, \quad \alpha^q \ge I(|g_k|^q).$$

We invoke Fatou's theorem and obtain

$$|g(x)|^q \overset{\text{a.e.}}{=} \liminf_{k\to\infty} |g_k(x)|^q \in L(X)$$

as well as

$$\alpha^q \ge I(|g|^q) \quad \text{and consequently} \quad \|g\|_q \le \alpha.$$

3. Now we assume $p = 1$. The quantity $\varepsilon > 0$ being given, we consider the set

$$E := \Big\{x \in X \,:\, |g(x)| \ge \alpha + \varepsilon\Big\}.$$

We insert the function $f = \chi_E \operatorname{sign} g$ into (3) and obtain

$$\alpha\mu(E) = \alpha\|f\|_1 \ge |I(fg)| \ge (\alpha + \varepsilon)\mu(E).$$

This implies $\mu(E) = 0$ for all $\varepsilon > 0$ and finally $\|g\|_\infty \le \alpha$. q.e.d.

Until now, we considered only one Daniell integral $I : M^\infty(X) \to \mathbb{R}$ as fixed, which we could extend onto the Lebesgue space $L^1(X)$. When a statement refers to this functional, we do not mention this functional I explicitly: We simplify $L^p(X) = L^p(X, I)$, for instance, or $f(x) = 0$ almost everywhere in X if and only if we have an I-null-set $N \subset X$ such that $f(x) = 0$ for all $x \in X \setminus N$ holds true. We already know that

$$M^\infty(X) \subset L^\infty(X) \subset L^p(X), \qquad 1 \le p \le +\infty$$

is correct. Additionally, we consider the Daniell integral J.

Definition 8.4. *We name a Daniell integral*

$$J \,:\, M^\infty(X) \longrightarrow \mathbb{R},$$

which satisfies the conditions (M1) to (M3) as well as (D1) to (D3) from Section 7 and is extendable onto $L^1(X, J) \supset L^\infty(X)$, as absolutely continuous with respect to I *if the following property is valid:*

(D4) Each I-null-set is a J-null-set.

With the aid of ideas of John v. Neumann (see L.H. Loomis: Abstract harmonic analysis), we prove the profound

Theorem 8.5. (Radon, Nikodym)
Let the Daniell integral J be absolutely continuous with respect to I. Then a uniquely determined function $g \in L^1(X)$ exists such that

$$J(f) = I(fg) \quad \textit{for all} \quad f \in M^\infty(X)$$

holds true.

Proof:

1. Let $f \in L^\infty(X)$ be given, then we have a null-set $N \subset X$ and a constant $c \in [0, +\infty)$ such that

$$|f(x)| \le c \qquad \text{for all} \quad x \in X \backslash N$$

is valid. We recall the property (D4), and see that N is a J-null-set as well, which implies $f \in L^\infty(X, J)$. A sequence $\{f_k\}_{k=1,2,\dots} \subset L^\infty(X)$ with $f_k \downarrow 0$ $(k \to \infty)$ a.e. on X fulfills the limit relation

$$f_k \downarrow 0 \quad \text{J-a.e. on} \quad X \quad \text{for} \quad k \to \infty$$

due to (D4). Now B.Levi's theorem on the space $L^1(X, J)$ yields

$$\lim_{k\to\infty} J(f_k) = 0.$$

Consequently, $J : L^\infty(X) \to \mathbb{R}$ represents a Daniell integral. Then we introduce the Daniell integral

$$K(f) := I(f) + J(f), \quad f \in L^\infty(X). \tag{4}$$

As in Section 2 we extend this functional onto the space $L^1(X, K)$; here the a.e.-properties are sufficient. We consider the inclusion $L^1(X, K) \supset L^p(X, K)$ for all $p \in [1, +\infty]$.

2. We take the exponents $p, q \in [1, +\infty]$ with $p^{-1} + q^{-1} = 1$ and obtain the following estimate for all $f \in M^\infty(X)$, namely

$$\begin{aligned} |J(f)| &\le J(|f|) \le K(|f|) \\ &\le \|f\|_{L^p(X,K)} \, \|1\|_{L^q(X,K)} \\ &= \big(I(1) + J(1)\big)^{\frac{1}{q}} \, \|f\|_{L^p(X,K)}. \end{aligned}$$

Therefore, J represents a bounded linear functional on the space $L^p(X, K)$ for an arbitrary exponent $p \in [1, +\infty)$. In the Hilbert space $L^2(X, K)$ we can apply the representation theorem of Fréchet-Riesz and obtain

$$J(f) = K(fh) \qquad \text{for all} \quad f \in M^\infty(X) \tag{5}$$

with an element $h \in L^2(X, K)$. Now Theorem 8.3 - in the case $p = 1$ - is utilized and we see the regularity improvement $h \in L^\infty(X, K)$. Since J is nonnegative, we infer $h(x) \ge 0$ K-a.e. on X. Furthermore, the relation (4) together with the assumption (D4) tell us that the K-null-sets coincide with the I-null-sets, and we arrive at

$$h(x) \ge 0 \qquad \text{a.e. in} \quad X.$$

3. Taking $f \in M^\infty(X)$, we can iterate (5) and (4) as follows

$$\begin{aligned} J(f) &= K(fh) = I(fh) + J(fh) \\ &= I(fh) + K(fh^2) \\ &= I(fh) + I(fh^2) + J(fh^2) = \ldots, \end{aligned}$$

and we obtain

$$J(f) = I\left(f \sum_{k=1}^{l} h^k\right) + J(fh^l), \qquad l = 1, 2, \ldots \tag{6}$$

Let us define

$$A := \left\{x \in X \,:\, h(x) \geq 1\right\}$$

and $f = \chi_A$. Via approximation, we immediately see that this element f can be inserted into (6). Then we observe

$$+\infty > J(f) \geq I\left(f \sum_{k=1}^{l} h^k\right) \geq l\, I(\chi_A) \qquad \text{for all} \quad l \in \mathbb{N}$$

and consequently $I(\chi_A) = 0$. Therefore, the inequality $0 \leq h(x) < 1$ a.e. in X is satisfied and, moreover,

$$h^l(x) \downarrow 0 \qquad \text{a.e. in} \quad X \quad \text{for} \quad l \to \infty. \tag{7}$$

Via transition to the limit $l \to \infty$ in (6), then B.Levi's theorem implies

$$J(f) = I\left(f \sum_{k=1}^{\infty} h^k\right) \qquad \text{for all} \quad f \in M^\infty(X),$$

when we note that $f = f^+ - f^-$ holds true. Taking $f(x) \equiv 1$ on X in particular, we infer that

$$g(x) = \sum_{k=1}^{\infty} h^k(x) \overset{\text{a.e.}}{=} \frac{h(x)}{1 - h(x)} \in L^1(X)$$

is fulfilled. q.e.d.

Theorem 8.6. (Decomposition theorem of Jordan and Hahn) *Let the bounded linear functional $A : M^\infty(X) \to \mathbb{R}$ be given on the linear normed space $\{M^\infty(X), \|\cdot\|_p\}$, where $1 \leq p < +\infty$ is fixed. Then we have two nonnegative bounded linear functionals $A^\pm : M^\infty(X) \to \mathbb{R}$ with $A = A^+ - A^-$; this means, more precisely,*

$$A(f) = A^+(f) - A^-(f) \qquad \textit{for all} \quad f \in M^\infty(X)$$

with

$$A^{\pm}(f) \geq 0 \qquad \text{for all} \quad f \in M^\infty(X) \quad \text{with} \quad f \geq 0.$$

Furthermore, we have the estimates

$$\|A^{\pm}\| \leq 2\|A\|, \quad \|A^-\| \leq 3\|A\|.$$

Here we define

$$\|A\| := \sup_{f \in M^\infty,\ \|f\|_p \leq 1} |A(f)|, \quad \|A^{\pm}\| := \sup_{f \in M^\infty,\ \|f\|_p \leq 1} |A^{\pm}(f)|.$$

Proof:

1. We take $f \in M^\infty(X)$ with $f \geq 0$ and set

$$A^+(f) := \sup\Big\{A(g) \,:\, g \in M^\infty(X),\ 0 \leq g \leq f\Big\}. \tag{8}$$

Evidently, we have $A^+(f) \geq 0$ for all $f \geq 0$. Moreover, the identity

$$\begin{aligned} A^+(cf) &= \sup\Big\{A(g) \,:\, 0 \leq g \leq cf\Big\} = \sup\Big\{A(cg) \,:\, 0 \leq g \leq f\Big\} \\ &= c \sup\Big\{A(g) \,:\, 0 \leq g \leq f\Big\} = cA^+(f) \end{aligned}$$

for all $f \geq 0$ and $c \geq 0$ holds true. When we take $f_j \in M^\infty(X)$ with $f_j \geq 0$ - for j=1,2 - we infer

$$\begin{aligned} &A^+(f_1) + A^+(f_2) \\ &\quad = \sup\Big\{A(g_1) \,:\, 0 \leq g_1 \leq f_1\Big\} + \sup\Big\{A(g_2) \,:\, 0 \leq g_2 \leq f_2\Big\} \\ &\quad = \sup\Big\{A(g_1 + g_2) \,:\, 0 \leq g_1 \leq f_1,\ 0 \leq g_2 \leq f_2\Big\} \\ &\quad \leq \sup\Big\{A(g) \,:\, 0 \leq g \leq f_1 + f_2\Big\} = A^+(f_1 + f_2). \end{aligned}$$

Given the function g with $0 \leq g \leq f_1 + f_2$, we introduce

$$g_1 := \min(g, f_1) \quad \text{and} \quad g_2 := (g - f_1)^+.$$

Then we observe $g_j \leq f_j$ for $j = 1, 2$ as well as $g_1 + g_2 = g$. Consequently, we obtain

$$A^+(f_1 + f_2) \leq A^+(f_1) + A^+(f_2)$$

and finally

$$A^+(f_1 + f_2) = A^+(f_1) + A^+(f_2).$$

Furthermore, the following inequality holds true for all $f \in M^\infty(X)$ with $f \geq 0$, namely

$$\begin{aligned}
|A^+(f)| &= \left|\sup\left\{A(g)\,:\, g\in M^\infty(X),\ 0\le g\le f\right\}\right| \\
&\le \sup\left\{|A(g)|\,:\, g\in M^\infty(X),\ 0\le g\le f\right\} \\
&\le \sup\left\{\|A\|\,\|g\|_p\,:\, g\in M^\infty(X),\ 0\le g\le f\right\} \\
&\le \|A\|\,\|f\|_p.
\end{aligned}$$

2. Now we extend $A^+ : M^\infty(X)\to\mathbb{R}$ via

$$M^\infty(X)\ni f(x) = f^+(x) - f^-(x) \qquad \text{with} \quad f^\pm(x)\ge 0$$

and define

$$A^+(f) := A^+(f^+) - A^+(f^-).$$

Consequently, we obtain with $A^+ : M^\infty(X)\to\mathbb{R}$ a bounded linear mapping. More precisely, we have the following estimate for all $f\in M^\infty(X)$:

$$\begin{aligned}
|A^+(f)| &\le |A^+(f^+)| + |A^+(f^-)| \\
&\le \|A\|\left(\|f^+\|_p + \|f^-\|_p\right) \le 2\|A\|\,\|f\|_p.
\end{aligned}$$

This implies $\|A^+\|\le 2\|A\|$.

3. Now we define

$$A^-(f) := A^+(f) - A(f) \qquad \text{for all} \quad f\in M^\infty(X).$$

Obviously, A^- represents a bounded linear functional. Here we observe

$$|A^-(f)| \le |A^+(f)| + |A(f)| \le 2\|A\|\cdot\|f\|_p + \|A\|\,\|f\|_p$$

and consequently $\|A^-\|\le 3\|A\|$. Finally, the inequality

$$A^-(f) = A^+(f) - A(f) = \sup\left\{A(g)\,:\, 0\le g\le f\right\} - A(f) \ge 0$$

for all $f\in M^\infty(X)$ with $f\ge 0$ is satisfied. q.e.d.

Theorem 8.7. (The Riesz representation theorem)
Let $1\le p<+\infty$ be fixed. For each bounded linear functional $A\in(\mathcal{L}^p(X))^*$ being given, there exists exactly one generating element $g\in\mathcal{L}^q(X)$ with the property

$$A(f) = I(fg) \quad \text{for all } f\in\mathcal{L}^p(X).$$

Here the identity $p^{-1}+q^{-1}=1$ holds true for the conjugate exponent $q\in(1,+\infty]$.

Proof: We perform our proof in two steps.

1. *Uniqueness:* Let the functions $g_1, g_2 \in \mathcal{L}^q(X)$ with
$$A(f) = I(fg_1) = I(fg_2) \qquad \text{for all} \quad f \in \mathcal{L}^p(X)$$
be given, and we deduce
$$0 = I\Big(f(g_1 - g_2)\Big) \qquad \text{for all} \quad f \in \mathcal{L}^p(X).$$
We recall Theorem 8.2 and obtain $0 = \|g_1 - g_2\|_{\mathcal{L}^q(X)}$, which implies $g_1 = g_2$ in $\mathcal{L}^q(X)$.
2. *Existence:* The functional $A : M^\infty(X) \to \mathbb{R}$ satisfies
$$|A(f)| \le \alpha \|f\|_p \qquad \text{for all} \quad f \in M^\infty(X) \tag{9}$$
with a bound $\alpha \in [0, +\infty)$. The decomposition theorem of Jordan-Hahn gives us nonnegative bounded linear functionals $A^\pm : M^\infty(X) \to \mathbb{R}$ satisfying
$$\|A^\pm\| \le 3\|A\| \le 3\alpha \quad \text{and} \quad A = A^+ - A^-.$$
Here the space $M^\infty(X)$ is endowed with the $\|\cdot\|_p$-norm. In particular, we observe $|A^\pm(f)| < +\infty$ for $f(x) = 1,\ x \in X$. A sequence $\{f_k\}_{k=1,2,\ldots} \subset M^\infty(X)$ with $f_k \downarrow 0$ in X converges uniformly on each compact set towards 0, due to Dini's theorem. Then we arrive at the estimate
$$|A^\pm(f_k)| \le 3\alpha \|f_k\|_p \longrightarrow 0 \qquad \text{for} \quad k \to \infty.$$
With $A^\pm$ we have two Daniell integrals, which are absolutely continuous with respect to I. When N namely is an I-null-set, we infer
$$|A^\pm(\chi_N)| \le 3\alpha \|\chi_N\|_p = 0.$$
Therefore, N is a null-set for the Daniell integrals $A^\pm$ as well. The Radon-Nikodym theorem provides elements $g^\pm \in \mathcal{L}^1(X)$ such that the representation
$$A^\pm(f) = I(fg^\pm) \qquad \text{for all} \quad f \in M^\infty(X)$$
holds true. This implies
$$\begin{aligned} A(f) &= A^+(f) - A^-(f) \\ &= I(fg^+) - I(fg^-) \\ &= I(fg) \qquad \text{for all} \quad f \in M^\infty(X), \end{aligned}$$
when we define $g := g^+ - g^- \in \mathcal{L}^1(X)$. On account of (9) our regularity theorem yields $g \in \mathcal{L}^q(X)$. When we extend the functional continuously onto $\mathcal{L}^p(X)$, we arrive at the representation
$$A(f) = I(fg) \qquad \text{for all} \quad f \in \mathcal{L}^p(X)$$
with a generating function $g \in \mathcal{L}^q(X)$. q.e.d.

Now we address the question of compactness in infinite-dimensional spaces of functions.

Definition 8.8. *A sequence $\{x_k\}_{k=1,2,\dots} \subset \mathcal{B}$ in a Banach space $\mathcal{B}$ is called weakly convergent towards an element $x \in \mathcal{B}$ - symbolically $x_k \rightharpoonup x$ - if the limit relations*

$$\lim_{k\to\infty} A(x_k) = A(x)$$

hold true for each continuous linear functional $A \in \mathcal{B}^*$.

Theorem 8.9. (Weak compactness of $L^p(X)$)
Let us take the exponent $1 < p < +\infty$. Furthermore, let $\{f_k\}_{k=1,2,\dots} \subset L^p(X)$ denote a bounded sequence with the property

$$\|f_k\|_p \leq c \quad \textit{for a constant} \quad c \in [0, +\infty) \quad \textit{and all indices} \quad k \in \mathbb{N}.$$

Then we have a subsequence $\{f_{k_l}\}_{l=1,2,\dots}$ and a limit element $f \in L^p(X)$ such that $f_{k_l} \rightharpoonup f$ in $L^p(X)$ holds true.

Proof:

1. We invoke the Riesz representation theorem and see the following: The relation $f_l \rightharpoonup f$ holds true if and only if $I(f_l g) \to I(fg)$ for all $g \in L^q(X)$ is correct; here we have $p^{-1}+q^{-1} = 1$ as usual. Theorem 7.14 from Section 7 tells us that the space $L^q(X)$ is separable. Therefore, we find a sequence $\{g_m\}_{m=1,2,\dots} \subset L^q(X)$ which lies densely in $L^q(X)$. From the bounded sequence $\{f_k\}_{k=1,2,\dots} \subset L^p(X)$ satisfying $\|f_k\|_p \leq c$ for all $k \in \mathbb{N}$, we now extract successively the subsequences

$$\{f_k\}_{k=1,2,\dots} \supset \{f_{k_l^{(1)}}\}_{l=1,2,\dots} \supset \{f_{k_l^{(2)}}\}_{l=1,2,\dots} \supset \cdots$$

such that

$$\lim_{l\to\infty} I(f_{k_l^{(m)}} g_m) =: \alpha_m \in \mathbb{R}, \qquad m = 1, 2, \dots .$$

Then we apply Cantor's diagonalization procedure, and we make the transition to the diagonal sequence $f_{k_l} := f_{k_l^{(l)}}$, $l = 1, 2, \dots$. Now we observe that

$$\lim_{l\to\infty} I(f_{k_l} g_m) = \alpha_m, \quad m = 1, 2, \dots$$

holds true.

2. By the symbol

$$\mathcal{D} := \left\{ g \in L^q(X) : \begin{array}{l} \text{There exist } N \in \mathbb{N} \text{ and } c_1, \dots, c_N \in \mathbb{R} \\ \text{and } 1 \leq i_1 < \dots < i_N < +\infty \text{ with } g = \sum_{k=1}^{N} c_k g_{i_k} \end{array} \right\}$$

we denote the vector space of finite linear combinations of $\{g_m\}_{m=1,2,\dots}$. Obviously, the limits

$$A(g) := \lim_{l\to\infty} I(f_{k_l} g) \qquad \text{for all} \quad g \in \mathcal{D}$$

exist. The linear functional $A : \mathcal{D} \to \mathbb{R}$ is bounded on the space $\mathcal{D}$ which lies densely in $L^q(X)$, and we have, more precisely,

$$|A(g)| \le c\|g\|_q \quad \text{for all} \quad g \in \mathcal{D}.$$

As described in Theorem 8.1, we continue our functional A from $\mathcal{D}$ onto the space $L^q(X)$, and the Riesz representation theorem provides an element $f \in L^p(X)$ such that

$$A(g) = I(fg) \quad \text{for all } g \in L^q(X).$$

3. Now we show that $f_{k_l} \rightharpoonup f$ in $L^p(X)$ holds true. For each element $g \in L^q(X)$ we find a sequence $\{\widetilde{g}_j\}_{j=1,2,\dots} \subset \mathcal{D}$ satisfying

$$g \overset{L^q}{=} \lim_{j\to\infty} \widetilde{g}_j \in L^q(X).$$

Then we obtain

$$\begin{aligned} |I(fg) - I(f_{k_l} g)| &\le |I(f(g - \widetilde{g}_j))| + |I((f - f_{k_l})\widetilde{g}_j)| + |I(f_{k_l}(\widetilde{g}_j - g))| \\ &\le 2C\|g - \widetilde{g}_j\|_q + |I((f - f_{k_l})\widetilde{g}_j)| \le \varepsilon \end{aligned}$$

for sufficiently large - but fixed - j and the indices $l \ge l_0$. q.e.d.

Remarks:

1. Similarly, we can introduce the notion of weak convergence in Hilbert spaces. Due to Hilbert's selection theorem, we can extract a weakly convergent subsequence from each bounded sequence in Hilbert spaces. However, it is not possible to extract a norm-convergent subsequence from an arbitrary bounded sequence in infinite-dimensional Hilbert spaces. Here we recommend the study of Section 6 in Chapter 8, in particular the first Definition and Example as well as Hilbert's selection theorem.
2. We assume $1 \le p_1 \le p_2 < +\infty$. Then the weak convergence $f_k \rightharpoonup f$ in $L^{p_2}(X)$ implies weak convergence $f_k \rightharpoonup f$ in $L^{p_1}(X)$, which is immediately inferred from the embedding relation $L^{p_2}(X) \hookrightarrow L^{p_1}(X)$.

Theorem 8.10. *The L^p-norm is* lower semicontinuous with respect to weak convergence, *which means:*

$$f_k \rightharpoonup f \text{ in } L^p(X) \quad \Longrightarrow \quad \|f\|_p \le \liminf_{k\to\infty} \|f_k\|_p.$$

Here we assume $1 < p < +\infty$ for the Hölder exponent.

Proof: We start with $f_k \rightharpoonup f$ in $L^p(X)$ and deduce

$$I(f_k g) \to I(fg) \qquad \text{for all} \quad g \in L^q(X).$$

When we choose

$$g(x) := |f(x)|^{\frac{p}{q}} \operatorname{sign} f(x) \in L^q(X),$$

we infer

$$I\Big(f_k |f|^{\frac{p}{q}} \operatorname{sign} f(x)\Big) \to I(|f|^p) = \|f\|_p^p$$

with $p^{-1} + q^{-1} = 1$. For all quantities $\varepsilon > 0$, we find an index $k_0 = k_0(\varepsilon) \in \mathbb{N}$ such that

$$\|f\|_p^p - \varepsilon \le I\Big(f_k |f|^{\frac{p}{q}} \operatorname{sign} f(x)\Big) \le I\Big(|f_k|\,|f|^{\frac{p}{q}}\Big)$$

$$\le \|f_k\|_p \Big(I(|f|^p)\Big)^{\frac{1}{q}} = \|f_k\|_p (\|f\|_p)^{\frac{p}{q}}$$

holds true for all indices $k \ge k_0(\varepsilon)$. When we assume $\|f\|_p > 0$ - without loss of generality - we find to each quantity $\varepsilon > 0$ an index $k_0(\varepsilon) \in \mathbb{N}$ such that

$$\|f_k\|_p \ge \|f\|_p - (\|f\|_p)^{-\frac{p}{q}} \varepsilon \quad \text{for all} \quad k \ge k_0(\varepsilon)$$

is correct. This implies

$$\liminf_{k \to \infty} \|f_k\|_p \ge \|f\|_p.$$

q.e.d.

9 Some Historical Notices to Chapter 2

The modern theory of partial differential equations requires to understand the class of Lebesgue integrable functions – extending the classical family of continuous functions. These more abstract concepts were only reluctantly accepted – even by some of the mathematical heroes of their time. A beautiful source of information, written within the *golden era for mathematics* in Poland between World War I and II, is the following textbook by
Stanisław Saks: Theory of the Integral; Warsaw 1933, Reprint by Hafner Publ. Co., New York (1937).

We would like to present a direct quotation from the preface of this monograph: "On several occasions attempts were made to generalize the old process of integration of Cauchy-Riemann, but it was Lebesgue who first made real progress in this matter. At the same time, Lebesgue's merit is not only to have created a new and more general notion of integral, nor even to have established its intimate connection with the theory of measure: the value of his work consists primarily in his theory of derivation which is parallel to that of integration. This enabled his discovery to find many applications in the most widely different branches of analysis and, from the point of view of method, made it possible to reunite the two fundamental conceptions of integral, namely that of definite integral and that of primitive, which appeared to be forever separated as soon as integration went outside the domain of continuous functions."

The integral of Lebesgue (1875–1941) was wonderfully combined with the abstract spaces created by D. Hilbert (1862–1943) and S. Banach (1892–1945). When we develop the modern theory of partial differential equations in the next volume of our textook, we shall highly appreciate the great vision of the words above by Stanisław Saks – written already in 1933.

Figure 1.2 PORTRAIT OF STEFAN BANACH (1892–1945)
taken from the *Lexikon bedeutender Mathematiker* edited by S. Gottwald, H.-J. Ilgauds, and K.-H. Schlote in Bibliographisches Institut Leipzig (1988).

Chapter 3

Brouwer's Degree of Mapping

Let the function $f : [a, b] \to \mathbb{R}$ be continuous with the property $f(a) < 0 < f(b)$. Due to the intermediate value theorem, there exists a number $\xi \in (a, b)$ satisfying $f(\xi) = 0$. When we assume that the function f is differentiable and each zero ξ of f is *nondegenerate* - this means $f'(\xi) \neq 0$ holds true - we name by

$$i(f, \xi) := \operatorname{sgn} f'(\xi)$$

the *index of f at the point ξ*. We easily deduce the following *index-sum formula*

$$\sum_{\xi \in (a,b):\; f(\xi)=0} i(f, \xi) = 1,$$

where this sum possesses only finitely many terms. In this chapter we intend to deduce corresponding results for functions in n variables. We start with the case $n = 2$, which is usually treated in a lecture on complex analysis.

1 The Winding Number

Let us begin with the following

Definition 1.1. *The number $k \in \mathbb{N}_0 := \mathbb{N} \cup \{0\}$ being prescribed, we define the* set of k-times continuously differentiable *(in the case $k \geq 1$) or* continuous *(in the case $k = 0$)* periodic complex-valued functions *by the symbol*

$$\Gamma_k := \Big\{\varphi = \varphi(t) : \mathbb{R} \to \mathbb{C} \in C^k(\mathbb{R}, \mathbb{C}) \;:\; \varphi(t + 2\pi) = \varphi(t) \quad \textit{for all}\; t \in \mathbb{R}\Big\}.$$

Now we note the following

Definition 1.2. *Let the function $\varphi \in \Gamma_1$ with $\varphi(t) \neq 0$ for all $t \in \mathbb{R}$ be given. Then we define the* winding number *of the closed curve $\varphi(t)$, $0 \leq t \leq 2\pi$ with respect to the point $z = 0$ as follows:*

F. Sauvigny, *Partial Differential Equations 1*, Universitext,
DOI 10.1007/978-1-4471-2981-3_3, © Springer-Verlag London 2012

$$W(\varphi) = W(\varphi, 0) := \frac{1}{2\pi i} \int_0^{2\pi} \frac{\varphi'(t)}{\varphi(t)}\, dt.$$

Remark: For the function $\varphi \in \Gamma_1$ we have the identity

$$\frac{1}{2\pi i} \int_0^{2\pi} \frac{\varphi'(t)}{\varphi(t)}\, dt = \frac{1}{2\pi i} \int_0^{2\pi} \frac{d}{dt}\Big(\log \varphi(t)\Big)\, dt$$

$$= \frac{1}{2\pi i} \int_0^{2\pi} \frac{d}{dt}\Big(\log |\varphi(t))| + i \arg \varphi(t)\Big)\, dt.$$

Therefore, we obtain

$$W(\varphi) = \frac{1}{2\pi} \int_0^{2\pi} \frac{d}{dt}\Big(\arg \varphi(t)\Big)\, dt = \frac{1}{2\pi}\Big(\arg \varphi(2\pi) - \arg \varphi(0)\Big),$$

where we have to extend the function $\arg \varphi(t)$ along the curve continuously. The integer $W(\varphi)$ consequently describes the number of rotations (or windings) of the curve φ about the origin.

Theorem 1.3. *Let the function $\varphi \in \Gamma_1$ with $\varphi(t) \neq 0$ for all $t \in \mathbb{R}$ be given. Then we have the statement $W(\varphi) \in \mathbb{Z}$.*

Proof: We consider the function

$$\Phi(t) := \varphi(t) \exp\left(-\int_0^t \frac{\varphi'(s)}{\varphi(s)}\, ds\right), \qquad 0 \leq t \leq 2\pi.$$

We observe

$$\Phi'(t) = \exp\left(-\int_0^t \frac{\varphi'(s)}{\varphi(s)}\, ds\right)\left\{\varphi'(t) + \varphi(t)\Big(-\frac{\varphi'(t)}{\varphi(t)}\Big)\right\} = 0$$

for all $0 \leq t \leq 2\pi$ and consequently $\Phi(t) = \text{const}$. In particular, we see

$$\varphi(0) = \Phi(0) = \Phi(2\pi) = \varphi(2\pi) \exp\left(-\int_0^{2\pi} \frac{\varphi'(s)}{\varphi(s)}\, ds\right)$$

and therefore

$$\exp\left(\int_0^{2\pi} \frac{\varphi'(s)}{\varphi(s)}\, ds\right) = 1 \quad \text{as well as} \quad \int_0^{2\pi} \frac{\varphi'(s)}{\varphi(s)}\, ds = 2\pi i k, \qquad k \in \mathbb{Z}.$$

This implies $W(\varphi) = k \in \mathbb{Z}$.

q.e.d.

Proposition 1.4. *For the functions* $\varphi_0, \varphi_1 \in \Gamma_1$ *we assume* $|\varphi_0(t)| > \varepsilon$ *and* $|\varphi_0(t) - \varphi_1(t)| < \varepsilon$, $t \in \mathbb{R}$ *with a number* $\varepsilon > 0$. *Then we have the identity*

$$W(\varphi_0) = W(\varphi_1).$$

Proof: With $t \in \mathbb{R}$, $0 \le \tau \le 1$ we consider the family of functions

$$\Phi_\tau(t) = \varphi(t,\tau) := (1-\tau)\varphi_0(t) + \tau\varphi_1(t) = \varphi_0(t) + \tau(\varphi_1(t) - \varphi_0(t)).$$

These have the properties

$$|\varphi(t,\tau)| \ge |\varphi_0(t)| - \tau|\varphi_1(t) - \varphi_0(t)| > \varepsilon - \tau\varepsilon \ge 0$$

as well as

$$\varphi(t,0) = \varphi_0(t), \quad \varphi(t,1) = \varphi_1(t) \qquad \text{for all} \quad t \in \mathbb{R}.$$

Furthermore, we note that

$$W(\Phi_\tau) = \frac{1}{2\pi i}\int_0^{2\pi} \frac{\Phi_\tau'(t)}{\Phi_\tau(t)}\,dt = \frac{1}{2\pi i}\int_0^{2\pi} \frac{(1-\tau)\varphi_0'(t) + \tau\varphi_1'(t)}{(1-\tau)\varphi_0(t) + \tau\varphi_1(t)}\,dt,$$

with an integrand which is continuous in the variables $(t,\tau) \in [0,2\pi] \times [0,1]$. Therefore, the winding number $W(\Phi_\tau)$ is continuous in the parameter $\tau \in [0,1]$ and gives an integer due to Theorem 1.3. Consequently, the identity $W(\varphi_\tau) = \text{const}$ holds true, and we arrive at the statement $W(\varphi_0) = W(\varphi_1)$ from above.

q.e.d.

We shall now define the winding number for continuous, closed curves as well. From Proposition 1.4 we immediately infer the subsequent

Proposition 1.5. *Let* $\{\varphi_k\}_{k=1,2,\ldots} \subset \Gamma_1$ *denote a sequence of curves with* $\varphi_k(t) \ne 0$ *for all* $t \in \mathbb{R}$ *and* $k \in \mathbb{N}$, *which converge uniformly on the interval* $[0,2\pi]$ *towards the continuous function* $\varphi \in \Gamma_0$. *Furthermore, we assume* $\varphi(t) \ne 0$ *for all* $t \in \mathbb{R}$. *Then we have a number* $k_0 \in \mathbb{N}$ *such that*

$$W(\varphi_k) = W(\varphi_l) \qquad \textit{for all} \quad k,l \ge k_0$$

holds true.

Definition 1.6. *Let us consider the function* $\varphi \in \Gamma_0$ *with* $\varphi(t) \ne 0$ *for all* $t \in \mathbb{R}$. *Furthermore, let the sequence of functions* $\{\varphi_k\}_{k=1,2,\ldots} \subset \Gamma_1$ *with* $\varphi_k(t) \ne 0$ *for all* $t \in \mathbb{R}$ *and* $k \in \mathbb{N}$ *be given, which converges uniformly on the interval* $[0,2\pi]$ *towards the function* φ *as follows:*

$$\lim_{k\to\infty} \varphi_k(t) = \varphi(t) \qquad \textit{for all} \quad t \in [0, 2\pi].$$

Then we define

$$W(\varphi) := \lim_{k\to\infty} W(\varphi_k).$$

Remark: The existence of such a sequence for each continuous function $\varphi \in \Gamma_0$ can be ascertained by the usual mollification process. We still have to show that the limit is independent of the choice of an approximating sequence $\{\varphi_k\}_{k=1,2,\dots} \subset \Gamma_1$. Taking two approximating sequences $\{\varphi_k\}_{k=1,2,\dots}$ and $\{\widetilde{\varphi}_k\}_{k=1,2,\dots}$, we make the transition to the mixed sequence

$$\varphi_1, \widetilde{\varphi}_1, \varphi_2, \widetilde{\varphi}_2, \dots =: \{\psi_k\}_{k=1,2,\dots}$$

and Proposition 1.5 yields

$$\lim_{k\to\infty} W(\widetilde{\varphi}_k) = \lim_{k\to\infty} W(\psi_k) = \lim_{k\to\infty} W(\varphi_k).$$

From Theorem 1.3 and Proposition 1.5 we infer the inclusion $W(\varphi) \in \mathbb{Z}$ for $\varphi \in \Gamma_0$.

Theorem 1.7. (Homotopy lemma)
Let the family of continuous curves $\Phi_\tau(t) = \varphi(t,\tau) \in \Gamma_0$ for $\tau^- \le \tau \le \tau^+$ be given. Furthermore, we have $\varphi(t,\tau) \in C^0([0,2\pi]\times[\tau^-,\tau^+],\mathbb{R}^2)$ and

$$\varphi(t,\tau) \neq 0 \qquad \textit{for all} \quad (t,\tau) \in [0,2\pi]\times[\tau^-,\tau^+].$$

Then the winding numbers $W(\Phi_\tau)$ in $[\tau^-,\tau^+]$ are constant.

Remark: The family of curves described in the theorem above is named a *homotopy.* Therefore, the winding number is homotopy-invariant.

Proof of Theorem 1.7: On account of the property $\varphi(t,\tau) \neq 0$ and the compactness of the set $[0,2\pi]\times[\tau^-,\tau^+]$, we have a number $\varepsilon > 0$ such that the inequality $|\varphi(t,\tau)| > \varepsilon$ for all $(t,\tau) \in [0,2\pi]\times[\tau^-,\tau^+]$ holds true. Since the function φ is uniformly continuous on the interval $[0,2\pi]\times[\tau^-,\tau^+]$, we have a number $\delta(\varepsilon) > 0$ with the property

$$|\varphi(t,\tau^*) - \varphi(t,\tau^{**})| < \varepsilon \qquad \text{for all} \quad t \in [0,2\pi], \quad \text{if } |\tau^* - \tau^{**}| < \delta(\varepsilon).$$

With the symbols $\{\varphi_k^*\}_{k=1,2,\dots} \subset \Gamma_1$ and $\{\varphi_k^{**}\}_{k=1,2,\dots} \subset \Gamma_1$, we consider two approximating sequences such that

$$\lim_{k\to\infty} \varphi_k^*(t) = \varphi(t,\tau^*) \quad \text{and} \quad \lim_{k\to\infty} \varphi_k^{**}(t) = \varphi(t,\tau^{**}) \qquad \text{for all} \quad t \in [0,2\pi]$$

hold true. Then we have an index $k_0 \in \mathbb{N}$ such that the following estimates are valid for all $k \ge k_0$:

$$|\varphi_k^*(t)| > \varepsilon, \quad |\varphi_k^{**}(t)| > \varepsilon, \quad |\varphi_k^*(t) - \varphi_k^{**}(t)| < \varepsilon \qquad \text{for all} \quad t \in [0, 2\pi].$$

Proposition 1.4 now yields $W(\varphi_k^*) = W(\varphi_k^{**})$ for all indices $k \geq k_0$, and we infer

$$W(\Phi_{\tau^*}) = W(\Phi_{\tau^{**}}) \qquad \text{for all} \quad \tau^*, \tau^{**} \in [\tau^-, \tau^+] \quad \text{with} \quad |\tau^* - \tau^{**}| < \delta(\varepsilon).$$

Since the quantity $\delta(\varepsilon)$ does not depend on τ^*, τ^{**} and the interval $[\tau^-, \tau^+]$ is compact, a continuation argument gives us the identity $W(\Phi_\tau) = \text{const}$ for all parameters $\tau \in [\tau^-, \tau^+]$.

q.e.d.

Theorem 1.8. *Let the disc*

$$B_R := \Big\{ z \in \mathbb{C} : |z| \leq R \Big\}$$

and the continuous function $f : B_R \to \mathbb{C}$ be given for a fixed radius $R > 0$. The boundary function $\varphi(t) := f(Re^{it})$ may fulfill the condition

$$\varphi(t) \neq 0 \qquad \text{for} \quad 0 \leq t \leq 2\pi,$$

and the winding number of φ satisfies $W(\varphi) \neq 0$. Then we have a point $z_* \in \overset{\circ}{B}_R$ with $f(z_*) = 0$.

Proof: We assume that f did not have any zero in B_R. Then we consider the following homotopy:

$$\Phi_\tau(t) := f(\tau e^{it}), \qquad 0 \leq t \leq 2\pi, \quad 0 \leq \tau \leq R.$$

With the aid of Theorem 1.7 and the identity $\Phi_0(t) = f(0) = \text{const}$, we infer

$$0 = W(\Phi_0) = W(\Phi_R)$$

in contradiction to the assumption $W(\Phi_R) = W(\varphi) \neq 0$. q.e.d.

Theorem 1.9. (Rouché)
The radius $R > 0$ being fixed, let $f_0, f_1 : B_R \to \mathbb{C}$ denote two continuous functions with the property

$$|f_1(z) - f_0(z)| < |f_0(z)| \qquad \text{for all} \quad z \in \partial B_R.$$

The curve $\varphi_0(t) := f_0(Re^{it})$ satisfies the condition

$$\varphi_0(t) \neq 0 \qquad \text{for} \quad 0 \leq t \leq 2\pi \quad \text{as well as} \quad W(\varphi_0) \neq 0.$$

Then we have a point $z_* \in \overset{\circ}{B}_R$ with $f_1(z_*) = 0$.

Proof: We set $\varphi_1(t) := f_1(Re^{it}),\ 0 \le t \le 2\pi$, and consider the homotopy

$$\Phi_\tau(t) = \varphi(t,\tau) := (1-\tau)\varphi_0(t) + \tau\varphi_1(t), \qquad 0 \le t \le 2\pi.$$

Note that

$$\begin{aligned} |\varphi(t,\tau)| &= |\varphi_0(t) + \tau(\varphi_1(t) - \varphi_0(t))| \\ &\ge |\varphi_0(t)| - |\varphi_1(t) - \varphi_0(t)| > 0 \end{aligned}$$

for all $(t,\tau) \in [0,2\pi] \times [0,1]$ holds true, and the homotopy lemma yields $W(\varphi_1) = W(\varphi_0) \neq 0$. According to Theorem 1.8, there exists a point $z_* \in \mathring{B}_R$ with $f_1(z_*) = 0$.

q.e.d.

Theorem 1.10. (Fundamental theorem of Algebra)
Each nonconstant complex polynomial

$$f(z) = z^n + a_{n-1}z^{n-1} + \ldots + a_0$$

of the degree $n \in \mathbb{N}$ possesses at least one complex zero.

Proof: (C.F. Gauß)
We set $f_0(z) := z^n,\ z \in \mathbb{C}$ and consider the following function for a fixed $R > 0$, namely

$$\varphi_0(t) := f(Re^{it}) = R^n e^{int}, \qquad 0 \le t \le 2\pi.$$

We calculate

$$W(\varphi_0) = \frac{1}{2\pi i}\int_0^{2\pi} \frac{\varphi_0'(t)}{\varphi_0(t)}\,dt = \frac{1}{2\pi i}\int_0^{2\pi} \frac{inR^n e^{int}}{R^n e^{int}}\,dt = n \ \in \mathbb{N}.$$

We choose the radius $R > 0$ so large that all points $z \in \mathbb{C}$ with $|z| = R$ fulfill the subsequent inequality:

$$|f_0(z)| = R^n > |f(z) - f_0(z)| = |a_{n-1}z^{n-1} + \ldots + a_0|.$$

Using the theorem of Rouché, we then find a point $z_* \in \mathbb{C}$ with $|z_*| < R$ such that $f(z_*) = 0$ is satisfied.

q.e.d.

Theorem 1.11. (Brouwer's fixed point theorem)
Let $f(z) : B_R \to B_R$ denote a continuous mapping. Then the function f possesses at least one fixed point*: This means that we have a point $z_* \in B_R$ satisfying $f(z_*) = z_*$.*

Proof: We consider the family of mappings

$$g(z,\tau) := z - \tau f(z), \qquad z \in B_R\,, \quad \tau \in [0,1).$$

For all points $z \in \partial B_R$ we have

$$|g(z,\tau)| \geq |z| - \tau |f(z)| \geq R(1-\tau) > 0.$$

Now we apply Rouché's theorem on the function $f_0(z) := z$, with the boundary function $\varphi_0(t) = Re^{it}$, and on the function $f_1(z) := g(z,\tau)$, for a fixed parameter $\tau \in [0,1)$. Then we find a point $z_\tau \in \overset{\circ}{B}_R$ - for each parameter $\tau \in [0,1)$ - with the following property:

$$0 = f_1(z_\tau) = z_\tau - \tau f(z_\tau).$$

For the parameters $\tau_n = 1 - \frac{1}{n}$ with $n = 1, 2, \ldots$, we obtain the relation

$$\Big(1 - \frac{1}{n}\Big) f(z_n) = z_n, \qquad n = 1, 2, \ldots$$

abbreviating $z_n := z_{\tau_n}$. Selecting a subsequence which converges in B_R, the continuity of the function f gives us the limit relation

$$\begin{aligned} z_* := \lim_{k\to\infty} z_{n_k} &= \lim_{k\to\infty} \tau_{n_k} f(z_{n_k}) \\ &= \lim_{k\to\infty} f(z_{n_k}) = f(z_*). \end{aligned}$$

q.e.d.

Definition 1.12. *Let $z \in \mathbb{C}$ denote an arbitrary point, and the function $\varphi(t) \in \Gamma_0$ may satisfy the condition $\varphi(t) \neq z$ for all $t \in \mathbb{R}$. Then we name*

$$W(\varphi, z) := W(\varphi(t) - z)$$

the winding number of the curve φ about the point z.

Theorem 1.13. *Let the function $\varphi \in \Gamma_0$ with the associate curve*

$$\gamma := \{\varphi(t) \in \mathbb{C} \,:\, 0 \leq t \leq 2\pi\}$$

be prescribed. Furthermore, let the domain $G \subset \mathbb{C} \setminus \gamma$ be given. Then the following function

$$\psi(z) := W(\varphi, z), \qquad z \in G$$

is constant. If the domain G contains a point z_0 with $|z_0| > \max\{|\varphi(t)| \,:\, 0 \leq t \leq 2\pi\}$, then we have the identity

$$\psi(z) \equiv 0, \qquad z \in G.$$

Proof:

1. Let z_0 and z_1 denote two points in G, which are connected by the continuous path
$$z = z(\tau) : [0,1] \to G \qquad \text{with} \quad z(0) = z_0, \; z(1) = z_1.$$
We consider the family of curves
$$\varphi_\tau(t) := \varphi(t) - z(\tau) \neq 0, \qquad t \in [0, 2\pi], \quad \tau \in [0,1].$$
The homotopy lemma implies
$$\text{const} = W(\varphi_\tau) = W(\varphi - z(\tau)) = W(\varphi, z(\tau)), \qquad \tau \in [0,1],$$
and consequently $W(\varphi, z_0) = W(\varphi, z_1)$ for arbitrary pairs $z_0, z_1 \in G$.
2. If we have a point $z_0 \in G$ with the property $|z_0| > \max\{|\varphi(t)| \, : \, 0 \le t \le 2\pi\}$, we consider the following path
$$z(\tau) := \frac{1}{1-\tau} z_0 \, , \qquad \tau \in [0,1)$$
satisfying the condition $z(\tau) \notin \gamma$ for all $\tau \in [0,1)$. Now we comprehend the identity $W(\varphi, z(\tau)) = \text{const}$ for $\tau \in [0,1)$. With the assumption $\varphi \in \Gamma_1$, we deduce the relation
$$\lim_{\tau \to 1-} W(\varphi, z(\tau)) = \lim_{\tau \to 1-} \left\{ \frac{1}{2\pi i} \int\limits_0^{2\pi} \frac{\varphi'(t)}{\varphi(t) - z(\tau)} \, dt \right\} = 0.$$
The functions $\varphi \in \Gamma_1$ consequently fulfill $W(\varphi, z(\tau)) = 0$ for all $\tau \in [0,1)$ and finally $W(\varphi, z_0) = 0$. Via approximation, we deduce the identity $W(\varphi, z_0) = 0$ for the functions $\varphi \in \Gamma_0$ as well.

q.e.d.

Definition 1.14. *Let the continuous function*
$$f = f(z) : \{z \in \mathbb{C} \, : \, |z - z_0| \le \varepsilon_0\} \to \mathbb{C} \quad \textit{with} \quad z_0 \in \mathbb{C} \quad \textit{and} \quad \varepsilon_0 > 0$$
be given, which possesses an isolated zero at the origin z_0 in the following sense: We have the relations
$$f(z_0) = 0 \quad \textit{and} \quad f(z) \neq 0 \qquad \textit{for all points} \quad 0 < |z - z_0| \le \varepsilon_0.$$
Then we define the index of f with respect to $z = z_0$ *as follows:*
$$i(f, z_0) := W(\varphi) \qquad \textit{with} \quad \varphi(t) := f(z_0 + \varepsilon e^{it}), \qquad 0 \le t \le 2\pi, \; 0 < \varepsilon \le \varepsilon_0.$$

Remark: On account of the homotopy lemma (Theorem 1.7), this definition is justified because $W(\varphi)$ does not depend on ε.

Example 1.15. Let the function $f(z)$ be holomorphic with an isolated zero at the point z_0. Then f admits the representation

$$f(z) = (z - z_0)^n g(z) \quad \text{with the integer} \quad n \in \mathbb{N},$$

where the function $g(z)$ is analytic and $g(z_0) \neq 0$ holds true. This implies the identity

$$i(f, z_0) = i((z - z_0)^n, z_0) = n \ \in \mathbb{N}.$$

Example 1.16. An antiholomorphic function $f(z)$ (that means $\overline{f}(z)$ is holomorphic) with the property $f(z_0) = 0$ admits the representation

$$f(z) = (\overline{z - z_0})^n g(\overline{z}).$$

Here the function $g(z)$ is analytic and $g(\overline{z_0}) \neq 0$ holds true. The index of f with respect to z_0 satisfies the identity

$$i(f, z_0) = -n \ \in -\mathbb{N}.$$

Theorem 1.17. (Index-sum formula)
The function $f \in C^2(B_R, \mathbb{C})$ has the boundary function $\varphi(t) := f(Re^{it}) \neq 0$, $t \in [0, 2\pi]$. Furthermore, this function f possesses in $\overset{\circ}{B}_R$ the mutually different zeroes z_k with their associate indices $i(f, z_k)$, $k = 1, \ldots, p$ and their total number $p \in \mathbb{N}_0$. Then we have the identity

$$W(\varphi) = \sum_{k=1}^{p} i(f, z_k).$$

Proof:

1. We set

$$F(x, y) := \log f(x, y), \qquad (x, y) \in B_R$$

and calculate

$$\begin{aligned} W(\varphi) &= \frac{1}{2\pi i} \int_0^{2\pi} \frac{\varphi'(t)}{\varphi(t)}\, dt \ = \ \frac{1}{2\pi i} \int_0^{2\pi} \frac{\frac{d}{dt} f(Re^{it})}{f(Re^{it})}\, dt \\ &= \frac{1}{2\pi i} \int_0^{2\pi} \frac{f_x(Re^{it})(-R\sin t) + f_y(Re^{it})(R\cos t)}{f(Re^{it})}\, dt \\ &= \frac{1}{2\pi i} \oint_{\partial B_R} \{F_x\, dx + F_y\, dy\} \ = \ \frac{1}{2\pi i} \oint_{\partial B_R} dF \end{aligned}$$

with the Pfaffian form $dF = F_x(x, y)dx + F_y(x, y)dy$. Here the boundary ∂B_R is described in the mathematically positive sense.

2. The sufficiently small quantity $\varepsilon > 0$ being given, we consider the domain

$$\Omega(\varepsilon) := \Big\{ z \in \overset{\circ}{B}_R : \ |z - z_k| > \varepsilon \ \text{ for } k = 1, \ldots, p \Big\}.$$

Setting

$$\varphi_k(t) := f(z_k + \varepsilon e^{it}), \qquad 0 \le t \le 2\pi, \quad k = 1, \ldots, p,$$

we deduce

$$W(\varphi_k) = \frac{1}{2\pi i} \oint\limits_{|z-z_k|=\varepsilon} dF, \qquad k = 1, \ldots, p,$$

similarly as in part 1 of our proof. Here the curves $|z - z_k| = \varepsilon$ are described in the mathematically positive sense. The Stokes integral theorem yields

$$\begin{aligned} W(\varphi) - \sum_{k=1}^{p} i(f, z_k) &= W(\varphi) - \sum_{k=1}^{p} W(\varphi_k) \\ &= \frac{1}{2\pi i} \oint\limits_{\partial B_R} dF - \frac{1}{2\pi i} \sum_{k=1}^{p} \oint\limits_{|z-z_k|=\varepsilon} dF \\ &= \frac{1}{2\pi i} \int\limits_{\partial \Omega(\varepsilon)} dF = \frac{1}{2\pi i} \int\limits_{\Omega(\varepsilon)} ddF \\ &= 0. \end{aligned}$$

q.e.d.

2 The Degree of Mapping in $\mathbb{R}^n$

J.L.E. Brouwer introduced the degree of mapping in $\mathbb{R}^n$ by simplicial approximation within combinatorial topology. When we intend to define the degree of mapping analytically, we have to replace the integral of the winding number by $(n-1)$-dimensional surface-integrals in $\mathbb{R}^n$ (compare G. de Rham: Varietés differentiables). E. Heinz transformed the boundary integral for the winding number into an area integral and thus created a possibility to define the degree of mapping in $\mathbb{R}^n$ in a natural way. We present the transition from the integral of the winding number to the area integral in $\mathbb{R}^2$ in the sequel:

Let the radius $R \in (0, +\infty)$ and the function $f = f(z) \in C^2(B_R, \mathbb{C})$ satisfying $\varphi(t) := f(Re^{it}) \neq 0$, $0 \le t \le 2\pi$ be given. We choose $\varepsilon > 0$ so small that $\varepsilon < |\varphi(t)|$ for all $t \in [0, 2\pi]$ holds true. Now we consider a function

$$\psi(r) = \begin{cases} 0, \ 0 \le r \le \delta \\ 1, \ \varepsilon \le r \end{cases} \qquad \in C^1([0, +\infty), \mathbb{R})$$

with $0 < \delta < \varepsilon$, and we investigate the integral of the winding number

$$2\pi i W(\varphi) = \oint\limits_{\partial B_R} \left\{ \frac{f_x}{f}\, dx + \frac{f_y}{f}\, dy \right\} = \oint\limits_{\partial B_R} dF$$

$$= \oint\limits_{\partial B_R} \psi(|f(z)|) dF(z) = \oint\limits_{\partial B_R} \psi(|f(x,y)|) dF(x,y)$$

with

$$F(x,y) = \log f(x,y) + 2\pi i k, \qquad k \in \mathbb{Z}.$$

We remark that F is only locally defined, whereas the differential dF is globally available. The 1-form

$$\psi(|f(x,y)|)\, dF(x,y), \qquad (x,y) \in B_R$$

belongs to the class $C^1(B_R)$, and we determine its exterior derivative. Via the identity

$$d\Big\{\psi(|f(x,y)|)\Big\} = \psi'(|f(x,y)|) \left\{ \left((f \cdot \overline{f})^{\frac{1}{2}} \right)_x dx + \left((f \cdot \overline{f})^{\frac{1}{2}} \right)_y dy \right\}$$

$$= \frac{\psi'(|f(x,y)|)}{2|f(x,y)|} \Big\{ f(\overline{f}_x\, dx + \overline{f}_y\, dy) + \overline{f}(f_x\, dx + f_y\, dy) \Big\}$$

we obtain

$$d\Big\{\psi(|f|)\, dF\Big\} = d\Big\{\psi(|f|)\Big\} \wedge dF$$

$$= \frac{\psi'(|f(x,y)|)}{2|f(x,y)|} \Big\{ f(\overline{f}_x\, dx + \overline{f}_y\, dy) + \overline{f}(f_x\, dx + f_y\, dy) \Big\}$$

$$\wedge \left\{ \frac{1}{f}(f_x\, dx + f_y\, dy) \right\}$$

$$= \frac{\psi'(|f(x,y)|)}{2|f(x,y)|} \{\overline{f}_x\, dx + \overline{f}_y\, dy\} \wedge \{f_x\, dx + f_y\, dy\}$$

$$= \frac{\psi'(|f(x,y)|)}{2|f(x,y)|} \Big\{ (\overline{f}_x\, dx \wedge f_y\, dy) - \overline{(\overline{f}_x\, dx \wedge f_y\, dy)} \Big\}$$

$$= i\, \frac{\psi'(|f(x,y)|)}{|f(x,y)|} \operatorname{Im}\big\{\overline{f}_x\, dx \wedge f_y\, dy\big\}.$$

When we set $f = u(x,y) + iv(x,y)$ as usual, we observe

$$\begin{aligned} d\Big\{\psi(|f|)\,dF\Big\} &= i\,\frac{\psi'(|f(x,y)|)}{|f(x,y)|}\,\mathrm{Im}\Big\{(u_x - iv_x)\,dx \wedge (u_y + iv_y)\,dy\Big\} \\ &= i\,\frac{\psi'(|f(x,y)|)}{|f(x,y)|}\,(u_x v_y - v_x u_y)\,dx \wedge dy \\ &= i\,\frac{\psi'(|f(x,y)|)}{|f(x,y)|}\,\frac{\partial(u,v)}{\partial(x,y)}\,dx \wedge dy. \end{aligned}$$

The Stokes integral theorem therefore yields

$$2\pi W(\varphi) = \iint\limits_{B_R} \frac{\psi'(|f(x,y)|)}{|f(x,y)|}\,\frac{\partial(u,v)}{\partial(x,y)}\,dxdy.$$

Now we define $\omega(t) := \dfrac{\psi'(t)}{t}$ with $t \geq 0$ and note that

$$\psi(t) = \int\limits_0^t \tau\omega(\tau)d\tau, \qquad t \geq 0$$

holds true. Let us choose a function $\omega(t) \in C^0([0,+\infty),\mathbb{R})$ with the following properties:

(a) We have $\omega(t) = 0$ for all $t \in [0,\delta] \cup [\varepsilon,+\infty)$;

(b) The condition $\int\limits_0^\infty \varrho\omega(\varrho)\,d\varrho = 1$ holds true.

Then we observe

$$W(\varphi) = \frac{1}{2\pi}\iint\limits_{B_R} \omega(|f(x,y)|)J_f(x,y)\,dxdy.$$

Via the transition

$$\tilde{\omega}(t) := \frac{1}{2\pi}\,\omega(t),$$

we obtain the normalization

(b') $\iint\limits_{\mathbb{R}^2} \tilde{\omega}(|z|)\,dxdy = 1$ with $z = x + iy$,

and we see

$$W(\varphi) = \iint\limits_{B_R} \tilde{\omega}(|f(x,y)|)J_f(x,y)\,dxdy.$$

These considerations propose the following definition for the degree of mapping in $\mathbb{R}^n$, namely

Definition 2.1. *Let $\Omega \subset \mathbb{R}^n$ denote a bounded open set in $\mathbb{R}^n$, and let us take the function*

$$f = (f_1(x_1, \dots, x_n), \dots, f_n(x_1, \dots, x_n)) \in C^k(\Omega, \mathbb{R}^n) \cap C^0(\overline{\Omega}, \mathbb{R}^n)$$

for $k \in \mathbb{N}$ with the property $f(x) \neq 0$ for all points $x \in \partial\Omega$. Given the inequality $0 < \varepsilon < \inf\{|f(x)| : x \in \partial\Omega\}$, we consider a function $\omega \in C^0([0, +\infty), \mathbb{R})$ with the following properties:

(a) We have $\omega(r) = 0$ for all $r \in [0, \delta] \cup [\varepsilon, +\infty)$, with $\delta \in (0, \varepsilon)$ chosen suitably;
(b) We require the condition

$$\int_{\mathbb{R}^n} \omega(|y|)\, dy = 1.$$

Then we define Brouwer's degree of mapping for f with respect to $y = 0$ *as follows:*

$$d(f, \Omega) = d(f, \Omega, 0) := \int_{\Omega} \omega\big(|f(x)|\big) J_f(x) dx.$$

Here we denote by

$$J_f(x) = \frac{\partial(f_1, \dots, f_n)}{\partial(x_1, \dots, x_n)}, \qquad x \in \Omega$$

the Jacobian of the mapping f.

Remarks:

1. Introducing n-dimensional spherical coordinates due to

$$y = r\eta = (r\eta_1, \dots, r\eta_n) \in \mathbb{R}^n \qquad \text{with} \quad r > 0,\ |\eta| = 1,$$

   we comprehend

$$\int_{\mathbb{R}^n} \omega(|y|)\, dy = \hat{\omega}_n \int_0^\infty r^{n-1} \omega(r)\, dr,$$

   where the symbol $\hat{\omega}_n$ means the area of the $(n-1)$-dimensional unit sphere in $\mathbb{R}^n$.
2. We still have to establish the independence of the quantity $d(f, \Omega)$ from the admissible test function ω chosen.

The subsequent result is fundamental.

Theorem 2.2. *Let $\Omega \subset \mathbb{R}^n$ denote a bounded open set - with $n \in \mathbb{N}$ - and consider the function $f \in C^1(\Omega, \mathbb{R}^n) \cap C^0(\overline{\Omega}, \mathbb{R}^n)$ satisfying $|f(x)| > \varepsilon > 0$ for all points $x \in \partial\Omega$. Furthermore, let $\omega(r) \in C^0\big([0, +\infty)\big)$ represent a test function with the following properties:*

(a) $\omega(r) = 0$ *for all* $r \in [0, \delta] \cup [\varepsilon, +\infty)$, $0 < \delta < \varepsilon$;

(b) $\int\limits_0^\infty r^{n-1}\omega(r)\,dr = 0.$

Then we have the identity

$$\int\limits_\Omega \omega(|f(x)|)\,J_f(x)\,dx = 0.$$

Proof:

1. It is sufficient to show the identity above only for those functions $f \in C^2(\Omega, \mathbb{R}^n) \cap C^0(\overline{\Omega}, \mathbb{R}^n)$. By approximation this relation pertains to all functions $f \in C^1(\Omega, \mathbb{R}^n) \cap C^0(\overline{\Omega}, \mathbb{R}^n)$.
2. Let us consider the function

$$f(x) = (f_1(x), \ldots, f_n(x)) \in C^2(\Omega, \mathbb{R}^n) \cap C^0(\overline{\Omega}, \mathbb{R}^n)$$

and an arbitrary vector-field

$$a(y) = (a_1(y), \ldots, a_n(y)) \in C^1(\mathbb{R}^n, \mathbb{R}^n).$$

Then we introduce the $(n-1)$-form

$$\lambda := \sum_{i=1}^n (-1)^{1+i} a_i(f(x))\,df_1 \wedge \ldots \wedge df_{i-1} \wedge df_{i+1} \wedge \ldots \wedge df_n.$$

With the aid of the identity

$$\begin{aligned} d\{a_i(f(x))\} &= \sum_{j=1}^n \frac{d}{dx_j}\Big(a_i(f(x))\Big)\,dx_j = \sum_{j,k=1}^n \frac{\partial a_i}{\partial y_k}(f(x))\frac{\partial f_k}{\partial x_j}\,dx_j \\ &= \sum_{k=1}^n \frac{\partial a_i}{\partial y_k}(f(x))\,df_k \end{aligned}$$

we determine the exterior derivative

$$\begin{aligned} d\lambda &= \sum_{i=1}^n (-1)^{i+1} d\{a_i(f(x))\} \wedge df_1 \wedge \ldots \wedge df_{i-1} \wedge df_{i+1} \wedge \ldots \wedge df_n \\ &= \sum_{i=1}^n \frac{\partial a_i}{\partial y_i}(f(x))\,df_1 \wedge \ldots \wedge f_n = \operatorname{div} a(f(x))\,J_f(x)\,dx_1 \wedge \ldots \wedge dx_n. \end{aligned}$$

3. Now we choose the vector-field $a(y)$ such that $\omega(|y|) = \operatorname{div} a(y)$ holds true. Via the function $\psi(r) \in C_0^1(0, +\infty)$, we propose the ansatz $a(y) := \psi(|y|)y$ and realize

$$\omega(|y|) = \operatorname{div} a(y) = n\psi(|y|) + \psi'(|y|)\Big(y \cdot \frac{y}{|y|}\Big) = n\psi(|y|) + |y|\psi'(|y|).$$

Using $r = |y|$, we obtain the differential equation

$$\frac{\omega(r)}{r} = \psi'(r) + n\frac{\psi(r)}{r} = \frac{(r^n\psi(r))'}{r^n}$$

with the solution

$$\psi(r) = r^{-n}\int_0^r \varrho^{n-1}\omega(\varrho)\,d\varrho.$$

We note that $\psi(r) = 0$ for all $r \in [0,\delta] \cup [\varepsilon, +\infty)$ holds true.

4. With the $(n-1)$-form

$$\lambda := \psi(|f(x)|)\sum_{i=1}^{n}(-1)^{i+1} f_i(x)\, df_1\wedge\ldots\wedge df_{i-1}\wedge df_{i+1}\wedge\ldots\wedge df_n \quad \in C_0^1(\Omega)$$

we consequently obtain

$$d\lambda = \omega(|f(x)|)\, J_f(x)\, dx_1 \wedge \ldots \wedge dx_n.$$

The Stokes integral theorem now yields

$$\int_\Omega \omega(|f(x)|)\, J_f(x)\, dx_1 \wedge \ldots \wedge dx_n = \int_\Omega d\lambda = 0.$$

q.e.d.

Corollary from Theorem 2.2: Definition 2.1 is independent from the choice of the test function: Let ω_1, ω_2 represent two admissible test functions: The function ω_1 may satisfy the condition (a) from Definition 2.1 with $\delta_1 \in (0,\varepsilon)$, and the function ω_2 may fulfill the condition (a) with $\delta_2 \in (0,\varepsilon)$. Then we have the identity

$$\int_0^\infty r^{n-1}(\omega_1(r) - \omega_2(r))dr = 0, \quad (\omega_1 - \omega_2)(r) = 0 \qquad \text{for} \quad r \in [0,\delta] \cup [\varepsilon, +\infty)$$

with $\delta := \min\{\delta_1, \delta_2\} \in (0,\varepsilon)$. Theorem 2.2 yields

$$\int_\Omega \Big(\omega_1(|f(x)|) - \omega_2(|f(x)|)\Big) J_f(x)dx = 0$$

and consequently

$$\int_\Omega \omega_1(|f(x)|)\, J_f(x)dx = \int_\Omega \omega_2(|f(x)|)\, J_f(x)dx.$$

q.e.d.

In order to prepare the homotopy lemma we prove

Proposition 2.3. *Let the two functions $f_1, f_2 \in C^1(\Omega,\mathbb{R}^n)\cap C^0(\overline{\Omega},\mathbb{R}^n)$ satisfy $|f_i(x)| > 5\varepsilon$ with $i = 1,2$ for all points $x \in \partial\Omega$. Furthermore, let the inequality*

$$|f_1(x) - f_2(x)| < \varepsilon \quad \textit{for all points} \quad x \in \overline{\Omega}$$

be valid. Then we have the identity

$$d(f_1,\Omega) = d(f_2,\Omega).$$

Proof: Let $\lambda = \lambda(r) \in C^1\big([0,+\infty),[0,1]\big)$ denote an auxiliary function such that

$$\lambda(r) = \begin{cases} 1,\ 0 \le r \le 2\varepsilon \\ 0,\ 3\varepsilon \le r \end{cases}.$$

Then we consider the function

$$f_3(x) := \Big(1 - \lambda\big(|f_1(x)|\big)\Big) f_1(x) + \lambda\big(|f_1(x)|\big) f_2(x), \quad x \in \overline{\Omega}.$$

We note that $f_3 \in C^1(\Omega,\mathbb{R}^n)\cap C^0(\overline{\Omega},\mathbb{R}^n)$ and

$$|f_3(x)| > 4\varepsilon \quad \text{for all points} \quad x \in \partial\Omega$$

as well as

$$\begin{aligned}|f_3(x) - f_i(x)| \le \Big(1 - \lambda\big(|f_1(x)|\big)\Big)|f_1(x) - f_i(x)| \\ +\lambda\big(|f_1(x)|\big)|f_2(x) - f_i(x)| < \varepsilon, \qquad x \in \overline{\Omega} \quad \text{with} \quad i = 1,2\end{aligned}$$

hold true. Now we observe

$$f_3(x) = \begin{cases} f_1(x) \text{ for all } x \in \Omega \text{ with } |f_1(x)| \ge 3\varepsilon \\ f_2(x) \text{ for all } x \in \Omega \text{ with } |f_2(x)| \le \varepsilon \end{cases}.$$

Let the symbols $\omega_1(r) \in C_0^0((3\varepsilon,4\varepsilon),\mathbb{R})$ and $\omega_2 \in C_0^0((0,\epsilon),\mathbb{R})$ denote two admissible test functions. Then we infer the identities

$$\omega_1(|f_1(x)|)J_{f_1}(x) = \omega_1(|f_3(x)|)J_{f_3}(x), \qquad x \in \Omega,$$

and

$$\omega_2(|f_2(x)|)J_{f_2}(x) = \omega_2(|f_3(x)|)J_{f_3}(x), \qquad x \in \Omega.$$

An integration immediately yields

$$\begin{aligned} d(f_1,\Omega) &= \int_\Omega \omega_1\big(|f_1(x)|\big)J_{f_1}(x)\,dx = \int_\Omega \omega_1\big(|f_3(x)|\big)J_{f_3}(x)\,dx \\ &= \int_\Omega \omega_2\big(|f_3(x)|\big)J_{f_3}(x)\,dx = \int_\Omega \omega_2\big(|f_2(x)|\big)J_{f_2}(x)\,dx = d(f_2,\Omega). \end{aligned}$$

q.e.d.

Proposition 2.3 directly implies

Proposition 2.4. *Let the function* $f : \Omega \to \mathbb{R}^n \in C^0(\overline{\Omega}, \mathbb{R}^n)$ *be given with the property* $f(x) \neq 0$ *for all* $x \in \partial\Omega$. *Furthermore, let*

$$\{f_k\}_{k=1,2,\ldots} \subset C^1(\Omega, \mathbb{R}^n) \cap C^0(\overline{\Omega}, \mathbb{R}^n)$$

denote a sequence of functions satisfying

$$f_k(x) \neq 0 \qquad \text{for all} \quad x \in \partial\Omega \quad \text{and all} \quad k \in \mathbb{N},$$

such that the convergence

$$\lim_{k\to\infty} f_k(x) = f(x)$$

is uniform in $\overline{\Omega}$. *Then we have an index* $k_0 \in \mathbb{N}$ *such that the identities*

$$d(f_k, \Omega) = d(f_l, \Omega) \qquad \text{for all} \quad k, l \geq k_0$$

are valid.

On account of Proposition 2.4 the following definition is justified.

Definition 2.5. *Let the function* $f(x) \in C^0(\overline{\Omega}, \mathbb{R}^n)$ *with* $f(x) \neq 0$ *for all* $x \in \partial\Omega$ *be given. Furthermore, let the sequence of functions* $\{f_k\}_{k=1,2,\ldots} \subset C^1(\Omega, \mathbb{R}^n) \cap C^0(\overline{\Omega}, \mathbb{R}^n)$ *be given with the property*

$$f_k(x) \neq 0 \qquad \text{for all points} \quad x \in \partial\Omega \quad \text{and all} \quad k \in \mathbb{N},$$

which converge uniformly in $\overline{\Omega}$ *as follows:*

$$f_k(x) \longrightarrow f(x) \qquad \text{for} \quad k \to \infty.$$

Then we define

$$d(f, \Omega) := \lim_{k\to\infty} d(f_k, \Omega)$$

and name this quantity Brouwer's degree of mapping for continuous functions.

Fundamental is the following result.

Theorem 2.6. (Homotopy lemma)
Let $f_\tau(x) \in C^0(\overline{\Omega}, \mathbb{R}^n)$ *for* $a \leq \tau \leq b$ *denote a family of continuous mappings with the following properties:*

(a) $f_\tau(x) = f(x, \tau) : \overline{\Omega} \times [a, b] \to \mathbb{R}^n \in C^0(\overline{\Omega} \times [a, b], \mathbb{R}^n)$,
(b) $f_\tau(x) \neq 0$ *for all points* $x \in \partial\Omega$ *and all parameters* $\tau \in [a, b]$.

Then we have the identity $d(f_\tau, \Omega) = const$ *in* $[a, b]$.

Proof: At first, we have a quantity $\varepsilon > 0$ such that $|f_\tau(x)| > 5\varepsilon$ for all points $x \in \partial\Omega$ and all parameters $\tau \in [a,b]$ is correct. Furthermore, there exists a number $\delta = \delta(\varepsilon) > 0$ such that all parameters $\tau^*, \tau^{**} \in [a,b]$ with $|\tau^* - \tau^{**}| < \delta(\epsilon)$ satisfy the inequality

$$|f(x,\tau^*) - f(x,\tau^{**})| < \varepsilon \qquad \text{for all points} \quad x \in \overline{\Omega}.$$

With

$$\{f_k^*\}_{k=1,2,\ldots} \quad \text{and} \quad \{f_k^{**}\}_{k=1,2,\ldots} \subset C^1(\Omega,\mathbb{R}^n) \cap C^0(\overline{\Omega},\mathbb{R}^n)$$

we consider admissible sequences of approximation for the functions $f_{\tau^*}(x)$ and $f_{\tau^{**}}(x)$, respectively. Then we have an index $k_0 \in \mathbb{N}$, such that the inequalities

$$|f_k^*(x)| > 5\varepsilon, \quad |f_k^{**}(x)| > 5\varepsilon \qquad \text{for all} \quad x \in \partial\Omega \quad \text{and all} \quad k \geq k_0$$

as well as

$$|f_k^*(x) - f_k^{**}(x)| < \varepsilon \qquad \text{for all points} \quad x \in \overline{\Omega} \quad \text{and all indices} \quad k \geq k_0$$

are valid. Now Proposition 2.3 yields

$$d(f_k^*, \Omega) = d(f_k^{**}, \Omega) \qquad \text{for all indices} \quad k \geq k_0,$$

and we observe

$$d(f_{\tau^*}, \Omega) = d(f_{\tau^{**}}, \Omega) \quad \text{for all } \tau^*, \tau^{**} \in [a,b] \text{ with } |\tau^* - \tau^{**}| < \delta(\varepsilon).$$

This implies the identity $d(f_\tau, \Omega) = \text{const}$ for $a \leq \tau \leq b$. q.e.d.

Theorem 2.7. *Let the function $f \in C^0(\overline{\Omega},\mathbb{R}^n)$ with $f(x) \neq 0$ for all points $x \in \partial\Omega$ be given, such that $d(f,\Omega) \neq 0$ holds true. Then we have a point $\xi \in \Omega$ satisfying $f(\xi) = 0$.*

Proof: If this statement were false, we would have a quantity $\varepsilon > 0$ with the property $|f(x)| > \varepsilon$ for all points $x \in \overline{\Omega}$. Let us denote the sequence of functions $\{f_k\}_{k=1,2,\ldots} \subset C^1(\Omega,\mathbb{R}^n) \cap C^0(\overline{\Omega},\mathbb{R}^n)$, where the convergence

$$f_k(x) \longrightarrow f(x) \qquad \text{for} \quad k \to \infty$$

is uniform in $\overline{\Omega}$. Now we have an index $k_0 \in \mathbb{N}$ such that $|f_k(x)| > \varepsilon$ in $\overline{\Omega}$ holds true for all indices $k \geq k_0$. When $\omega = \omega(r) \in C_0^0((0,\varepsilon),\mathbb{R})$ denotes an admissible test function satisfying

$$\int\limits_{\mathbb{R}^n} \omega(|y|)dy = 1,$$

we deduce the relation

$$d(f_k, \Omega) = \int_{\Omega} \omega(|f_k(x)|) J_{f_k}(x) dx = 0 \quad \text{for all indices} \quad k \geq k_0$$

and consequently

$$d(f, \Omega) = \lim_{k \to \infty} d(f_k, \Omega) = 0$$

in contradiction to the assumption. Therefore, we have a point $\xi \in \Omega$ satisfying $f(\xi) = 0$. q.e.d.

Theorem 2.8. *Let the functions $f_0, f_1 \in C^0(\overline{\Omega}, \mathbb{R}^n)$ with $|f_0(x) - f_1(x)| < |f_1(x)|$ for all $x \in \partial\Omega$ be given. Then we have the identity*

$$d(f_0, \Omega) = d(f_1, \Omega).$$

Proof: We utilize the linear homotopy

$$f_\tau(x) = \tau f_0(x) + (1 - \tau) f_1(x), \qquad x \in \overline{\Omega}, \quad \tau \in [0, 1].$$

On account of $f_\tau(x) \neq 0$ for all $x \in \partial\Omega$ and all $\tau \in [0, 1]$, Theorem 2.6 yields the identity

$$d(f_0, \Omega) = d(f_1, \Omega).$$

q.e.d.

Definition 2.9. *Let $\Omega \subset \mathbb{R}^n$ denote a bounded open set, and let the function $f(x) : \partial\Omega \to \mathbb{R}^n \setminus \{0\}$ be continuous. Furthermore, let the function $\hat{f}(x) : \mathbb{R}^n \to \mathbb{R}^n \in C^0(\mathbb{R}^n, \mathbb{R}^n)$ with $\hat{f}(x) = f(x)$ for all $x \in \partial\Omega$ constitute a continuous extension of f onto the entire space $\mathbb{R}^n$. Then we set*

$$v(f, \partial\Omega) := d(\hat{f}, \Omega)$$

for the order of the function f with respect to the point $z = 0$.

Remarks:

1. Due to Tietze's extension theorem, there always exists such a continuation $\hat{f}$ of f.
2. Theorem 2.8 tells us that $v(f, \partial\Omega)$ is independent of the extension chosen.

Using Definition 2.9, we obtain the following corollary from the homotopy lemma:

Theorem 2.10. *Let $f_\tau(x) = f(x, \tau) : \partial\Omega \times [a, b] \to \mathbb{R}^n \setminus \{0\} \in C^0(\partial\Omega \times [a, b])$ constitute a continuous family of zero-free mappings. Then we have the identity $v(f_\tau, \partial\Omega) = const$ in $[a, b]$.*

3 Topological Existence Theorems

We begin with the fundamental

Proposition 3.1. *Let $\Omega \subset \mathbb{R}^n$ denote a bounded open set and define the function $f(x) = \varepsilon(x - \xi)$, $x \in \Omega$; here we choose $\varepsilon = \pm 1$ and $\xi \in \Omega$. Then we have the identity $d(f, \Omega) = \varepsilon^n$.*

Proof: We take a number $\eta > 0$ such that $|f(x)| > \eta$ for all points $x \in \partial\Omega$ holds true. Let $\omega \in C_0^0((0,\eta),\mathbb{R})$ denote an arbitrary test function satisfying

$$\int\limits_{\mathbb{R}^n} \omega(|x|)\,dx = 1.$$

Then we have

$$d(f,\Omega) = \int\limits_{\Omega} \omega(|f(x)|)\,J_f(x)\,dx = \int\limits_{\Omega} \omega(|x-\xi|)\varepsilon^n\,dx = \varepsilon^n.$$

q.e.d.

Theorem 3.2. *Let $f_\tau(x) = f(x,\tau) : \overline{\Omega} \times [a,b] \to \mathbb{R}^n \in C^0(\overline{\Omega} \times [a,b], \mathbb{R}^n)$ denote a family of mappings with*

$$f_\tau(x) \neq 0 \qquad \textit{for all} \quad x \in \partial\Omega \quad \textit{and all} \quad \tau \in [a,b].$$

Furthermore, we have the function

$$f_a(x) = (x - \xi), \qquad x \in \Omega$$

with a point $\xi \in \Omega$. For each parameter $\tau \in [a,b]$ we find a point $x_\tau \in \Omega$ satisfying $f(x_\tau,\tau) = 0$.

Proof: The homotopy lemma and Proposition 3.1 yield

$$d(f_\tau,\Omega) = d(f_a,\Omega) = 1 \qquad \text{for all} \quad \tau \in [a,b].$$

Consequently, there exists a point $x_\tau \in \Omega$ with $f(x_\tau,\tau) = 0$ for each parameter $\tau \in [a,b]$, due to Theorem 2.7 in Section 2. q.e.d.

Theorem 3.3. (Brouwer's fixed point theorem)
Each continuous mapping $f(x) : B \to B$ of the unit ball $B := \{x \in \mathbb{R}^n : |x| \leq 1\}$ into itself possesses a fixed point $\xi \in B$, which satisfies the condition $\xi = f(\xi)$.

Proof: We consider the mapping

$$f_\tau(x) = x - \tau f(x), \qquad x \in B$$

for all parameters $\tau \in [0,1)$, which fulfills the following boundary condition

$$|f_\tau(x)| \geq |x| - \tau|f(x)| \geq 1 - \tau > 0 \quad \text{for all} \quad x \in \partial B \quad \text{and all} \quad \tau \in [0,1).$$

According to Theorem 3.2, each $\tau \in [0,1)$ admits a point $x_\tau \in \overset{\circ}{B}$ satisfying $f_\tau(x_\tau) = 0$ and equivalently $\tau f(x_\tau) = x_\tau$. We now choose a sequence $\tau_n \uparrow 1$ for $n \to \infty$, such that $\{x_{\tau_n}\}_{n=1,2,\ldots}$ in B converges. This implies

$$\xi := \lim_{n\to\infty} x_{\tau_n} = \lim_{n\to\infty} \tau_n f(x_{\tau_n}) = \lim_{n\to\infty} f(x_{\tau_n}) = f(\xi).$$

q.e.d.

Remark: Brouwer's fixed point theorem remains true for all those sets which are homeomorphic to the ball B.

Theorem 3.4. (H. Poincaré and L.E.J. Brouwer)
Let the dimension $n \in \mathbb{N}$ be even. By the symbol

$$S^n := \Big\{x \in \mathbb{R}^{n+1} \,:\, |x| = 1\Big\}$$

we denote the n-dimensional sphere in $\mathbb{R}^{n+1}$. Then there do not exist tangential, zero-free, continuous vector-fields on the sphere S^n.

Proof: If $\varphi : S^n \to \mathbb{R}^{n+1}$ were such a vector-field, we would have the properties $|\varphi(x)| > 0$ and $(\varphi(x), x) = 0$ for all $x \in S^n$. Given the sign factor $\varepsilon = \pm 1$, we consider the mapping $f(x) := \varepsilon x$, $x \in S^n$ and the homotopy

$$f_\tau(x) = (1-\tau) f(x) + \tau\varphi(x), \qquad x \in S^n.$$

We observe

$$|f_\tau(x)|^2 = (1-\tau)^2|f(x)|^2 + \tau^2|\varphi(x)|^2 > 0$$

for all points $x \in S^n$ and all parameters $\tau \in [0,1]$. With the aid of Theorem 2.10 in Section 2 we comprehend

$$v(\varphi, S^n) = v(f_1, S^n) = v(f_0, S^n) = v(f, S^n) = \varepsilon^{n+1}$$

where we use Proposition 3.1. When the dimension n is even, we deduce the relation

$$-1 = v(\varphi, S^n) = +1.$$

This reveals an evident contradiction! q.e.d.

4 The Index of a Mapping

In this section we transfer the index-sum formula from the case $n = 2$ to the situation of arbitrary dimensions. In this context we derive that the degree of mapping gives us an integer. We begin with the easy

Proposition 4.1. *Let $\Omega_j \subset \mathbb{R}^n$ for $j = 1, 2$ denote two bounded open disjoint sets and $\Omega := \Omega_1 \cup \Omega_2$ their union. Furthermore, let $f(x) \in C^0(\overline{\Omega}, \mathbb{R}^n)$ represent a continuous mapping with the property*

$$f(x) \neq 0 \qquad \textit{for all points} \quad x \in \partial\Omega_1 \cup \partial\Omega_2.$$

Then we have the identity

$$d(f, \Omega) = d(f, \Omega_1) + d(f, \Omega_2).$$

Proof: When we choose the quantity $\varepsilon > 0$ sufficiently small, we obtain $|f(x)| > \varepsilon$ for all points $x \in \partial\Omega_1 \cup \partial\Omega_2$. Furthermore, we have a sequence of functions $\{f_k\}_{k=1,2,\ldots} \subset C^1(\Omega, \mathbb{R}^n) \cap C^0(\overline{\Omega}, \mathbb{R}^n)$ satisfying $f_k \to f$ uniformly on $\overline{\Omega}$ as well as $|f_k(x)| > \varepsilon$ for all points $x \in \partial\Omega_1 \cup \partial\Omega_2$ and all indices $k \geq k_0$. Now we utilize the admissible test function $\omega \in C_0^0((0, \varepsilon), \mathbb{R})$ with the property $\int\limits_{\mathbb{R}^n} \omega(|y|)dy = 1$, and we easily see for all indices $k \geq k_0$ the following equation:

$$\begin{aligned} d(f_k, \Omega) &= \int\limits_{\Omega} \omega\big(|f_k(x)|\big) J_{f_k}(x)\,dx \\ &= \int\limits_{\Omega_1} \omega\big(|f_k(x)|\big) J_{f_k}(x)\,dx + \int\limits_{\Omega_2} \omega\big(|f_k(x)|\big) J_{f_k}(x)\,dx \\ &= d(f_k, \Omega_1) + d(f_k, \Omega_2). \end{aligned}$$

This implies the desired identity $d(f, \Omega) = d(f, \Omega_1) + d(f, \Omega_2)$. q.e.d.

Proposition 4.2. *Let the function $f \in C^0(\overline{\Omega}, \mathbb{R}^n)$ be defined on the bounded open set $\Omega \subset \mathbb{R}^n$, with the associate set of zeroes*

$$F := \Big\{x \in \overline{\Omega} : \; f(x) = 0\Big\}.$$

Furthermore, choose an open set with $\Omega_0 \subset \Omega$ such that $F \subset \Omega_0$ holds true. Then we have the conclusion

$$d(f, \Omega) = d(f, \Omega_0).$$

Proof: We set $\Omega_1 := \overset{\circ}{(\Omega \setminus \Omega_0)}$ and observe $\Omega \setminus \partial\Omega_1 = \Omega_0 \,\dot{\cup}\, \Omega_1$. On account of the property $f(x) \neq 0$ for all points $x \in \overline{\Omega_1}$, Theorem 2.7 from Section 2 yields the statement $d(f, \Omega_1) = 0$. Now Proposition 4.1 implies

$$d(f, \Omega) = d(f, \Omega_0) + d(f, \Omega_1) = d(f, \Omega_0).$$

q.e.d.

Definition 4.3. *We consider the function $f(x) \in C^0(\overline{\Omega}, \mathbb{R}^n)$. With a sufficiently small quantity $\varepsilon > 0$, the point $z \in \Omega$ satisfies the conditions $f(z) = 0$ and $f(x) \neq 0$ for all $0 < |x - z| \leq \varepsilon$. Then we name*

$$i(f, z) := d(f, B_\varepsilon(z))$$

the index of f at the point $x = z$. *Here we abbreviate $B_\varepsilon(z) := \{x \in \mathbb{R}^n : |x - z| < \varepsilon\}$.*

Theorem 4.4. *Consider the function $f \in C^0(\overline{\Omega}, \mathbb{R}^n)$, and let the equation $f(x) = 0$, $x \in \overline{\Omega}$ possess $p \in \mathbb{N}_0$ mutually different solutions $x^{(1)}, \ldots, x^{(p)} \in \Omega$. Then we have the identity*

$$d(f, \Omega) = \sum_{j=1}^{p} i(f, x^{(j)}).$$

Proof: We choose a sufficiently small quantity $\varepsilon > 0$ such that the open sets

$$\Omega_j := \left\{x \in \mathbb{R}^n : |x - x^{(j)}| < \varepsilon\right\}$$

are mutually disjoint. Now Proposition 4.1 and 4.2 yield

$$d(f, \Omega) = d\Big(f, \bigcup_{j=1}^{p} \Omega_j\Big) = \sum_{j=1}^{p} d(f, \Omega_j) = \sum_{j=1}^{p} i(f, x^{(j)}).$$

q.e.d.

Proposition 4.5. *With $A = (a_{ij})_{i,j=1,\ldots,n}$ let us consider a real $n \times n$-matrix satisfying* $\det A \neq 0$. *Then we have an orthogonal matrix $S = (s_{ij})_{i,j=1,\ldots,n}$ and a symmetric positive-definite matrix $P = (p_{ij})_{i,j=1,\ldots,n}$ such that $A = S \circ P$ holds true.*

Proof: On account of $\det A \neq 0$ we have a positive-definite matrix P with $P^2 = A^t A$. The matrix $A^t A$ is namely symmetric and positive-definite, due to

$$(A^t A x, x) = |Ax|^2 > 0 \qquad \text{for all vectors} \quad x \in \mathbb{R}^n \setminus \{0\}.$$

Via the principal axes transformation theorem, we find an orthogonal matrix U and positive eigenvalues $\lambda_1, \ldots, \lambda_n \in (0, +\infty)$ such that

$$A^t A = U^t \circ \Lambda \circ U \qquad \text{with} \quad \Lambda = \begin{pmatrix} \lambda_1 & & 0 \\ & \ddots & \\ 0 & & \lambda_n \end{pmatrix} =: \mathrm{Diag}(\lambda_1, \ldots, \lambda_n)$$

holds true. Setting

$$P := U^t \circ \Lambda^{1/2} \circ U, \qquad \Lambda^{1/2} := \mathrm{Diag}(\sqrt{\lambda_1}, \ldots, \sqrt{\lambda_n}\,),$$

we obtain a symmetric positive-definite matrix with the property

$$P^2 = U^t \circ \Lambda \circ U = A^t A.$$

This implies

$$|Px|^2 = (Px, Px) = (P^2 x, x) = (A^t A x, x) = |Ax|^2$$

and consequently

$$|Px| = |Ax| \qquad \text{for all} \quad x \in \mathbb{R}^n.$$

Now we introduce the matrix $S := A \circ P^{-1}$: For all vectors $x \in \mathbb{R}^n$ we infer

$$|Sx| = |A \circ P^{-1} x| = |P \circ P^{-1} x| = |x|.$$

Therefore, the matrix S is orthogonal, and we arrive at the desired representation $A = S \circ P$. q.e.d.

Theorem 4.6. *Let the quantity $\varepsilon > 0$ and the function $f \in C^1(B_\varepsilon(z), \mathbb{R}^n)$ be given with $f(z) = 0$ as well as $J_f(z) \neq 0$. Then we have the identity*

$$i(f, z) = sgn\ J_f(z) \ \in \{\pm 1\}.$$

Proof: There exists a real $n \times n$-matrix A, such that the representation

$$f(x) = A(x - z) + R(x) \qquad \text{for all} \quad |x - z| \leq \varrho_0 \quad \text{with} \quad 0 < \varrho_0 < \varepsilon$$

holds true. Here the condition $\det A = J_f(z) \neq 0$ is fulfilled, and we have the behavior

$$|R(x)| \leq \eta(\varrho)|x - z| \qquad \text{for all} \quad |x - z| \leq \varrho \leq \varrho_0 \quad \text{with} \quad \lim_{\varrho \to 0} \eta(\varrho) = 0$$

for the remainder term. Due to Proposition 4.5, we have the decomposition $A = S \circ P$ with an orthogonal matrix S and a positive-definite symmetric matrix P. To the matrix

$$P = U^t \circ \mathrm{Diag}(\lambda_1, \ldots, \lambda_n) \circ U$$

we associate the family of positive-definite symmetric matrices

$$P_\tau := U^t \circ \mathrm{Diag}\Big(\tau + (1-\tau)\lambda_1, \ldots, \tau + (1-\tau)\lambda_n\Big) \circ U,$$

which satisfies $P_0 = P$ and $P_1 = E$; here the symbol E denotes the unit matrix. When $\lambda_{\min} > 0$ gives us the least eigenvalue of P and

$$\lambda := \min(1, \lambda_{\min}) > 0$$

is defined, we deduce

$$\begin{aligned}
|P_\tau x|^2 = (P_\tau x, P_\tau x) &= (P_\tau^2 x, x) \\
&= \Big(U^t \circ \mathrm{Diag}\Big([\tau + (1-\tau)\lambda_1]^2, \ldots, [\tau + (1-\tau)\lambda_n]^2\Big) \circ Ux, x\Big) \\
&\geq \Big(U^t \circ \mathrm{Diag}(\lambda^2, \ldots, \lambda^2) \circ Ux, x\Big) \\
&= \lambda^2 (x, x) = \lambda^2 |x|^2,
\end{aligned}$$

and consequently

$$|P_\tau x| \geq \lambda |x| \qquad \text{for all} \quad x \in \mathbb{R}^n \quad \text{and all} \quad \tau \in [0,1].$$

Now we consider the family of mappings

$$f_\tau(x) = f(x, \tau) = S \circ P_\tau(x - z) + (1-\tau)R(x), \quad x \in B_\varepsilon(z), \quad \tau \in [0,1].$$

Evidently, we infer

$$f_0(x) = S \circ P_0(x-z) + R(x) = S \circ P(x-z) + R(x) = A(x-z) + R(x) = f(x)$$

as well as

$$f_1(x) = S \circ P_1(x - z) = S(x - z) =: g(x), \qquad x \in B_\varepsilon(z).$$

Furthermore, we estimate for all points $x \in \mathbb{R}^n$ with $|x - z| = \varrho \in (0, \varrho_0]$ as follows:

$$\begin{aligned}
|f_\tau(x)| &\geq |S \circ P_\tau(x-z)| - (1-\tau)|R(x)| \\
&\geq |P_\tau(x-z)| - |R(x)| \\
&\geq (\lambda - \eta(\varrho))|x - z| \ \geq \ \frac{\lambda}{2}|x - z| \ > \ 0.
\end{aligned}$$

Here we have chosen the quantity $\varrho_0 > 0$ sufficiently small. The homotopy lemma implies

$$d(f_\tau, B_\varrho(z)) = \text{const} \qquad \text{for all parameters} \quad \tau \in [0,1]$$

and finally

$$i(f,z) = d(f, B_\varrho(z)) = d(f_0, B_\varrho(z)) = d(f_1, B_\varrho(z)) = d(g, B_\varrho(z)).$$

When $\omega \in C_0^0((0,\varrho),\mathbb{R})$ denotes an admissible test function with the property

$$\int\limits_{\mathbb{R}^n} \omega(|y|)dy = 1,$$

we infer

$$d(g, B_\varrho(z)) = \int\limits_{|x-z|<\varrho} \omega\big(|g(x)|\big) J_g(x)dx = (\det S) \int\limits_{\mathbb{R}^n} \omega(|x-z|)dx = \det S.$$

We summarize our considerations to the identity $i(f,z) = \det S = \operatorname{sgn} J_f(z)$.
q.e.d.

Theorem 4.7. *The mapping $f : \overline{\Omega} \to \mathbb{R}^n$ may be continuous and the equation*

$$f(x) = 0, \qquad x \in \overline{\Omega},$$

possesses only finitely many solutions $x^{(1)}, \ldots, x^{(N)} \in \Omega$. Let the function f be continuously differentiable in each neighborhood of the zeroes $x^{(\nu)}$, and we assume

$$J_f\big(x^{(\nu)}\big) \neq 0 \quad for \quad \nu = 1, \ldots, N.$$

Then we have the identity

$$d(f,\Omega) = \sum_{\nu=1}^{N} sgn\, J_f\big(x^{(\nu)}\big) = N^+ - N^-.$$

Here the symbols N^+ and N^- give us the numbers of zeroes with $sgn\, J_f = +1$ and $sgn\, J_f = -1$, respectively.

Proof: Theorem 4.4 combined with Theorem 4.6 immediately provide the statement above. q.e.d.

With the assumptions of the theorem above for the function f, we obtain that the degree of mapping is an integer. In the sequel, the latter property will be shown for arbitrary functions $f \in C^0(\overline{\Omega}, \mathbb{R}^n)$ satisfying $f|_{\partial\Omega} \neq 0$.

Proposition 4.8. *With $a = (a_1, \ldots, a_n) \in \mathbb{R}^n$ and $h > 0$, we define the cube*

$$W := \Big\{x \in \mathbb{R}^n \;:\; a_i \leq x_i \leq a_i + h, \; i = 1, \ldots, n\Big\}$$

and consider a function

$$f(x) = (f_1(x_1, \ldots, x_n), \ldots, f_n(x_1, \ldots, x_n)) : W \to \mathbb{R}^n \in C^1(W, \mathbb{R}^n).$$

The associate image-set is denoted by the symbol $W^* := f(W)$. The functional matrix

$$\partial f(x) = \left(\frac{\partial f_i}{\partial x_j}(x)\right)_{i,j=1,\dots,n} = (f_{x_1}(x), \dots, f_{x_n}(x)), \qquad x \in W$$

possesses the following norm:

$$\|\partial f(x)\| := \left(\sum_{i,j=1}^{n} \left(\frac{\partial f_i}{\partial x_j}(x)\right)^2\right)^{\frac{1}{2}} = \left(\sum_{i=1}^{n} |f_{x_i}(x)|^2\right)^{\frac{1}{2}}, \quad x \in W.$$

Furthermore, we have a constant $M \in [0, +\infty)$ and a quantity $\varepsilon \in (0, +\infty)$, such that the inequalities

$$\|\partial f(x')\| \le M \quad \text{and} \quad \|\partial f(x') - \partial f(x'')\| \le \varepsilon \qquad \text{for all pairs} \quad x', x'' \in W$$

hold true. Finally, we have a point $\xi \in W$ satisfying $J_f(\xi) = 0$.
Then we have a function $\varphi = \varphi(y) \in C_0^0(\mathbb{R}^n, [0,1])$ with the property $\varphi(y) = 1$ for all $y \in W^*$, such that

$$\int_{\mathbb{R}^n} \varphi(y)dy \le K(M,n)h^n\varepsilon$$

is correct with the constant $K(M,n) := 4^n \sqrt{n}^n M^{n-1}$.

Remark: Therefore, we can estimate the exterior measure of the set W^* by $K(M,n)h^n\varepsilon$.

Proof of Proposition 4.8:

1. We easily comprehend the invariance of the statement above with respect to translations and rotations. Therefore, we can assume $f(\xi) = 0$ without loss of generality. On acount of the condition $J_f(\xi) = 0$, we find a point $z \in \mathbb{R}^n \setminus \{0\}$ satisfying $z \circ \partial f(\xi) = 0$. Via an adequate rotation, we can assume the condition $z = e_n = (0, \dots, 0, 1) \in \mathbb{R}^n$ without loss of generality, and consequently
$$0 = e_n \circ \partial f(\xi) = \nabla f_n(\xi).$$
2. The intermediate value theorem, applied to each component function, yields
$$f_i(x) = f_i(x) - f_i(\xi) = \sum_{j=1}^{n} \frac{\partial f_i}{\partial x_j}(z^{(i)})(x_j - \xi_j) = \nabla f_i(z^{(i)}) \cdot (x - \xi)$$
with an individual point $z^{(i)} = \xi + t_i(x - \xi)$ and $t_i \in (0,1)$ for each $i \in \{1, \dots, n\}$. This implies
$$|f_i(x)| \le |\nabla f_i(z^{(i)})||x - \xi| \le M\sqrt{n}h, \qquad i = 1, \dots, n-1,$$
$$|f_n(x)| \le |\nabla f_n(z^{(n)})||x - \xi| = |\nabla f_n(z^{(n)}) - \nabla f_n(\xi)||x - \xi| \le \varepsilon\sqrt{n}h$$

for arbitrary points $x \in W$. We obtain

$$W^* \subset W^{**} := \Big\{ y \in \mathbb{R}^n \,:\, |y_i| \le M\sqrt{n}h,\; i = 1, \ldots, n-1;\; |y_n| \le \varepsilon\sqrt{n}h \Big\}.$$

3. Let the function $\varrho \in C_0^0(\mathbb{R}, [0,1])$ with

$$\varrho(t) = \begin{cases} 1, & |t| \le 1 \\ 0, & |t| \ge 2 \end{cases}$$

be given. We set

$$\varphi = \varphi(y) := \varrho\left(\frac{y_1}{M\sqrt{n}h}\right) \cdot \ldots \cdot \varrho\left(\frac{y_{n-1}}{M\sqrt{n}h}\right) \cdot \varrho\left(\frac{y_n}{\varepsilon\sqrt{n}h}\right), \qquad y \in \mathbb{R}^n.$$

Then we observe $\varphi \in C_0^0(\mathbb{R}^n, [0,1])$ and $\varphi(y) = 1$ for all $y \in W^{**} \supset W^*$. Furthermore, we deduce

$$\begin{aligned} &\int\limits_{\mathbb{R}^n} \varphi(y)\,dy \\ &= \int\limits_{-\infty}^{+\infty} \varrho\left(\frac{y_1}{M\sqrt{n}h}\right) dy_1 \cdot \ldots \cdot \int\limits_{-\infty}^{+\infty} \varrho\left(\frac{y_{n-1}}{M\sqrt{n}h}\right) dy_{n-1} \cdot \int\limits_{-\infty}^{+\infty} \varrho\left(\frac{y_n}{\varepsilon\sqrt{n}h}\right) dy_n \\ &= \left(\int\limits_{-\infty}^{+\infty} \varrho(t)\,dt \right)^n M^{n-1} \sqrt{n}^{\,n} h^n \varepsilon \\ &\le (4^n M^{n-1} \sqrt{n}^{\,n}) h^n \varepsilon \;=\; K(M,n) h^n \varepsilon. \end{aligned}$$

q.e.d.

Theorem 4.9. (Sard's lemma)
Let $\Omega \subset \mathbb{R}^n$ denote an open set and $f : \Omega \to \mathbb{R}^n \in C^1(\Omega, \mathbb{R}^n)$ a continuously differentiable mapping. Furthermore, let the set $F \subset \Omega$ be compact and

$$F^* := \Big\{ y = f(x) \,:\, x \in F,\; J_f(x) = 0 \Big\}$$

describe the set of its critical values. *Then F^* is an n-dimensional Lebesgue null-set.*

Proof: Without loss of generality, we can assume that F represents a cube:

$$F = W = \Big\{ x \in \mathbb{R}^n \,:\, a_i \le x_i \le a_i + h,\; i = 1, \ldots, n \Big\}.$$

Now we consider a uniform decomposition of the cube W into N^n subcubes, with the lateral lenghts $\frac{h}{N}$ and an arbitrary number $N \in \mathbb{N}$. This is achieved by decomposing the axes via $a_i + j\frac{h}{N}$ with $i = 1, \ldots, n$ and $j = 0, 1, \ldots, N$ and by a subsequent Cartesian multiplication. In this way we obtain the subcubes W_α for $\alpha = 1, \ldots, N^n$ with the following properties:

$$W = \bigcup_{\alpha=1}^{N^n} W_\alpha, \qquad \overset{\circ}{W}_\alpha \cap \overset{\circ}{W}_\beta = \emptyset \ (\alpha \neq \beta).$$

The diameter of a subcube W_α is determined by

$$\operatorname{diam}(W_\alpha) = \sqrt{n}\,\frac{h}{N}.$$

Now we set

$$M := \sup_{x \in W} \|\partial f(x)\| \qquad \text{and} \qquad \varepsilon_N := \sup_{\substack{x', x'' \in W \\ |x' - x''| \leq \frac{\sqrt{n}h}{N}}} \|\partial f(x') - \partial f(x'')\|.$$

Let $\mathbf{N} \subset \{1, \ldots, N^n\}$ describe the index set belonging to those subcubes W_α, which possess at least one point $\xi \in W_\alpha$ with $J_f(\xi) = 0$. Then we infer the inclusion

$$W^* \subset \bigcup_{\alpha \in \mathbf{N}} W_\alpha^* \qquad \text{with} \quad W_\alpha^* := \Big\{ y = f(x) \ : \ x \in W_\alpha \Big\}.$$

According to Proposition 4.8, we obtain a function to each index $\alpha \in \mathbf{N}$ as follows:

$$\varphi_\alpha = \varphi_\alpha(y) \in C_0^0(\mathbb{R}^n, [0,1]) \quad \text{with} \quad \varphi_\alpha(y) \geq \chi_{W_\alpha^*}(y), y \in \mathbb{R}^n$$

and

$$\int_{\mathbb{R}^n} \varphi_\alpha(y) dy \leq K(M,n) \Big(\frac{h}{N}\Big)^n \varepsilon_N.$$

Here χ_A means the characteristic function of a set A. We infer the estimate

$$\chi_{W^*}(y) \leq \sum_{\alpha \in \mathbf{N}} \chi_{W_\alpha^*}(y) \leq \sum_{\alpha \in \mathbf{N}} \varphi_\alpha(y), \qquad y \in \mathbb{R}^n,$$

and the function $\sum\limits_{\alpha \in \mathbf{N}} \varphi_\alpha(y) \in C_0^0(\mathbb{R}^n, [0, +\infty))$ satisfies

$$\int_{\mathbb{R}^n} \Big(\sum_{\alpha \in \mathbf{N}} \varphi_\alpha(y) \Big)\, dy \leq \sum_{\alpha \in \mathbf{N}} \left(K(M,n) \left(\frac{h}{N}\right)^n \varepsilon_N \right) \leq |W| K(M,n) \varepsilon_N$$

for all $N \in \mathbb{N}$. Letting $N \to \infty$ we observe $\varepsilon_N \downarrow 0$. Therefore, W^* represents an n-dimensional Lebesgue null-set.

q.e.d.

Theorem 4.10. (Generic finiteness)
Let $\Omega \subset \mathbb{R}^n$ denote a bounded open set and $f \in C^1(\Omega, \mathbb{R}^n) \cap C^0(\overline{\Omega}, \mathbb{R}^n)$ a function satisfying $\inf_{x \in \partial\Omega} |f(x)| > \varepsilon > 0$.
Then we have a point $z \in \mathbb{R}^n$ with $|z| \le \varepsilon$ such that the following properties hold true:

(1) The equation $f(x) = z$, $x \in \overline{\Omega}$ possesses at most finitely many solutions $x^{(1)}, \ldots, x^{(N)} \in \Omega$.
(2) The conditions $J_f\big(x^{(\nu)}\big) \neq 0$ are correct for the indices $\nu = 1, \ldots, N$.

Proof: Let us consider the set

$$F := \Big\{x \in \overline{\Omega} \,:\, |f(x)| \le \varepsilon\Big\},$$

and we observe that $F \subset \mathbb{R}^n$ is compact as well as $F \subset \Omega$. The set

$$F^* := \Big\{y = f(x) \,:\, x \in F,\ J_f(x) = 0\Big\}$$

of the critical values for f is a Lebesgue null-set, due to Sard's lemma. Therefore, we find a point $z \in \mathbb{R}^n$ satisfying $|z| \le \varepsilon$ and $z \notin F^*$. Now we show that this point z realizes the property (1): Assuming on the contrary that the equation $f(x) = z$ had infinitely many solutions $x^1, x^2, \ldots \in \overline{\Omega}$, we easily achieve the convergence $x^\nu \to \xi$ for $\nu \to \infty$. When we observe the property $f(x^\nu) = f(\xi) = z$ for all $\nu \in \mathbb{N}$, the preimages of the point z with respect to the mapping f would accumulate at the point ξ. Because $\xi \in \Omega$ and $J_f(\xi) \neq 0$ are correct, the mapping f is there locally injective, and we attain an obvious contradiction. Consequently, only finitely many solutions exist for the equation $f(x) = z$, $x \in \overline{\Omega}$ and each of them has the property (2). q.e.d.

Theorem 4.11. *Let $\Omega \subset \mathbb{R}^n$ denote a bounded open set and $f \in C^0(\overline{\Omega}, \mathbb{R}^n)$ a continuous mapping satisfying $f(x) \neq 0$ for all points $x \in \partial\Omega$. Then the statement $d(f, \Omega) \in \mathbb{Z}$ is correct.*

Proof: We have only to consider mappings $f \in C^1(B, \mathbb{R}^n) \cap C^0(\overline{\Omega}, \mathbb{R}^n)$. Let us choose a sequence of points $\{z^\nu\}_{\nu=1,2,\ldots} \subset \mathbb{R}^n \setminus f(\partial\Omega)$, which do not represent critical values of the function f and fulfill the asymptotic condition

$$\lim_{\nu \to \infty} z^\nu = 0.$$

The functions

$$f_\nu(x) := f(x) - z^\nu, \qquad x \in \overline{\Omega}, \quad \nu \in \mathbb{N}$$

satisfy the condition $d(f_\nu, \Omega) \in \mathbb{Z}$, due to Theorem 4.10 and Theorem 4.7. Consequently, we have an index $\nu_0 \in \mathbb{N}$ such that

$$d(f, \Omega) = d(f_\nu, \Omega) \qquad \text{for all indices} \quad \nu \ge \nu_0$$

is correct. Therefore, the statement $d(f, \Omega) \in \mathbb{Z}$ is established. q.e.d.

5 The Product Theorem

Let the function $f \in C^1(\Omega, \mathbb{R}^n) \cap C^0(\overline{\Omega}, \mathbb{R}^n)$ with $0 < \varepsilon < \inf_{x \in \partial\Omega} |f(x)|$ be given. Furthermore, we take an admissible test function $\omega \in C_0^0((0,\varepsilon), \mathbb{R})$ satisfying

$$\int_{\mathbb{R}^n} \omega(|y|)\, dy = 1.$$

Then we have the identity

$$\int_{\Omega} \omega(|f(x)|)\, J_f(x)\, dx = d(f,\Omega) \int_{\mathbb{R}^n} \omega(|y|)\, dy.$$

Now we shall generalize this identity to the class of arbitrary test functions $\varphi \in C_0^0(\mathbb{R}^n \setminus f(\partial\Omega), \mathbb{R})$. Then we utilize this result to determine the degree of mapping $d(g \circ f, \Omega, z)$ for a composed function $g \circ f$ with the generators $f, g \in C^0(\mathbb{R}^n, \mathbb{R}^n)$, and we obtain the so-called product theorem.

Definition 5.1. *Let $\mathcal{O} \subset \mathbb{R}^n$ denote an open set and assume $x \in \mathcal{O}$. Then we call the following set*

$$G_x := \left\{ y \in \mathcal{O} : \begin{array}{l} \textit{There exists a path } \varphi(t) : [0,1] \to \mathcal{O} \in C^0([0,1]) \\ \textit{satisfying } \varphi(0) = x,\ \varphi(1) = y. \end{array} \right\}$$

the connected component of x in $\mathcal{O}$.

Remarks:

1. The connected component G_x represents the largest open connected subset of $\mathcal{O}$ which contains the point x.
2. When we consider two connected components with G_x and G_y, only the alternative $G_x \cap G_y = \emptyset$ or $G_x = G_y$ is possible.

We easily establish the following

Proposition 5.2. *Each open set $\mathcal{O} \subset \mathbb{R}^n$ can be decomposed into countably many connected components. Therefore, we have open connected sets $\{G_i\}_{i \in I}$ - with the index-set $I \subset \mathbb{N}$ - such that $G_i \cap G_j = \emptyset$ for all $i, j \in I$ with $i \neq j$ as well as*

$$\mathcal{O} = \dot{\bigcup_{i \in I}} G_i$$

hold true. This decomposition is, apart from rearrangements, uniquely determined.

Definition 5.3. *When the function $\varphi \in C_0^0(\mathbb{R}^n)$ is given, we name*

$$supp\,\varphi = \overline{\left\{x \in \mathbb{R}^n \,:\, \varphi(x) \neq 0\right\}}$$

the support of φ.

Proposition 5.4. *Let the open set $\mathcal{O} \subset \mathbb{R}^n$ be decomposable into the connected components $\{G_i\}_{i=1,2,\dots}$, which means $\mathcal{O} = \bigcup\limits_{i=1}^{\infty} G_i$, and take the function $\varphi \in C_0^0(\mathcal{O})$. Then we have the identity*

$$\int\limits_{\mathcal{O}} \varphi(x)dx = \sum_{i=1}^{\infty} \int\limits_{G_i} \varphi(x)\,dx,$$

where the series above possesses only finitely many nonvanishing terms.

Proof: We define the following functions

$$\varphi_i(x) := \begin{cases} \varphi(x), \; x \in G_i \\ 0, \qquad x \in \mathbb{R}^n \setminus G_i \end{cases}, \qquad i = 1, 2, \dots$$

Now we have an index $N_0 \in \mathbb{N}$, such that $\varphi_i(x) \equiv 0$, $x \in \mathbb{R}^n$ is correct for all indices $i \geq N_0$. If this were not true, we could find points $x^{(i_j)} \in G_{i_j}$ for $j = 1, 2, \dots$ satisfying $i_1 < i_2 < \dots$ and $\varphi(x^{(i_j)}) \neq 0$. Since the inclusion

$$\{x^{(i_j)}\}_{j=1,2,\dots} \subset \operatorname{supp} \varphi$$

holds true and $\operatorname{supp} \varphi$ is compact, the selection of a subsequence $x^{(i_j)} \to \xi(j \to \infty)$ allows us to achieve $\xi \in \operatorname{supp} \varphi \subset \mathcal{O}$. When we denote by $G_{i^*} = G_\xi$ the connected component of ξ in $\mathcal{O}$, we can find an index $j_0 \in \mathbb{N}$ such that $x^{(i_j)} \in G_{i^*}$ for all $j \geq j_0$ holds true. This reveals a contradiction to the property $x^{(i_j)} \in G_{i_j}$ for $j = 1, 2, \dots$. Consequently, we see

$$\begin{aligned} \int\limits_{\mathbb{R}^n} \varphi(x)\,dx = \int\limits_{\mathbb{R}^n} \sum_{i=1}^{N_0} \varphi_i(x)\,dx &= \sum_{i=1}^{N_0} \int\limits_{\mathbb{R}^n} \varphi_i(x)\,dx \\ = \sum_{i=1}^{N_0} \int\limits_{G_i} \varphi(x)\,dx &= \sum_{i=1}^{\infty} \int\limits_{G_i} \varphi(x)\,dx. \end{aligned}$$

q.e.d.

Definition 5.5. *Let us consider $f \in C^0(\overline{\Omega}, \mathbb{R}^n)$ and $z \in \mathbb{R}^n \setminus f(\partial\Omega)$. Then we set*

$$d(f, \Omega, z) := d(f(x) - z, \Omega, 0)$$

for the degree of mapping for the function f with respect to the point z.

Proposition 5.6. *If $G \subset \mathbb{R}^n \setminus f(\partial\Omega)$ denotes a domain, we infer*

$$d(f, \Omega, z) = const \qquad \text{for all points} \quad z \in G.$$

Proof: Given the two arbitrary points $z_0, z_1 \in G$, we consider the connecting path

$$\varphi(t) : [0,1] \to G \in C^0([0,1], G), \qquad \varphi(0) = z_0, \quad \varphi(1) = z_1.$$

Now the family of functions $f(x) - \varphi(t)$ with $x \in \Omega$ and $t \in [0,1]$ describes a homotopy. This implies

$$d(f, \Omega, \varphi(t)) = d(f - \varphi(t), \Omega, 0) = \text{const}, \qquad t \in [0,1],$$

and we obtain $d(f, \Omega, z_0) = d(f, \Omega, z_1)$, in particular. q.e.d.

Definition 5.7. *When $G \subset \mathbb{R}^n \setminus f(\partial\Omega)$ represents a domain, we define*

$$d(f, \Omega, G) := d(f, \Omega, z) \qquad \text{for a point} \quad z \in G.$$

Remark: Let $\Omega \subset \mathbb{R}^n$ denote an open bounded set and $f \in C^0(\overline{\Omega}, \mathbb{R}^n)$ a continuous function. Then the set $f(\partial\Omega) \subset \mathbb{R}^n$ is compact. When $\{G_i\}_{i=1,\ldots,N_0}$ with $N_0 \in \{0, 1, \ldots, +\infty\}$ constitute the bounded connected components of $\mathbb{R}^n \setminus f(\partial\Omega)$ and G_∞ the unbounded connected component, we have the representation

$$\mathbb{R}^n \setminus f(\partial\Omega) = \bigcup_{i=1}^{N_0} G_i \,\cup\, G_\infty.$$

Since we can find a point $z \notin f(\overline{\Omega})$, we infer the identity

$$d(f, \Omega, G_\infty) = d(f, \Omega, z) = 0.$$

Theorem 5.8. *Let $\{f_k\}_{k=1,2,\ldots} \subset C^1(\Omega, \mathbb{R}^n) \cap C^0(\overline{\Omega}, \mathbb{R}^n)$ denote a sequence of functions, which converge uniformly on the set $\overline{\Omega}$ towards the function $f \in C^0(\overline{\Omega}, \mathbb{R}^n)$. Furthermore, let*

$$\mathbb{R}^n \setminus f(\partial\Omega) = \bigcup_{i=1}^{N_0} G_i \,\cup\, G_\infty, \quad N_0 \in \{0, 1, \ldots, +\infty\}$$

describe the decomposition into their connected components. To each function $\varphi \in C_0^0(\mathbb{R}^n \setminus f(\partial\Omega))$ we find a number $k^* = k^*(\varphi) \in \mathbb{N}$, such that the identity

$$\int_\Omega \varphi(f_k(x)) J_{f_k}(x)\, dx = \sum_{i=1}^{N_0} d(f, \Omega, G_i) \int_{G_i} \varphi(z)\, dz$$

for all indices $k \geq k^*$ is correct. Here the series above possesses only finitely many nonvanishing terms - even in the case $N_0 = +\infty$.

Proof:

1. We observe that $\operatorname{supp}\varphi \cap f(\partial\Omega) = \emptyset$ holds true and both sets are compact. Therefore, we find a quantity $\varepsilon_0 > 0$ such that the estimate

$$|f(x) - z| > \varepsilon_0 \quad \text{for all} \quad x \in \partial\Omega \quad \text{and all} \quad z \in \operatorname{supp}\varphi$$

is correct. Because the convergence $f_k \to f$ is uniform on $\overline{\Omega}$, we find an index $k^* = k^*(\varphi) \in \mathbb{N}$ such that

$$|f_k(x) - z| > \varepsilon_0 \qquad \text{for all points} \quad x \in \partial\Omega, \quad z \in \operatorname{supp}\varphi, \quad k \geq k^*$$

holds true. We then take an admissible test function $\omega \in C_0^0((0,1),\mathbb{R})$ satisfying $\int\limits_{\mathbb{R}^n} \omega(|y|)\,dy = 1$. With the number $\varepsilon \in (0, \varepsilon_0]$, we set

$$\omega_\varepsilon(r) := \frac{1}{\varepsilon^n}\,\omega\Big(\frac{r}{\varepsilon}\Big) \in C_0^0((0,\varepsilon),\mathbb{R}) \quad \text{satisfying} \quad \int\limits_{\mathbb{R}^n} \omega_\varepsilon(|y|)\,dy = 1.$$

Finally, we define the function

$$\vartheta(z) := \begin{cases} d(f,\Omega,z), & \text{if } z \in \mathbb{R}^n \setminus f(\partial\Omega) \\ 0, & \text{if } z \in f(\partial\Omega) \end{cases}.$$

2. For all points $z \in \operatorname{supp}\varphi$ and all indices $k \geq k^*(\varphi)$ we observe

$$\vartheta(z) = d(f,\Omega,z) = \int\limits_{\Omega} \omega_\varepsilon(|f_k(x) - z|) J_{f_k}(x)\,dx, \qquad 0 < \varepsilon \leq \varepsilon_0.$$

Now the integration of $\varphi(z)\vartheta(z) \in C_0^0(\mathbb{R}^n \setminus f(\partial\Omega))$ yields

$$\begin{aligned}\int\limits_{\mathbb{R}^n} \varphi(z)\vartheta(z)\,dz &= \int\limits_{\mathbb{R}^n}\Big(\int\limits_{\Omega} \varphi(z)\omega_\varepsilon(|f_k(x) - z|) J_{f_k}(x)\,dx\Big)\,dz \\ &= \int\limits_{\Omega}\Big(\int\limits_{\mathbb{R}^n} \varphi(z)\omega_\varepsilon(|f_k(x) - z|)\,dz\Big) J_{f_k}(x)\,dx.\end{aligned}$$

On the other hand, Proposition 5.4 implies

$$\begin{aligned}\int\limits_{\mathbb{R}^n} \varphi(z)\vartheta(z)\,dz &= \Big(\int\limits_{G_\infty} \varphi(z)\,dz\Big) \underbrace{d(f,\Omega,G_\infty)}_{=0} \\ &\quad + \sum_{i=1}^{N_0}\Big(\int\limits_{G_i} \varphi(z)\,dz\Big)\,d(f,\Omega,G_i),\end{aligned}$$

where the series above possesses only finitely many nonvanishing terms. The transition $\varepsilon \to 0+$ to the limit gives us the identity

$$\sum_{i=1}^{N_0} d(f,\Omega,G_i) \int\limits_{G_i} \varphi(z)\,dz = \int\limits_{\Omega} \varphi(f_k(x)) J_{f_k}(x)\,dx \quad \text{for all} \quad k \geq k^*(\varphi).$$

Here we take the convergence

$$\lim_{\varepsilon \to 0+} \int\limits_{\mathbb{R}^n} \varphi(z)\omega_\varepsilon(|f_k(x)-z|)\,dz = \varphi(f_k(x)) \qquad \text{for} \quad x \in \Omega \quad \text{uniformly}$$

into account. q.e.d.

Theorem 5.9. (Product theorem for the degree of mapping)
Let us consider the functions $f, g \in C^0(\mathbb{R}^n,\mathbb{R}^n)$, and let the set $\Omega \subset \mathbb{R}^n$ be open and bounded. We set $E := f(\partial\Omega)$. With the symbols $\{D_i\}_{i=1,\ldots,N_0}$, where $N_0 \in \{0,1,\ldots,+\infty\}$ holds true, we denote the bounded connected components of the set $\mathbb{R}^n \setminus E$. Finally, we choose a point $z \in \mathbb{R}^n \setminus g(E)$. Then we have the identity

$$d(g \circ f, \Omega, z) = \sum_{i=1}^{N_0} d(f,\Omega,D_i)\, d(g,D_i,z),$$

where this series possesses only finitely many nonvanishing terms.

Proof: (L.Bers)

1. We define

$$h(x) := g \circ f(x).$$

According to the Weierstraß approximation theorem from Section 1 in Chapter 1, we can choose sequences

$$\{f_l(x)\}_{l=1,2,\ldots} \subset C^1(\mathbb{R}^n,\mathbb{R}^n) \quad \text{and} \quad \{g_k(y)\}_{k=1,2,\ldots} \subset C^1(\mathbb{R}^n,\mathbb{R}^n)$$

which converge uniformly on each compact set towards the function $f(x)$ and $g(y)$, respectively. In addition, we define the functions

$$h_k(x) := g_k \circ f(x), \quad h_{kl}(x) := g_k \circ f_l(x), \qquad k,l \in \mathbb{N}.$$

This implies the uniform convergence

$$h_k(x) \longrightarrow h(x) \qquad \text{for} \quad k \to \infty$$

as well as

$$h_{kl}(x) \longrightarrow h_k(x) \qquad \text{for} \quad l \to \infty$$

on each compact set.

2. There exists a quantity $\varepsilon > 0$, such that the estimate

$$|h(x) - z| > \varepsilon, \quad |h_k(x) - z| > \varepsilon \quad \text{for all} \quad x \in \partial\Omega \quad \text{and all} \quad k \geq k_0(\varepsilon)$$

is correct. Now we choose an admissible test function $\omega \in C_0^0((0, \varepsilon), \mathbb{R})$ with the property $\int\limits_{\mathbb{R}^n} \omega(|u|)du = 1$. Then we obtain the following identity for all indices $k \geq k_0(\varepsilon)$ and $l \geq l_0(k)$, namely

$$\begin{aligned} d(h_k, \Omega, z) = d(h_{kl}, \Omega, z) &= \int\limits_{\Omega} \omega(|h_{kl}(x) - z|) J_{h_{kl}}(x)\, dx \\ &= \int\limits_{\Omega} \omega\big(|g_k(f_l(x)) - z|\big) J_{g_k}(f_l(x)) J_{f_l}(x)\, dx. \end{aligned}$$

When we define

$$\varphi_k(y) := \omega(|g_k(y) - z|) J_{g_k}(y) \in C_0^0(\mathbb{R}^n \setminus E) \qquad \text{for} \quad k \geq k_0,$$

Theorem 5.8 yields

$$d(h_k, \Omega, z) = \int\limits_{\Omega} \varphi_k(f_l(x)) J_{f_l}(x)\, dx = \sum_{i=1}^{N_0} d(f, \Omega, D_i) \int\limits_{D_i} \varphi_k(y)\, dy$$

for $k \geq k_0$. Here only finitely many terms of the sum are nonvanishing. When we note that

$$\int\limits_{D_i} \varphi_k(y)\, dy = \int\limits_{D_i} \omega(|g_k(y) - z|) J_{g_k}(y)\, dy = d(g_k, D_i, z), \qquad k \geq k_0,$$

we immediately infer

$$d(h_k, \Omega, z) = \sum_{i=1}^{N_0} d(f, \Omega, D_i)\, d(g_k, D_i, z).$$

Now we have an index $k_1 \geq k_0$ such that

$$d(h_k, \Omega, z) = d(h, \Omega, z) \qquad \text{for all} \quad k \geq k_1$$

holds true. Furthermore, we find an index $k_2 \geq k_1$ such that

$$d(g_k, D_i, z) = d(g, D_i, z) \qquad \text{for all} \quad k \geq k_2 \quad \text{and all} \quad i = 1, \ldots, N_0$$

is valid. We summarize our results to

$$d(h, \Omega, z) = \sum_{i=1}^{N_0} d(f, \Omega, D_i)\, d(g, D_i, z).$$

q.e.d.

6 Theorems of Jordan-Brouwer

Let us consider the compact set $F \subset \mathbb{R}^n$, and we denote the number of connected components for the open set $\mathbb{R}^n \setminus F$ by the symbol $N(F) \in \{0, 1, \ldots, +\infty\}$.

Theorem 6.1. (C. Jordan and J.L.E. Brouwer)
Let two homeomorphic compact sets F and F^* in $\mathbb{R}^n$ be given. Then we have the identity $N(F) = N(F^*)$.

Proof: (J. Leray)
Since the sets F and F^* are homeomorphic, we have a topological mapping $\hat{f} : F \to F^*$ with its inverse mapping $\hat{f}^{-1} : F^* \to F$. With the aid of Tietze's extension theorem we construct mappings $f, g \in C^0(\mathbb{R}^n)$ satisfying $f(x) = \hat{f}(x)$ for all points $x \in F$ and $g(y) = \hat{f}^{-1}(y)$ for all points $y \in F^*$. On the contrary, we assume that

$$N := N(F) \neq N(F^*) =: N^*$$

was correct: Then we depart from the inequality $N^* < N$, without loss of generality. Consequently, the number N^* is finite. We denote by the symbols $\{D_i\}_{i=1,\ldots,N}$ and $\{D_i^*\}_{i=1,\ldots,N^*}$ the bounded connected components of $\mathbb{R}^n \setminus F$ and $\mathbb{R}^n \setminus F^*$, respectively. When we take $z \in D_k$ and $k \in \{1, \ldots, N^* + 1\}$, the product theorem yields

$$\begin{aligned} \delta_{ik} = d(g \circ f, D_i, D_k) &= d(g \circ f, D_i, z) \\ &= \sum_{j=1}^{N^*} \underbrace{d(f, D_i, D_j^*)}_{:=a_{ij}} \underbrace{d(g, D_j^*, z)}_{:=b_{jk}} = \sum_{j=1}^{N^*} a_{ij} b_{jk} \qquad \text{for} \quad i, k = 1, \ldots, N^* + 1. \end{aligned}$$

Now we have a point $\xi = (\xi_1, \ldots, \xi_{N^*+1}) \in \mathbb{R}^{N^*+1} \setminus \{0\}$ satisfying

$$\sum_{k=1}^{N^*+1} b_{jk} \xi_k = 0 \qquad \text{for} \quad j = 1, \ldots, N^*.$$

We obtain with the relation

$$\xi_i = \sum_{j=1}^{N^*} \sum_{k=1}^{N^*+1} a_{ij} b_{jk} \xi_k = \sum_{j=1}^{N^*} a_{ij} \left(\sum_{k=1}^{N^*+1} b_{jk} \xi_k \right) = 0, \qquad i = 1, \ldots, N^* + 1,$$

which constitutes an evident contradiction. The assumption $N \neq N^*$ was incorrect and equality follows.

q.e.d.

Theorem 6.2. (C. Jordan and J.L.E. Brouwer)
Let $S^* \subset \mathbb{R}^n$ denote a set which is homeomorphic to the unit sphere $S = \{x \in \mathbb{R}^n : |x| = 1\}$ via the topological mapping $\hat{f} : S \to S^*$. Then this topological sphere *S^* decomposes the space $\mathbb{R}^n$ into a bounded domain G_1, which we call the* inner domain, *and an unbounded domain G_2, which we call the* outer domain. *We have the following property for the mapping $\hat{f}$, namely*

$$v(\hat{f}, S, z) = \begin{cases} \pm 1, & \text{for } z \in G_1 \\ 0, & \text{for } z \in G_2 \end{cases}.$$

Proof: As in the proof of Theorem 6.1, we extend the mappings $\hat{f} : S \to S^*$ and $\hat{f}^{-1} : S^* \to S$ to continuous mappings f and g, respectively, onto the whole space $\mathbb{R}^n$. Since the sphere S decomposes $\mathbb{R}^n$ into an inner domain and an outer domain, we infer the identity

$$N(S^*) = N(S) = 1$$

from Theorem 6.1. The mapping $g \circ f$ satisfies $g \circ f(x) = x$ for all $x \in S$. Now the product theorem implies

$$1 = d(g \circ f, B, 0) = d(f, B, G_1)\, d(g, G_1, 0), \qquad B := B_1(0).$$

Since the degree of mapping gives us an integer, the point $z \in G_1$ possesses the order

$$v(\hat{f}, S, z) = d(f, B, G_1) = \pm 1.$$

q.e.d.

Remark: In the special case $n = 2$ of our theorem above, we call the topological sphere a *Jordan curve* and the associate inner domain a *Jordan domain.* We obtain *Jordan's curve theorem* in this plane situation. However, we have proved *Brouwer's sphere theorem* in the higher-dimensional case $n \in \mathbb{N}$ with $n \geq 3$.

Theorem 6.3. *Let U_z denote an n-dimensional neighborhood of the point $z \in \mathbb{R}^n$, and the mapping $f : U_z \to \mathbb{R}^n$ may be injective and continuous satisfying $f(z) = 0$. Then we have $i(f, z) = \pm 1$.*

Proof: At first, we choose the quantity $\varrho > 0$ so small that

$$B_\varrho(z) := \{x \in \mathbb{R}^n : |x - z| < \varrho\}$$

is subject to the condition $\overline{B_\varrho(z)} \subset U_z$. Then we consider the sphere $S := \partial B_\varrho(z)$ and the topological sphere $S^* := f(S)$, where G_1 denotes the inner domain of S^*. Now Theorem 6.2 implies

$$d(f, B_\varrho(z), G_1) = \pm 1.$$

When we choose the point $y' \in G_1$ with its preimage $x' \in B_\varrho(z)$, we observe $f(x') = y'$. The connecting straight line

$$\mathbf{s} := \{(1-t)z + tx' \,:\, 0 \leq t \leq 1\} \subset B_\varrho(z)$$

possesses the image arc $\mathbf{s}^* := f(\mathbf{s}) \subset \mathbb{R}^n \setminus S^*$. Since the point $y' = f(x') \in \mathbf{s}^*$ is situated within G_1, we infer $0 = f(z) \in G_1$. Consequently, we obtain

$$i(f,z) = d(f, B_\varrho(z), 0) = d(f, B_\varrho(z), G_1) = \pm 1.$$

q.e.d.

Theorem 6.4. (Invariance of domains in $\mathbb{R}^n$)
Let $G \subset \mathbb{R}^n$ denote a domain and $f: G \to \mathbb{R}^n$ a continuous injective mapping. Then the image $G^* := f(G)$ is a domain as well.

Proof: Since the set G is connected and the function f is continuous, we obtain that $G^* = f(G)$ is connected. We now show that the set G^* is open: Let the point $z \in G$ be arbitrary, and the quantity $\varrho > 0$ may be chosen so small that $\overline{B_\varrho(z)} \subset G$ is fulfilled. The continuous injective mapping

$$g(x) := f(x) - f(z), \qquad x \in \overline{B_\varrho(z)}$$

satisfies $i(g,z) = \pm 1$, due to Theorem 6.3. This implies

$$d(f, B_\varrho(z), f(z)) = d(g, B_\varrho(z), 0) = \pm 1.$$

Taking the quantity $\varepsilon > 0$ sufficiently small, we infer the estimate

$$|f(x) - f(z)| > \varepsilon \quad \text{for all} \quad x \in \partial B_\varrho(z).$$

The homotopy theorem gives us the identity

$$d(f, B_\varrho(z), \zeta) = d(f, B_\varrho(z), f(z)) = \pm 1 \quad \text{for} \quad |\zeta - f(z)| < \frac{\varepsilon}{2}.$$

For all $\zeta \in \mathbb{R}^n$ with $|\zeta - f(z)| < \frac{\varepsilon}{2}$ we have a point $x \in B_\varrho(z)$ satisfying $f(x) = \zeta$. This means $B_{\frac{\varepsilon}{2}}(f(z)) \subset f(G)$. Consequently, the function f represents an open mapping, and the image $G^* = f(G)$ is a domain.

q.e.d.

We supplement the following result to Theorem 6.2.

Theorem 6.5. (Jordan, Brouwer) *Each topological sphere $S^* \subset \mathbb{R}^n$ decomposes the space $\mathbb{R}^n$ into an inner domain G_1 and an outer domain G_2, which means*

$$\mathbb{R}^n = G_1 \,\dot{\cup}\, S^* \,\dot{\cup}\, G_2.$$

Furthermore, we have the property $\partial G_1 = S^* = \partial G_2$.

Proof: We only have to show the property $\partial G_i = S^*$ for $i = 1, 2$. Let $f: S \to S^*$ represent the topological mapping, and take with $\tilde{x} \in S^*$ an arbitrary point. Then we define $\xi := f^{-1}(\tilde{x}) \in S$, and we consider the following sets

$$E := \{x \in S : |x - \xi| \leq \varepsilon\} \quad \text{and} \quad F := \{x \in S : |x - \xi| \geq \varepsilon\}$$

with $S = E \cup F$. The transition to their images $E^* := f(E)$ and $F^* := f(F)$ yields the relation $S^* = E^* \cup F^*$. Since the set $\mathbb{R}^n \setminus F$ is connected, our Theorem 6.1 tells us that $\mathbb{R}^n \setminus F^*$ is connected as well. Consequently, there exists a continuous path π connecting two arbitrarily chosen points $a_1 \in G_1$ and $a_2 \in G_2$, which does not meet the set F^*. Since the image S^* separates the domains G_1 and G_2, we see $\pi \cap S^* \neq \emptyset$ and therefore $\pi \cap E^* \neq \emptyset$. When $a_1' \in \pi$ is the first point starting from a_1, which meets the set E^*, and $a_2' \in \pi$ is the first point starting from a_2, which meets the set E^*, we select two points $a_i'' \in G_i$ for $i = 1, 2$ on the path π satisfying $|a_i'' - a_i'| \leq \varepsilon$. Now we observe $\varepsilon \downarrow 0$ and obtain two sequences of points $\{a_{i,j}''\}_{j=1,2,\ldots} \subset G_i$ for $i = 1, 2$ such that

$$\lim_{j \to \infty} a_{i,j}'' = \tilde{x} \quad \text{for } i = 1, 2.$$

Thus we arrive at the conclusion $\partial G_1 = S^* = \partial G_2$. q.e.d.

Figure 1.3 PORTRAIT OF BERNHARD RIEMANN (1826–1866) taken from page 848 of his *Gesammelte Werke*, Springer-Verlag (1990).

Chapter 4

Generalized Analytic Functions

The theory of analytic functions in one and several complex variables has been founded by Cauchy, Riemann and Weierstraß and belongs to the most beautiful mathematical creations of modern times. We recommend the textbooks of Behnke-Sommer [BS], Grauert-Fritzsche [Gr], [GF], Hurwitz-Courant [HC] and Vekua [V]. The investigations of analytic functions with respect to their differentiable properties will be founded on the integral theorems from Chapter 1 and with respect to their topological properties will be based on the winding number from Chapter 3. We additionally obtain a direct approach to the solutions of the inhomogeneous Cauchy-Riemann differential equations in this chapter. In the last section we investigate the discontinuous behavior of Cauchy's integral across the boundary.

1 The Cauchy-Riemann Differential Equation

We begin with the following

Definition 1.1. *Let the function $f = f(z) : \Omega \to \mathbb{C}$ be defined on the open set $\Omega \subset \mathbb{C}$, and $z_0 \in \Omega$ denotes an arbitrary point. Then we name the function f* complex differentiable at the point z_0 *if the following limit*

$$\lim_{\substack{z \to z_0 \\ z \neq z_0}} \frac{f(z) - f(z_0)}{z - z_0} =: f'(z_0)$$

exists. We call $f'(z_0)$ the complex derivative *of the function f at the point z_0. When $f'(z)$ exists for all $z \in \Omega$ and the function $f' : \Omega \to \mathbb{C}$ is continuous, we name the function f* holomorphic in Ω.

We note the well-known

F. Sauvigny, *Partial Differential Equations 1*, Universitext,
DOI 10.1007/978-1-4471-2981-3_4,

Theorem 1.2. *If the power series*

$$f(z) = \sum_{n=0}^{\infty} a_n z^n$$

converges for all points $|z| < R$ *with the radius of convergence* $R > 0$ *being fixed, then the function* $f(z)$ *is holomorphic in the disc* $\{z \in \mathbb{C} : |z| < R\}$ *and we have*

$$f'(z) = \sum_{n=1}^{\infty} n a_n z^{n-1}.$$

Proof:

1. At first, we show convergence of the series

$$\sum_{n=1}^{\infty} n a_n z^{n-1}$$

for all points $|z| < R$. According to Cauchy's convergence criterion for series, the given series converges if and only if the series

$$\sum_{n=1}^{\infty} n a_n z^n = \sum_{n=1}^{\infty} b_n z^n \qquad \text{with} \quad b_n := n a_n$$

converges. Now we observe

$$\limsup_{n\to\infty} \sqrt[n]{|b_n|} = \limsup_{n\to\infty} \left(\sqrt[n]{n}\, \sqrt[n]{|a_n|} \right) = \limsup_{n\to\infty} \sqrt[n]{|a_n|}.$$

Consequently, this series possesses the same radius of convergence $R > 0$ as the series $\sum\limits_{n=0}^{\infty} a_n z^n$.

2. Choosing a fixed point $z \in \mathbb{C}$ with $|z| \le R_0 < R$ we take a point $w \ne z$ satisfying $|w| \le R_0$ and calculate

$$\begin{aligned} \frac{f(w) - f(z)}{w - z} &= \sum_{n=0}^{\infty} a_n \frac{w^n - z^n}{w - z} \\ &= \sum_{n=1}^{\infty} a_n \left(w^{n-1} + w^{n-2} z + \ldots + z^{n-1} \right) \qquad (1) \\ &= \sum_{n=1}^{\infty} a_n g_n(w, z). \end{aligned}$$

Here we have set $g_n(w, z) := w^{n-1} + w^{n-2} z + \ldots + z^{n-1}$ for $n \in \mathbb{N}$. We note that

$$|a_n g_n(w, z)| \le n |a_n| R_0^{n-1} \qquad \text{for all} \quad |w| \le R_0, \ |z| \le R_0.$$

Now part 1 of our proof yields

$$\sum_{n=1}^{\infty} n|a_n|R_0^{n-1} < +\infty.$$

Due to the Weierstraß majorant test, the uniform convergence of the series in (1) for $|w| \le R_0$, $|z| \le R_0$ follows. Performing the transition to the limit $w \to z$ in (1), we obtain

$$f'(z) = \sum_{n=0}^{\infty} a_n g_n(z,z) = \sum_{n=1}^{\infty} n a_n z^{n-1}.$$

q.e.d.

Now we shall study the relationship between complex differentiation and partial differentiation.

Theorem 1.3. *Let the function* $w = f(z) = f(x,y) = u(x,y) + iv(x,y) : \Omega \to \mathbb{C}$ *be holomorphic in the open set* $\Omega \subset \mathbb{C}$*. Then we have* $f \in C^1(\Omega, \mathbb{C})$ *and the following two equivalent conditions are satisfied:*

$$f_x + i f_y = 0 \qquad in \quad \Omega, \tag{2}$$

or

$$u_x = v_y, \quad u_y = -v_x \qquad in \quad \Omega. \tag{3}$$

The equations (3) are named the Cauchy-Riemann differential equations.

Remark: The functions $u = \operatorname{Re} f(z) : \Omega \to \mathbb{R}$ and $v = \operatorname{Im} f(z) : \Omega \to \mathbb{R}$ denote the *real* and *imaginary part of the function* f, respectively.

Proof of Theorem 1.3: Since the function f is holomorphic in Ω, the complex derivative exists:

$$f'(z) = \lim_{\substack{|\Delta z| \to 0 \\ \Delta z \in \mathbb{C} \setminus \{0\}}} \frac{f(z + \Delta z) - f(z)}{\Delta z}.$$

Setting $\Delta z = \varepsilon > 0$, we find in particular

$$\begin{aligned} f'(z) &= \lim_{\substack{\varepsilon \to 0 \\ \varepsilon \in \mathbb{R} \setminus \{0\}}} \frac{f(z+\varepsilon) - f(z)}{\varepsilon} \\ &= \lim_{\substack{\varepsilon \to 0 \\ \varepsilon \in \mathbb{R} \setminus \{0\}}} \frac{f(x+\varepsilon, y) - f(x,y)}{\varepsilon} = f_x(x,y). \end{aligned}$$

Passing to the limit with $\Delta z = i\varepsilon$, we deduce

$$\begin{aligned} f'(z) &= \lim_{\substack{\varepsilon \to 0 \\ \varepsilon \in \mathbb{R} \setminus \{0\}}} \frac{f(z+i\varepsilon) - f(z)}{i\varepsilon} \\ &= \lim_{\substack{\varepsilon \to 0 \\ \varepsilon \in \mathbb{R} \setminus \{0\}}} \frac{f(x, y+\varepsilon) - f(x,y)}{i\varepsilon} = \frac{1}{i} f_y(x,y). \end{aligned}$$

Consequently, $f \in C^1(\Omega, \mathbb{C})$ holds true, and (2) is immediately inferred from $f_x = f' = \frac{1}{i} f_y$ in Ω. On account of the identity

$$f_x + i f_y = (u + iv)_x + i(u + iv)_y = (u_x - v_y) + i(v_x + u_y)$$

the relation (2) is satisfied if and only if (3) is correct. q.e.d.

Remark: The property (2) for holomorphic functions implies that the oriented angles are preserved with respect to the mapping $w = f(z)$ at all points $z \in \Omega$ with $f'(z) \neq 0$. Mercator has already discovered this property *conformal* - so important for geography - proposing his well-known stereographic projection of the sphere onto the plane.

Theorem 1.4. *Let the function $f(z) = u(x, y) + iv(x, y) \in C^1(\Omega, \mathbb{C})$ be defined on the open set $\Omega \subset \mathbb{R}^2 \cong \mathbb{C}$, and we assume (2) or alternatively (3). Then the function f is holomorphic in Ω.*

Proof: We apply the mean value theorem separately on the functions $u = \operatorname{Re} f$ and $v = \operatorname{Im} f$. Utilizing $z = x+iy \in \Omega$ and $\Delta z = \Delta x+i\Delta y \in \mathbb{C}$ with $|\Delta z| < \varepsilon$, we obtain the identities

$$\begin{aligned} u(z + \Delta z) - u(z) &= u(x + \Delta x, y + \Delta y) - u(x, y) \\ &= u_x(\xi_1, \eta_1)\Delta x + u_y(\xi_1, \eta_1)\Delta y \\ &= u_x(\xi_1, \eta_1)\Delta x - v_x(\xi_1, \eta_1)\Delta y \end{aligned}$$

and

$$\begin{aligned} v(z + \Delta z) - v(z) &= v(x + \Delta x, y + \Delta y) - v(x, y) \\ &= v_x(\xi_2, \eta_2)\Delta x + v_y(\xi_2, \eta_2)\Delta y \\ &= v_x(\xi_2, \eta_2)\Delta x + u_x(\xi_2, \eta_2)\Delta y \end{aligned}$$

at the intermediate points $(\xi_1, \eta_1), (\xi_2, \eta_2) \in \Omega$ satisfying $|z - (\xi_k + i\eta_k)| < \varepsilon$ for $k = 1, 2$. We now compose

$$\begin{aligned} &f(z + \Delta z) - f(z) \\ &\quad= \Big\{u(x + \Delta x, y + \Delta y) - u(x, y)\Big\} + i\Big\{v(x + \Delta x, y + \Delta y) - v(x, y)\Big\} \\ &\quad= \Big\{u_x(\xi_1, \eta_1) + iv_x(\xi_2, \eta_2)\Big\}\Delta x + i\Big\{u_x(\xi_2, \eta_2) + iv_x(\xi_1, \eta_1)\Big\}\Delta y \\ &\quad= \Big\{u_x(\xi_1, \eta_1) + iv_x(\xi_2, \eta_2)\Big\}(\Delta x + i\Delta y) \\ &\qquad+i\Big\{\Big[u_x(\xi_2, \eta_2) - u_x(\xi_1, \eta_1)\Big] + i\Big[v_x(\xi_1, \eta_1) - v_x(\xi_2, \eta_2)\Big]\Big\}\Delta y. \end{aligned}$$

Abbreviating

$$g(z,\Delta z) := \Big[u_x(\xi_2,\eta_2) - u_x(\xi_1,\eta_1)\Big] + i\Big[v_x(\xi_1,\eta_1) - v_x(\xi_2,\eta_2)\Big]$$

we find

$$\frac{f(z+\Delta z)-f(z)}{\Delta z} = u_x(\xi_1,\eta_1) + iv_x(\xi_2,\eta_2) + ig(z,\Delta z)\frac{\Delta y}{\Delta z}.$$

The transition to the limit $|\Delta z| \to 0$ yields

$$\lim_{\substack{|\Delta z|\to 0\\ \Delta z\in\mathbb{C}\setminus\{0\}}} \frac{f(z+\Delta z)-f(z)}{\Delta z} = f_x(z) + \lim_{\substack{|\Delta z|\to 0\\ \Delta z\in\mathbb{C}\setminus\{0\}}} \Big\{ig(z,\Delta z)\frac{\Delta y}{\Delta z}\Big\} = f_x(z)$$

utilizing $f \in C^1(\Omega,\mathbb{C})$. Therefore, the function $f:\Omega\to\mathbb{C}$ is holomophic in Ω due to Definition 1.1.

q.e.d.

2 Holomorphic Functions in $\mathbb{C}^n$

Our present considerations are based on the theory of curvilinear integrals from Section 6 in Chapter 1.

We choose the domain $\Omega \subset \mathbb{C}$ and denote by

$$w = f(z) = u(x,y) + iv(x,y), \qquad (x,y)\in\Omega$$

a complex-valued function with $u, v \in C^1(\Omega,\mathbb{R})$. Let the points $P, Q \in \Omega$ and the curve $X \in \mathcal{C}(\Omega, P, Q)$ be given. Then we consider the curvilinear integral

$$\begin{aligned}\int_X f(z)\,dz &= \int_X \Big\{u(x,y)+iv(x,y)\Big\}(dx+idy)\\ &= \int_X (u\,dx - v\,dy) + i\int_X (v\,dx + u\,dy)\\ &= \int_X \omega_1 + i\int_X \omega_2\end{aligned}$$

with the real differential forms

$$\omega_1 := u\,dx - v\,dy, \quad \omega_2 := v\,dx + u\,dy.$$

Now the differential forms ω_1 and ω_2 are closed if and only if the identities

$$0 = d\omega_1 = -\left(\frac{\partial u}{\partial y} + \frac{\partial v}{\partial x}\right) dx \wedge dy,$$

$$0 = d\omega_2 = \left(\frac{\partial u}{\partial x} - \frac{\partial v}{\partial y}\right) dx \wedge dy$$

hold true in Ω. We infer the following equations:

$$\frac{\partial u(x,y)}{\partial x} = \frac{\partial v(x,y)}{\partial y}, \quad \frac{\partial u(x,y)}{\partial y} = -\frac{\partial v(x,y)}{\partial x} \quad \text{in} \quad \Omega. \tag{1}$$

This *Cauchy-Riemann system of differential equations* is equivalent to the property that the function $f : \Omega \to \mathbb{C}$ is holomorphic, and therefore possesses a complex derivative at each point $z \in \Omega$ which represents a continuous function.

Theorem 2.1. (Cauchy, Riemann)
Let $\Omega \subset \mathbb{C}$ denote a simply connected domain, and let $f \in C^1(\Omega, \mathbb{C})$ be a continuously differentiable function. Then the following four statements are equivalent:

(a) The function f is holomorphic in Ω;
(b) The real part and the imaginary part of $f(x,y) = u(x,y) + iv(x,y)$ satisfy the Cauchy-Riemann system of differential equations (1);
(c) For each closed curve $X \in \mathcal{C}(\Omega, P, P)$ with $P \in \Omega$ we have the identity

$$\int_X f(z)\, dz = 0;$$

(d) There exists a holomorphic function $F : \Omega \to \mathbb{C}$ satisfying

$$F'(z) = f(z), \qquad z \in \Omega,$$

which represents a primitive function F of f.

Proof:

1. The equivalence of $(a) \Leftrightarrow (b)$ has been shown in Section 1.
2. Now we prove $(b) \Leftrightarrow (c)$: Evidently, the condition

$$\int_X f(z)\, dz = 0 \qquad \text{for all} \quad X \in \mathcal{C}(\Omega)$$

is fulfilled if and only if

$$\int_X \omega_1 = 0, \quad \int_X \omega_2 = 0 \qquad \text{for all} \quad X \in \mathcal{C}(\Omega)$$

holds true. This is equivalent to

$$d\omega_1 = 0, \quad d\omega_2 = 0 \qquad \text{in} \quad \Omega$$

and finally to (1).

3. We now prove $(c) \Rightarrow (d)$: Here we apply Theorem 6.5 from Section 6 in Chapter 1. The statement (c) then is equivalent to the existence of functions $U, V \in C^1(\Omega, \mathbb{R})$ with the properties

$$dU(x,y) = \omega_1(x,y), \qquad dV(x,y) = \omega_2(x,y) \qquad \text{in} \quad \Omega$$

giving us the conditions

$$\begin{aligned} U_x(x,y) &= u(x,y), \quad U_y(x,y) = -v(x,y), \\ V_x(x,y) &= v(x,y), \quad V_y(x,y) = u(x,y). \end{aligned} \tag{2}$$

Now the equations (2) are equivalent to

$$\begin{aligned} &\frac{\partial}{\partial x}\Big(U(x,y) + iV(x,y)\Big) = u(x,y) + iv(x,y) = f(x,y), \\ &\frac{1}{i}\frac{\partial}{\partial y}\Big(U(x,y) + iV(x,y)\Big) = u(x,y) + iv(x,y) = f(x,y). \end{aligned} \tag{3}$$

By the definition $F = U + iV$ we obtain a holomorphic function in Ω satisfying

$$F'(z) = \frac{\partial}{\partial x} F(x,y) = f(z), \qquad z \in \Omega.$$

4. Finally, we show the direction $(d) \Rightarrow (c)$: When the curve $X \in \mathcal{C}(\Omega)$ is given, we deduce

$$\begin{aligned} \int_X f(z)\,dz &= \int_a^b f\Big(X(t)\Big)X'(t)\,dt = \int_a^b \frac{d}{dt} F\Big(X(t)\Big)\,dt \\ &= F\Big(X(b)\Big) - F\Big(X(a)\Big) = 0 \end{aligned}$$

on account of $X(a) = X(b)$. q.e.d.

Remark: The statement in the direction $(a) \Rightarrow (c)$ is known as *Cauchy's integral theorem.*

The subsequent statements, which can be taken from Theorems 6.9 and 6.12 in Section 6 of Chapter 1, are valid for arbitrary domains $\Omega \subset \mathbb{C}$.

Theorem 2.2. *Let $\Omega \subset \mathbb{C}$ denote a domain where the two closed curves $X, Y \in \mathcal{C}(\Omega)$ are homotopic to each other. Furthermore, let $w = f(z)$, $z \in \Omega$, represent a holomorphic function in Ω. Then we have the identity*

$$\int_X f(z)\,dz = \int_Y f(z)\,dz.$$

When we fix the end points of our curve we obtain the following

Theorem 2.3. (Monodromy)
Let $\Omega \subset \mathbb{C}$ be a domain where we choose two arbitrary points $P, Q \in \Omega$. Furthermore, we consider two curves $X, Y \in \mathcal{C}(\Omega, P, Q)$ homotopic to each other, which both have the start-point $P \in \Omega$ and the end-point $Q \in \Omega$. If the function $f : \Omega \to \mathbb{C}$ is holomorphic, we then have the identity

$$\int_X f(z)\, dz = \int_Y f(z)\, dz.$$

A set $\Theta \subset \mathbb{R}^n$ is denoted as *compactly contained* in a set $\Omega \subset \mathbb{R}^n$ - symbolically $\Theta \subset\subset \Omega$ - if the set $\overline{\Theta}$ is compact and $\overline{\Theta} \subset \Omega$ holds true.

Theorem 2.4. (Cauchy, Weierstraß)
Let $\Omega \subset \mathbb{C}$ denote a domain, and let the center $z_0 \in \Omega$ and the radius $r > 0$ be prescribed such that the open disc

$$K = K_r(z_0) := \left\{ z \in \mathbb{C} : |z - z_0| < r \right\}$$

realizes the inclusion $K \subset\subset \Omega$. Furthermore, let us consider the function $f \in C^1(\Omega, \mathbb{C})$. Then the following statements are equivalent:

(a) The function $f(z)$ is holomorphic in K;
(b) We have the validity of Cauchy's integral formula

$$f(z) = \frac{1}{2\pi i} \oint_{\partial K} \frac{f(\zeta)}{\zeta - z}\, d\zeta$$

for all points $z \in K$ - with $\zeta = \xi + i\eta$ - where the curvilinear integral is evaluated over the positive-oriented circumference;
(c) The series expansion

$$f(z) = \sum_{k=0}^{\infty} a_k (z - z_0)^k, \qquad z \in K$$

with the coefficients

$$a_k := \frac{1}{k!} f^{(k)}(z_0), \qquad k = 0, 1, 2, \ldots$$

holds true.

Proof:

1. At first, we show the direction $(a) \Rightarrow (b)$. We observe that the function

$$g(\zeta) := \frac{f(\zeta)}{\zeta - z}, \qquad \zeta \in K \setminus \{z\}$$

is holomorphic on its domain of definition. For all sufficiently small quantities $\varepsilon > 0$, the curves

$$X(t) := z + \varepsilon\, e^{it}, \qquad 0 \le t \le 2\pi$$

and

$$Y(t) := z_0 + r\, e^{i\varphi}, \qquad 0 \le \varphi \le 2\pi$$

are homotopic to each other in the set $\overline{K} \setminus \{z\}$. This implies

$$\begin{aligned}
\oint_{\partial K} \frac{f(\zeta)}{\zeta - z}\, d\zeta = \int_Y g(\zeta)\, d\zeta &= \int_X g(\zeta)\, d\zeta \\
&= \int_0^{2\pi} \frac{f(z + \varepsilon\, e^{it})}{\varepsilon\, e^{it}}\, i\varepsilon\, e^{it}\, dt \\
&= i \int_0^{2\pi} f(z + \varepsilon\, e^{it})\, dt.
\end{aligned}$$

Via transition to the limit $\varepsilon \to 0+$ we obtain

$$\oint_{\partial K} \frac{f(\zeta)}{\zeta - z}\, d\zeta = 2\pi i f(z)$$

and

$$f(z) = \frac{1}{2\pi i} \oint_{\partial K} \frac{f(\zeta)}{\zeta - z}\, d\zeta \qquad \text{for all} \quad z \in K.$$

2. We secondly deduce the direction $(b) \Rightarrow (c)$: For all points $z \in K$, $\zeta \in \partial K$ we have the identity

$$\frac{1}{\zeta - z} = \frac{1}{(\zeta - z_0) - (z - z_0)} = \frac{1}{\zeta - z_0} \frac{1}{1 - \dfrac{z - z_0}{\zeta - z_0}}.$$

Now we observe

$$\left| \frac{z - z_0}{\zeta - z_0} \right| < 1,$$

such that we can expand this fraction into the uniformly convergent geometric series

$$\frac{1}{\zeta - z_0} \sum_{k=0}^{\infty} \left(\frac{z - z_0}{\zeta - z_0} \right)^k = \sum_{k=0}^{\infty} \frac{1}{(\zeta - z_0)^{k+1}} (z - z_0)^k .$$

This implies

$$\begin{aligned} f(z) &= \frac{1}{2\pi i} \oint\limits_{\partial K} \frac{f(\zeta)}{\zeta - z}\, d\zeta \\ &= \frac{1}{2\pi i} \sum_{k=0}^{\infty} \left(\oint\limits_{\partial K} \frac{f(\zeta)}{(\zeta - z_0)^{k+1}}\, d\zeta \right) (z - z_0)^k \\ &= \sum_{k=0}^{\infty} a_k (z - z_0)^k \end{aligned}$$

with the coefficients

$$a_k := \frac{1}{2\pi i} \oint\limits_{\partial K} \frac{f(\zeta)}{(\zeta - z_0)^{k+1}}\, d\zeta = \frac{f^{(k)}(z_0)}{k!}, \qquad k = 0, 1, 2, \ldots$$

3. The direction $(c) \Rightarrow (a)$ has already been shown in Section 1. q.e.d.

Remark: In the sequel, we tacitly mean uniform convergence of power series in the interior of their domain of convergence.

Theorem 2.5. (Identity theorem for holomorphic functions)
Let two holomorphic functions $f, g : \Omega \to \mathbb{C}$ be given on the domain $\Omega \subset \mathbb{C}$. Furthermore, let $\{z_k\}_{k=1,2,\ldots} \subset \Omega \setminus \{z_0\}$ denote a convergent sequence with the limit

$$\lim_{k\to\infty} z_k = z_0 \in \Omega .$$

Finally, the coincidence

$$f(z_k) = g(z_k), \qquad k = 1, 2, \ldots$$

may hold true. Then we have the identity

$$f(z) \equiv g(z) \qquad in \quad \Omega .$$

Proof: On the contrary, we assume that the holomorphic function $h(z) := f(z) - g(z)$ did not vanish identically. At the point $z_0 \in \Omega$ we expand the function $h = h(z)$ into the following power series

$$h(z) = \sum_{k=0}^{\infty} a_k (z - z_0)^k , \qquad z \in K_\varrho(z_0), \quad \varrho := \operatorname{dist}(z_0, \partial\Omega).$$

On account of $h(z_0) = 0$ we have a positive integer $n \in \mathbb{N}$ with $a_n \neq 0$, such that

$$h(z) = a_n(z - z_0)^n \Big\{1 + \alpha(z)\Big\} \qquad \text{with} \quad \lim_{z \to z_0} \alpha(z) = 0$$

holds true. Taking the quantity $\varrho > 0$ sufficiently small we obtain

$$|h(z)| \geq |a_n||z - z_0|^n \left(1 - \frac{1}{2}\right) = \frac{|a_n|}{2} |z - z_0|^n, \qquad z \in K_\varrho(z_0).$$

This implies the relation

$$h(z) \neq 0 \qquad \text{for all} \quad z \in K_\varrho(z_0) \setminus \{z_0\}$$

contradicting the assumption

$$h(z_k) = f(z_k) - g(z_k) = 0, \qquad k = 1, 2, 3, \ldots$$

q.e.d.

We now define *Wirtinger's differential operators*

$$\frac{\partial}{\partial z} := \frac{1}{2}\left(\frac{\partial}{\partial x} - i\frac{\partial}{\partial y}\right), \quad \frac{\partial}{\partial \overline{z}} := \frac{1}{2}\left(\frac{\partial}{\partial x} + i\frac{\partial}{\partial y}\right).$$

The function $f(z) = u(x, y) + iv(x, y)$ fulfills the Cauchy-Riemann system of differential equations if the identity

$$\begin{aligned}\frac{\partial}{\partial \overline{z}} f(z) &= \frac{1}{2}(f_x + if_y) = \frac{1}{2}\Big(u_x + iv_x + i(u_y + iv_y)\Big) \\ &= \frac{1}{2}(u_x - v_y) + \frac{i}{2}(v_x + u_y) = 0\end{aligned}$$

holds true. Therefore, holomorphic functions satisfy the partial differential equation

$$\frac{\partial}{\partial \overline{z}} f(z) = 0 \qquad \text{in} \quad \Omega. \tag{4}$$

We can comprehend the function $f(z) = u(x, y) + iv(x, y) : \Omega \to \mathbb{C} \in C^1(\Omega, \mathbb{C})$ alternatively as a function depending on the variables z and $\overline{z}$; then we see $\frac{\partial}{\partial z}$ and $\frac{\partial}{\partial \overline{z}}$ as partial derivatives. These differentiators are $\mathbb{C}$-linear and the product as well as the quotient rule are valid. Furthermore, we have the following complex version of the

Chain rule: Let the symbols $\Omega, \Theta \subset \mathbb{C}$ denote two domains with the associate C^1-functions $w = f(z) : \Omega \to \Theta$ and $\alpha = g(w) : \Theta \to \mathbb{C}$. Then the composition

$$h(z) := g\Big(f(z)\Big), \qquad z \in \Omega$$

represents a C^1-function as well, and we have the differentiation rules

$$\begin{aligned} h_z(z) &= g_w(f(z))f_z(z) + g_{\overline{w}}(f(z))\overline{f}_z(z), \\ h_{\overline{z}}(z) &= g_w(f(z))f_{\overline{z}}(z) + g_{\overline{w}}(f(z))\overline{f}_{\overline{z}}(z) \qquad \text{in} \quad \Omega. \end{aligned} \tag{5}$$

Furthermore, we have the following

Calculus rules: For a C^1-function $f(z) : \Omega \to \mathbb{C}$ the identities

$$\overline{(f_z(z))} = \overline{f}_{\overline{z}}(z), \quad \overline{(f_{\overline{z}}(z))} = \overline{f}_z(z)$$

and

$$J_f(z) = \begin{vmatrix} u_x & u_y \\ v_x & v_y \end{vmatrix} = \begin{vmatrix} f_z & f_{\overline{z}} \\ \overline{f}_z & \overline{f}_{\overline{z}} \end{vmatrix} = |f_z|^2 - |f_{\overline{z}}|^2$$

are correct. In particular, a holomorphic function $f : \Omega \to \mathbb{C}$ satisfies the condition

$$J_f(z) = |f'(z)|^2 \geq 0 \qquad \text{for all} \quad z \in \Omega.$$

When we take (4) into account, holomorphic functions are exactly those which are independent of the variable $\overline{z}$. These statements are proved in Chapter 1 of the book [GF] by H. Grauert and K. Fritsche or in the monograph by [Re] by R. Remmert on pp. 52-56.

When we finally consider a function $f : \Omega \to \mathbb{C} \in C^2(\Omega, \mathbb{C})$, we infer

$$\frac{\partial}{\partial z}\frac{\partial}{\partial \overline{z}} f(z) = \frac{\partial}{\partial \overline{z}}\frac{\partial}{\partial z} f(z) = \frac{1}{4}\Delta f(z), \qquad z \in \Omega.$$

Now we investigate holomorphic functions in several complex variables.

Definition 2.6. *We name the function*

$$w = f(z) = f(z_1, \ldots, z_n) : \Omega \longrightarrow \mathbb{C}, \qquad (z_1, \ldots, z_n) \in \Omega$$

- defined on the domain $\Omega \subset \mathbb{C}^n$ with $n \in \mathbb{N}$ - holomorphic *if the following conditions are satisfied:*

(a) We have the regularity $f \in C^0(\Omega, \mathbb{C})$;
(b) For each fixed vector $(z_1, \ldots, z_n) \in \Omega$ and index $k \in \{1, \ldots, n\}$ being given, the function

$$\Phi(t) := f(z_1, \ldots, z_{k-1}, t, z_{k+1}, \ldots, z_n), \qquad t \in K_{\varepsilon_k}(z_k)$$

is holomorphic in the disc

$$K_{\varepsilon_k}(z_k) := \Big\{ t \in \mathbb{C} : |t - z_k| < \varepsilon_k \Big\}$$

with a sufficiently small radius $\varepsilon_k = \varepsilon_k(z) > 0$.

Theorem 2.7. (Cauchy's integral formula in $\mathbb{C}^n$)
Let the function $f = f(z_1, \dots, z_n) : \Omega \to \mathbb{C}$ be holomorphic in the domain $\Omega \subset \mathbb{C}^n$. With the data $z^0 = (z_1^0, \dots, z_n^0) \in \Omega$ and $R_1 > 0, \dots, R_n > 0$ let the polycylinder

$$P := \left\{ z = (z_1, \dots, z_n) : |z_k - z_k^0| < R_k,\ k = 1, \dots, n \right\}$$

be compactly contained in Ω, which means $\overline{P} \subset \Omega$. Then we have the following integral representation for all points $z = (z_1, \dots, z_n) \in P$, namely

$$\begin{aligned} f(z_1, \dots, z_n) &= \frac{1}{(2\pi i)^n} \oint\limits_{|\zeta_1 - z_1^0| = R_1} \cdots \oint\limits_{|\zeta_n - z_n^0| = R_n} \frac{f(\zeta_1, \dots, \zeta_n)}{(\zeta_1 - z_1) \cdot \ldots \cdot (\zeta_n - z_n)}\, d\zeta_1 \dots d\zeta_n \\ &= \frac{1}{(2\pi i)^n} \int\limits_0^{2\pi} \cdots \int\limits_0^{2\pi} \frac{f(z_1^0 + R_1 e^{it_1}, \dots, z_n^0 + R_n e^{it_n})}{(z_1^0 + R_1 e^{it_1} - z_1) \cdot \ldots \cdot (z_n^0 + R_n e^{it_n} - z_n)} \cdot \\ &\quad \cdot (iR_1 e^{it_1}) \cdot \ldots \cdot (iR_n e^{it_n})\, dt_1 \dots dt_n. \end{aligned}$$

Proof: The function $f = f(z)$ is holomorphic with respect to the variables $z_1, \dots, z_n$. Therefore, we deduce

$$\begin{aligned} f(z_1, \dots, z_n) &= \frac{1}{2\pi i} \oint\limits_{|\zeta_1 - z_1^0| = R_1} \frac{f(\zeta_1, z_2, \dots, z_n)}{\zeta_1 - z_1}\, d\zeta_1 \\ &= \frac{1}{(2\pi i)^2} \oint\limits_{|\zeta_1 - z_1^0| = R_1} \frac{d\zeta_1}{\zeta_1 - z_1} \oint\limits_{|\zeta_2 - z_2^0| = R_2} \frac{f(\zeta_1, \zeta_2, z_3, \dots, z_n)}{\zeta_2 - z_2}\, d\zeta_2 \\ &\ \vdots \\ &= \frac{1}{(2\pi i)^n} \oint\limits_{|\zeta_1 - z_1^0| = R_1} \cdots \oint\limits_{|\zeta_n - z_n^0| = R_n} \frac{f(\zeta_1, \dots, \zeta_n)}{(\zeta_1 - z_1) \cdot \ldots \cdot (\zeta_n - z_n)}\, d\zeta_1 \dots d\zeta_n. \end{aligned}$$

The second representation is revealed via introduction of polar coordinates.

q.e.d.

Theorem 2.8. *Let the sequence of holomorphic functions $f_k(z_1, \dots, z_n) : \Omega \to \mathbb{C}$, $k = 1, 2, \dots$, be given on the domain $\Omega \subset \mathbb{C}^n$, which converges uniformly on each compact subset of $\Omega \subset \mathbb{C}^n$. Then the limit function*

$$f(z_1, \dots, z_n) := \lim_{k \to \infty} f_k(z_1, \dots, z_n), \quad z = (z_1, \dots, z_n) \in \Omega$$

is holomorphic in $\Omega \subset \mathbb{C}^n$.

Proof: We apply Cauchy's integral formula in $\mathbb{C}^n$. Here we choose the polycylinder P as in Theorem 2.7 and infer the following representation for all points $z \in P$:

$$\begin{aligned} f(z_1,\dots,z_n) &= \lim_{k\to\infty} f_k(z_1,\dots,z_n) \\ &= \lim_{k\to\infty} \frac{1}{(2\pi i)^n} \oint\limits_{|\zeta_1-z_1^0|=R_1} \cdots \oint\limits_{|\zeta_n-z_n^0|=R_n} \frac{f_k(\zeta_1,\dots,\zeta_n)}{(\zeta_1-z_1)\cdot\ldots\cdot(\zeta_n-z_n)}\,d\zeta_1\ldots d\zeta_n \\ &= \frac{1}{(2\pi i)^n} \oint\limits_{|\zeta_1-z_1^0|=R_1} \cdots \oint\limits_{|\zeta_n-z_n^0|=R_n} \frac{f(\zeta_1,\dots,\zeta_n)}{(\zeta_1-z_1)\cdot\ldots\cdot(\zeta_n-z_n)}\,d\zeta_1\ldots d\zeta_n. \end{aligned}$$

Therefore, the limit function $f = f(z)$ is holomorphic in P. q.e.d.

Theorem 2.9. *With the assumptions of Theorem 2.7, we have the following power-series-expansion for all points $z = (z_1,\dots,z_n)$ satisfying $|z_k - z_k^0| < R_k$ with $k = 1,\dots,n$, namely*

$$f(z_1,\dots,z_n) = \sum_{k_1,\dots,k_n=0}^{\infty} a_{k_1\ldots k_n}(z_1-z_1^0)^{k_1}\cdot\ldots\cdot(z_n-z_n^0)^{k_n}.$$

Here the coefficients fulfill

$$\begin{aligned} a_{k_1\ldots k_n} &= \frac{1}{(2\pi i)^n} \oint\limits_{|\zeta_1-z_1^0|=R_1} \cdots \oint\limits_{|\zeta_n-z_n^0|=R_n} \frac{f(\zeta_1,\dots,\zeta_n)}{(\zeta_1-z_1^0)^{k_1+1}\cdot\ldots\cdot(\zeta_n-z_n^0)^{k_n+1}}\,d\zeta_1\ldots d\zeta_n \\ &= \frac{1}{k_1!\cdot\ldots\cdot k_n!}\left\{\left(\frac{\partial}{\partial\zeta_1}\right)^{k_1}\cdots\left(\frac{\partial}{\partial\zeta_n}\right)^{k_n} f(\zeta_1,\dots,\zeta_n)\right\}_{\zeta=z^0} \end{aligned}$$

for $k_1,\dots,k_n = 0,1,2,\dots$. Setting

$$M := \max_{\substack{|\zeta_k-z_k^0|=R_k \\ k=1,\dots,n}} |f(\zeta_1,\dots,\zeta_n)|$$

we have Cauchy's estimates

$$|a_{k_1\ldots k_n}| \le \frac{M}{R_1^{k_1}\cdot\ldots\cdot R_n^{k_n}}, \qquad k_1,\dots k_n = 0,1,2,\dots$$

Proof: As in the proof of Theorem 2.4 we deduce

$$\frac{1}{\zeta_k - z_k} = \sum_{l=0}^{\infty} \frac{(z_k-z_k^0)^l}{(\zeta_k-z_k^0)^{l+1}}$$

for $k = 1, \ldots, n$. The absolute convergence of this series yields

$$\frac{1}{(\zeta_1 - z_1) \cdot \ldots \cdot (\zeta_n - z_n)} = \sum_{k_1, \ldots, k_n = 0}^{\infty} \frac{(z_1 - z_1^0)^{k_1}}{(\zeta_1 - z_1^0)^{k_1+1}} \cdot \ldots \cdot \frac{(z_n - z_n^0)^{k_n}}{(\zeta_n - z_n^0)^{k_n+1}}.$$

Now Theorem 2.7 implies the following identity for all points $z = (z_1, \ldots, z_n)$ satisfying $|z_k - z_k^0| < R_k$ with $k = 1, \ldots, n$, namely

$$f(z_1, \ldots, z_n) = \sum_{k_1, \ldots, k_n = 0}^{\infty} a_{k_1 \ldots k_n} (z_1 - z_1^0)^{k_1} \cdot \ldots \cdot (z_n - z_n^0)^{k_n}.$$

The further statements are evident. q.e.d.

Theorem 2.10. (Liouville)
Let the entire holomorphic function $f(z_1, \ldots, z_n) : \mathbb{C}^n \to \mathbb{C}$ be given. Furthermore, we have a constant $M \in [0, +\infty)$ such that the estimate

$$|f(z_1, \ldots, z_n)| \leq M \qquad \textit{for all} \quad (z_1, \ldots, z_n) \in \mathbb{C}^n$$

holds true. Then we have a constant $c \in \mathbb{C}$ such that

$$f(z_1, \ldots, z_n) \equiv c \qquad \textit{on the space} \quad \mathbb{C}^n$$

is correct. Therefore, each bounded entire holomorphic function is constant.

Proof: We can expand the function $f = f(z)$ into the power series

$$f(z_1, \ldots, z_n) = \sum_{k_1, \ldots, k_n = 0}^{\infty} a_{k_1 \ldots k_n} z_1^{k_1} \cdot \ldots \cdot z_n^{k_n}$$

on the whole space $\mathbb{C}^n$ about the origin $z_1 = 0, \ldots, z_n = 0$. Choosing the polycylinder

$$P := \Big\{ (z_1, \ldots, z_n) \in \mathbb{C}^n \; : \; |z_j| < R \;\text{ for } j = 1, \ldots, n \Big\} \subset \mathbb{C}^n,$$

the Cauchy estimates yield

$$|a_{k_1 \ldots k_n}| \leq \frac{M}{R^{k_1 + \ldots + k_n}} \longrightarrow 0 \qquad \text{for} \quad R \to \infty$$

for all $(k_1, \ldots, k_n) \in \mathbb{N}^n$ with $k_1 + \ldots + k_n > 0$. This implies

$$f(z_1, \ldots, z_n) = a_{0 \ldots 0} =: c \in \mathbb{C} \quad \text{for all points} \quad (z_1, \ldots, z_n) \in \mathbb{C}^n.$$

q.e.d.

Theorem 2.11. (Identity theorem in $\mathbb{C}^n$)
Let the functions $f(z) : \Omega \to \mathbb{C}$ and $g(z) : \Omega \to \mathbb{C}$ be holomorphic on the domain $\Omega \subset \mathbb{C}^n$. Furthermore, let the point $z^0 = (z_1^0, \ldots, z_n^0) \in \Omega$ be fixed where the coincidence

$$\left(\frac{\partial}{\partial \zeta_1}\right)^{k_1} \cdots \left(\frac{\partial}{\partial \zeta_n}\right)^{k_n} f(\zeta_1, \ldots, \zeta_n)\Big|_{\zeta = z^0}$$
$$= \left(\frac{\partial}{\partial \zeta_1}\right)^{k_1} \cdots \left(\frac{\partial}{\partial \zeta_n}\right)^{k_n} g(\zeta_1, \ldots, \zeta_n)\Big|_{\zeta = z^0}$$

for all indices $k_1, \ldots, k_n = 0, 1, 2, \ldots$ is valid. Then we infer the identity

$$f(z) \equiv g(z) \qquad \text{for all} \quad z \in \Omega.$$

Proof: We consider the function

$$h(z) := f(z) - g(z), \qquad z \in \Omega$$

and the nonvoid set

$$\Theta := \left\{ z \in \Omega : \begin{array}{c} \left(\frac{\partial}{\partial \zeta_1}\right)^{k_1} \cdots \left(\frac{\partial}{\partial \zeta_n}\right)^{k_n} h(\zeta)\Big|_{\zeta = z} = 0 \\ \text{for } k_1, \ldots, k_n = 0, 1, 2, \ldots \end{array} \right\}.$$

Evidently, this set is closed and open as well - because the function $h = h(z)$ can be expanded into a vanishing power series at each point $z \in \Theta$. When we connect an arbitrary point $z^1 \in \Omega$ with the point $z^0 \in \Theta$ by a continuous path $\varphi : [0,1] \to \Omega \in C^0([0,1], \Omega)$ satisfying $\varphi(0) = z^0$ and $\varphi(1) = z^1$, then a continuation argument yields the inclusion $\varphi([0,1]) \subset \Theta$. The set Θ is namely simultaneously open and closed. This implies $z^1 = \varphi(1) \in \Theta$ and consequently $\Theta = \Omega$. Therefore, we obtain $h(z) \equiv 0$ in Ω and finally $f(z) \equiv g(z)$ in Ω.
q.e.d.

Remarks:

1. When two functions $f = f(z)$ and $g = g(z)$ coincide on an open set, they are identical on the whole domain of definition due to Theorem 2.11.
2. If two functions $f = f(z)$ and $g = g(z)$ coincide only on a sequence of points which converges in the domain of holomorphy, they are not necessarily identical! For instance, we consider the holomorphic function

$$f(z_1, \ldots, z_n) = z_1 \cdot \ldots \cdot z_n, \qquad z = (z_1, \ldots, z_n) \in \mathbb{C}^n$$

vanishing on the coordinate axes.

The following result provides a powerful tool (see Theorem 1.9 in Chapter 5, Section 1) for the investigation of analyticity for solutions of partial differential equations.

Theorem 2.12. (Holomorphic parameter integrals)
Assumptions: *Let $\Theta \subset \mathbb{R}^m$ and $\Omega \subset \mathbb{C}^n$ denote domains in the respective spaces of dimensions $m, n \in \mathbb{N}$. Furthermore, let*

$$f = f(t,z) = f(t_1, \ldots, t_m, z_1, \ldots, z_n) \,:\, \Theta \times \Omega \longrightarrow \mathbb{C} \in C^0(\Theta \times \Omega, \mathbb{C})$$

represent a continuous function with the following properties:

(a) For each fixed vector $t \in \Theta$ the function

$$\Phi(z) := f(t,z), \qquad z \in \Omega$$

is holomorphic.

(b) We have a continuous integrable function $F(t) : \Theta \longrightarrow [0, +\infty) \in C^0(\Theta, \mathbb{R})$ satisfying

$$\int\limits_{\Theta} F(t)\,dt < +\infty,$$

which represents a uniform majorant to our function $f = f(t,z)$ - that means

$$|f(t,z)| \leq F(t) \qquad \text{for all} \quad (t,z) \in \Theta \times \Omega.$$

Statement: *Then the function*

$$\varphi(z) := \int\limits_{\Theta} f(t,z)\,dt, \qquad z \in \Omega$$

is holomorphic in Ω.

Proof:

1. We consider a closed n-dimensional rectangle Q satisfying $Q \subset \Theta$, and we show that the function

$$\Psi(z) := \int\limits_{Q} f(t,z)\,dt, \qquad z \in \Omega$$

is holomorphic. Here we decompose the rectangle Q via the formula

$$\mathcal{Z}_k \,:\, Q = \bigcup_{l=1}^{N_k} Q_l$$

into subrectangles whose measure of fineness satisfies: $\delta(\mathcal{Z}_k) \to 0$ for $k \to \infty$. Then we consider an arbitrary compact set $K \subset \Omega$: For each $\varepsilon > 0$ we have an index $k_0 = k_0(\varepsilon) \in \mathbb{N}$ such that the following estimate holds true for all $k \geq k_0$, namely

$$\left|\Psi(z)-\sum_{l=1}^{N_k} f\left(t^{(l)},z\right)|Q_l|\right| = \left|\int\limits_Q f(t,z)\,dt - \sum_{l=1}^{N_k} f\left(t^{(l)},z\right)|Q_l|\right| \le \varepsilon$$

for all $z \in K$ with $t^{(l)} \in Q_l$. On a compact set the continuous function $f = f(t,z)$ is namely uniformly continuous! The sequence of holomorphic functions

$$\Psi_k(z) := \sum_{l=1}^{N_k} f\left(t^{(l)},z\right)|Q_l|, \qquad z \in \Omega, \quad k = 1,2,3,\ldots$$

converges uniformly on each compact set $K \subset \Omega$ towards the holomorphic function

$$\Psi(z) := \int\limits_Q f(t,z)\,dt, \qquad z \in \Omega$$

due to Theorem 2.8.

2. Now we exhaust the open set Θ by a sequence of compact sets

$$R_1 \subset R_2 \subset R_3 \subset \ldots \subset \Theta$$

where each set R_k is a union of finitely many closed rectangles in Θ. From the first part of our proof we infer that the function

$$\varphi_k(z) := \int\limits_{R_k} f(t,z)\,dt, \qquad z \in \Omega$$

is holomorphic for each index $k \in \mathbb{N}$. With an arbitrarily given quantity $\varepsilon > 0$, we have the relation

$$\int\limits_{\Theta\setminus R_k} F(t)\,dt \le \varepsilon \qquad \text{for all} \quad k \ge k_0(\varepsilon).$$

This implies the following inequality for all $z \in \Omega$, namely

$$|\varphi(z)-\varphi_k(z)| = \left|\int\limits_{\Theta\setminus R_k} f(t,z)\,dt\right| \le \int\limits_{\Theta\setminus R_k} F(t)\,dt \le \varepsilon$$

with the index $k \ge k_0(\varepsilon)$. Therefore, the sequence of holomorphic functions $\varphi_k = \varphi_k(z)$ with $k = 1,2,3,\ldots$ converges uniformly towards the holomorphic function

$$\varphi(z) = \int\limits_\Theta f(t,z)\,dt, \qquad z \in \Omega.$$

This completes the proof. q.e.d.

Remarks:

1. The transition from the equation $f_{\overline{z}}(z) = 0$ to the system
$$\frac{\partial}{\partial \overline{z}_i} f(z_1, \ldots, z_n) = 0, \qquad i = 1, \ldots, n$$
is easy, since the latter constitutes a linear system.
2. We refer the reader to the excellent book [GF] by Grauert and Fritzsche for further studies in the theory of holomorphic functions with several complex variables.

3 Geometric Behavior of Holomorphic Functions in $\mathbb{C}$

We begin with the surprising

Theorem 3.1. *Let the function $f : G \to \mathbb{C}$ be holomorphic on the domain $G \subset \mathbb{C}$, and take a point $z_0 \in G$. Then the following statements are equivalent:*

(a) The function f is locally injective about the point z_0;
(b) The function f is locally bijective about the point z_0;
(c) We have the condition $J_f(z_0) > 0$.

Proof:

1. The direction $(a) \Rightarrow (b)$ is contained in the Theorems of Jordan-Brouwer in $\mathbb{R}^n$ for the special case $n = 2$.
2. We now show the direction $(b) \Rightarrow (c)$: Here we consider the disc
$$K := \Big\{ z \in \mathbb{C} : |z - z_0| < \varrho \Big\} \subset\subset G$$
with a sufficiently small radius $\varrho > 0$. Then we define
$$F(z) := f(z) - f(z_0), \qquad z \in \overline{K}$$
and
$$\varphi(t) := F(\varrho e^{it}) \neq 0, \qquad 0 \leq t \leq 2\pi.$$
Now the index-sum formula yields
$$\pm 1 = W(\varphi) = i(F, z_0) = n$$
if the expansion
$$F(z) = a_n (z - z_0)^n + o(|z - z_0|^n), \qquad z \to z_0$$
with $a_n \neq 0$ and $n \in \mathbb{N}$ holds true. Here the function
$$\psi(z) := o(|z - z_0|^n)$$

satisfies

$$\lim_{\substack{z\to z_0\\ z\neq z_0}} \frac{\psi(z)}{|z-z_0|^n} = 0.$$

Then we infer the statements $n = 1$ and $F'(z_0) \neq 0$. Finally, we arrive at the condition

$$J_f(z_0) = J_F(z_0) = |F'(z_0)|^2 > 0.$$

3. From the fundamental theorem on the inverse mapping we infer the implication $(c) \Rightarrow (a)$.

q.e.d.

Example 3.2. Theorem 3.1 becomes false for functions which are only real-differentiable. In this context, we consider the bijective function

$$f(x) = x^3\,, \qquad x \in \mathbb{R},$$

whose derivative $J_f(x) = 2x^2$ possesses a zero at the origin $x = 0$.

Problem: Generalize the results of Theorem 3.1 to holomorphic functions with n complex variables

$$f = \Big(f_1(z_1,\ldots,z_n),\ldots,f_n(z_1,\ldots,z_n)\Big) \,:G \longrightarrow \mathbb{C}^n$$

in the domain $G \subset \mathbb{C}^n$ (compare [GF] Chapter 1)!

Even for holomorphic mappings, which are not necessarily injective, we have the following

Theorem 3.3. (Invariance of domains in $\mathbb{C}$)
Let us consider a domain $G \subset \mathbb{C}$, where a nonconstant holomorphic function $w = f(z) : G \to \mathbb{C}$, $z \in G$ is defined. Then the image

$$G^* := f(G) = \Big\{w = f(z)\,:\,z \in G\Big\}$$

is a domain in $\mathbb{C}$ as well.

Proof: We transfer the proof from Theorem 6.4 in Chapter 3, Section 6 to the plane situation, and we note that the function $f = f(z)$ locally possesses the expansion

$$f(z) = f(z_0) + a_n(z-z_0)^n + o(|z-z_0|^n) \qquad \text{with} \quad a_n \in \mathbb{C}\setminus\{0\}$$

at an arbitrary point $z_0 \in G$. Consequently, the function

$$g(z) := f(z) - f(z_0), \qquad |z-z_0| \leq \varrho$$

satisfies the conditions

$$i(g, z_0) = n \neq 0 \quad \text{and} \quad g(z) \neq 0$$

for all points $z \in \mathbb{C}$ with $|z - z_0| = \varrho$; here we have chosen the radius $\varrho > 0$ sufficiently small. The arguments of the proof quoted above yield the statement of our theorem.

q.e.d.

Theorem 3.4. (Maximum principle for holomorphic functions)
Let the nonconstant holomorphic function $f : G \to \mathbb{C}$ be given on the domain $G \subset \mathbb{C}$. Then we have the following inequality for all points $z \in G$, namely

$$|f(z)| < \sup_{\zeta \in G} |f(\zeta)| =: M.$$

Proof: If $M = +\infty$ holds true, nothing has to be shown. Therefore, we assume the condition $M < +\infty$. Choosing the point $z \in G$ arbitrarily, we find a quantity $\delta = \delta(z) > 0$ such that the disc

$$B_\delta(f(z)) := \Big\{ w \in \mathbb{C} : |w - f(z)| < \delta \Big\}$$

fulfills the inclusion

$$B_\delta(f(z)) \subset G^*$$

according to Theorem 3.3. Consequently, we infer the statement above from the following inequality:

$$M := \sup_{\zeta \in G} |f(\zeta)| \geq \sup_{w \in B_\delta(f(z))} |w| = |f(z)| + \delta > |f(z)|.$$

q.e.d.

Remarks:

1. When we additionally assume that the domain G is bounded and the function $f : \overline{G} \to \mathbb{C}$ is continuous, we find a point $z_0 \in \partial G$ with the property

$$\sup_{\zeta \in G} |f(\zeta)| = |f(z_0)| > |f(z)| \qquad \text{for all} \quad z \in G.$$

2. The transition from f to $\frac{1}{f}$ reveals the
Minimum principle for holomorphic functions: The nonconstant holomorphic function $f : G \to \mathbb{C} \setminus \{0\}$ on the domain $G \subset \mathbb{C}$ satisfies the estimate

$$|f(z)| > \inf_{\zeta \in G} |f(\zeta)| \qquad \text{for all} \quad z \in G.$$

3. In the minimum principle we cannot renounce the assumption $f \neq 0$, as demonstrated by the subsequent

Example 3.5. On the domain

$$G := \Big\{ z \in \mathbb{C} \,:\, |z| < 1 \Big\}$$

we consider the holomorphic function

$$f(z) := z, \qquad z \in G.$$

Here the function $|f(z)|$ attains its minimum at the interior point $z = 0$.

4. Let the function $f(z_1, \ldots, z_n) : G \to \mathbb{C}$ be holomorphic on the domain $G \subset \mathbb{C}^n$. Then we infer the identities

$$f_{\overline{z}_j}(z) = 0 \qquad \text{in} \quad G \quad \text{for} \quad j = 1, \ldots, n.$$

We consider the square of the modulus for this function, namely

$$\Phi = \Phi(z) = \Phi(z_1, \ldots, z_n) := |f(z)|^2 = f(z)\overline{f}(z), \qquad z \in G.$$

For the indices $j = 1, \ldots, n$ we evaluate its derivatives

$$\Phi_{z_j} = f_{z_j}\overline{f} + f\overline{f}_{z_j} = f_{z_j}\overline{f} + f\overline{(f_{\overline{z}_j})} = f_{z_j}\overline{f} \qquad \text{in} \quad G$$

and

$$\Phi_{z_j\overline{z}_j} = f_{z_j\overline{z}_j}\overline{f} + f_{z_j}\overline{f}_{\overline{z}_j} = |f_{z_j}|^2 \qquad \text{in} \quad G.$$

Therefore, we arrive at the inequality

$$\Delta\Phi(z) = 4\sum_{j=1}^{n} \Phi_{z_j\overline{z}_j}(z) = 4\sum_{j=1}^{n} |f_{z_j}(z)|^2 \geq 0, \qquad z \in G. \tag{1}$$

Those functions are subharmonic and consequently subject to the maximum principle, as we shall show in Chapter 5.

Now we consider the *reflection at the real axis*

$$\tau(z) := \overline{z}, \qquad z \in \mathbb{C}. \tag{2}$$

This function is continuous on the complex plane $\mathbb{C}$, and we observe

$$\tau(z) = z \quad \Longleftrightarrow \quad z \in \mathbb{R}. \tag{3}$$

Denoting the upper and lower half-plane in $\mathbb{C}$ by

$$\mathbb{H}^{\pm} := \Big\{ z = x + iy \in \mathbb{C} \,:\, \pm y > 0 \Big\},$$

respectively, we obtain the topological mappings

$$\tau : \mathbb{H}^+ \to \mathbb{H}^-, \quad \tau : \mathbb{R} \to \mathbb{R}, \quad \tau : \mathbb{H}^- \to \mathbb{H}^+.$$

The function $\tau = \tau(z)$ is antiholomorphic in the following sense.

Definition 3.6. *On the open set $\Omega \subset \mathbb{C}$ we call the function $f : \Omega \to \mathbb{C}$* antiholomorphic *if the associate function*

$$g(z) := \overline{f(z)}, \qquad z \in \Omega$$

is holomorphic in Ω.

Theorem 3.7. *Each holomorphic function $f : \Omega \to \mathbb{C}$ is orientation-preserving, which means*

$$J_f(z) \geq 0 \qquad \text{for all} \quad z \in \Omega.$$

Each antiholomorphic function $f : \Omega \to \mathbb{C}$ is orientation-reversing, which means

$$J_f(z) \leq 0 \qquad \text{for all} \quad z \in \Omega.$$

Proof: If the function $f = f(z)$ is holomorphic, we infer

$$J_f(z) = \begin{vmatrix} f_z & f_{\overline{z}} \\ \overline{f}_z & \overline{f}_{\overline{z}} \end{vmatrix} = |f_z|^2 \geq 0 \qquad \text{in} \quad \Omega.$$

When the function $f = f(z)$ is antiholomorphic, we consider the holomorphic function $g(z) := \overline{f(z)}$ with $z \in \Omega$ and calculate

$$J_f(z) = \begin{vmatrix} f_z & f_{\overline{z}} \\ \overline{f}_z & \overline{f}_{\overline{z}} \end{vmatrix} = \begin{vmatrix} \overline{g}_z & \overline{g}_{\overline{z}} \\ g_z & g_{\overline{z}} \end{vmatrix} = -g_z \overline{g}_{\overline{z}} = -|g_z|^2 \leq 0 \qquad \text{in} \quad \Omega.$$

Consequently, all statements are proved. q.e.d.

The basic tool for the investigation of the boundary behavior for solutions of two-dimensional partial differential equations is provided by the following

Theorem 3.8. (Schwarzian reflection principle)
In the upper half-plane we consider the open set $\Omega^+ \subset \mathbb{H}^+$ such that

$$\Gamma := \partial\Omega^+ \cap \mathbb{R} \subset \mathbb{R}$$

represents a nonvoid open set. Furthermore, we define the open set

$$\Omega^- := \left\{ z \in \mathbb{C} : \overline{z} \in \Omega^+ \right\} \subset \mathbb{H}^-$$

and consider the disjoint union

$$\Omega := \Omega^+ \dot{\cup} \Gamma \dot{\cup} \Omega^-.$$

Finally, let the function $f : \Omega^+ \cup \Gamma \to \mathbb{C} \in C^1(\Omega^+) \cap C^0(\Omega^+ \cup \Gamma)$ be holomorphic in Ω^+ and satisfy $f(\Gamma) \subset \mathbb{R}$. Then the function

$$F(z) := \begin{cases} f(z), \ z \in \Omega^+ \cup \Gamma \\ \overline{f(\overline{z})}, \ z \in \Omega^- \end{cases} \tag{4}$$

is holomorphic in the set Ω.

Proof:

1. Obviously, we have the regularity $F \in C^1(\Omega^+ \cup \Omega^-)$. We consider all points $z \in \Omega^-$ and derive as follows:

$$\begin{aligned}
F_{\overline{z}}(z) &= \frac{\partial}{\partial \overline{z}} \{\tau \circ f \circ \tau\}(z) \\
&= (\tau \circ f)_w \Big|_{\tau(z)} \tau_{\overline{z}} + (\tau \circ f)_{\overline{w}} \Big|_{\tau(z)} \overline{\tau}_{\overline{z}} \\
&= (\tau \circ f)_w \Big|_{\tau(z)} \\
&= \tau_\zeta \Big|_{f \circ \tau(z)} f_w \Big|_{\tau(z)} + \tau_{\overline{\zeta}} \Big|_{f \circ \tau(z)} \overline{f}_w \Big|_{\tau(z)} \\
&= 0.
\end{aligned}$$

Consequently, the function $F = F(z)$ is holomorphic in $\Omega^+ \cup \Omega^-$.

2. Furthermore, the function $F = F(z)$ is continuous in Ω and, in particular, on the real line in Γ. We choose the limit point $z_0 \in \Gamma$ arbitrarily and consider a sequence of points $\{z_k\}_{k=1,2,\ldots} \subset \Omega^-$ with the property

$$\lim_{k \to \infty} z_k = z_0.$$

We infer the relation

$$\begin{aligned}
\lim_{k \to \infty} F(z_k) = \lim_{k \to \infty} \overline{f(\overline{z}_k)} &= \overline{f(\overline{z}_0)} = \overline{f(z_0)} \\
= f(z_0) &= F(z_0),
\end{aligned}$$

where we note that $f = f(z)$ is continuous in $\Omega^+ \cup \Gamma$.

3. We still have to show the holomorphy of the function $F = F(z)$ on the set Ω: Let $z_0 \in \Gamma$ be an arbitrary point, and we consider the semidiscs

$$H_\varepsilon^\pm := \Big\{ z \in \mathbb{C} : |z - z_0| < \varrho, \;\; \pm \mathrm{Im}\, z > \varepsilon \Big\} \subset \Omega^\pm$$

with a sufficiently small radius $\varrho > 0$ being fixed and the parameter $\varepsilon \to 0+$. With the aid of Cauchy's integral theorem and the Cauchy integral formula we deduce: For each point $z \in \mathbb{C} \setminus \mathbb{R}$ with $|z - z_0| < \varrho$ we have a sufficiently small quantity $\varepsilon = \varepsilon(z) > 0$ with the property

$$F(z) = \frac{1}{2\pi i} \oint_{\partial H_\varepsilon^+} \frac{F(\zeta)}{\zeta - z}\, d\zeta + \frac{1}{2\pi i} \oint_{\partial H_\varepsilon^-} \frac{F(\zeta)}{\zeta - z}\, d\zeta. \tag{5}$$

In the transition to the limit $\varepsilon \to 0+$ the integrals on the real axis cancel out, and we obtain

$$F(z) = \frac{1}{2\pi i} \oint_{|\zeta - z_0| = \varrho} \frac{F(\zeta)}{\zeta - z}\, d\zeta, \qquad |z - z_0| < \varrho. \tag{6}$$

This representation formula reveals the holomorphy of the function $F = F(z)$ about the point $z_0 \in \Gamma$.

q.e.d.

The *reflection at the unit circle* is important:

$$\sigma(z) := \frac{1}{z}, \qquad z \in \mathbb{C} \setminus \{0\}. \tag{7}$$

This function is holomorphic with the complex derivative

$$\sigma'(z) = -\frac{1}{z^2}, \qquad z \in \mathbb{C} \setminus \{0\}.$$

Combined with the reflection at the real axis, the function

$$f(z) := \tau \circ \sigma(z) = \frac{1}{\overline{z}}, \qquad z \in \mathbb{C} \setminus \{0\}$$

in polar coordinates satisfies the identity

$$f(re^{i\varphi}) = \frac{1}{r} e^{i\varphi}, \qquad 0 < r < +\infty, \quad 0 \le \varphi < 2\pi. \tag{8}$$

Obviously, the unit circle line $|z| = 1$ remains fixed with respect to the mapping $f = f(z)$.

We add a further element to the Gaussian plane $\mathbb{C}$, namely the *infinitely distant point* $\infty \notin \mathbb{C}$, and we obtain *Riemann's sphere*

$$\overline{\mathbb{C}} := \mathbb{C} \cup \{\infty\}.$$

Now we define the *ε-disc about the point* ∞ by

$$K_\varepsilon(\infty) := \Big\{ z \in \mathbb{C} \, : \, |z| > \frac{1}{\varepsilon} \Big\} \cup \{\infty\}, \qquad 0 < \varepsilon < +\infty. \tag{9}$$

When we use the familiar discs

$$K_\varepsilon(0) := \Big\{ z \in \mathbb{C} \, : \, |z| < \varepsilon \Big\},$$

we obtain the topological mapping

$$\sigma \, : \, K_\varepsilon(0) \setminus \{0\} \longrightarrow K_\varepsilon(\infty) \setminus \{\infty\} \tag{10}$$

for all $0 < \varepsilon < +\infty$.

Definition 3.9. *We call a set $O \subset \overline{\mathbb{C}}$* open *- in Riemann's sphere - if each point $z_0 \in O$ possesses a disc $K_\varepsilon(z_0)$, with a sufficiently small radius $\varepsilon > 0$ about the respective center, such that*

$$K_\varepsilon(z_0) \subset O$$

is fulfilled. Here we mean, as usual,

$$K_\varepsilon(z_0) := \Big\{ z \in \mathbb{C} \, : \, |z - z_0| < \varepsilon \Big\}$$

for all points $z_0 \in \mathbb{C}$ and radii $\varepsilon > 0$.

Theorem 3.10. *The system of open sets*

$$\mathcal{T}(\overline{\mathbb{C}}) := \Big\{ O \subset \overline{\mathbb{C}} : O \text{ is open} \Big\}$$

constitutes a topological space.

Proof: Exercise.

The limit point $z_0 \in \overline{\mathbb{C}}$ being given, we define the notion *limit* for a sequence of points $\{z_k\}_{k=1,2,\ldots} \subset \overline{\mathbb{C}}$ tending to z_0 as follows:

$$\lim_{k\to\infty} z_k = z_0 \quad \Longleftrightarrow \quad \begin{cases} \text{For all } \varepsilon > 0 \text{ we have an index } k_0 = k_0(\varepsilon) \in \mathbb{N} \\ \text{such that } z_k \in K_\varepsilon(z_0) \text{ for all } k \geq k_0(\varepsilon) \text{ holds true.} \end{cases} \tag{11}$$

We obtain the usual notion of convergence for the points $z_0 \in \mathbb{C}$, however, the infinitely distant point $z_0 = \infty$ as limit point means:

$$\lim_{k\to\infty} z_k = z_0 \quad \Longleftrightarrow \quad \begin{cases} \text{For all } \varepsilon > 0 \text{ we have an index } k_0 = k_0(\varepsilon) \in \mathbb{N} \\ \text{such that } |z_k| > \dfrac{1}{\varepsilon} \text{ for all } k \geq k_0(\varepsilon) \text{ holds true} \end{cases} . \tag{12}$$

As an exercise one proves the subsequent

Theorem 3.11. *Riemann's sphere $\{\overline{\mathbb{C}}, \mathcal{T}(\overline{\mathbb{C}})\}$ is compact in the following sense:*

(a) To each sequence of points $\{z_k\}_{k=1,2,\ldots} \subset \overline{\mathbb{C}}$ there exists a convergent subsequence $\{z_{k_l}\}_{l=1,2,\ldots} \subset \{z_k\}_{k=1,2,\ldots}$ with the property

$$z_0 := \lim_{l\to\infty} z_{k_l} \in \overline{\mathbb{C}}.$$

(b) Each open covering $\{O_\iota\}_{\iota\in J}$ of Riemann's sphere $\overline{\mathbb{C}}$ contains a finite subcovering.

Definition 3.12. *Let $\Omega \subset \overline{\mathbb{C}}$ denote an open set, where $f : \Omega \to \overline{\mathbb{C}}$ may be defined. Then the function $f = f(z)$ is called* continuous at the point $z_0 \in \Omega$, *if each quantity $\varepsilon > 0$ admits a number $\delta = \delta(\varepsilon, z_0) > 0$ such that the inclusion*

$$f\Big(K_\delta(z_0)\Big) \subset K_\varepsilon\Big(f(z_0)\Big)$$

holds true. When the function $f = f(z)$ is continuous at each point $z_0 \in \Omega$, we call the function continuous in Ω.

Theorem 3.13. *The reflection at the unit circle*

$$\sigma(z) := \begin{cases} \infty, \; z = 0 \\ \dfrac{1}{z}, \; z \in \mathbb{C} \setminus \{0\} \\ 0, \quad z = \infty \end{cases}$$

gives us a continuous bijective mapping $\sigma : \overline{\mathbb{C}} \to \overline{\mathbb{C}}$. This function is holomorphic in $\mathbb{C} \setminus \{0\}$ with the derivative

$$\sigma'(z) = -\frac{1}{z^2}, \qquad z \in \mathbb{C} \setminus \{0\}.$$

Proof: Exercise.

We define the *stereographic projection* from the unit sphere

$$\begin{aligned} S^2 &:= \Big\{ x = (x_1, x_2, x_3) \in \mathbb{R}^3 \; : \; |x| = 1 \Big\} \\ &= \Big\{ (\sin\vartheta \cos\varphi, \sin\vartheta \sin\varphi, \cos\vartheta) \; : \; 0 \le \vartheta \le \pi, \; 0 \le \varphi < 2\pi \Big\} \end{aligned}$$

onto the plane $\mathbb{R}^2 = \mathbb{C}$ according to

$$\pi \; : \; S^2 \longrightarrow \mathbb{R}^2 \cup \{\infty\}, \quad S^2 \ni (x_1, x_2, x_3) \mapsto (p_1, p_2) \in \mathbb{R}^2 \cup \{\infty\} \tag{13}$$

with

$$x_1 = \sin\vartheta \cos\varphi, \quad x_2 = \sin\vartheta \sin\varphi, \quad x_3 = \cos\vartheta,$$

$$p_1 = \frac{\sin\vartheta \cos\varphi}{1 - \cos\vartheta}, \quad p_2 = \frac{\sin\vartheta \sin\varphi}{1 - \cos\vartheta}.$$

This mapping ist bijective, and its restriction to the *sphere without north-pole* $S^2 \setminus \{(0,0,1)\}$ is conformal in the following sense: Oriented angles between two intersecting curves are preserved with respect to the mapping

$$\pi : S^2 \setminus \{(0,0,1)\} \to \mathbb{R}^2.$$

Here we refer the reader to the brilliant *Grundlehren* [BL] by W. Blaschke and K. Leichtweiß.

We observe the following behavior for a sequence of points

$$\{x^{(k)}\}_{k=1,2,3,\ldots} \subset S^2 \setminus \{(0,0,1)\}$$

satisfying $x^{(k)} \to (0,0,1)$ with $k \to \infty$, namely

$$\pi\Big(x^{(k)}\Big) \longrightarrow \infty \qquad \text{for} \quad k \to \infty.$$

Therefore, the definition $\pi((0,0,1)) := \infty$ makes sense, in order to extend the mapping π continuously onto the whole sphere S^2.

4 Isolated Singularities and the General Residue Theorem

On the basis of the Gaussian integral theorem in the plane, we establish the fundamental

Theorem 4.1. (General residue theorem)
Assumptions:

I. *Let $G \subset \mathbb{C}$ denote a bounded domain whose boundary points $\dot{G}$* are accessible from the exterior *as follows: For all points $z_0 \in \dot{G}$ there exists a sequence $\{z_k\}_{k=1,2,\ldots} \subset \mathbb{C} \setminus \overline{G}$ satisfying $\lim_{k\to\infty} z_k = z_0$. Furthermore, we have $J \in \mathbb{N}$ regular C^1-curves*

$$X^{(j)}(t) : [a_j, b_j] \longrightarrow \mathbb{C} \in C^1([a_j, b_j], \mathbb{C}), \qquad j = 1, \ldots, J$$

with the following properties:

$$X^{(j)}\Big((a_j, b_j)\Big) \cap X^{(k)}\Big((a_k, b_k)\Big) = \emptyset, \quad j, k \in \{1, \ldots, J\}, \quad j \neq k$$

and

$$\dot{G} = \bigcup_{j=1}^{J} X^{(j)}\Big([a_j, b_j]\Big).$$

Finally, the domain G is situated at the left-hand side of the respective curves: More precisely, the function

$$-i\left|\frac{d}{dt}X^{(j)}(t)\right|^{-1}\frac{d}{dt}X^{(j)}(t), \quad t \in (a_j, b_j)$$

represents the exterior normal vector to the domain G for $j = 1, \ldots, J$. The entire curvilinear integral - over these J oriented curves - will be addressed by the symbol $\int\limits_{\partial G} \cdots$.

II. *Furthermore, let $N \in \mathbb{N} \cup \{0\}$ singular points - where $N = 0$ describes the case that no singular point exists - be given, which are denoted by $\zeta_j \in G$ for $j = 1, \ldots, N$. Now we define the* punctured domain

$$G' := G \setminus \{\zeta_1, \ldots, \zeta_N\} \quad \textit{and} \quad \overline{G}' := \overline{G} \setminus \{\zeta_1, \ldots, \zeta_N\}.$$

III. *Let the function $f = f(z) : \overline{G}' \to \mathbb{C} \in C^1(G', \mathbb{C}) \cap C^0(\overline{G}', \mathbb{C})$ satisfy the* inhomogeneous Cauchy-Riemann equation

$$\frac{\partial}{\partial \overline{z}} f(z) = g(z) \qquad \textit{for all} \quad z \in G'. \tag{1}$$

IV. Finally, let the right-hand side of our differential equation (1) fulfill the following integrability condition:

$$\iint\limits_{G'} |g(z)|\, dxdy < +\infty.$$

Statement: *Then the limits*

$$\operatorname{Res}(f,\zeta_k) := \lim_{\varepsilon\to 0+} \left\{ \frac{\varepsilon}{2\pi} \int_0^{2\pi} f(\zeta_k + \varepsilon e^{i\varphi}) e^{i\varphi}\, d\varphi \right\} \tag{2}$$

exist for $k = 1, \ldots, N$, *and we have the identity*

$$\int\limits_{\partial G} f(z)\, dz - 2i \iint\limits_{G'} g(z)\, dxdy = 2\pi i \sum_{k=1}^{N} \operatorname{Res}(f,\zeta_k). \tag{3}$$

Proof: We apply the Gaussian integral theorem to the domain

$$G_\varepsilon := \Big\{ z \in G : |z - \zeta_k| > \varepsilon_k \;\text{ for }\; k = 1, \ldots, N \Big\}.$$

Here the vector $\varepsilon = (\varepsilon_1, \ldots, \varepsilon_N)$ consists of the entries $\varepsilon_1 > 0, \ldots, \varepsilon_N > 0$.

Figure 1.4 ILLUSTRATION OF THE RESIDUE THEOREM

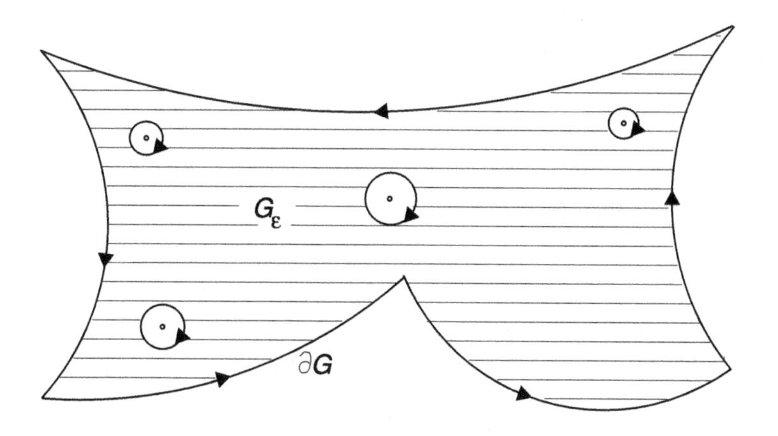

With the notations $f(z) = u(x,y) + iv(x,y)$ and

$$\partial G_\varepsilon : z(t) = x(t) + iy(t), \quad t \in [a_k, b_k], \quad k = 1, \ldots, K = J + N$$

we obtain

$$\begin{aligned}\int\limits_{\partial G_\varepsilon} f(z)\,dz &= \int\limits_{\partial G_\varepsilon} (u+iv)\,(dx+idy)\\ &= \int\limits_{\partial G_\varepsilon} (u\,dx - v\,dy) + i\int\limits_{\partial G_\varepsilon} (v\,dx + u\,dy)\\ &= \sum_{k=1}^{K}\int\limits_{a_k}^{b_k} (ux' - vy')\,dt + i\sum_{k=1}^{K}\int\limits_{a_k}^{b_k} (vx' + uy')\,dt.\end{aligned}$$

The exterior normal to the domain G_ε satisfies

$$\begin{aligned}\xi(z(t)) &= -i\left\{x'(t)^2 + y'(t)^2\right\}^{-\frac{1}{2}}\left\{x'(t) + iy'(t)\right\}\\ &= \left\{x'(t)^2 + y'(t)^2\right\}^{-\frac{1}{2}}\Big(y'(t), -x'(t)\Big)\end{aligned}$$

with $t \in (a_k, b_k)$ for $k = 1, \ldots, K$. Consequently, the Gaussian integral theorem implies

$$\begin{aligned}\int\limits_{\partial G_\varepsilon} f(z)\,dz &= \sum_{k=1}^{K}\int\limits_{a_k}^{b_k}\Big\{(-v,-u)\cdot\xi\Big\}\Big|_{z(t)} d\sigma(t) + i\sum_{k=1}^{K}\int\limits_{a_k}^{b_k}\Big\{(u,-v)\cdot\xi\Big\}\Big|_{z(t)} d\sigma(t)\\ &= \iint\limits_{G_\varepsilon} (-v_x - u_y + iu_x - iv_y)\,dxdy\end{aligned}$$

with the line element

$$d\sigma(t) = \sqrt{x'(t)^2 + y'(t)^2}\,dt.$$

We observe

$$2if_{\overline{z}} = i(f_x + if_y) = -f_y + if_x = -u_y - iv_y + iu_x - v_x$$

and infer the identity

$$\int\limits_{\partial G} f(z)\,dz - 2i\iint\limits_{G_\varepsilon} f_{\overline{z}}(z)\,dxdy = \sum_{k=1}^{N}\oint\limits_{|z-\zeta_k|=\varepsilon_k} f(z)\,dz. \tag{4}$$

Here we integrate over the positive-oriented circular lines on the right-hand side. On the left-hand side in (4) we can implement the transition to the limit $\varepsilon_k \to 0_+$ for each $k \in \{1, \ldots, N\}$ separately, and therefore the limit on the right-hand side exists:

$$\lim_{\varepsilon_k \to 0_+}\oint\limits_{|z-\zeta_k|=\varepsilon_k} f(z)\,dz \in \mathbb{C}.$$

In particular, we can evaluate

$$\begin{aligned}
\lim_{\varepsilon_k \to 0+} \oint\limits_{|z-\zeta_k|=\varepsilon_k} f(z)\,dz &= \lim_{\varepsilon_k \to 0+} \left\{ \varepsilon_k \int\limits_0^{2\pi} f(\zeta_k + \varepsilon_k e^{i\varphi}) i e^{i\varphi}\,d\varphi \right\} \\
&= 2\pi i \lim_{\varepsilon_k \to 0+} \left\{ \frac{\varepsilon_k}{2\pi} \int\limits_0^{2\pi} f(\zeta_k + \varepsilon_k e^{i\varphi}) e^{i\varphi}\,d\varphi \right\} \\
&= 2\pi i \operatorname{Res}(f, \zeta_k)
\end{aligned}$$

for $k = 1, \ldots, N$. The transition to the limit $\varepsilon \to 0$ in (4) implies

$$\int\limits_{\partial G} f(z)\,dz - 2i \iint\limits_{G'} g(z)\,dxdy = 2\pi i \sum_{k=1}^{N} \operatorname{Res}(f, \zeta_k),$$

and finally the statement above follows. q.e.d.

Definition 4.2. *We name* $\operatorname{Res}(f, \zeta_k)$ *from (2) the* residue of f at the point ζ_k.

Definition 4.3. *We denote those domains* $G \subset \mathbb{C}$*, which satisfy Assumption I. in Theorem 4.1, as* normal domains.

Remarks to Theorem 4.1:

1. In the case $N = 0$ - without interior singular points - we obtain the *Gaussian integral theorem in the complex form*

$$\iint\limits_{G} \frac{\partial}{\partial \overline{z}} f(z)\,dxdy = \frac{1}{2i} \int\limits_{\partial G} f(z)\,dz. \tag{5}$$

2. In the case that $g(z) \equiv 0$ in G' holds true, the function $f = f(z)$ is holomorphic in G', and we infer the classical *residue theorem* due to Liouville:

$$\int\limits_{\partial G} f(z)\,dz = 2\pi i \sum_{k=1}^{N} \operatorname{Res}(f, \zeta_k). \tag{6}$$

3. If the function $f = f(z)$ remains bounded about the point ζ_k, which means

$$\sup_{0<|z-\zeta_k|<\varepsilon_k} |f(z)| < +\infty,$$

we infer $\operatorname{Res}(f, \zeta_k) = 0$.

4. Let the following representation

$$f(z) = \frac{\Phi(z)}{z - \zeta_k}, \qquad 0 < |z - \zeta_k| < \varepsilon_k, \tag{7}$$

with a continuous function $\Phi = \Phi(z)$ at the point ζ_k, hold true. Then we have the following identity

$$\begin{aligned}\operatorname{Res}(f, \zeta_k) &= \lim_{\varepsilon \to 0+} \left\{ \frac{\varepsilon}{2\pi} \int_0^{2\pi} \frac{\Phi(\zeta_k + \varepsilon e^{i\varphi})}{\varepsilon e^{i\varphi}} e^{i\varphi} \, d\varphi \right\} \\ &= \lim_{\varepsilon \to 0+} \left\{ \frac{1}{2\pi} \int_0^{2\pi} \Phi(\zeta_k + \varepsilon e^{i\varphi}) \, d\varphi \right\}\end{aligned}$$

for the residue, and consequently

$$\operatorname{Res}(f, \zeta_k) = \Phi(\zeta_k).$$

Theorem 4.4. (Integral representation)
Let the assumptions I. to IV. of Theorem 4.1 be fulfilled. Additionally, the function $f = f(z)$ satisfies the condition

$$\sup_{z \in G'} |f(z)| < +\infty. \tag{8}$$

Then we have the integral representation

$$f(z) = \frac{1}{2\pi i} \int_{\partial G} \frac{f(\zeta)}{\zeta - z} \, d\zeta - \frac{1}{\pi} \iint_{G''} \frac{g(\zeta)}{\zeta - z} \, d\xi d\eta, \qquad z \in G', \tag{9}$$

where we abbreviate $G'' := G' \setminus \{z\}$ and $\zeta = \xi + i\eta$.

Proof: Choosing the point $z \in G'$ as fixed, we apply Theorem 4.1 to the following function

$$h(\zeta) := \frac{f(\zeta)}{\zeta - z}, \qquad \zeta \in G''.$$

Now we calculate

$$\begin{aligned}\int_{\partial G} h(\zeta) \, d\zeta - 2i \iint_{G''} h_{\overline{\zeta}}(\zeta) \, d\xi d\eta &= 2\pi i \sum_{k=1}^{N} \operatorname{Res}(h, \zeta_k) + 2\pi i \operatorname{Res}(h, z) \\ &= 2\pi i f(z).\end{aligned}$$

This implies

$$f(z) = \frac{1}{2\pi i} \int\limits_{\partial G} \frac{f(\zeta)}{\zeta - z}\, d\zeta - \frac{1}{\pi} \iint\limits_{G''} \frac{g(\zeta)}{\zeta - z}\, d\xi d\eta, \qquad z \in G'$$

which corresponds to the statement above. q.e.d.

As a corollary we obtain the following

Theorem 4.5. (Riemann's theorem on removable singularities)
In the punctured disc

$$\Omega := \Big\{ z \in \mathbb{C} : 0 < |z - z_0| \le r \Big\}$$

with the center $z_0 \in \mathbb{C}$ and the radius $r \in (0, +\infty)$, let the function $f : \Omega \to \mathbb{C}$ be holomorphic and bounded, which means

$$\sup_{z \in \Omega} |f(z)| < +\infty.$$

Then $f = f(z)$ can be continued onto the full disc

$$\hat{\Omega} := \Big\{ z \in \mathbb{C} : |z - z_0| \le r \Big\}$$

as a holomorphic function.

Proof: We apply Theorem 4.4 to the set Ω and the holomorphic function $f = f(z)$. Now we infer the statement above from the following integral representation:

$$f(z) = \frac{1}{2\pi i} \oint\limits_{|\zeta - z_0| = r} \frac{f(\zeta)}{\zeta - z}\, d\zeta, \quad z \in \Omega. \tag{10}$$

q.e.d.

In the sequel, we investigate holomorphic functions in the neighborhood of singular points.

Theorem 4.6. (Laurent expansion)
Let the function $f = f(z)$ be holomorphic in the punctured disc

$$\Omega := \Big\{ z \in \mathbb{C} : 0 < |z - z_0| < r \Big\},$$

where $z_0 \in \mathbb{C}$ and $r \in (0, +\infty)$ holds true. Then we have the representation

$$f(z) = \sum_{n=-\infty}^{+\infty} a_n (z - z_0)^n \qquad \textit{for all} \quad z \in \Omega \tag{11}$$

with the coefficients

$$a_n := \frac{1}{2\pi i} \oint\limits_{|\zeta - z_0| = \varrho} \frac{f(\zeta)}{(\zeta - z_0)^{n+1}}\, d\zeta \qquad \textit{for} \quad n \in \mathbb{Z},$$

where the radius $\varrho \in (0, r)$ *has been chosen arbitrarily. The convergence of this so-called* Laurent series, *with its* principal part

$$g(z) := \sum_{n=-1}^{-\infty} a_n (z - z_0)^n\,, \qquad z \in \Omega$$

and its subordinate part

$$h(z) := \sum_{n=0}^{+\infty} a_n (z - z_0)^n\,, \qquad z \in \Omega,$$

is uniform on each compact set contained in Ω*. Finally, we have the identity*

$$Res\,(f, z_0) = a_{-1}. \tag{12}$$

Remark: The determination of the Laurent series gives us the residue as its coefficient a_{-1}, and then we can evaluate integrals by the residue theorem. The coefficients of the Laurent series are uniquely determined.

Proof of the theorem: Without loss of generality, we can assume $z_0 = 0$. We consider the point $z \in \Omega$, fix the numbers $0 < \varepsilon < |z| < \delta < r$, and apply Theorem 4.4 to the domain

$$G := \Big\{ z \in \mathbb{C} : \varepsilon < |z| < \delta \Big\}.$$

Now we obtain

$$f(z) = \frac{1}{2\pi i} \oint\limits_{|\zeta| = \delta} \frac{f(\zeta)}{\zeta - z}\, d\zeta - \frac{1}{2\pi i} \oint\limits_{|\zeta| = \varepsilon} \frac{f(\zeta)}{\zeta - z}\, d\zeta \qquad \text{for all points} \quad z \in G.$$

The expansion of the power series as usual, namely

$$\frac{1}{2\pi i} \oint\limits_{|\zeta| = \delta} \frac{f(\zeta)}{\zeta - z}\, d\zeta = \sum_{n=0}^{\infty} a_n z^n \qquad \text{for all} \quad |z| < \delta,$$

gives us the subordinate part of the Laurent series. Now we expand the expression below for all $|\zeta| = \varepsilon$ and $|z| > \varepsilon$ as follows:

$$-\frac{1}{\zeta - z} = \frac{1}{z} \frac{1}{1 - \frac{\zeta}{z}} = \frac{1}{z} \sum_{n=0}^{\infty} \frac{\zeta^n}{z^n} = \sum_{n=0}^{\infty} \zeta^n z^{-n-1}.$$

Here the convergence of the series is uniform on each compact set. For all points satisfying $|z| > \varepsilon$ we obtain the following expansion

$$-\frac{1}{2\pi i}\oint\limits_{|\zeta|=\varepsilon}\frac{f(\zeta)}{\zeta-z}\,d\zeta=\sum_{n=0}^{\infty}\left(\frac{1}{2\pi i}\oint\limits_{|\zeta|=\varepsilon}\frac{f(\zeta)}{\zeta^{-n}}\,d\zeta\right)z^{-n-1}$$
$$=\sum_{n=0}^{\infty}a_{-n-1}z^{-n-1}$$
$$=\sum_{n=-1}^{-\infty}a_n z^n.$$

This gives us the principal part of our Laurent series. Therefore, we have shown the uniform convergence of the series

$$f(z)=\sum_{n=-\infty}^{+\infty}a_n z^n \qquad \text{for} \quad \varepsilon<|z|<\delta,$$

where $0<\varepsilon<\delta<r$ has been chosen arbitrarily. q.e.d.

Definition 4.7. *Let the holomorphic function $f=f(z)$ be represented by its Laurent series (11), in the neighborhood of the point $z_0\in\mathbb{C}$, due to Theorem 4.6.*

i) When we find a nonvanishing coefficient $a_n\neq 0$ with its index $n\leq N$ for each integer $N\in\mathbb{Z}$, we say that the function f possesses an essential singularity *at the point z_0.*
ii) When we have a negative integer $N\in\mathbb{Z}$ with $N<0$, such that $a_n=0$ for all indices $n<N$ holds true and $a_N\neq 0$, we say that f possesses a pole of the order $(-N)\in\mathbb{N}$ *at the point z_0.*
iii) When the condition $a_n=0$ for all $n\in\mathbb{Z}$ with $n<0$ is correct, we say that f possesses a removable singularity *at the point z_0.*

Theorem 4.8. (Casorati, Weierstraß)
Let the assumptions and notations of Theorem 4.6 be valid. Additionally, let the function $f:\overline{\Omega}\to\overline{\mathbb{C}}$ be continuous. Then this function $f=f(z)$ does not possess an essential singularity at the point z_0. It has a pole at this point if and only if $f(z_0)=\infty$ is correct; and it has a removable singularity at the point z_0 if and only if $f(z_0)\in\mathbb{C}$ holds true.

Proof: Since $f:\Omega\to\mathbb{C}$ is extendable into the point z_0 as a continuous function, we have a constant $c\in\mathbb{C}$ and a quantity $\varepsilon>0$ such that

$$f(z)\neq c \qquad \text{for all} \quad z\in K_\varepsilon(z_0)$$

holds true. Now we consider the holomorphic function

$$g(z) := \frac{1}{f(z) - c}, \qquad z \in K_\varepsilon(z_0) \setminus \{z_0\}.$$

On account of the condition

$$\sup_{0<|z-z_0|<\varepsilon} |g(z)| < +\infty,$$

we can continue $g = g(z)$ as a holomorphic function into the point z_0, due to Riemann's theorem on removable singularities. Consequently, we have a holomorphic function $h = h(z)$, $z \in K_\varepsilon(z_0)$ satisfying $h(z_0) \neq 0$ and a nonnegative integer $n \in \mathbb{N}_0$ such that

$$\frac{1}{f(z) - c} = g(z) = (z - z_0)^n h(z), \quad z \in K_\varepsilon(z_0) \setminus \{z_0\}$$

holds true. This implies the representation

$$f(z) = c + (z - z_0)^{-n} h(z)^{-1} = \sum_{k=-n}^{+\infty} b_k (z - z_0)^k = (z - z_0)^N \psi(z)$$

for all points $z \in K_\varepsilon(z_0) \setminus \{z_0\}$. Here we have used the integer $N \in \mathbb{Z}$ and the holomorphic function

$$\psi = \psi(z), \quad z \in K_\varepsilon(z_0)$$

satisfying $\psi(z_0) \neq 0$.
Now the function $f = f(z)$ possesses a pole at the point z_0 if and only if the relation

$$\lim_{\substack{z\to z_0\\ z\neq z_0}} |f(z)| = \lim_{\substack{z\to z_0\\ z\neq z_0}} \Big\{ |z - z_0|^N |\psi(z)| \Big\} = |\psi(z_0)| \lim_{\substack{z\to z_0\\ z\neq z_0}} |z - z_0|^N = +\infty$$

or equivalently

$$f(z_0) = \infty$$

holds true.
Correspondingly, the function possesses a removable singularity at the point z_0 if and only if

$$\lim_{\substack{z\to z_0\\ z\neq z_0}} |f(z)| = \lim_{\substack{z\to z_0\\ z\neq z_0}} \Big\{ |z - z_0|^N |\psi(z)| \Big\} = |\psi(z_0)| \lim_{\substack{z\to z_0\\ z\neq z_0}} |z - z_0|^N < +\infty$$

or equivalently

$$f(z_0) \in \mathbb{C}$$

holds true. Thus the statement above is established. q.e.d.
Remark: We could show by the method described that a function with an essential singularity z_0 approaches each value in $\mathbb{C}$ as closely as possible in each neighborhood of this singular point.

We consider the holomorphic function $f : \Omega \to \mathbb{C}$ defined on the punctured disc

$$\Omega := \Big\{ z \in \mathbb{C} : 0 < |z - z_0| < r \Big\},$$

with no essential singularity at the point z_0. Then we have the representation

$$f(z) = (z - z_0)^n \varphi(z), \qquad z \in \Omega, \quad n \in \mathbb{Z}, \tag{13}$$

with the holomorphic function $\varphi : \Omega \cup \{z_0\} \to \mathbb{C}$ satisfying $\varphi(z_0) \neq 0$.

Definition 4.9. *We name the integer $n \in \mathbb{Z}$ from the representation (13) the* order of the zero z_0.

Remark: If $n \in \mathbb{N}$ holds true, we infer $f(z_0) = 0$. In the case $n = 0$, the condition $f(z_0) \neq 0$ is fulfilled. If $n \in -\mathbb{N}$ holds true, the function $f = f(z)$ possesses a pole of the order $N = -n \in \mathbb{N}$ at the point z_0, and we note that $f(z_0) = \infty$.

Theorem 4.10. (Principle of the argument)
Let the assumptions I. and II. of Theorem 4.1 be fulfilled. The function

$$f = f(z) : \overline{G}' \to \mathbb{C} \setminus \{0\}$$

may be holomorphic in $\overline{G}'$ and extendable into the singular points as a continuous function $f : \overline{G} \to \overline{\mathbb{C}}$. We denote the order of the zeroes for the singular points ζ_k by the symbols $n_k = n_k(\zeta_k) \in \mathbb{Z}$ for $k = 1, \ldots, N$. Then we have the index-sum formula

$$\sum_{k=1}^{N} n_k = \frac{1}{2\pi i} \int\limits_{\partial G} \frac{1}{f(\zeta)} \Big\{ f_\xi(\zeta)\, d\xi + f_\eta(\zeta)\, d\eta \Big\}. \tag{14}$$

Proof: We apply the residue theorem on the holomorphic function

$$F(z) := \frac{f'(z)}{f(z)}, \qquad z \in \overline{G}'.$$

Here we consider the expansions

$$f(z) = (z - \zeta_k)^{n_k} \varphi_k(z), \qquad z \in G \setminus \{\zeta_k\}, \quad z \to \zeta_k, \tag{15}$$

with the holomorphic functions $\varphi_k = \varphi_k(z)$ satisfying $\varphi_k(\zeta_k) \neq 0$. These imply the representations

$$\begin{aligned} F(z) &= \frac{n_k (z - \zeta_k)^{n_k - 1} \varphi_k(z) + (z - \zeta_k)^{n_k} \varphi_k'(z)}{(z - \zeta_k)^{n_k} \varphi_k(z)} \\ &= \frac{n_k}{z - \zeta_k} + \frac{\varphi_k'(z)}{\varphi_k(z)} \end{aligned}$$

for

$$z \in G \setminus \{\zeta_k\}, \quad z \to \zeta_k,$$

and we obtain

$$\text{Res}\,(F, \zeta_k) = n_k\,, \qquad k = 1, \ldots, N. \tag{16}$$

The residue theorem yields

$$\begin{aligned}
\sum_{k=1}^{N} n_k = \sum_{k=1}^{N} \text{Res}\,(F, \zeta_k) &= \frac{1}{2\pi i} \int\limits_{\partial G} F(\zeta)\, d\zeta \\
= \frac{1}{2\pi i} \int\limits_{\partial G} \frac{f'(\zeta)}{f(\zeta)}\, d\zeta &= \frac{1}{2\pi i} \int\limits_{\partial G} \frac{f_\xi\, d\xi + i f_\xi\, d\eta}{f} \\
= \frac{1}{2\pi i} \int\limits_{\partial G} \frac{f_\xi\, d\xi + f_\eta\, d\eta}{f}&,
\end{aligned}$$

and the statement above is proved. q.e.d.

Remark: We refer the reader to Section 1 in Chapter 3, presenting the winding number, in order to comprehend the notation *principle of the argument.*

Now we shall investigate properties of the singular double integral from representation (9).

Definition 4.11. *Let $\Omega \subset \mathbb{C}$ denote a bounded open set, and the bounded continuous function*

$$g \in L^\infty(\Omega, \mathbb{C}) \cap C^0(\Omega, \mathbb{C})$$

may be given. Then we name

$$T_\Omega[g](z) := -\frac{1}{\pi} \iint\limits_{\Omega} \frac{g(\zeta)}{\zeta - z}\, d\xi d\eta, \qquad z \in \Omega, \tag{17}$$

the Hadamard integral operator; *here we use $\zeta = \xi + i\eta$ again.*

The following result is fundamental for the two-dimensional potential theory.

Theorem 4.12. (Hadamard's estimate)
Let $\Omega \subset \mathbb{C}$ denote a bounded open set, and let $g \in C^0(\Omega, \mathbb{C})$ be a continuous function with the property

$$\|g\|_\infty := \sup_{\zeta \in \Omega} |g(\zeta)| < +\infty.$$

Then we have a constant $\gamma \in (0, +\infty)$ such that the function

$$\psi(z) := T_\Omega[g](z), \qquad z \in \mathbb{C}$$

satisfies the inequality

$$|\psi(z_1) - \psi(z_2)| \le 2\gamma \|g\|_\infty |z_1 - z_2| \log \frac{\vartheta(z_1)}{|z_1 - z_2|}$$
$$\text{for all points } z_1, z_2 \in \mathbb{C} \text{ with } |z_1 - z_2| \le \frac{1}{2}\vartheta(z_1). \tag{18}$$

Here we have defined the quantity

$$\vartheta(z_1) := \sup_{z \in \Omega} |z - z_1|.$$

Proof: We take the points $z_1, z_2 \in \mathbb{C}$ with $z_1 \ne z_2$ and observe

$$\begin{aligned} \psi(z_1) - \psi(z_2) &= \frac{1}{\pi} \iint\limits_{\Omega} \left(\frac{g(\zeta)}{\zeta - z_2} - \frac{g(\zeta)}{\zeta - z_1} \right) d\xi d\eta \\ &= \frac{1}{\pi} \iint\limits_{\Omega} \frac{z_2 - z_1}{(\zeta - z_2)(\zeta - z_1)} \, g(\zeta) \, d\xi d\eta. \end{aligned} \tag{19}$$

We utilize the transformation

$$\zeta = z_1 + z(z_2 - z_1), \qquad z \in \mathbb{C},$$

which satisfies $0 \mapsto z_1$ and $1 \mapsto z_2$ and has the Jacobian $|z_2 - z_1|^2$. Then we estimate as follows:

$$\begin{aligned} &|\psi(z_1) - \psi(z_2)| \\ &\le \frac{1}{\pi} |z_2 - z_1| \|g\|_\infty \iint\limits_{\Omega} \frac{1}{|\zeta - z_1||\zeta - z_2|} \, d\xi d\eta \\ &\le \frac{1}{\pi} |z_2 - z_1| \|g\|_\infty \iint\limits_{\zeta : |\zeta - z_1| \le \vartheta(z_1)} \frac{1}{|\zeta - z_1||\zeta - z_2|} \, d\xi d\eta \\ &= \frac{1}{\pi} |z_2 - z_1| \|g\|_\infty \iint\limits_{z : |z| \le \frac{\vartheta(z_1)}{|z_2 - z_1|}} \frac{1}{|z(z_2 - z_1)||(z-1)(z_2 - z_1)|} \, |z_2 - z_1|^2 \, dxdy \\ &= \frac{1}{\pi} |z_2 - z_1| \|g\|_\infty \iint\limits_{z : |z| \le \frac{\vartheta(z_1)}{|z_1 - z_2|}} \frac{1}{|z||z-1|} \, dxdy. \end{aligned}$$

Now we have a constant $\gamma \in (0, +\infty)$ such that

$$\iint\limits_{|z| \le R} \frac{1}{|z||z-1|} \, dxdy \le \gamma \iint\limits_{1 \le |z| \le R} \frac{1}{|z|^2} \, dxdy \qquad \text{for all} \quad R \in [2, +\infty) \tag{20}$$

holds true. For the points $z_1, z_2 \in \mathbb{C}$ with $0 < |z_1 - z_2| \le \frac{1}{2}\vartheta(z_1)$ we have

$$2 \le \frac{\vartheta(z_1)}{|z_1 - z_2|}.$$

Therefore, the following estimate

$$\begin{aligned} |\psi(z_1) - \psi(z_2)| &\le \frac{\gamma}{\pi}|z_2 - z_1| \|g\|_\infty \iint\limits_{1 \le |z| \le \frac{\vartheta(z_1)}{|z_2 - z_1|}} \frac{1}{|z|^2}\,dxdy \\ &= \frac{\gamma}{\pi}\|g\|_\infty |z_2 - z_1|\, 2\pi \int\limits_1^{\frac{\vartheta(z_1)}{|z_1 - z_2|}} \frac{1}{r^2}\, r\, dr \\ &= 2\gamma \|g\|_\infty |z_1 - z_2| \log \frac{\vartheta(z_1)}{|z_1 - z_2|} \end{aligned}$$

yields the statement above. q.e.d.

Definition 4.13. *We consider the function $f : \Omega \to \mathbb{R}^m$ on the set $\Omega \subset \mathbb{R}^n$, where $m, n \in \mathbb{N}$ holds true. Furthermore, let $\omega : [0, +\infty) \to [0, +\infty)$ denote a continuous function - with $\omega(0) = 0$ - which prescribes a* modulus of continuity*. Then the function f is named* Dini continuous *if the estimate*

$$|f(x) - f(y)| \le \omega(|x - y|) \qquad \textit{for all points} \quad x, y \in \Omega \tag{21}$$

holds true.
In the special case

$$\omega(t) = Lt, \qquad t \in [0, +\infty),$$

the function f is called Lipschitz continuous *with the* Lipschitz constant $L \in [0, +\infty)$.
In the special case

$$\omega(t) = Ht^\alpha\,, \qquad t \in [0, +\infty),$$

we name f Hölder continuous *with the* Hölder constant $H \in [0, +\infty)$ *and the* Hölder exponent $\alpha \in (0, 1)$.

Corollary of Theorem 4.12: The function $\psi(z) = T_\Omega[g](z)$, $z \in \overline{\Omega}$ is Dini continuous with the following modulus of continuity

$$\omega(t) = 2\gamma \|g\|_\infty t \log \frac{\vartheta}{t}\,, \qquad t \in [0, +\infty), \quad \vartheta := \operatorname{diam} \Omega. \tag{22}$$

Therefore, the function $\psi = \psi(z)$ is Hölder continuous in $\overline{\Omega}$ with each Hölder exponent $\alpha \in (0, 1)$.

Theorem 4.14. (Removable singularities)
Let the assumptions I. to IV. of Theorem 4.1 be satisfied. Furthermore, the function $f = f(z)$ may be subject to the condition

$$\sup_{z \in G'} |f(z)| < +\infty.$$

Finally, let the right-hand side $g = g(z)$ of the inhomogeneous Cauchy-Riemann differential equation (1) fulfill

$$\sup_{z \in G'} |g(z)| < +\infty.$$

Then the function $f = f(z)$ is extendable into the singular points $\zeta_1, \ldots, \zeta_N \in G$ as a Hölder continuous function, with an arbitrary Hölder exponent $\alpha \in (0,1)$.

Proof: We use Theorem 4.4 and Theorem 4.12. q.e.d.

5 The Inhomogeneous Cauchy-Riemann Differential Equation

We recommend the study of the excellent monograph [V] of I.N.Vekua and, moreover, the interesting treatise by

I. N. Vekua: *Systeme von Differentialgleichungen erster Ordnung vom elliptischen Typus und Randwertaufgaben.* Deutscher Verlag der Wissenschaften, Berlin, 1956.

Definition 5.1. *Let the continuous function $\Phi : \Omega \to \mathbb{C}$ be given on the open set $\Omega \subset \mathbb{C}$, and choose a point $z_0 \in \Omega$ as fixed. We consider the normal domains G_k for $k = 1, 2, \ldots$ of the topological type of the disc - with area $|G_k|$ and length $|\partial G_k|$ of their boundary curves - which satisfy the inclusions*

$$z_0 \in G_k \subset \Omega, \qquad k \in \mathbb{N} \tag{1}$$

and the asymptotic condition

$$\lim_{k \to \infty} |\partial G_k| = 0. \tag{2}$$

When all these sequences of domains $\{G_k\}_{k=1,2,\ldots}$ possess the uniquely determined limit

$$\lim_{k \to \infty} \frac{1}{2i|G_k|} \int\limits_{\partial G_k} \Phi(z)\, dz =: \frac{d}{d\overline{z}}\, \Phi(z_0), \tag{3}$$

we call the function $\Phi = \Phi(z)$ (weakly) differentiable *at the point z_0* in the sense of Pompeiu.

Remark: The Gaussian integral theorem in the complex form yields the identity

$$\frac{d}{d\overline{z}}\,\Phi(z_0) = \Phi_{\overline{z}}(z_0) \qquad \text{for all points} \quad z_0 \in \Omega \tag{4}$$

for all functions $\Phi \in C^1(\Omega, \mathbb{C})$; here we used Wirtinger's derivative on the right-hand side and Pompeiu's derivative on the left-hand side.

Definition 5.2. *We take the open set* $\Omega \subset \mathbb{C}$ *and define* Vekua's class of functions *as follows:*

$$C_{\overline{z}}(\Omega) := \left\{ \Phi \in C^0(\Omega, \mathbb{C}) \, : \begin{array}{l} \textit{There exists} \quad \dfrac{d}{d\overline{z}}\,\Phi(z) =: g(z) \\ \textit{for all} \quad z \in \Omega \quad \textit{with} \quad g \in C^0(\Omega, \mathbb{C}) \end{array} \right\}.$$

Proposition 5.3. *The rules of differentiation for the class* $C^1(\Omega)$ *remain valid even in the class* $C_{\overline{z}}(\Omega)$ *- if the formula contains only the functions* Φ *and* $\Phi_{\overline{z}}$.

Proof: Consider the unit disc

$$B := \Big\{ \zeta = \xi + i\eta \in \mathbb{C} \, : \, |\zeta| < 1 \Big\}$$

and the mollifier $\chi = \chi(\zeta) \in C_0^\infty(B, [0, +\infty))$ with the property

$$\iint\limits_B \chi(\xi, \eta)\, d\xi d\eta = 1.$$

With an arbitrary function $\Phi = \Phi(z) \in C_{\overline{z}}(\Omega)$, we associate the mollified function

$$\Phi^\varepsilon(z) := \iint\limits_\Omega \frac{1}{\varepsilon^2}\,\chi\left(\frac{\zeta - z}{\varepsilon}\right) \Phi(\zeta)\, d\xi d\eta, \qquad z \in \Omega \quad \text{with} \quad \operatorname{dist}(z, \mathbb{C} \setminus \Omega) \geq \varepsilon, \tag{5}$$

where $0 < \varepsilon < \varepsilon_0$ is valid. One easily shows that the Pompeiu derivation commutes with the mollification process (compare the Friedrichs theorems in Section 1 of Chapter 10 of Volume 2 for Sobolev spaces). Thus the statements

$$\Phi^\varepsilon(z) \longrightarrow \Phi(z) \qquad \text{for} \quad \varepsilon \to 0+ \quad \text{uniformly in} \quad \Theta \tag{6}$$

and

$$\Phi^\varepsilon_{\overline{z}}(z) \longrightarrow \frac{d}{d\overline{z}}\,\Phi(z) \qquad \text{for} \quad \varepsilon \to 0+ \quad \text{uniformly in} \quad \Theta \tag{7}$$

for each compact set $\Theta \subset \Omega$ hold true. Therefore, we can transfer the rules of differentiation into the class $C_{\overline{z}}(\Omega)$.

q.e.d.

Proposition 5.4. *Let $\Omega \subset \mathbb{C}$ denote a bounded open set and $g \in C^0(\Omega, \mathbb{C}) \cap L^\infty(\Omega, \mathbb{C})$ a function. Then the integral*

$$\Psi(z) := T_\Omega[g](z), \qquad z \in \Omega$$

is differentiable - in the sense of Pompeiu - with respect to $\overline{z}$ at each point $z_0 \in \Omega$, and we have

$$\frac{d}{d\overline{z}} \Psi(z_0) = g(z_0), \qquad z_0 \in \Omega. \tag{8}$$

Proof: As described in Definition 5.1, let us consider a sequence $\{G_k\}_{k=1,2,\ldots}$ of domains contracting to the point $z_0 \in \Omega$. We utilize the characteristic function

$$\chi_{G_k}(z) := \begin{cases} 1, \; z \in G_k \\ 0, \; z \in \mathbb{C} \setminus G_k \end{cases}.$$

Now we deduce

$$\begin{aligned}
\frac{1}{2i|G_k|} \int\limits_{\partial G_k} \Psi(z)\,dz &= \frac{1}{2i|G_k|} \int\limits_{\partial G_k} \left(-\frac{1}{\pi} \iint\limits_{\Omega} \frac{g(\zeta)}{\zeta - z}\,d\xi d\eta \right) dz \\
&= \frac{1}{2\pi i|G_k|} \iint\limits_{\Omega} \left(g(\zeta) \int\limits_{\partial G_k} \frac{1}{z-\zeta}\,dz \right) d\xi d\eta \\
&= \frac{1}{2\pi i|G_k|} \iint\limits_{\Omega} \Big(g(\zeta) 2\pi i \chi_{G_k}(\zeta) \Big)\,d\xi d\eta \\
&= \frac{1}{|G_k|} \iint\limits_{G_k} g(\zeta)\,d\xi d\eta
\end{aligned}$$

for $k = 1, 2, 3, \ldots$. Finally, we obtain the identity

$$\frac{d}{d\overline{z}} \Psi(z_0) = \lim_{k \to \infty} \left\{ \frac{1}{2i|G_k|} \int\limits_{\partial G_k} \Psi(z)\,dz \right\} = g(z_0)$$

for all points $z_0 \in \Omega$. q.e.d.

Remark: In general, the function $\Psi = \Psi(z)$ does not belong to the class $C^1(\Omega)$; however, Ψ lies only in the class $C_{\overline{z}}(\Omega)$.

Theorem 5.5. (Pompeiu, Vekua)
On the open set $\Omega \subset \mathbb{C}$ we consider the continuous function $g \in C^0(\Omega, \mathbb{C})$. Then the following statements are equivalent:

(a) *The element $f = f(z)$ belongs to Vekua's class of functions $C_{\overline{z}}(\Omega)$ and satisfies the partial differential equation*

$$\frac{d}{d\overline{z}} f(z) = g(z), \qquad z \in \Omega \tag{9}$$

in the sense of Pompeiu;

(b) *The element $f = f(z)$ belongs to the class $C^0(\Omega, \mathbb{C})$, and we have the following integral representation for each normal domain $G \subset\subset \Omega$:*

$$f(z) = \frac{1}{2\pi i} \int\limits_{\partial G} \frac{f(\zeta)}{\zeta - z}\, d\zeta - \frac{1}{\pi} \iint\limits_{G} \frac{g(\zeta)}{\zeta - z}\, d\xi d\eta, \qquad z \in G. \tag{10}$$

Proof: We show the direction $(a) \Rightarrow (b)$: Let us consider the element $f \in C_{\overline{z}}(\Omega)$ with

$$\frac{d}{d\overline{z}} f(z) = g(z), \qquad z \in \Omega.$$

Then we have a sequence of functions $f_k(z) \in C^1(\Omega, \mathbb{C})$ for $k = 1, 2, \ldots$ satisfying

$$\begin{cases} f_k(z) \longrightarrow f(z), & z \in \Theta \\ f_{k\overline{z}}(z) \longrightarrow \dfrac{d}{d\overline{z}} f(z), & z \in \Theta \end{cases} \qquad \text{uniformly for} \quad k \to \infty \tag{11}$$

in each compact set $\Theta \subset \Omega$. Based on Theorem 4.4 from Section 4, we comprehend the following identity for each normal domain $G \subset\subset \Omega$:

$$f_k(z) = \frac{1}{2\pi i} \int\limits_{\partial G} \frac{f_k(\zeta)}{\zeta - z}\, d\zeta - \frac{1}{\pi} \iint\limits_{G} \frac{\frac{\partial}{\partial \overline{\zeta}} f_k(\zeta)}{\zeta - z}\, d\xi d\eta, \qquad z \in G, \quad k \in \mathbb{N}.$$

Therefore, we obtain the integral representation (10) via transition $k \to \infty$ to the limit.

We now show the direction $(b) \Rightarrow (a)$: The curvilinear integral in (10) represents an analytic function in the domain G. Furthermore, the parameter integral $T_G[g]$ is continuous in G and weakly differentiable with respect to $\overline{z}$ in Pompeiu's sense. Therefore, we obtain

$$\frac{d}{d\overline{z}} f(z) = g(z), \qquad z \in G,$$

according to Proposition 5.4. q.e.d.

Definition 5.6. *We name a function $g : \Omega \to \mathbb{C}$, defined on the open set $\Omega \subset \mathbb{C}$,* Hölder continuous *if each compact set $\Theta \subset \Omega$ admits a constant $H = H(\Theta) \in [0, +\infty)$ and an exponent $\alpha = \alpha(\Theta) \in (0, 1]$ such that the estimate*

$$|g(z_1) - g(z_2)| \le H(\Theta)|z_1 - z_2|^{\alpha(\Theta)} \qquad \text{for all points} \quad z_1, z_2 \in \Theta \tag{12}$$

holds true.

Definition 5.7. *Let $G \subset \mathbb{C}$ be a normal domain, take the point $z \in G$ as fixed, and let*

$$f : \overline{G} \setminus \{z\} \to \mathbb{C} \in C^0(\overline{G} \setminus \{z\})$$

denote a continuous function. We consider the domains

$$G_\varepsilon(z) := \Big\{\zeta \in G : |\zeta - z| > \varepsilon\Big\}$$

for all numbers $0 \le \varepsilon < dist\{z, \mathbb{C} \setminus G\}$. Then we call the expression

$$\underset{G_0(z)}{\iint\!\!\!\!\!\!\!\bigcirc} f(\zeta)\, d\xi d\eta := \lim_{\varepsilon \to 0+} \iint\limits_{G_\varepsilon(z)} f(\zeta)\, d\xi d\eta \tag{13}$$

Cauchy's principal value *of the integral*

$$\iint\limits_{G_0(z)} f(\zeta)\, d\xi d\eta$$

if the limit (13) exists.

Remark: When the improper Riemannian integral $\iint\limits_{G_0(z)} f(\zeta)\, d\xi d\eta$ exists, we infer

$$\underset{G_0(z)}{\iint\!\!\!\!\!\!\!\bigcirc} f(\zeta)\, d\xi d\eta = \iint\limits_{G_0(z)} f(\zeta)\, d\xi d\eta. \tag{14}$$

Example 5.8. We consider the function

$$\Lambda(z) := \Lambda_G(z) := T_G[1](z) = -\frac{1}{\pi} \iint\limits_{G} \frac{1}{\zeta - z}\, d\xi d\eta = -\frac{1}{\pi} \underset{G_0(z)}{\iint\!\!\!\!\!\!\!\bigcirc} \frac{1}{\zeta - z}\, d\xi d\eta$$

for all $z \in G$. The Gaussian integral theorem in the complex form yields

$$\begin{aligned}
\Lambda(z) &= -\frac{1}{\pi} \lim_{\varepsilon \to 0+} \iint\limits_{G_\varepsilon(z)} \frac{1}{\zeta - z}\, d\xi d\eta \\
&= -\frac{1}{\pi} \lim_{\varepsilon \to 0+} \iint\limits_{G_\varepsilon(z)} \frac{d}{d\overline{\zeta}} \frac{\overline{\zeta}}{\zeta - z}\, d\xi d\eta \\
&= -\frac{1}{\pi} \lim_{\varepsilon \to 0+} \frac{1}{2i} \int\limits_{\partial G_\varepsilon(z)} \frac{\overline{\zeta}}{\zeta - z}\, d\zeta \\
&= -\frac{1}{2\pi i} \int\limits_{\partial G} \frac{\overline{\zeta}}{\zeta - z}\, d\zeta + \frac{1}{2\pi i} \lim_{\varepsilon \to 0+} \oint\limits_{|\zeta - z| = \varepsilon} \frac{\overline{\zeta}}{\zeta - z}\, d\zeta
\end{aligned}$$

for all $z \in G$. With the aid of the transformation $\zeta = z + \varepsilon e^{i\varphi}$ and $d\zeta = i\varepsilon e^{i\varphi}\,d\varphi$, let us calculate as follows:

$$\begin{aligned}
\lim_{\varepsilon\to 0+} \oint\limits_{|\zeta-z|=\varepsilon} \frac{\overline{\zeta}}{\zeta - z}\,d\zeta &= \lim_{\varepsilon\to 0+} \int\limits_0^{2\pi} \frac{\overline{(z+\varepsilon e^{i\varphi})}}{\varepsilon e^{i\varphi}}\, i\varepsilon e^{i\varphi}\,d\varphi \\
&= i \lim_{\varepsilon\to 0+} \int\limits_0^{2\pi} \overline{(z+\varepsilon e^{i\varphi})}\,d\varphi \\
&= i \lim_{\varepsilon\to 0+} \left(2\pi\overline{z} + \varepsilon \int\limits_0^{2\pi} e^{-i\varphi}\,d\varphi \right) \\
&= 2\pi i\overline{z}.
\end{aligned}$$

This implies

$$\Lambda(z) = \overline{z} - \frac{1}{2\pi i} \int\limits_{\partial G} \frac{\overline{\zeta}}{\zeta - z}\,d\zeta, \qquad z \in G. \tag{15}$$

The function $\Lambda = \Lambda(z)$ belongs to the class $C^\infty(G)$, and we observe

$$\Lambda_{\overline{z}}(z) = 1, \qquad z \in G, \tag{16}$$

as well as

$$\Lambda_z(z) = -\frac{1}{2\pi i} \int\limits_{\partial G} \frac{\overline{\zeta}}{(\zeta - z)^2}\,d\zeta, \quad z \in G. \tag{17}$$

Now we deduce the identity

$$\begin{aligned}
\lim_{\varepsilon\to 0+} \oint\limits_{|\zeta-z|=\varepsilon} \frac{\overline{\zeta}}{(\zeta - z)^2}\,d\zeta &= \lim_{\varepsilon\to 0+} \int\limits_0^{2\pi} \frac{\overline{z}+\varepsilon e^{-i\varphi}}{\varepsilon^2 e^{2i\varphi}}\, i\varepsilon e^{i\varphi}\,d\varphi \\
&= i \lim_{\varepsilon\to 0+} \left\{ \frac{\overline{z}}{\varepsilon} \int\limits_0^{2\pi} e^{-i\varphi}\,d\varphi + \int\limits_0^{2\pi} e^{-2i\varphi}\,d\varphi \right\},
\end{aligned}$$

where we utilize the substitution $\zeta = z + \varepsilon e^{i\varphi}$ again. Both integrals in the last line vanish, which implies

$$\lim_{\varepsilon\to 0+} \oint\limits_{|\zeta-z|=\varepsilon} \frac{\overline{\zeta}}{(\zeta - z)^2}\,d\zeta = 0.$$

Therefore, we obtain

$$\begin{aligned}\Lambda_z(z) &= -\frac{1}{2\pi i}\int\limits_{\partial G}\frac{\overline{\zeta}}{(\zeta-z)^2}\,d\zeta+\lim_{\varepsilon\to 0+}\frac{1}{2\pi i}\oint\limits_{|\zeta-z|=\varepsilon}\frac{\overline{\zeta}}{(\zeta-z)^2}\,d\zeta\\ &= -\frac{1}{2\pi i}\lim_{\varepsilon\to 0+}\int\limits_{\partial G_\varepsilon(z)}\frac{\overline{\zeta}}{(\zeta-z)^2}\,d\zeta,\end{aligned}$$

and the Gaussian integral theorem yields

$$\begin{aligned}\Lambda_z(z) &= -\frac{1}{\pi}\lim_{\varepsilon\to 0+}\iint\limits_{G_\varepsilon(z)}\frac{d}{d\overline{\zeta}}\frac{\overline{\zeta}}{(\zeta-z)^2}\,d\xi d\eta\\ &= -\frac{1}{\pi}\oiint\limits_{G_0(z)}\frac{1}{(\zeta-z)^2}\,d\xi d\eta.\end{aligned}$$

Finally, we arrive at the formula

$$\Lambda_z(z) = -\frac{1}{\pi}\oiint\limits_{G_0(z)}\frac{1}{(\zeta-z)^2}\,d\xi d\eta,\qquad z\in G. \tag{18}$$

Here we have presented Cauchy's principal value of an integral which does not converge absolutely.

Proposition 5.9. *On the normal domain $G\subset\mathbb{C}$, let the function $g:\overline{G}\to\mathbb{C}\in C^0(\overline{G},\mathbb{C})$ be Hölder continuous. Then Cauchy's principal value of the following integral exists for all points $z\in G$, namely*

$$\begin{aligned}\chi(z)=\Pi_G[g](z) &:= -\frac{1}{\pi}\oiint\limits_{G_0(z)}\frac{g(\zeta)}{(\zeta-z)^2}\,d\xi d\eta\\ &= \lim_{\varepsilon\to 0+}\left\{-\frac{1}{\pi}\iint\limits_{G_\varepsilon(z)}\frac{g(\zeta)}{(\zeta-z)^2}\,d\xi d\eta\right\}.\end{aligned} \tag{19}$$

The function $\chi:G\to\mathbb{C}$ is continuous in G.

Definition 5.10. *We call Π_G from (19) the* Vekua integral operator.

Proof of Proposition 5.9: For all points $z\in G$ and all numbers $0<\varepsilon<\operatorname{dist}(z,\mathbb{C}\setminus G)$ we observe

$$-\frac{1}{\pi}\iint\limits_{G_\varepsilon(z)}\frac{g(\zeta)}{(\zeta-z)^2}\,d\xi d\eta = -\frac{1}{\pi}\iint\limits_{G_\varepsilon(z)}\frac{g(\zeta)-g(z)}{(\zeta-z)^2}\,d\xi d\eta-\frac{g(z)}{\pi}\iint\limits_{G_\varepsilon(z)}\frac{1}{(\zeta-z)^2}\,d\xi d\eta. \tag{20}$$

Now the integral

$$\Phi(z) := -\frac{1}{\pi} \iint\limits_{G_0(z)} \frac{g(\zeta) - g(z)}{(\zeta - z)^2} \, d\xi d\eta, \qquad z \in G$$

converges absolutely, and the function $\Phi : G \to \mathbb{C}$ is continuous. We utilize the function $\Lambda = \Lambda(z)$, $z \in G$ from Example 5.8, and we infer from (20) the following identity for $\varepsilon \to 0+$, namely

$$\chi(z) = \Phi(z) + g(z)\Lambda_z(z), \qquad z \in G. \tag{21}$$

Furthermore, the function $\chi = \chi(z)$ is continuous in G. q.e.d.

Proposition 5.11. *Let $G \subset \mathbb{C}$ denote a normal domain, and let the function $g \in C^0(\overline{G}, \mathbb{C})$ be Hölder continuous in G. Then the function*

$$\Psi(z) := T_G[g](z) = -\frac{1}{\pi} \iint\limits_{G} \frac{g(\zeta)}{\zeta - z} \, d\xi d\eta, \qquad z \in G$$

belongs to the regularity class $C^1(G, \mathbb{C})$, and we have the identities

$$\Psi_{\overline{z}}(z) = g(z), \quad \Psi_z(z) = \Pi_G[g](z) \qquad \text{for all points} \quad z \in G. \tag{22}$$

Proof:

1. Let the point $z_0 \in G$ be fixed and $g(z_0) = 0$ hold true. With the aid of formula (19) from the proof of Theorem 4.12 in Section 4, we determine the difference quotient for the points $z \in G \setminus \{z_0\}$ as follows:

$$\frac{\Psi(z) - \Psi(z_0)}{z - z_0} = -\frac{1}{\pi} \iint\limits_{G} \frac{g(\zeta) - g(z_0)}{(\zeta - z_0)(\zeta - z)} \, d\xi d\eta.$$

This implies

$$\lim_{\substack{z \to z_0 \\ z \neq z_0}} \frac{\Psi(z) - \Psi(z_0)}{z - z_0} = -\frac{1}{\pi} \iint\limits_{G} \frac{g(\zeta) - g(z_0)}{(\zeta - z_0)^2} \, d\xi d\eta = \Pi_G[g - g(z_0)](z_0). \tag{23}$$

Choosing the arguments $z = z_0 + \delta$ and $z = z_0 + i\delta$, respectively, with $\delta \to 0$ and $\delta \neq 0$, the limit (23) reveals the relation

$$\Psi_x(z_0) = \Pi_G[g - g(z_0)](z_0) = -i\Psi_y(z_0), \tag{24}$$

and consequently

$$\Psi_{\overline{z}}(z_0) = 0, \quad \Psi_z(z_0) = \Pi_G[g - g(z_0)](z_0). \tag{25}$$

2. If the point $z_0 \in G$ is fixed, we consider the function

$$g(z) = \Big\{ g(z) - g(z_0) \Big\} + g(z_0) =: \tilde{g}(z) + g(z_0).$$

With the aid of Example 5.8, we obtain the identity

$$\begin{aligned}\Psi(z) = T_G[g](z) &= T_G[\tilde{g} + g(z_0)](z) \\ &= T_G[\tilde{g}](z) + g(z_0)T_G[1](z) \\ &= T_G[\tilde{g}](z) + g(z_0)\Lambda(z)\end{aligned}$$

for all points $z \in G$. Due to part 1 of our proof, we can differentiate the first summand at the point z_0 with respect to z and $\overline{z}$ using the result (25); the second summand is infinitely often differentiable in G according to the results (16) and (18). Therefore, we obtain

$$\Psi_{\overline{z}}(z_0) = 0 + g(z_0) \cdot 1$$

and

$$\Psi_z(z_0) = \Pi_G[\tilde{g}](z_0) + g(z_0)\Pi_G[1](z_0) = \Pi_G[g](z_0).$$

This implies the identities

$$\Psi_z(z_0) = \Pi_G[g](z_0), \quad \Psi_{\overline{z}}(z_0) = g(z_0) \qquad \text{for all} \quad z_0 \in G. \tag{26}$$

Since the right-hand sides are continuous in G, the function $\Psi = \Psi(z)$ belongs to the class $C^1(G, \mathbb{C})$.

q.e.d.

Proposition 5.12. *Let the function $g : \overline{G} \to \mathbb{C}$ of the class $C^1(\overline{G}, \mathbb{C})$ be given on the normal domain $G \subset \mathbb{C}$. Then we have the identity*

$$\Pi_G[g](z) = T_G\left[\frac{\partial}{\partial \zeta} g\right](z) - \frac{1}{2\pi i} \int\limits_{\partial G} \frac{g(\zeta)}{\zeta - z}\, d\overline{\zeta} \tag{27}$$

for all $z \in G$.

Proof: We observe

$$\Pi_G[g](z) = \lim_{\varepsilon \to 0+} \left\{ -\frac{1}{\pi} \iint\limits_{G_\varepsilon(z)} \frac{g(\zeta)}{(\zeta - z)^2}\, d\xi d\eta \right\}, \qquad z \in G.$$

Let us calculate with the aid of the Gaussian theorem in the complex form as follows:

$$\begin{aligned}\lim_{\varepsilon \to 0+} \iint\limits_{G_\varepsilon(z)} \frac{\overline{g}(\zeta)}{(\overline{\zeta} - \overline{z})^2}\, d\xi d\eta &= -\lim_{\varepsilon \to 0+} \iint\limits_{G_\varepsilon(z)} \frac{\partial}{\partial \overline{\zeta}} \left(\frac{\overline{g}(\zeta)}{\overline{\zeta} - \overline{z}} \right) d\xi d\eta \\ &\quad + \lim_{\varepsilon \to 0+} \iint\limits_{G_\varepsilon(z)} \frac{1}{\overline{\zeta} - \overline{z}} \frac{\partial}{\partial \overline{\zeta}} \overline{g}(\zeta)\, d\xi d\eta \\ &= -\lim_{\varepsilon \to 0+} \frac{1}{2i} \int\limits_{\partial G_\varepsilon(z)} \frac{\overline{g}(\zeta)}{\overline{\zeta} - \overline{z}}\, d\zeta + \iint\limits_{G_0(z)} \frac{\frac{\partial}{\partial \overline{\zeta}} \overline{g}(\zeta)}{\overline{\zeta} - \overline{z}}\, d\xi d\eta.\end{aligned}$$

We set $\zeta = z + \varepsilon e^{i\varphi}$ again and obtain

$$\lim_{\varepsilon\to 0+} \oint_{|\zeta - z|=\varepsilon} \frac{\overline{g}(\zeta)}{\overline{\zeta} - \overline{z}}\, d\zeta = \lim_{\varepsilon\to 0+} \int_0^{2\pi} \frac{\overline{g}(z+\varepsilon e^{i\varphi})}{\varepsilon e^{-i\varphi}}\, i\varepsilon e^{i\varphi}\, d\varphi$$

$$= i \lim_{\varepsilon\to 0+} \left\{ \int_0^{2\pi} \overline{g}(z+\varepsilon e^{i\varphi}) e^{2i\varphi}\, d\varphi \right\}$$

and consequently

$$\lim_{\varepsilon\to 0+} \oint_{|\zeta - z|=\varepsilon} \frac{\overline{g}(\zeta)}{\overline{\zeta} - \overline{z}}\, d\zeta = 0.$$

This implies

$$\lim_{\varepsilon\to 0+} \iint_{G_\varepsilon(z)} \frac{\overline{g}(\zeta)}{(\overline{\zeta} - \overline{z})^2}\, d\xi d\eta = -\frac{1}{2i} \int_{\partial G} \frac{\overline{g}(\zeta)}{\overline{\zeta} - \overline{z}}\, d\zeta + \iint_{G_0(z)} \frac{\frac{\partial}{\partial \overline{\zeta}} \overline{g}(\zeta)}{\overline{\zeta} - \overline{z}}\, d\xi d\eta$$

for all $z \in G$. Therefore, we deduce the following identity for all points $z \in G$:

$$\varPi_G[g](z) = \lim_{\varepsilon\to 0+} \left\{ -\frac{1}{\pi} \iint_{G_\varepsilon(z)} \frac{g(\zeta)}{(\zeta - z)^2}\, d\xi d\eta \right\}$$

$$= \overline{-\frac{1}{\pi} \left\{ \lim_{\varepsilon\to 0+} \iint_{G_\varepsilon(z)} \frac{\overline{g}(\zeta)}{(\overline{\zeta} - \overline{z})^2}\, d\xi d\eta \right\}}$$

$$= -\frac{1}{\pi} \left(\frac{1}{2i} \int_{\partial G} \frac{g(\zeta)}{\zeta - z}\, d\overline{\zeta} + \iint_{G_0(z)} \frac{g_\zeta(\zeta)}{\zeta - z}\, d\xi d\eta \right)$$

$$= T_G\left[\frac{\partial}{\partial \zeta} g\right](z) - \frac{1}{2\pi i} \int_{\partial G} \frac{g(\zeta)}{\zeta - z}\, d\overline{\zeta}.$$

This corresponds to the statement above. q.e.d.

We summarize our considerations to the important

Theorem 5.13. (Regularity for the inhomogeneous Cauchy-Riemann equation)
Let the set $\varOmega \subset \mathbb{C}$ be open, where the function $g \in C^k(\varOmega, \mathbb{C})$ with $k \in \mathbb{N} \cup \{0\}$ is defined. Furthermore, the element $f = f(z)$ belongs to Vekua's class of

functions $C_{\overline{z}}(\Omega)$ and satisfies the inhomogeneous Cauchy-Riemann differential equation

$$\frac{d}{d\overline{z}} f(z) = g(z), \qquad z \in \Omega \tag{28}$$

in the sense of Pompeiu. Then the function f is contained in the regularity class $C^k(\Omega, \mathbb{C})$, and its derivatives of the order k are Dini continuous - with the modulus of continuity described in Section 4, Theorem 4.12. If additionally all k-th derivatives of the right-hand side $g = g(z)$ are Hölder continuous functions in Ω, the regularity $f \in C^{k+1}(\Omega, \mathbb{C})$ follows.

Proof:

1. According to Theorem 5.5, the differential equation (28) is equivalent to the integral equation

$$f(z) = \frac{1}{2\pi i} \int\limits_{\partial G} \frac{f(\zeta)}{\zeta - z}\, d\zeta + T_G[g](z), \qquad z \in G,$$

in arbitrary normal domains $G \subset\subset \Omega$. The first summand on the right-hand side represents a holomorphic function in G and the regularity of $f = f(z)$ is consequently determined by the Hadamard integral operator

$$\Psi(z) := T_G[g](z), \qquad z \in G.$$

In the basic situation $k = 0$, Theorem 4.12 from Section 4 reveals that the function $\Psi = \Psi(z)$, and $f = f(z)$ as well, is Dini continuous in G - with the modulus of continuity described there. If the right-hand side $g = g(z)$ is additionally Hölder continuous in Ω, Propositions 5.9 and 5.11 yield

$$\Psi \in C^1(G); \quad \Psi_{\overline{z}}(z) = g(z), \quad \Psi_z(z) = \Pi_G[g](z), \qquad z \in G. \tag{29}$$

2. In the case $k = 1$ we have $g \in C^1(\Omega)$, and the relation (29) implies $\Psi_{\overline{z}} \in C^1(\Omega)$. Furthermore, Proposition 5.12 yields

$$\Psi_z(z) = \Pi_G[g](z) = T_G\left[\frac{\partial}{\partial \zeta} g\right](z) - \frac{1}{2\pi i} \int\limits_{\partial G} \frac{g(\zeta)}{\zeta - z}\, d\overline{\zeta}, \qquad z \in G \subset\subset \Omega. \tag{30}$$

Here the second summand on the right-hand side is again holomorphic in G, and the function

$$\Phi(z) := T_G\left[\frac{\partial}{\partial \zeta} g\right](z), \qquad z \in G$$

is Dini continuous. If g_z and $g_{\overline{z}}$ or equivalently g_x and g_y are additionally Hölder continuous in Ω, the relation (30) combined with Proposition 5.11 imply the regularity $\Psi_z \in C^1(\Omega)$ and the identity

$$\begin{aligned}\Psi_{zz} &= \frac{\partial}{\partial z}\left\{T_G\left[\frac{\partial}{\partial \zeta}g\right](z) - \frac{1}{2\pi i}\int\limits_{\partial G}\frac{g(\zeta)}{\zeta - z}\,d\overline{\zeta}\right\} \\ &= \Pi_G\left[\frac{\partial}{\partial \zeta}g\right](z) - \frac{1}{2\pi i}\int\limits_{\partial G}\frac{g(\zeta)}{(\zeta - z)^2}\,d\overline{\zeta}\end{aligned} \tag{31}$$

for all points $z \in G$. Furthermore, we have

$$\Psi_{z\overline{z}}(z) = g_z(z) = \Psi_{\overline{z}z}(z) \qquad \text{in} \quad G, \tag{32}$$

and

$$\Psi_{\overline{z}\overline{z}}(z) = g_{\overline{z}}(z) \qquad \text{in} \quad G. \tag{33}$$

Therefore, the regularity $\Psi \in C^2(\Omega)$ is proved, and the derivatives are determined by the formulas above.

3. In the cases $k = 2, 3, \ldots$ one continues the procedure in the way described. Here one essentially utilizes the formula

$$\Pi_G\left[\frac{\partial^{k-1}}{\partial \zeta^{k-1}}g\right](z) = T_G\left[\frac{\partial^k}{\partial \zeta^k}g\right](z) - \frac{1}{2\pi i}\int\limits_{\partial G}\frac{\frac{d^{k-1}}{d\zeta^{k-1}}g(\zeta)}{\zeta - z}\,d\overline{\zeta}$$

for all $z \in G$. q.e.d.

6 Pseudoholomorphic Functions

Let $\Omega \subset \mathbb{C}$ denote an open set, and let us define the linear *space of complex potentials*

$$\begin{aligned}\mathcal{B}(\Omega) := \Big\{a : \Omega \to \mathbb{C} \;:\; &\text{There exists a bounded open set } \Theta \subset \Omega \text{ such that} \\ &a \in C^0(\Theta, \mathbb{C}) \cap L^\infty(\Theta, \mathbb{C}) \text{ and } a(z) = 0 \text{ for all } z \in \Omega \setminus \Theta \\ &\text{holds true}\Big\}.\end{aligned}$$

Definition 6.1. *The function $f = f(z) = u(x,y) + iv(x,y)$, $(x,y) \in \Omega$ of the class $C^0(\Omega, \mathbb{C}) \cap C_{\overline{z}}(\Omega)$ is called* pseudoholomorphic *in Ω, if we have a complex potential $a \in \mathcal{B}(\Omega)$ such that the differential equation*

$$\frac{d}{d\overline{z}}f(z) = a(z)f(z), \quad z \in \Omega \tag{1}$$

is satisfied in the Pompeiu sense.

Example 6.2. Let the function $f \in C_{\overline{z}}(\Omega)$ satisfy the differential inequality

$$|f_{\overline{z}}(z)| \le M|f(z)|, \qquad z \in \Omega \tag{2}$$

in the bounded open set $\Omega \subset \mathbb{C}$, with the constant $M \in [0, +\infty)$. Now we define the open set

$$\Theta := \Big\{z \in \Omega : f(z) \neq 0\Big\}$$

and choose the potential

$$a(z) := \begin{cases} \dfrac{f_{\overline{z}}(z)}{f(z)}, & z \in \Theta \\ 0, & z \in \Omega \setminus \Theta \end{cases} \tag{3}$$

satisfying $\|a\|_\infty \leq M$. Consequently, the inequality (1) is fulfilled and the function $f = f(z)$ is pseudoholomorphic in Ω.

If the potential

$$a(z) := \frac{1}{2}\Big\{\alpha(x,y) + i\beta(x,y)\Big\}$$

is Hölder continuous in Ω, the solution of (1) belongs to the class $C^1(\Omega)$ according to the regularity theorem from Section 5. Then we can transform (1) into a real system of differential equations. At first, we note

$$2f_{\overline{z}} = f_x + if_y = (u_x - v_y) + i(v_x + u_y)$$

and simultaneously

$$2af = (\alpha + i\beta)(u + iv) = (\alpha u - \beta v) + i(\alpha v + \beta u).$$

Therefore, the equation (1) is equivalent to the system

$$\begin{aligned} u_x - v_y &= \alpha u - \beta v \\ v_x + u_y &= \alpha v + \beta u \end{aligned} \qquad \text{in} \quad \Omega \tag{4}$$

and finally to

$$\begin{pmatrix} \dfrac{\partial}{\partial x} & -\dfrac{\partial}{\partial y} \\ \dfrac{\partial}{\partial y} & \dfrac{\partial}{\partial x} \end{pmatrix} \begin{pmatrix} u(x,y) \\ v(x,y) \end{pmatrix} = \begin{pmatrix} \alpha(x,y) & -\beta(x,y) \\ \beta(x,y) & \alpha(x,y) \end{pmatrix} \begin{pmatrix} u(x,y) \\ v(x,y) \end{pmatrix} \qquad \text{in} \quad \Omega. \tag{5}$$

Theorem 6.3. (Similarity principle of Bers and Vekua)
On the open set $\Omega \subset \mathbb{C}$ we consider the pseudoholomorphic function $f = f(z)$, with the associate potential $a \in \mathcal{B}(\Omega)$ and the associate open set $\Theta \subset \Omega$ being given. Furthermore, we introduce the parameter integral

$$\Psi(z) := -\frac{1}{\pi} \iint\limits_{\Theta} \frac{a(\zeta)}{\zeta - z}\, d\xi d\eta, \qquad z \in \Omega, \tag{6}$$

representing a Dini continuous function - according to Theorem 4.12 from Section 4. Then the following function

$$\Phi(z) := f(z)\, e^{-\Psi(z)}\,, \qquad z \in \Omega,$$

is holomorphic in Ω, and we have Vekua's representation formula

$$f(z) = e^{\Psi(z)}\, \Phi(z), \qquad z \in \Omega. \tag{7}$$

Proof: Let the symbols $\chi_n \in C_0^\infty(\Theta, [0,1])$, $n = 1, 2, \dots$ denote a sequence of functions such that

$$\lim_{n\to\infty} \chi_n(z) = \chi(z) := \begin{cases} 1, \ z \in \Theta \\ 0, \ z \in \mathbb{C} \setminus \Theta \end{cases}.$$

Then we consider the functions

$$\Psi_n(z) := T_{\mathbb{C}}[a\chi_n](z) = -\frac{1}{\pi} \iint\limits_{\mathbb{C}} \frac{a(\zeta)\chi_n(\zeta)}{\zeta - z}\, d\xi d\eta, \qquad z \in \mathbb{C}, \tag{8}$$

for $n = 1, 2, \dots$ of the class $C_{\overline{z}}(\mathbb{C})$ satisfying

$$\frac{d}{d\overline{z}} \Psi_n(z) = a(z)\chi_n(z), \qquad z \in \mathbb{C} \quad \text{with} \quad n \in \mathbb{N}. \tag{9}$$

We now investigate the sequence

$$\Phi_n(z) := f(z)\, e^{-\Psi_n(z)}\,, \qquad z \in \Omega \quad \text{for} \quad n = 1, 2, 3, \dots \tag{10}$$

of the class $C_{\overline{z}}(\Omega)$. With the aid of (1), we calculate

$$\begin{aligned} \frac{d}{d\overline{z}} \Phi_n(z) &= e^{-\Psi_n(z)} \left\{ \frac{d}{d\overline{z}} f(z) - f(z) \frac{d}{d\overline{z}} \Psi_n(z) \right\} \\ &= e^{-\Psi_n(z)} \left\{ a(z)f(z) - f(z)a(z)\chi_n(z) \right\} \\ &= e^{-\Psi_n(z)}\, a(z)f(z) \left\{ 1 - \chi_n(z) \right\} \end{aligned} \tag{11}$$

for all points $z \in \Omega$ and the indices $n = 1, 2, \dots$. We apply Theorem 5.5 from Section 5 and obtain the following identity for each normal domain $G \subset\subset \Omega$, namely

$$\Phi_n(z) = \frac{1}{2\pi i} \int\limits_{\partial G} \frac{\Phi_n(\zeta)}{\zeta - z}\, d\zeta - \frac{1}{\pi} \iint\limits_{G} \frac{e^{-\Psi_n(\zeta)} a(\zeta) f(\zeta)\{1 - \chi_n(\zeta)\}}{\zeta - z}\, d\xi d\eta \tag{12}$$

for all $z \in G$ and $n = 1, 2, \dots$. Via Lebesgue's convergence theorem we easily verify

$$
\begin{aligned}
\lim_{n\to\infty} \Phi_n(z) &= \lim_{n\to\infty} \left\{ f(z) \exp\left(\frac{1}{\pi} \iint\limits_{\mathbb{C}} \frac{a(\zeta)\chi_n(\zeta)}{\zeta - z}\, d\xi d\eta \right) \right\} \\
&= f(z) \exp\left\{ \frac{1}{\pi} \iint\limits_{\Theta} \frac{a(\zeta)}{\zeta - z}\, d\xi d\eta \right\} \\
&= f(z) \exp\Big\{ -\Psi(z) \Big\} \\
&= \Phi(z)
\end{aligned}
\tag{13}
$$

for the point $z \in \Omega$ being fixed. The transition to the limit in (12) yields the identity

$$
\Phi(z) = \frac{1}{2\pi i} \int\limits_{\partial G} \frac{\Phi(\zeta)}{\zeta - z}\, d\zeta, \qquad z \in G, \tag{14}
$$

for each normal domain $G \subset\subset \Omega$. Consequently, the function $\Phi = \Phi(z)$ is holomophic in Ω.

q.e.d.

On the basis of Vekua's representation formula, we can immediately transfer various properties of holomorphic functions to the class of pseudoholomorphic functions.

Theorem 6.4. (Carleman)
We have given the pseudoholomorphic function $f : \Omega \to \mathbb{C}$ on the open set $\Omega \subset \mathbb{C}$. Furthermore, let us consider the limit point $z_0 \in \Omega$ and the sequence of points $\{z_k\}_{k=1,2,\ldots} \subset \Omega \setminus \{z_0\}$ with the following properties

$$
\lim_{k\to\infty} z_k = z_0 \quad \textit{and} \quad f(z_k) = 0 \quad \textit{for all} \quad k \in \mathbb{N}.
$$

Then we infer the identity

$$
f(z) \equiv 0 \qquad \textit{in} \quad \Omega.
$$

Proof: Combine the identity theorem for holomorphic functions with Theorem 6.3 from above.

q.e.d.

In the same way we transfer the principle of the argument to pseudoholomorphic functions.

Theorem 6.5. (Uniqueness theorem of Vekua)
Let the function $f : \mathbb{C} \to \mathbb{C}$ be pseudoholomorphic with the asymptotic property

$$
\lim_{\varepsilon\to 0+} \sup_{|z|\geq \frac{1}{\varepsilon}} |f(z)| = 0. \tag{15}
$$

Then we have the identity

$$f(z) \equiv 0 \qquad in \quad \mathbb{C}.$$

Proof: We denote the complex potential belonging to the function $f = f(z)$ by $a \in \mathcal{B}(\mathbb{C})$, and we mean by $\Theta \subset \mathbb{C}$ the associate bounded open set. According to Theorem 6.3 we have the representation

$$f(z) = e^{\Psi(z)}\, \Phi(z), \qquad z \in \mathbb{C},$$

with a holomorphic function $\Phi = \Phi(z)$, $z \in \mathbb{C}$. Furthermore, the function

$$\Psi(z) := -\frac{1}{\pi} \iint\limits_{\Theta} \frac{a(\zeta)}{\zeta - z}\, d\xi d\eta, \qquad z \in \mathbb{C}$$

is bounded: We have a fixed constant $C \in (0, +\infty)$ such that the estimate

$$|\Psi(z)| \leq \frac{1}{\pi}\, \|a\|_\infty \iint\limits_{\Theta} \frac{1}{|\zeta - z|}\, d\xi d\eta \leq \frac{1}{\pi}\, \|a\|_\infty\, C, \qquad z \in \mathbb{C}$$

is valid. Consequently, the holomorphic function

$$\Phi(z) = f(z)\, e^{-\Psi(z)}, \qquad z \in \Omega$$

is bounded and finally constant - due to Liouville's theorem. We take (15) into account, and we infer

$$\lim_{\varepsilon \to 0+} \sup_{|z| \geq \frac{1}{\varepsilon}} |\Phi(z)| = 0$$

and consequently

$$f(z)\, e^{-\Psi(z)} = \Phi(z) \equiv 0 \qquad \text{in} \quad \mathbb{C}.$$

Finally, we obtain

$$f(z) \equiv 0 \qquad \text{in} \quad \mathbb{C},$$

and the statement above is established. q.e.d.

Remark: An entire pseudoholomorphic function, vanishing at the infinitely distant point, is identically zero.

7 Conformal Mappings

We begin with the central

Definition 7.1. *Let* $\Omega_j \subset \mathbb{C}$ *for* $j = 1, 2$ *denote two domains, and we call the mapping* $w = f(z) : \Omega_1 \to \Omega_2$ conformal *if the following properties are fulfilled:*

(a) The function $f : \Omega_1 \to \Omega_2$ *is bijective;*
(b) The function $f : \Omega_1 \to \Omega_2$ *is holomorphic;*
(c) The Jacobian satisfies $J_f(z) = |f'(z)|^2 > 0$ *for all points* $z \subset \Omega_1$.

Remark: On account of Theorem 3.1 in Section 3, we can deduce the condition (c) from the properties (a) and (b).

Remark: A conformal mapping preserves the oriented angles between two intersecting arcs.

Definition 7.2. *Two domains* Ω_1 *and* Ω_2 *in the complex plane* $\mathbb{C}$ *are called* conformally equivalent, *if there exists a conformal mapping* $f : \Omega_1 \to \Omega_2$ *between them.*

Definition 7.3. *For a domain* $\Omega \subset \mathbb{C}$ *we name the set*

$$Aut(\Omega) := \Big\{ f : \Omega \to \Omega \, : \, f \text{ is conformal} \Big\}$$

the automorphism group of the domain Ω.

Remark: An easy exercise reveals that Aut (Ω) represents a group with respect to the composition

$$f_1, f_2 \in \operatorname{Aut}(\Omega), \quad \text{then} \quad f := f_2 \circ f_1 \in \operatorname{Aut}(\Omega)$$

with the unit element $f = id_\Omega$.

Definition 7.4. *Let the complex parameters* $a, b, c, d \in \mathbb{C}$ *be given such that*

$$det \begin{pmatrix} a & b \\ c & d \end{pmatrix} = ad - bc \neq 0$$

holds true, and we define

$$\mathbb{C}^* := \Big\{ z \in \mathbb{C} \, : \, cz + d \neq 0 \Big\}.$$

Then we name the mapping

$$w = f(z) := \frac{az + b}{cz + d}, \qquad z \in \mathbb{C}^*$$

a Möbius transformation *or alternatively a* fractional linear transformation.

With the coefficient matrix

$$\begin{pmatrix} 1 & b \\ 0 & 1 \end{pmatrix}$$

we obtain the *translation*

$$f(z) = z + b, \qquad z \in \mathbb{C}$$

by the vector $b \in \mathbb{C}$. The coefficient matrix

$$\begin{pmatrix} a & 0 \\ 0 & 1 \end{pmatrix}$$

yields a *rotational dilation*

$$f(z) = az, \qquad z \in \mathbb{C}$$

- using a complex parameter $a \in \mathbb{C} \setminus \{0\}$ - about the angle $\varphi = \arg a$ with the modulus $|a|$. Both mappings are conformal on the complex plane $\mathbb{C}$, they can be continued onto the closure $\overline{\mathbb{C}} = \mathbb{C} \cup \{\infty\}$, and we have the fixed point $f(\infty) = \infty$.

With the coefficient matrix

$$\begin{pmatrix} 0 & 1 \\ 1 & 0 \end{pmatrix}$$

we obtain the *reflection at the unit circle*

$$f(z) = \frac{1}{z}, \qquad z \in \mathbb{C} \setminus \{0\}$$

which is conformal on the set $\mathbb{C} \setminus \{0\}$; this mapping can be continued onto the extended complex plane $\overline{\mathbb{C}} = \mathbb{C} \cup \{\infty\}$ setting $f(0) = \infty$.

We speak of an *elementary mapping* when we jointly refer to a translation, a rotational dilation, or a reflection at the unit circle.

Theorem 7.5. *Each Möbius transformation*

$$f(z) = \frac{az+b}{cz+d}, \qquad z \in \mathbb{C}^*$$

possesses finitely many elementary mappings $f_1(z), \ldots, f_n(z)$ - with $n \in \mathbb{N}$ - such that the representation

$$f(z) = f_n \circ \ldots \circ f_2 \circ f_1(z), \qquad z \in \mathbb{C}^*$$

is valid. The domain $\mathbb{C}^*$ is conformally mapped onto the image $f(\mathbb{C}^*)$ by the function $f = f(z)$. Each circle in $\mathbb{C}$ is transformed into a circle or a straight line by the mapping $f = f(z)$; this function additionally transfers each straight line in $\mathbb{C}$ into a straight line or a circle in $\mathbb{C}$.

Remark: When we comprehend a straight line as a circle extending to the infinitely distant point, then the Möbius transformations are *circle-preserving*.

Proof of Theorem 7.5:

1. Given the *linear transformation* $f(z) = az + b$, $z \in \mathbb{C}$ with the complex parameters $a \in \mathbb{C} \setminus \{0\}$ and $b \in \mathbb{C}$, we choose the elementary mappings

$$f_1(z) := az, \quad f_2(z) := z + b$$

and obtain

$$f_2 \circ f_1(z) = az + b = f(z), \qquad z \in \mathbb{C}.$$

2. With an arbitrary *fractional linear transformation*

$$f(z) = \frac{az+b}{cz+d}, \qquad z \in \mathbb{C}^*$$

- for the parameter $c \neq 0$ - we choose the following mappings

$$f_1(z) := cz + d, \quad f_2(z) := \frac{1}{z}, \quad f_3(z) := \frac{bc - ad}{c} z + \frac{a}{c}$$

and obtain

$$f_3 \circ f_2 \circ f_1(z) = f_3\left(\frac{1}{cz+d}\right) = \frac{bc-ad}{c} \frac{1}{cz+d} + \frac{a}{c}$$

$$= \frac{bc - ad + acz + ad}{c(cz+d)} = \frac{az+b}{cz+d} = f(z)$$

for all points $z \in \mathbb{C}^*$. Observing part 1. of our proof, the mappings f_1, f_2 and f_3 can be represented as a composition of elementary mappings, and this remains true for the mapping $f = f(z)$ as well.

3. Since the elementary mappings transform the extended complex plane $\overline{\mathbb{C}}$ topologically onto itself, the mapping $f : \overline{\mathbb{C}} \to \overline{\mathbb{C}}$ is topological as well. Furthermore, the function $f : \mathbb{C}^* \to f(\mathbb{C}^*)$ is analytic, and the mapping

$$f(z) = \frac{az+b}{cz+d}, \qquad z \in \mathbb{C}^*$$

satisfies the identity

$$f'(z) = \frac{acz + ad - caz - cb}{(cz+d)^2} = \frac{ad - bc}{(cz+d)^2} \neq 0 \quad \text{for all points} \quad z \in \mathbb{C}^*.$$

Consequently, the mapping $f : \mathbb{C}^* \to f(\mathbb{C}^*)$ is conformal.

4. Evidently, linear transformations transfer circles into circles - and straight lines into straight lines. In order to show the *circle-preserving property* of the Möbius transformations, we establish this feature only for the reflection at the unit circle:

Circles and straight lines in the $z = x + iy$ - plane are described by the following equation:

$$0 = \alpha(x^2 + y^2) + \beta x + \gamma y + \delta \tag{1}$$

with suitable real numbers $\alpha, \beta, \gamma, \delta \in \mathbb{R}$. We now define the number

$$a := \frac{1}{2}(\beta - i\gamma) \in \mathbb{C}$$

and transform (1) into the complex equation

$$0 = \alpha z\overline{z} + 2\,\mathrm{Re}\,(az) + \delta = \alpha z\overline{z} + az + \overline{az} + \delta. \tag{2}$$

On the set $\mathbb{C} \setminus \{0\}$ we multiply (2) by $\frac{1}{z}\frac{1}{\overline{z}}$ and obtain

$$0 = \alpha + a\frac{1}{\overline{z}} + \overline{a}\,\frac{1}{z} + \delta\,\frac{1}{z}\frac{1}{\overline{z}}, \qquad z \in \mathbb{C} \setminus \{0\}.$$

Setting $w = \frac{1}{z}$ and $\overline{w} = \frac{1}{\overline{z}}$, we arrive at the *joint equation for circles and straight lines*:

$$0 = \alpha + a\overline{w} + \overline{a}w + \delta w\overline{w} = \delta w\overline{w} + 2\,\mathrm{Re}\,(\overline{a}w) + \alpha. \tag{3}$$

Therefore, the function $z \to \frac{1}{z}$ maps circles/lines into circles/lines. q.e.d.

Remarks:

1. Given the two Möbius transformations

$$f(z) = \frac{az+b}{cz+d} \quad \text{and} \quad \varphi(z) = \frac{\alpha z+\beta}{\gamma z+\delta},$$

their composition

$$F(z) = f \circ \varphi(z)$$

represents the following Möbius transformation

$$F(z) = \frac{Az+B}{Cz+D}$$

with the coefficient matrix

$$\begin{pmatrix} A & B \\ C & D \end{pmatrix} = \begin{pmatrix} a & b \\ c & d \end{pmatrix} \circ \begin{pmatrix} \alpha & \beta \\ \gamma & \delta \end{pmatrix}. \tag{4}$$

2. The Möbius transformation

$$f(z) = \frac{az+b}{cz+d}$$

possesses the following inverse Möbius transformation

$$g(z) = \frac{-dz+b}{cz-a}.$$

We leave the proof of these statements as an exercise to the reader.

Example 7.6. Let us denote the unit disc by

$$B := \Big\{ z = x + iy \in \mathbb{C} \, : \, |z| < 1 \Big\}$$

and the upper half-plane by the symbol

$$H^+ := \Big\{ w = u + iv \in \mathbb{C} \, : \, v > 0 \Big\}.$$

We then consider the Möbius transformation

$$f(z) = \frac{z+i}{iz+1}, \qquad z \in \mathbb{C} \setminus \{i\}, \tag{5}$$

and evaluate

$$f(0) = i, \quad f(i) = \lim_{z \to i} f(z) = \infty, \quad f(1) = \frac{1+i}{i+1} = 1$$

as well as

$$f(-i) = 0, \quad f(-1) = \frac{-1+i}{-i+1} = -1.$$

Therefore, the domains H^+ and B are conformally equivalent via the mapping $f : B \to H^+$.

Example 7.7. Let the point $z_0 \in B$ be chosen as fixed. We now consider the Möbius transformation

$$w = f(z) = \frac{z - z_0}{\overline{z}_0 z - 1}, \qquad z \in \overline{B}, \tag{6}$$

with the coefficient matrix

$$\begin{pmatrix} 1 & -z_0 \\ \overline{z}_0 & -1 \end{pmatrix}.$$

Here, we observe

$$\det \begin{pmatrix} 1 & -z_0 \\ \overline{z}_0 & -1 \end{pmatrix} = -1 + |z_0|^2 < 0$$

as well as $f(z_0) = 0$. We calculate as follows:

$$|f(1)| = \left| \frac{1 - z_0}{\overline{z}_0 - 1} \right| = \left| \frac{1 - z_0}{1 - z_0} \right| = 1,$$

$$|f(-1)| = \left| \frac{-1 - z_0}{-1 - \overline{z}_0} \right| = \left| \frac{-1 - z_0}{-1 - z_0} \right| = 1,$$

$$|f(i)| = \left| \frac{i - z_0}{i\overline{z}_0 - 1} \right| = \left| \frac{i - z_0}{-\overline{z}_0 - i} \right| = \left| \frac{i - z_0}{\overline{z}_0 + i} \right| = \left| \frac{i - z_0}{z_0 - i} \right| = 1.$$

Therefore, we have the mapping properties $f : \partial B \to \partial B$ and $f(z_0) = 0$. Consequently, the function $f = f(z)$ represents a conformal mapping from the unit disc B onto itself.

Definition 7.8. *Let us consider a continuous mapping $f : \Omega \to \Omega$ of the domain $\Omega \subset \overline{\mathbb{C}}$ into itself. We call $z_0 \in \Omega$ a* fixed point *of the mapping $f = f(z)$ if the identity $f(z_0) = z_0$ holds true. When the property $0 \in \Omega$ is valid, and 0 provides a fixed point of the mapping, we name this function* origin-preserving.

The automorphism group Aut (B) can be explicitly determined with the aid of the following

Theorem 7.9. (Schwarzian lemma)
Let $w = f(z) : B \to B$ denote a holomorphic, origin-preserving function. Then we have the estimate

$$|f(z)| \le |z| \qquad \text{for all points} \quad z \in B.$$

If there exists a point $z_0 \in B \setminus \{0\}$ satisfying $|f(z_0)| = |z_0|$, the function $f = f(z)$ admits the representation

$$f(z) = e^{i\vartheta} z, \qquad z \in B$$

with a certain angle $\vartheta \in [0, 2\pi)$.

Proof: The function

$$g(z) := \frac{f(z)}{z}, \quad z \in B \setminus \{0\}$$

can be holomorphically continued onto the disc B, and we have the boundary behavior

$$\limsup_{z \to \partial B} |g(z)| \le 1.$$

From Theorem 3.4 in Section 3 we infer

$$\sup_{z \in B} |g(z)| \le \limsup_{z \to \partial B} |g(z)| \le 1$$

and therefore

$$|f(z)| \le |z| \qquad \text{for all points} \quad z \in B.$$

If there exists a point $z_0 \in B \setminus \{0\}$ with $|f(z_0)| = |z_0|$, we observe $|g(z_0)| = 1$. Consequently, the mapping $g = g(z)$ is constant - due to the theorem quoted above. This implies

$$g(z) = e^{i\vartheta}, \qquad z \in B$$

or equivalently

$$f(z) = e^{i\vartheta} z, \qquad z \in B,$$

with an angle $\vartheta \in [0, 2\pi)$. q.e.d.

Theorem 7.10. (Automorphism group of the unit disc)
An automorphism $w = f(z) : B \to B$ of the unit disc is necessarily of the following form:

$$w = f(z) = e^{i\vartheta}\frac{z - z_0}{\overline{z}_0 z - 1}, \qquad z \in B, \tag{7}$$

with $z_0 := f^{-1}(0) \in B$ and $\vartheta \in [0, 2\pi)$. On the other hand, each mapping of the form (7) - with $z_0 \in B$ and $\vartheta \in [0, 2\pi)$ - represents an automorphism of the unit disc B. In particular, the origin-preserving automorphisms of B are of the form

$$f(z) = e^{i\vartheta} z, \qquad z \in B, \tag{8}$$

with an angle $\vartheta \in [0, 2\pi)$.

Proof:

1. From Example 7.7 we see that all Möbius transformations of the form (7) represent automorphisms of the unit disc.
2. When the function $w = f(z)$, $z \in B$ gives us an origin-preserving automorphism of B, Theorem 7.9 yields the estimate

$$|w| = |f(z)| \le |z| \qquad \text{for all} \quad z \in B.$$

Now the inverse mapping $z = g(w)$, $w \in B$ represents an origin-preserving automorphism of B as well, and we deduce

$$|z| = |g(w)| \le |w| \qquad \text{for all} \quad w \in B.$$

We combine our estimates to

$$|z| \le |w| = |f(z)| \le |z|, \qquad z \in B$$

and obtain

$$|f(z)| = |z|, \qquad z \in B.$$

Our Theorem 7.9 provides an angle $\vartheta \in [0, 2\pi)$ such that

$$f(z) = e^{i\vartheta} z, \qquad z \in B$$

holds true.
3. If $w = f(z) : B \to B$ represents an arbitrary automorphism of B, we set $z_0 := f^{-1}(0)$. Then we consider the Möbius transformation

$$w = g(z) := \frac{z - z_0}{\overline{z}_0 z - 1}, \qquad z \in B,$$

and we obtain the following origin-preserving automorphism of B, namely

$$h(w) := f \circ g^{-1}(w), \qquad w \in B.$$

Recalling the second part of our proof, we infer

$$f \circ g^{-1}(w) = e^{i\vartheta} w, \qquad w \in B,$$

with an angle $\vartheta \in [0, 2\pi)$. The mapping $w = g(z)$ then satisfies

$$f(z) = e^{i\vartheta} g(z) = e^{i\vartheta} \frac{z - z_0}{\overline{z}_0 z - 1}, \qquad z \in B,$$

and the statement above is proved. q.e.d.

Remarks:

1. With the aid of Theorem 7.10 we can investigate *Poincaré's half-plane*, which provides a model for the non-Euclidean geometry.
2. As an exercise, we show the subsequent representation:

$$\begin{aligned}\text{Aut}(B) &= \left\{ f(z) = e^{i\vartheta} \frac{z - z_0}{\overline{z}_o z - 1} \ : \ z_0 \in B, \ \vartheta \in [0, 2\pi) \right\} \\ &= \left\{ f = \frac{az + b}{\overline{b}z + \overline{a}} \ : \ a, b \in \mathbb{C}, \ a\overline{a} - b\overline{b} = 1 \right\}.\end{aligned}$$

3. With the aid of Example 7.7, one then deduces the following statement:

$$\text{Aut}\,(H^+) = \left\{ f(z) = \frac{\alpha z + \beta}{\gamma z + \delta} \ : \ \alpha, \beta, \gamma, \delta \in \mathbb{R} \ \text{ mit } \ \alpha\delta - \beta\gamma = 1 \right\}.$$

When two domains are given in the complex plane, which are bounded by a circle or a straight line, we then can map them conformally onto each other via a Möbius transformation. This fact is contained in the following

Theorem 7.11. *Let the points $z_\nu \in \overline{\mathbb{C}}$ and $w_\nu \in \overline{\mathbb{C}}$ for $\nu = 1, 2, 3$ with $z_\nu \neq z_\mu$ and $w_\nu \neq w_\mu$ for $\nu \neq \mu$ be arbitrarily given. Then we have a uniquely determined Möbius transformation*

$$f(z) = \frac{az + b}{cz + d} \quad \textit{satisfying} \quad f(z_\nu) = w_\nu \quad \textit{with} \quad \nu = 1, 2, 3.$$

Remark: In particular, each Möbius transformation with at least three fixed points reduces to the identical mapping.
Proof of Theorem 7.11:

1. We establish the existence of our mapping: If the inclusion $z_1, z_2, z_3 \in \mathbb{C}$ is correct, we consider the transformation

$$f(z) := \frac{z - z_1}{z - z_3} : \frac{z_2 - z_1}{z_2 - z_3}, \qquad z \in \mathbb{C} \setminus \{z_3\}$$

and observe

$$f(z_1) = 0, \quad f(z_2) = 1, \quad f(z_3) = \infty.$$

If one of the points z_1, z_2, z_3 coincides with the infinitely distant point ∞, i.e. $z_3 = \infty$ without loss of generality, we define

$$f(z) = \frac{z - z_1}{z_2 - z_1}, \qquad z \in \mathbb{C}$$

and obtain

$$f(z_1) = 0, \quad f(z_2) = 1, \quad f(z_3) = \infty.$$

Correspondingly, we construct a mapping $g = g(w)$ with the property

$$g(w_1) = 0, \quad g(w_2) = 1, \quad g(w_3) = \infty.$$

With the function $h(z) := g^{-1} \circ f(z)$, we then get a Möbius transformation satisfying

$$h(z_1) = g^{-1}(f(z_1)) = g^{-1}(0) = w_1, \quad h(z_2) = w_2, \quad h(z_3) = w_3.$$

2. We now show the uniqueness: With $f_j(z)$ for $j = 1, 2$ let us consider two Möbius transformations satisfying

$$f_j(z_\nu) = w_\nu\,, \qquad \nu = 1, 2, 3 \quad \text{for} \quad j = 1, 2.$$

Then the Möbius transformation $f_2^{-1} \circ f_1(z) : \overline{\mathbb{C}} \to \overline{\mathbb{C}}$ possesses three fixed points z_ν with $\nu = 1, 2, 3$. Choosing a Möbius transformation $g = g(z)$ with

$$g(0) = z_1, \quad g(1) = z_2, \quad g(\infty) = z_3,$$

the resulting mapping

$$h(z) := g^{-1} \circ f_2^{-1} \circ f_1 \circ g(z), \qquad z \in \overline{\mathbb{C}}$$

possesses the fixed points 0, 1, and ∞. When we observe

$$h(z) = \frac{az + b}{cz + d}, \qquad z \in \overline{\mathbb{C}},$$

we deduce

$$0 = h(0) = \frac{b}{d}$$

and therefore $b = 0$. Furthermore, we have

$$\infty = \lim_{z \to \infty} h(z) = \lim_{z \to \infty} \frac{az}{cz + d}$$

and consequently $c = 0$. Finally, we note that

$$1 = h(1) = \frac{a}{d} \cdot 1$$

and infer $\frac{a}{d} = 1$. This implies

$$h(z) = z \qquad \text{for all points} \quad z \in \overline{\mathbb{C}}$$

and, moreover,

$$g^{-1} \circ f_2^{-1} \circ f_1 \circ g(z) = z, \qquad z \in \overline{\mathbb{C}}.$$

This is equivalent to the statement

$$f_2^{-1} \circ f_1(z) = z, \qquad z \in \overline{\mathbb{C}},$$

and we arrive at the identity

$$f_1(z) = f_2(z) \qquad \text{for all points} \quad z \in \overline{\mathbb{C}}.$$

q.e.d.

We shall now determine those domains $\Omega \subset \overline{\mathbb{C}}$, which are conformally equivalent to the unit disc B. Since the extended complex plane $\overline{\mathbb{C}}$ is compact in contrast to the unit disc B, these two domains cannot be conformally equivalent. A conformal mapping being topological in particular, the topological properties of conformally equivalent domains have to coincide! Now the function

$$f(z) := \frac{z}{1 + |z|}, \qquad z \in \mathbb{C}$$

represents a topological mapping of $\mathbb{C}$ onto B. Such a conformal mapping - between those domains - cannot exist, since this function has to be constant due to Liouville's theorem! Consequently, the domains $\mathbb{C}$ and B are not conformally equivalent. However, we provide the fundamental

Theorem 7.12. (Riemannian mapping theorem)
Let $\Omega \subset \mathbb{C}$ with $\Omega \neq \mathbb{C}$ denote a simply connected domain. Then there exists a conformal mapping $f : \Omega \to B$.

Remark: The whole class of conformal mappings from Ω onto the unit disc B is given in the form $g \circ f$ with $g \in \mathrm{Aut}\,(B)$.

Before we provide a proof of our theorem, we need the following preparatory lemmas.

Proposition 7.13. (Arzelà, Ascoli)
Let the numbers $m, n \in \mathbb{N}$ and the compact set $K \subset \mathbb{R}^n$ be chosen; and let the set of functions

$$\mathcal{F} := \Big\{ f_\iota : K \to \mathbb{R}^m \; : \; \iota \in J \Big\}$$

be given - with the index set J - satisfying the following properties:

(1) The set $\mathcal{F}$ is uniformly bounded: Consequently, we have a constant $\mu > 0$ such that

$$|f_\iota(x)| \leq \mu \qquad \text{for all points} \quad x \in K \quad \text{and all indices} \quad \iota \in J.$$

(2) The set $\mathcal{F}$ is equicontinuous: Therefore, each quantity $\varepsilon > 0$ admits a number $\delta = \delta(\varepsilon) > 0$ such that all points $x', x'' \in K$ with $|x' - x''| < \delta$ and all indices $\iota \in J$ satisfy the inequality

$$|f_\iota(x') - f_\iota(x'')| < \varepsilon.$$

Statement: *Then the set $\mathcal{F}$ contains a uniformly convergent subsequence in K, namely $g^{(k)} \in \mathcal{F}$ for $k = 1, 2, 3, \ldots$, which converges uniformly towards a continuous function $g \in C^0(K, \mathbb{R}^m)$.*

Usually this result is proved in connection with Peano's existence theorem in the theory of ordinary differential equations.

Proposition 7.14. (Root lemma)
Let $G \subset \mathbb{C} \setminus \{0\}$ denote a simply connected domain such that $z_1 = r_1 e^{i\varphi_1} \in G$ with $r_1 \in (0, +\infty)$ and $\varphi_1 \in [0, 2\pi)$ holds true, and define $w_1 = \sqrt{r_1} e^{\frac{i}{2}\varphi_1}$. Then we have exactly one conformal mapping

$$f(z) = \sqrt{z}, \qquad z \in G$$

onto the simply connected domain $\widetilde{G} := f(G) \subset \mathbb{C} \setminus \{0\}$ with the following properties:

$$f^2(z) = z, \qquad f'(z) = \frac{1}{2f(z)} \quad \text{for all points} \quad z \in G, \tag{9}$$

$$f(z_1) = w_1, \tag{10}$$

and

$$\widetilde{G} \cap (-\widetilde{G}) = \emptyset. \tag{11}$$

Proof: We consider the holomorphic function in G, namely

$$g(z) := \int_{z_1}^{z} \frac{1}{\zeta}\, d\zeta = \log z - \log z_1\,, \qquad z \in G.$$

Here the curvilinear integral has to be evaluated along an arbitrary curve from z_1 to z - within G - and the logarithm function has to be continued along this path. Then the function

$$f(z) := w_1 \exp\left\{\frac{1}{2} g(z)\right\}, \qquad z \in G$$

is holomorphic and satisfies the conditions

$$f(z_1) = w_1 \exp 0 = w_1$$

as well as

$$\begin{aligned} f^2(z) &= w_1^2 \exp g(z) = r_1 e^{i\varphi_1} e^{\log z} e^{-\log z_1} \\ &= \frac{z_1 z}{z_1} = z \qquad \text{for all points} \quad z \in G. \end{aligned}$$

The property (11) follows from the construction. q.e.d.

Proposition 7.15. (Hurwitz)
Given the domain $\Omega \subset \mathbb{C}$, let the holomorphic functions $f_k : \Omega \to \mathbb{C}$ for $k = 1, 2, 3, \ldots$ converge uniformly on each compact set in Ω towards the non-constant holomorphic function $f : \Omega \to \mathbb{C}$. Furthermore, let the functions $f_k = f_k(z)$ be injective for all indices $k \in \mathbb{N}$. Then the limit function $f = f(z)$ is injective as well.

Proof: If the function $f = f(z)$ were not injective, we would have two different points $z_1, z_2 \in \Omega$ with the property

$$f(z_1) = w_1 = f(z_2).$$

We then consider the function

$$g(z) := f(z) - w_1$$

which possesses the two zeroes z_1 and z_2. At these points we determine their topological indices $i(g, z_j) = n_j \in \mathbb{N}$ for $j = 1, 2$. We now consider the functions

$$g_k(z) := f_k(z) - w_1\,, \qquad z \in K_j := \Big\{z \in \mathbb{C} : |z - z_j| \le \varepsilon_j\Big\},$$

with a sufficiently small number $\varepsilon_j > 0$ for $j = 1, 2$ and $k = 1, 2, 3, \ldots$. Then their winding numbers fulfill

$$W(g_k, K_j) = i(g, z_j) = n_j \in \mathbb{N} \quad \text{with} \quad j = 1, 2 \quad \text{for all} \quad k \ge k_0\,; \tag{12}$$

here the index $k_0 \in \mathbb{N}$ has to be chosen sufficiently large. On account of (12), the functions $g_k = g_k(z)$ possess at least two zeroes for all indices $k \ge k_0$ - in contradiction to the injectivity of $f_k = f_k(z)$ assumed above. q.e.d.

We are now prepared to establish the *Proof of Theorem 7.12:*

1. Let $\Omega \subset \mathbb{C}$ with the property $\Omega \ne \mathbb{C}$ denote a simply connected domain. At first, we find a point $z_0 \in \mathbb{C} \setminus \Omega$. Via the conformal mapping

$$f(z) := z - z_0\,, \qquad z \in \Omega$$

we make the transition to the conformally equivalent domain

$$\Omega \subset \mathbb{C} \setminus \{0\}. \tag{13}$$

With the aid of the conformal mapping

$$f(z) = \sqrt{z}, \qquad z \in \Omega$$

from Proposition 7.14, we construct a conformally equivalent domain such that

$$\Omega \cap (-\Omega) = \emptyset. \tag{14}$$

2. We now start from a simply connected domain with the property (13) as well as (14), and choose a point $z_0 \in \Omega$ as fixed. We consider the following *set of admissible functions*

$$\mathcal{F} := \Big\{ f : \Omega \to B \, : \, f \text{ is holomorphic and injective in } \Omega, \;\; f(z_0) = 0 \Big\}.$$

With the aid of the *extremal principle by P. Koebe*, we are looking for those mappings $f \in \mathcal{F}$ which realize the following condition:

$$|f'(z_0)| = \sup_{\Phi \in \mathcal{F}} |\Phi'(z_0)|. \tag{15}$$

At first, we verify that the class $\mathcal{F}$ is nonvoid. On account of (14) we find a point $z_1 \in \mathbb{C}$ and a radius $\varrho > 0$, such that the statement $z \notin \Omega$ is satisfied for all $z \in \mathbb{C}$ with $|z - z_1| \leq \varrho$. The function

$$f_1(z) := \frac{1}{z - z_1}, \qquad z \in \Omega$$

is bounded according to

$$|f_1(z)| \leq \frac{1}{\varrho}, \qquad z \in \Omega.$$

Application of the conformal mapping

$$f_2(w) := r\{w - f_1(z_0)\}, \qquad w \in \mathbb{C}$$

- with a sufficiently small radius $r > 0$ - finally gives us the admissible function

$$f := f_2 \circ f_1 \in \mathcal{F}.$$

3. Let us consider an arbitrary function with $f \in \mathcal{F}$ and observe its Dirichlet integral

$$D(f) := \iint\limits_{\Omega} \Big\{ |f_x|^2 + |f_y|^2 \Big\} \, dxdy = 2 \iint\limits_{\Omega} |f_x \wedge f_y| \, dxdy \leq 2\pi.$$

Let $z_1 \in \Omega$ denote an arbitrary point, and $\delta > 0$ may be chosen so small that the disc

$$B_\delta(z_1) := \Big\{ z \in \mathbb{C} \, : \, |z - z_1| < \delta \Big\}$$

fulfills the inclusion $B_{\sqrt{\delta}}(z_1) \subset\subset \Omega$. Then the oscillation lemma of Courant and Lebesgue provides a number $\delta^* \in [\delta, \sqrt{\delta}]$ with the following property:

$$\int\limits_{z:|z-z_1|=\delta^*} |df(z)| \le \frac{2\sqrt{2\pi}}{\sqrt{-\log\delta}}. \tag{16}$$

When we observe the injectivity of the mapping $f = f(z)$, the diameter of the corresponding domains is estimated as follows:

$$\operatorname{diam} f\Big(B_\delta(z_1)\Big) \le \operatorname{diam} f\Big(B_{\delta^*}(z_1)\Big) \le \frac{\sqrt{2\pi}}{\sqrt{-\log\delta}}. \tag{17}$$

For each compact set $K \subset \Omega$, the class of functions

$$\mathcal{F}_K := \Big\{f : K \to \mathbb{C} \,:\, f \in \mathcal{F}\Big\}$$

is consequently equicontinuous and uniformly bounded. With the aid of Proposition 7.13, we can select a subsequence - converging uniformly on each compact set $K \subset \Omega$ - from all sequences of functions $\{f_k\}_{k=1,2,\ldots} \subset \mathcal{F}$.

4. Invoking Proposition 7.15 we obtain *compactness of the class of functions* $\mathcal{F}$ in the following sense: From each sequence $\{f_k\}_{k=1,2,\ldots} \subset \mathcal{F}$ satisfying

$$0 < |f_k'(z_0)| \le |f_{k+1}'(z_0)| \quad \text{for} \quad k \in \mathbb{N},$$

we can select a subsequence $\{f_{k_l}\}_{l=1,2,\ldots}$ converging uniformly on each compact set $K \subset \Omega$ towards a function $f \in \mathcal{F}$. In this way, we find a function $f \in \mathcal{F}$ with the extremal property (15).

Finally, we have to show the surjectivity

$$f(\Omega) = B. \tag{18}$$

5. If the statement $G \ne B$ for the image $G := f(\Omega) \subset B$ was correct on the contrary, we could find a point $z_1 \in B \setminus G$. The mapping

$$w = \psi_1(z) := \frac{z - z_1}{\overline{z}_1 z - 1}, \qquad z \in B$$

belongs to the class Aut (B) and fulfills

$$\psi_1(z_1) = 0, \quad \psi_1(0) = z_1.$$

On the simply connected domain

$$G_1 := \psi_1(G) \subset B \setminus \{0\}$$

we consider the conformal root-function from Proposition 7.14, namely

$$w = \psi_2(z) := \sqrt{z}\,, \qquad z \in G_1$$

with $z_2 := \sqrt{z_1}$. We then obtain the simply connected domain

$$G_2 := \psi_2(G_1) \subset B \setminus \{0\}$$

with $z_2 \in G_2$. Finally, we utilize the automorphism

$$w = \psi_3(z) = \frac{z - z_2}{\overline{z}_2 z - 1}\,, \qquad z \in B$$

with the property

$$\psi_3(z_2) = 0$$

and define the domain

$$G_3 := \psi_3(G_2) \subset B.$$

The composition

$$\psi := \psi_3 \circ \psi_2 \circ \psi_1 : G \longrightarrow G_3$$

is conformal, and we note that

$$\psi(0) = \psi_3 \circ \psi_2 \circ \psi_1(0) = \psi_3 \circ \psi_2(z_1) = \psi_3(z_2) = 0.$$

Then we observe $\psi \circ f \in \mathcal{F}$ on account of

$$\psi \circ f(z_0) = \psi(0) = 0.$$

We now evaluate as follows:

$$\begin{aligned}
(\psi \circ f)'(z_0) &= \psi'(0) f'(z_0) \\
&= \psi_3'(z_2)\psi_2'(z_1)\psi_1'(0) f'(z_0) \\
&= \frac{1}{\overline{z}_2 z_2 - 1}\,\frac{1}{2\sqrt{z_1}}\,(\overline{z}_1 z_1 - 1) f'(z_0) \\
&= \frac{1}{\overline{z}_2 z_2 - 1}\,\frac{1}{2 z_2}\,\Big\{(\overline{z}_2 z_2)^2 - 1\Big\} f'(z_0) \\
&= \frac{|z_2|^2 + 1}{2 z_2}\, f'(z_0).
\end{aligned}$$

Here we take $z_2 = \sqrt{z_1}$ into account. From $0 < |z_2| < 1$ we infer

$$(1 - |z_2|)^2 > 0 \quad \text{as well as} \quad |z_2|^2 - 2|z_2| + 1 > 0$$

and consequently

$$\frac{|z_2|^2+1}{2|z_2|} > 1.$$

With the subsequent inequality

$$|(\psi \circ f)'(z_0)| = \frac{|z_2|^2+1}{2|z_2|}\,|f'(z_0)| > |f'(z_0)| = \sup_{\Phi\in\mathcal{F}} |\Phi'(z_0)|$$

we arrive at a contradiction. Therefore, the proof is complete. q.e.d.

8 Boundary Behavior of Conformal Mappings

We begin with the fundamental

Definition 8.1. *A bounded domain $\Omega \subset \mathbb{C}$ is called* Jordan domain, *if its boundary $\partial\Omega = \Gamma$ represents a Jordan curve with the topological positive-oriented representation $\gamma : \partial B \to \Gamma$ and the parametrization*

$$\beta(t) := \gamma(e^{it}), \qquad t \in \mathbb{R}.$$

The number $k \in \mathbb{N}$ being given, we name Γ k-times continuously differentiable and regular *at the point $z_1 = \beta(t_1) \in \Gamma$ with $t_1 \in [0, 2\pi)$, if we have a quantity $\varepsilon = \varepsilon(t_1) > 0$ such that*

$$\beta \in C^k((t_1-\varepsilon, t_1+\varepsilon), \mathbb{C})$$

as well as

$$\beta'(t) \neq 0 \qquad \textit{for all} \quad t \in (t_1-\varepsilon, t_1+\varepsilon)$$

holds true. If we have additionally an expansion into the power series

$$\beta(t) = \sum_{k=0}^{\infty} \frac{1}{k!}\,\beta^{(k)}(t_1)(t-t_1)^k \qquad \textit{for all} \quad t_1-\varepsilon < t < t_1+\varepsilon, \tag{1}$$

we call $z_1 = \beta(t_1)$ a regular analytic boundary point. *We speak of a* C^k-Jordan curve *(and an* analytic Jordan curve*) Γ, if each boundary point $z_1 \in \Gamma$ is regular and k-times continuously differentiable (or analytic, respectively).*

Theorem 8.2. (Carathéodory, Courant)
Let $\Omega \subset \mathbb{C}$ denote a Jordan domain. Then the conformal mapping $f : \Omega \to B$ can be continuously extended onto the closure $\overline{\Omega}$ as a topological mapping $f : \overline{\Omega} \to \overline{B}$.

Proof:

1. We take the point $z_1 = \beta(t_1) \in \Gamma$ as fixed, and for all numbers $0 < \delta < \delta_0$ we consider that connected component $G_\delta(z_1)$ of the open set $\{z \in \Omega : |z - z_1| < \delta\}$ satisfying $z_1 \in \partial G_\delta(z_1)$. With the parameters $t_2 < t_3$ we denote by
$$\beta[t_2, t_3] := \Big\{\beta(t) \,:\, t_2 \le t \le t_3\Big\}$$
the oriented Jordan arc on the curve Γ from the point $z_2 = \beta(t_2)$ to the point $z_3 = \beta(t_3)$. The boundary of the set $G_\delta(z_1)$ consists of the circular arc $S_\delta(z_1) \subset \Omega$ and the Jordan arc
$$\Gamma_\delta(z_1) := \beta[t_2, t_3] \qquad \text{with} \quad t_2 < t_1 < t_3.$$
This implies
$$\partial G_\delta(z_1) = \Gamma_\delta(z_1) \,\dot{\cup}\, S_\delta(z_1).$$
The Courant-Lebesgue oscillation lemma from Section 5 in Chapter 1 can be transferred to the present situation. To each quantity $\delta > 0$ prescribed, this gives us a number $\delta^* \in [\delta, \sqrt{\delta}]$ with the following property:
$$\int\limits_{z \in S_{\delta^*}(z_1)} |df(z)| \le \frac{2\sqrt{2}\,\pi}{\sqrt{-\log \delta}}. \tag{2}$$
Now the image $f(S_{\delta^*}(z_1)) \subset B$ represents a Jordan arc of finite length, whose end points - being continuously extended - are situated on ∂B. Since the mapping $f : \Omega \to B$ is injective, we infer
$$\operatorname{diam} f(G_\delta(z_1)) \le \operatorname{diam} f(G_{\delta^*}(z_1)) \le \frac{2\sqrt{2}\,\pi}{\sqrt{-\log \delta}}. \tag{3}$$
Therefore, the function $f = f(z)$ is uniformly continuous on Ω and can be continuously extended onto the closure $\overline{\Omega}$.
2. In the same way we prove the continuous extendability of the inverse function
$$g(w) := f^{-1}(w), \qquad w \in B,$$
onto the closure $\overline{B}$. Here we utilize the modulus of continuity for the Jordan curve Γ in the following sense: For each quantity $\varepsilon > 0$ we have a number $\delta = \delta(\varepsilon) > 0$, such that each pair of consecutive points $z_j = \beta(t_j) \in \Gamma$ for $j = 1, 2$ with $t_1 < t_2$ and $|z_1 - z_2| \le \delta(\varepsilon)$ fulfills the following estimate:
$$\operatorname{diam} \beta[t_1, t_2] := \sup_{t_1 \le \tau_1 < \tau_2 \le t_2} |\beta(\tau_1) - \beta(\tau_2)| \le \varepsilon. \tag{4}$$
3. Since the function $f = f(z)$ is continuously extendable onto $\overline{\Omega}$ and the function $g = g(w)$ onto $\overline{B}$ as well, the mapping $f : \overline{\Omega} \to \overline{B}$ is topological.

q.e.d.

Theorem 8.3. (Analytic boundary behavior)
Let $z = g(w) : B \to \Omega$ denote a conformal mapping onto the Jordan domain $\Omega \subset \mathbb{C}$, which is - via the prescription $g : \overline{B} \to \overline{\Omega}$ - topologically extendable. Let the boundary Γ be regular and analytic at the point $z_1 = g(w_1) \in \Gamma = \partial\Omega$ with $w_1 \in \partial B$. Then we have a convergent power series

$$\sum_{k=0}^{\infty} a_k (w - w_1)^k \qquad \text{for all points} \quad w \in \mathbb{C} \quad \text{satisfying} \quad |w - w_1| < \varepsilon$$

- with the coefficients $a_k \in \mathbb{C}$ for $k \in \mathbb{N}_0$ and $a_1 \neq 0$ - choosing $\varepsilon > 0$ sufficiently small, such that the following representation holds true:

$$g(w) = \sum_{k=0}^{\infty} a_k (w - w_1)^k \qquad \text{for all points} \quad w \in B \quad \text{with} \quad |w - w_1| < \varepsilon. \quad (5)$$

Therefore, the function $g = g(w)$ can be analytically extended across the boundary ∂B at the point $w_1 \in \partial B$.

Proof:

1. Since $z_1 = g(w_1) = \beta(t_1) \in \Gamma$ represents a regular analytic boundary point of the curve Γ, we observe

$$\beta(t) = \sum_{k=0}^{\infty} \frac{1}{k!} \beta^{(k)}(t_1)(t - t_1)^k \quad \text{for all} \quad t_1 - \varepsilon < t < t_1 + \varepsilon, \qquad (6)$$

with $\beta'(t_1) \neq 0$. We now can extend the convergent power series into a complex neighborhood using the variable $r = (t + is) \in \mathbb{C}$, and we obtain the following function:

$$h(r) := \sum_{k=0}^{\infty} \frac{1}{k!} \beta^{(k)}(t_1)(r - t_1)^k \qquad \text{for all} \quad r \in \mathbb{C} \quad \text{with} \quad |r - t_1| < \varepsilon. \qquad (7)$$

On account of the condition $\beta'(t_1) \neq 0$, the holomorphic inverse mapping h^{-1} exists in a neighborhood of the point $z_1 = h(t_1)$.

2. We now use the following Möbius transformation:

$$\ell : H^+ \longrightarrow B \qquad \text{is conformal and satisfies} \quad \ell(0) = w_1.$$

To the holomorphic mapping

$$\Psi(\zeta) := h^{-1} \circ g \circ \ell(\zeta), \qquad \zeta \in H^+ \quad \text{with} \quad |\zeta| < \varepsilon \qquad (8)$$

we can apply the Schwarzian reflection principle and obtain the holomorphic function

$$\Psi(\zeta), \qquad |\zeta| < \varepsilon \qquad (9)$$

on the full disc about the origin. Now the function

$$h \circ \Psi \circ \ell^{-1}(w) = \sum_{k=0}^{\infty} a_k (w - w_1)^k \quad \text{for all} \quad |w - w_1| < \varepsilon \tag{10}$$

is holomorphic as well; and we consider their expansion about the point $w_1 \in \partial B$ into a convergent power series. From (8) and (10), we finally infer

$$g(w) = \sum_{k=0}^{\infty} a_k (w - w_1)^k \qquad \text{for all points} \quad w \in B \quad \text{with} \quad |w - w_1| < \varepsilon. \tag{11}$$

Since the mapping $g : \overline{B} \to \overline{\Omega}$ is topological, the coefficient a_1 in the expansion (11) satisfies the condition $a_1 \neq 0$.

q.e.d.

Remark: If the boundary Γ denotes a polygon, we can represent the conformal mapping $g : B \to \Omega$ with $\Gamma = \partial\Omega$ via the *Schwarz-Christoffel formulas* by a curvilinear integral in a nearly explicit way.

Theorem 8.4. (Boundary point lemma in $\mathbb{C}$)
On the disc

$$B_\varrho(z_1) := \Big\{ z \in \mathbb{C} : |z - z_1| < \varrho \Big\} \quad \text{with} \quad z_1 \in \mathbb{C} \quad \text{and} \quad \varrho > 0$$

let the holomorphic function

$$w = f(z) : B_\varrho(z_1) \longrightarrow B \in C^1(\overline{B_\varrho(z_1)}, \overline{B}) \tag{12}$$

be given, such that the condition

$$|f(z_1)| \leq 1 - \varepsilon \quad \text{with a quantity} \quad \varepsilon > 0$$

is satisfied. Furthermore, let $z_2 \in \partial B_\varrho(z_1)$ denote a boundary point with $|f(z_2)| = 1$. Then we have the inequality

$$|f'(z_2)| \geq \frac{\varepsilon^2}{\varrho}. \tag{13}$$

Proof: We consider the function

$$\ell(w) := z_1 + (z_2 - z_1)w, \qquad w \in \overline{B}$$

satisfying

$$\ell(0) = z_1\,, \quad \ell(1) = z_2\,, \quad |\ell'(w)| = |z_2 - z_1| = \varrho \qquad \text{for all} \quad w \in \overline{B}. \tag{14}$$

We set $w_1 = f(z_1) \in B$ as well as $w_2 = f(z_2) \in \partial B$ and use the Möbius transformation

$$h(w) := e^{i\vartheta}\,\frac{w - w_1}{\overline{w}_1 w - 1}, \qquad w \in \overline{B},$$

with a suitable angle $\vartheta \in [0, 2\pi)$. Then we obtain

$$h(w_1) = 0, \quad h(w_2) = 1 \tag{15}$$

and calculate

$$\begin{aligned} |h'(w_2)| &= \frac{|(\overline{w}_1 w_2 - 1) - \overline{w}_1(w_2 - w_1)|}{|\overline{w}_1 w_2 - 1|^2} = \frac{|1 - |w_1|^2|}{|1 - \overline{w}_1 w_2|^2} \\ &\le \frac{1}{(1 - |\overline{w}_1 w_2|)^2} = \frac{1}{(1 - |w_1|)^2} \\ &\le \frac{1}{(1 - (1 - \varepsilon))^2} = \frac{1}{\varepsilon^2}. \end{aligned} \tag{16}$$

We now consider the origin-preserving holomorphic mapping

$$\Phi(w) := h \circ f \circ \ell(w), \qquad w \in \overline{B}$$

of the class $C^1(\overline{B}, \overline{B})$. The Schwarzian lemma yields

$$|\Phi(w)| \le |w|, \qquad w \in \overline{B}. \tag{17}$$

Therefore, we arrive at the inequality

$$\left|\frac{\Phi(r) - \Phi(1)}{r - 1}\right| \ge \frac{|\Phi(1)| - |\Phi(r)|}{1 - r} \ge \frac{1 - r}{1 - r} = 1$$

for all $r \in (0, 1)$, and we infer

$$|\Phi'(1)| \ge 1. \tag{18}$$

The combination of (14), (16), and (18) yields

$$1 \le |\Phi'(1)| = |h'(w_2) f'(z_2) \ell'(1)| \le \frac{1}{\varepsilon^2}\,|f'(z_2)|\varrho$$

and consequently

$$|f'(z_2)| \ge \frac{\varepsilon^2}{\varrho},$$

which implies the statement above. q.e.d.

Theorem 8.5. (Lipschitz estimate)
The C^2-Jordan-domain $\Omega \subset \mathbb{C}$ is conformally transformed by the mapping

$f : \Omega \to B$ onto the unit disc with the inverse mapping $z = g(w) : B \to \Omega$. Then we have the estimate

$$\sup_{w \in B} |g'(w)| < +\infty, \tag{19}$$

and consequently the mapping $g = g(w)$ is Lipschitz continuous on the closure $\overline{B}$.

Proof: According to the Weierstraß approximation theorem, we can approximate the domain Ω by the Jordan domains Ω_n - with $n \in \mathbb{N}$ - such that their bounding analytic Jordan curves $\Gamma_n = \partial\Omega_n$ converge for $n \to \infty$ towards the C^2-Jordan-curve $\Gamma = \partial\Omega$ inclusive of their derivatives up to the second order. Based on Theorem 8.3, we now consider the conformal mappings

$$g_n : \overline{B} \longrightarrow \overline{\Omega}_n \in C^1(\overline{B}, \overline{\Omega}_n)$$

with their inverse mappings

$$f_n : \overline{\Omega}_n \longrightarrow \overline{B} \in C^1(\overline{\Omega}_n, \overline{B})$$

for all indices $n \in \mathbb{N}$: They converge for $n \to \infty$ uniformly - and in the interior together with their derivatives - towards the function $g \in C^0(\overline{B}, \overline{\Omega})$ and its inverse function $f \in C^0(\overline{\Omega}, \overline{B})$, respectively. Now we have a fixed radius $\varrho > 0$ independent of $n \in \mathbb{N}$, such that each domain Ω_n possesses a support circle

$$B_\varrho(z_1) \subset \Omega_n \qquad \text{with} \quad z_1 \in \Omega_n, \quad z_2 \in \partial B_\varrho(z_1) \cap \Gamma_n$$

at each boundary point $z_2 \in \Gamma_n = \partial\Omega_n$ given. Observing the relation $f_n \to f$ for $n \to \infty$, we find a quantity $\varepsilon > 0$ independent of $n \in \mathbb{N}$ such that the following estimate holds true:

$$|f_n(z_1)| \le |f(z_1)| + |f_n(z_1) - f(z_1)| \le 1 - \varepsilon \qquad \text{for all indices} \quad n \ge n_0(\varepsilon). \tag{20}$$

Here we have chosen the index $n_0(\varepsilon)$ so large that the inequalities

$$|f(z_1)| \le 1 - 2\varepsilon, \qquad |f_n(z_1) - f(z_1)| \le \varepsilon$$

are satisfied. From Theorem 8.4 we infer the estimate

$$|f_n'(z_2)| \ge \frac{\varepsilon^2}{\varrho}, \qquad n \ge n_0(\varepsilon).$$

Setting $w_2 = f_n(z_2)$, we obtain the following statements for the inverse mappings:

$$|g_n'(w_2)| \le \frac{\varrho}{\varepsilon^2} \qquad \text{for all} \quad w_2 \in \partial B \quad \text{and} \quad n \ge n_0(\varepsilon). \tag{21}$$

The maximum principle for holomorphic functions yields the estimate

$$\sup_{B} |g'_n(w)| \leq \frac{\varrho}{\varepsilon^2}, \qquad n \geq n_0(\varepsilon). \tag{22}$$

For $n \to \infty$, we finally obtain the statement above with the following inequality:

$$\sup_{B} |g'(w)| < +\infty. \tag{23}$$

q.e.d.

Remarks to the reflection of the mapping over the differentiable boundary:

For an arbitrary boundary point $z_1 = \beta(t_1) \in \Gamma$ we consider the plane C^1-mapping

$$h(r) = h(t + is) = h(t, s) := \beta(t) + is\beta'(t), \quad r = t + is, \, |r - t_1| < \varepsilon$$

and evaluate

$$\begin{aligned}
\frac{\partial}{\partial \overline{r}} h(r) &= \frac{1}{2} \Big\{ h_t(r) + ih_s(r) \Big\} \\
&= \frac{1}{2} \Big\{ \beta'(t) + is\beta''(t) + i^2\beta'(t) \Big\} \\
&= \frac{i}{2} s\beta''(t) = \frac{i}{2} \beta''(Re\, r) Im\, r, \quad |r - t_1| < \varepsilon.
\end{aligned}$$

For a sufficiently small $\varepsilon > 0$ the Jacobian satisfies

$$J_h(r) = det \begin{pmatrix} \beta'(t) + is\beta''(t) \\ i\beta'(t) \end{pmatrix} = \begin{vmatrix} h_r(r) & h_{\overline{r}}(r) \\ \overline{h}_r(r) & \overline{h}_{\overline{r}}(r) \end{vmatrix} \geq \lambda, \quad |r - t_1| < \varepsilon \tag{24}$$

with a constant $\lambda > 0$. Now we make the transition from the function $z = h(r)$ to its inverse mapping $r = h^{-1}(z)$, $|z - z_1| < \varepsilon^*$ with a sufficiently small $\varepsilon^* > 0$. We differentiate the identity

$$h^{-1} \circ h(r) = r, \quad |r - t_1| < \varepsilon$$

with respect to r and $\overline{r}$ and obtain in

$$h_z^{-1}(h(r))h_r(r) + h_{\overline{z}}^{-1}(h(r))\overline{h}_r(r) = 1$$

$$h_z^{-1}(h(r))h_{\overline{r}}(r) + h_{\overline{z}}^{-1}(h(r))\overline{h}_{\overline{r}}(r) = 0$$

a nonsingular linear system of equations for the unknowns $h_z^{-1}(h(r))$ and $h_{\overline{z}}^{-1}(h(r))$. Via Cramer's rule we determine

$$h_{\overline{z}}^{-1}(h(r)) = \frac{-h_{\overline{r}}(r)}{J_h(r)} = \frac{-i\beta''(Re\, r)}{2J_h(r)} Im\, r, \quad |r - t_1| < \varepsilon.$$

Inserting $r = h^{-1}(z)$ we obtain the equation

$$\frac{\partial}{\partial \overline{z}} h^{-1}(z) = \frac{-i\beta''(Re\, h^{-1}(z))}{2J_h(h^{-1}(z))} Im\, h^{-1}(z), \quad |z - z_1| < \varepsilon^*. \tag{25}$$

On account of (24) and since the function $|\beta''(t)|$, $|t - t_1| < \varepsilon$ is bounded, we obtain the pseudoholomorphic function

$$\left| \frac{\partial}{\partial \overline{z}} h^{-1}(z) \right| \le c_1 \left| Im\, h^{-1}(z) \right|, \quad |z - z_1| < \varepsilon^* \tag{26}$$

with a constant $c_1 > 0$. As in the proof of Theorem 8.3, we now insert the holomorphic function $g \circ \ell(\zeta)$ with $\zeta \in H^+$ and $|\zeta| < \varepsilon$ into the function $h^{-1} = h^{-1}(z)$. Then we obtain a pseudoholomorphic function with

$$\Psi(\zeta) := h^{-1} \circ g \circ \ell(\zeta), \qquad \zeta \in H^+, \quad |\zeta| < \varepsilon.$$

Due to Theorem 8.5, we obtain further constants $c_2 > 0$ and $c_3 > 0$ such that the estimate

$$\begin{aligned} |\Psi_{\overline{\zeta}}(\zeta)| &= \left| \frac{\partial}{\partial z} h^{-1} \Big|_{g\circ\ell(\zeta)} (g \circ \ell)_{\overline{\zeta}} + \frac{\partial}{\partial \overline{z}} h^{-1} \Big|_{g\circ\ell(\zeta)} (\overline{g \circ \ell})_{\overline{\zeta}} \right| \\ &= \left| \frac{\partial}{\partial \overline{z}} h^{-1}\Big(g \circ \ell(\zeta)\Big) \right| \left| g'(\ell(\zeta)) \right| \left| \ell'(\zeta) \right| \\ &\le c_1 c_2 c_3 \left| Im\, h^{-1} \circ g \circ \ell(\zeta) \right| \\ &= c_1 c_2 c_3 \left| Im\, \Psi(\zeta) \right| \end{aligned}$$

for all points $\zeta \in H^+$ with $|\zeta| < \varepsilon$ holds true. With the aid of an integral representation from Theorem 5.5 in Section 5, we can immediately derive a reflection principle for the pseudoholomorphic function $\Psi = \Psi(\zeta)$, which possesses real values and a vanishing Pompeiu derivative on the interval $(-\varepsilon, +\varepsilon)$. Then we obtain the pseudoholomorphic function

$$\left| \frac{d}{d\overline{\zeta}} \Psi(\zeta) \right| \le c\, |\Psi(\zeta)|, \quad |\zeta| < \varepsilon \tag{27}$$

with the constant $c := c_1 c_2 c_3$. We now apply the similarity principle of Bers und Vekua to the function Ψ and obtain asymptotic expansions for our original function g on the boundary ∂B. The functions, which appear in this context, do not belong to the regularity class C^1, in general.

Now we shall prove that the regularity property $g \in C^1(\overline{B}, \mathbb{C})$ holds true and that its derivative $g'(w) : \overline{B} \to \mathbb{C} \backslash \{0\}$ is subject to a Hölder condition for each exponent $\alpha \in (0, 1)$. Here we need the following statement which is shown via the theory of harmonic functions presented in Chapter 5.

Proposition 8.6. (Hardy, Littlewood)
The holomorphic function $G(w) = x(w) + iy(w) \in C^1(\overline{B})$ may satisfy the condition

$$\left|\frac{d}{dt}y(e^{it})\right| \leq l < +\infty \qquad \textit{for all parameters} \quad t \in \mathbb{R}. \tag{28}$$

Then we have a constant $L = L(l) \in (0, +\infty)$ *such that the estimate*

$$|G'(w)| \leq L \qquad \textit{for all points} \quad w \in \overline{B} \tag{29}$$

is correct.

Proof: We consider the Jordan curve

$$\Gamma := \Big\{(\cos t, \sin t, y(e^{it})) \in \mathbb{R}^3 \;:\; 0 \leq t \leq 2\pi\Big\}.$$

According to (28), each point of this curve

$$(u_0, v_0, y_0) = (u_0, v_0, y(u_0, v_0)) \in \Gamma$$

admits a lower and an upper support plane

$$y^{\pm}(u, v) := y_0 + \alpha^{\pm}(u - u_0) + \beta^{\pm}(v - v_0), \qquad (u, v) \in \mathbb{R}^2, \tag{30}$$

which are both situated entirely at one side of the curve Γ. For their measure of ascent we have a constant $L = L(l) \in (0, +\infty)$ such that the real coefficients $\alpha^{\pm}, \beta^{\pm}$ satisfy the following conditions

$$\sqrt{(\alpha^{\pm})^2 + (\beta^{\pm})^2} \leq L. \tag{31}$$

On account of the maximum principle for harmonic functions, we derive

$$\begin{aligned} &y^-(u, v) \leq y(u, v) \leq y^+(u, v) \qquad \text{for all points} \quad (u, v) \in \overline{B}, \\ &y^-(u_0, v_0) = y_0 = y^+(u_0, v_0). \end{aligned} \tag{32}$$

This implies

$$\left|\frac{\partial}{\partial r}y(re^{it})\right|_{r=1} \leq L \qquad \text{for all} \quad t \in \mathbb{R}, \tag{33}$$

and together with (28) we infer the inequality

$$|y_w(w)| \leq L \qquad \text{for all points} \quad w \in \partial B. \tag{34}$$

The maximum principle for the holomorphic function y_w yields

$$|y_w(w)| \leq L \qquad \text{for all points} \quad w \in \overline{B}. \tag{35}$$

Now we invoke the Cauchy-Riemann differential equations for the function $G(w) = x(w) + iy(w)$ with $w \in \overline{B}$, and we arrive at the estimate (29). q.e.d.

Theorem 8.7. **($C^{1,1}$-regularity)**
Let $g : B \to \Omega$ denote a conformal mapping onto the C^2-Jordan-domain $\Omega \subset \mathbb{C}$ with the bounding C^2-Jordan-curve $\Gamma = \partial\Omega$. Then we have the regularity $g \in C^1(\overline{B}, \overline{\Omega})$ and the condition

$$g'(w) \neq 0 \qquad \textit{for all points} \quad w \in \overline{B}.$$

Furthermore, we have a Lipschitz constant $L = L(g) \in (0, +\infty)$ such that the estimate

$$|g'(w_1) - g'(w_2)| \leq L|w_1 - w_2| \quad \textit{for all points} \quad w_1, w_2 \in \overline{B}$$

is satisfied.

Proof: As in the proof of Theorem 8.5, we approximate the function $g : \overline{B} \to \overline{\Omega}$ uniformly in $\overline{B}$ by conformal mappings $g_n : \overline{B} \to \overline{\Omega}_n$ for $n = 1, 2, \ldots$ such that

$$\sup_B |g_n'(w)| \leq c_1, \qquad n \in \mathbb{N}.$$

When we define the functions

$$G_n(w) := \log g_n'(w) = \log |g_n'(w)| + i \arg g_n'(w), \qquad w \in \overline{B} \quad \text{for} \quad n \in \mathbb{N}, \tag{36}$$

we observe

$$\lim_{n\to\infty} G_n(0) = \lim_{n\to\infty} \log g_n'(0) = \log g'(0) \in \mathbb{C}. \tag{37}$$

We still have to verify the estimate

$$\sup_{w\in B} |G_n'(w)| \leq c_2, \qquad n \in \mathbb{N}. \tag{38}$$

Now we associate the subsequent Gaussian metric with the mapping $g_n = g_n(w)$, namely

$$ds_n^2 = E_n(w)\,(du^2 + dv^2) = |g_n'(w)|^2(du^2 + dv^2). \tag{39}$$

We invoke the following *formula by F. Minding* from a lecture of differential geometry for the geodesic curvature κ_n of the boundary curve $\Gamma_n = \partial\Omega_n$:

$$\frac{\partial}{\partial r} \log \sqrt{E_n(r\cos t, r\sin t)}\,\Big|_{r=1} = \kappa_n \sqrt{E_n(\cos t, \sin t)} - 1, \qquad t \in \mathbb{R}. \tag{40}$$

Here we recommend the first volume in the *Grundlehren der mathematischen Wissenschaften* [BL] by W. Blaschke and K. Leichtweiss, where especially § 77, § 78, and § 92 are relevant. Minding's formula is explicitly derived in the treatise on *Minimal Surfaces* [DHS] by U. Dierkes, S. Hildebrandt, and F. Sauvigny with the identity (48) of Section 1.3.

On account of (40) and Theorem 8.5, the mapping

$$G_n(w) = x_n(w) + iy_n(w), \quad w \in \overline{B}$$

from (36) then satisfies the following estimate:

$$\left|\frac{\partial}{\partial r}x_n(re^{it})\right|_{r=1} \leq \tilde{c}_2 \qquad \text{for all} \quad t \in \mathbb{R} \quad \text{and} \quad n \in \mathbb{N} \tag{41}$$

with a constant $\tilde{c}_2$. The Cauchy-Riemann differential equations yield

$$\left|\frac{d}{dt}y_n(e^{it})\right| \leq \tilde{c}_2 \qquad \text{for all} \quad t \in \mathbb{R} \quad \text{and} \quad n \in \mathbb{N}. \tag{42}$$

Via Proposition 8.6 we arrive at the estimate (38).

Therefore, the sequence of functions $\{G_n\}_{n=1,2,\ldots}$ is equicontinuous and uniformly bounded. Via the theorem of Arzelà-Ascoli, we achieve the transition to a uniformly convergent subsequence on $\overline{B}$ with $\{G_{n_k}\}_{k=1,2,\ldots}$, and we obtain the continuous function

$$G(w) := \lim_{k\to\infty} G_{n_k}(w), \qquad w \in \overline{B}.$$

Now we observe

$$G(w) = \lim_{k\to\infty} G_{n_k}(w) = \lim_{k\to\infty} \log g'_{n_k}(w) = \log g'(w), \qquad w \in B.$$

Consequently, the function

$$\Phi(w) := \log g'(w), \qquad w \in B$$

can be continuously extended on the closure $\overline{B}$, and we deduce the continuity of the derivative $g'(w) : \overline{B} \to \mathbb{C}\setminus\{0\}$. Since the functions $\{G_n\}_{n=1,2,\ldots}$ satisfy a joint Lipschitz condition in $\overline{B}$, this remains true for the limit function $G = G(w)$ and consequently for $g = g(w)$, $w \in \overline{B}$.

q.e.d.

Remark: In order to obtain the statement $g \in C^2(\overline{B})$, we have to assume higher regularity for the boundary curve $\Gamma = \partial\Omega$.

9 Behavior of Cauchy's Integral across the Boundary

Let $G \subset \mathbb{C}$ denote a bounded regular C^2-domain, such that its boundary $\Gamma = \partial G$ represents a regular C^2-Jordan-curve. When the length of the curve Γ is given by $L = |\partial G| = |\Gamma| \in (0, +\infty)$, we imagine this contour being parametrized with unit-velocity – such that the domain lies to the left – by the following function:

$$\zeta = \zeta(t) : \mathbb{R} \to \Gamma \in C^2(\mathbb{R}, \mathbb{C}), \tag{1}$$

$$\zeta(t+kL) = \zeta(t) \text{ for all } t \in \mathbb{R} \text{ and } k \in \mathbb{Z} \quad \text{(periodicity with the period L)}, \tag{2}$$

$$|\zeta'(t)| = 1 \quad \text{for all} \quad t \in \mathbb{R} \quad (unit - velocity), \tag{3}$$

$$\nu(t) := -i\zeta'(t),\ t \in \mathbb{R} \quad \text{is the exterior unit normal to G.} \tag{4}$$

For a sufficiently small $s_0 > 0$, the 2-dimensional vector-field

$$Z(t,s) := \zeta(t) + s\nu(t), \quad t \in \mathbb{R},\ s \in (-s_0, +s_0) \tag{5}$$

yields a C^1-diffeomorphism of

$$\textit{the stripe} \quad \mathbb{R} \times (-s_0, +s_0) \quad \text{onto } \textit{the tube} \quad G_0^- \cup \Gamma \cup G_0^+,$$

which is periodic in the variable t with the period L. Here we have chosen the domains

$$G_0^{\pm} := \{z = Z(t,s) \in \mathbb{C} : t \in \mathbb{R}, \pm s \in (0, +s_0)\},$$

which are not simply connected.

Example:
For the unit disc $G = B := \{z \in \mathbb{C} : |z| < 1\}$ we take the parametrization $\zeta(t) = e^{it}$, $t \in \mathbb{R}$ and the exterior unit normal $\nu(t) = -i\zeta'(t)$, $t \in \mathbb{R}$. When we set $\Gamma := \{z \in \mathbb{C} : |z| = 1\}$ and $G_1^- := \{z \in G : 0 < |z| < 1\}$ as well as $G_1^+ := \{z \in G : 1 < |z| < 2\}$, the vector-field

$$Z(t,s) := e^{it} + se^{it}, \quad t \in \mathbb{R},\ s \in (-1, +1) \tag{6}$$

yields the diffeomorphism

$$Z : \mathbb{R} \times (-1, +1) \to G_1^- \cup \Gamma \cup G_1^+, \tag{7}$$

which is periodic in the variable t with the period 2π.

We remark that $G_0^- \subset G$ lies within the domain G with Γ as its exterior boundary, while $G_0^+ \subset \mathbb{C} \setminus \overline{G}$ lies in the the exterior domain of G with Γ as its interior boundary. For each point $z = Z(t,s) \in G_0^{\pm} \cup \Gamma$ with the uniquely determined parameters $t \in [0, L)$ and $\pm s \in [0, s_o)$ respectively, we have the unique *projection point* $\hat{z} = Z(t, 0) \in \Gamma$ *onto the curve* Γ; we observe that the correspondence $G_0^- \cup \Gamma \cup G_0^+ \ni z \to \hat{z} \in \Gamma$ is continuous!

For each point $z = Z(t,s) \in G_0^- \cup \Gamma \cup G_0^+$, we consider the *complex logarithm function*

$$\log_z(\zeta) := \log_t(\zeta - z) = \ln|\zeta - z| + i \arg_{\nu(t)}(\zeta - z), \quad \zeta \in \mathbb{C} \setminus \{z\}, \tag{8}$$

on the leaf $\mathbb{C} \setminus \{z\}$ *sliced along the semi-line* $z + R_t$. Here we have introduced the ray

$$R_t := \{s\nu(t) \in \mathbb{C} : s > 0\}$$

where the argument function $\arg_{\nu(t)}$ is discontinuous, with a jump of the size $2\pi > 0$ for an approach in the clockwise to the counter-clockwise direction $+$ or $-$, respectively.

Finally, we define the *characteristic function*

$$\chi(z) := \begin{cases} 1, & z \in G \\ \frac{1}{2}, & z \in \partial G \\ 0, & z \in \mathbb{C} \setminus \overline{G} \end{cases} \tag{9}$$

Now we prescribe a *boundary function*

$$f = f(\zeta) : \Gamma \to \mathbb{C} \in C^k(\Gamma) \tag{10}$$

for $k = 1$ or $k = 2$, meaning that the function

$$F(t) := f(\zeta(t)), \, t \in \mathbb{R} \quad \text{belongs to the class} \quad C_L^k(\mathbb{R}, \mathbb{C}) \tag{11}$$

of those $C^k(\mathbb{R}, \mathbb{C})$-functions, which are periodic in $\mathbb{R}$ with the period $L > 0$.

We shall study the behavior of Cauchy's integral

$$\Phi(z) := \frac{1}{2\pi i} \oint\limits_{\partial G} \frac{f(\zeta)}{\zeta - z} \, d\zeta, \quad z \in \mathbb{C} \setminus \Gamma \tag{12}$$

near the boundary contour Γ. To achieve this aim, we perform real partial integrations of Cauchy's integral in the subsequent Propositions 9.1 and 9.2.

Proposition 9.1. *For boundary functions $f \in C^1(\Gamma)$, Cauchy's integral appears in the form*

$$\Phi(z) = \chi(z) f(\hat{z}) - \frac{1}{2\pi i} \int_0^L F'(\tau) \log_z(\zeta(\tau) - z) \, d\tau, \quad z \in G_0^- \cup G_0^+ \tag{13}$$

with the complex logarithmic kernel (8) above.

Proof: For all $z = Z(t, s) \in G_0^- \cup G_0^+$ with $t \in [0, L)$ and $\pm s \in (0, s_0)$ we calculate as follows:

$$\begin{aligned}\Phi(z) &= \frac{1}{2\pi i}\oint\limits_{\partial G}\frac{f(\zeta)}{\zeta - z}\,d\zeta = \frac{1}{2\pi i}\int_0^L \frac{F(\tau)}{\zeta(\tau)-z}\,\zeta'(\tau)\,d\tau \\
&= \frac{1}{2\pi i}\int_t^{t+L}\Big(\log_z(\zeta(\tau)-z)\Big)'F(\tau)\,d\tau \\
&= \frac{1}{2\pi i}\Big[F(\tau)\log_z(\zeta(\tau)-z)\Big]_{t+}^{(t+L)-} - \frac{1}{2\pi i}\int_t^{t+L} F'(\tau)\log_z(\zeta(\tau)-z)\,d\tau \\
&= \chi(z)f(\hat{z}) - \frac{1}{2\pi i}\int_0^L F'(\tau)\log_z(\zeta(\tau)-z)\,d\tau.\end{aligned}$$

q.e.d.

In order to study the behavior of the complex derivative for the holomorphic Cauchy integral, we provide

Proposition 9.2. *For boundary functions $f \in C^1(\Gamma)$, the complex derivative of Cauchy's integral appears in the form*

$$\frac{d}{dz}\Phi(z) = \frac{1}{2\pi i}\oint\limits_{\partial G}\frac{\dot{f}(\zeta)}{\zeta - z}\,d\zeta, \quad z \in G_0^- \cup G_0^+ \tag{14}$$

as Cauchy integral for the directional derivative

$$\dot{f}(\zeta) := \frac{F'(t)}{\zeta'(t)} \quad \textit{for} \quad \zeta = \zeta(t) \in \Gamma \quad \textit{and} \quad t \in [0, L). \tag{15}$$

Proof: For all $z \in G_0^- \cup G_0^+$ let us calculate:

$$\begin{aligned}\frac{d}{dz}\Phi(z) &= \frac{1}{2\pi i}\oint\limits_{\partial G}\frac{f(\zeta)}{(\zeta - z)^2}\,d\zeta = \frac{1}{2\pi i}\int_0^L \frac{F(\tau)}{(\zeta(\tau)-z)^2}\,\zeta'(\tau)\,d\tau \\
&= \frac{1}{2\pi i}\int_0^L \Big(\frac{-1}{\zeta(\tau)-z}\Big)'F(\tau)\,d\tau \\
&= \frac{1}{2\pi i}\Big[F(\tau)\cdot\frac{-1}{\zeta(\tau)-z}\Big]_0^L + \frac{1}{2\pi i}\int_0^L \frac{F'(\tau)}{\zeta(\tau)-z}\,d\tau \\
&= \frac{1}{2\pi i}\int_0^L \frac{F'(\tau)}{\zeta'(\tau)}\cdot\frac{\zeta'(\tau)}{\zeta(\tau)-z}\,d\tau = \frac{1}{2\pi i}\oint\limits_{\partial G}\frac{\dot{f}(\zeta)}{\zeta - z}\,d\zeta.\end{aligned}$$

q.e.d.

We are now prepared to establish the interesting

Theorem 9.3. (Cauchy's integral across the boundary)
For Cauchy's integral $\Phi(z)$ from (12) with the boundary function $f \in C^k(\Gamma)$ from (10) with $k = 1$ or $k = 2$, we have the following statements:

(a) If $k = 1$ is assumed, Cauchy's integral $\Phi(z)$ can be extended to a continuous function onto the set $G_0^- \cup \Gamma$ from the interior and to a continuous function onto the set $\Gamma \cup G_0^+$ from the exterior. Along the curve Γ, Cauchy's integral $\Phi(z)$ possesses a jump of the size $f(\hat{z})$ due to the representation (13) above. Here the parametric integral of (13) with the complex logarithmic kernel is continuous on $G_0^- \cup \Gamma \cup G_0^+$.

(b) If $k = 2$ is assumed, the complex derivative $\frac{d}{dz}\Phi(z)$, $z \in \mathbb{C} \setminus \Gamma$ can be extended continuously onto $G_0^- \cup \Gamma$ from the interior and onto $\Gamma \cup G_0^+$ from the exterior. Furthermore, we have the jump relation

$$\frac{d}{dz}\Phi(z) = \chi(z)\dot{f}(\hat{z}) - \frac{1}{2\pi i}\int_0^L \left(\frac{F'(\tau)}{\zeta'(\tau)}\right)' \log_z(\zeta(\tau) - z)\, d\tau\,,\ z \in G_0^- \cup G_0^+ \tag{16}$$

for the complex derivative of Cauchy's integral.

Proof:

(a) Since $\int_0^1 -\ln r\, dr < \infty$ is correct, the parametric integral with the complex logarithmic kernel

$$\frac{1}{2\pi i}\int_0^L F'(\tau)\log_z(\zeta(\tau) - z)\, d\tau, \quad z \in G_0^- \cup \Gamma \cup G_0^+ \tag{17}$$

possesses an integrable majorant. With the aid of the convergence theorem for – absolutely convergent – improper Riemannian integrals, we comprehend the continuity of the parametric integral (17) in dependence of the variable z on the domain $G_0^- \cup \Gamma \cup G_0^+$. Now the identity (13) of Proposition 9.1 reveals the statement (a).

(b) We integrate via Proposition 9.2 and Proposition 9.1. The identity in (16) is derived for all $z \in G_0^- \cup G_0^+$ as follows:

$$\begin{aligned}\frac{d}{dz}\Phi(z) &= \frac{1}{2\pi i}\oint\limits_{\partial G} \frac{\dot{f}(\zeta)}{\zeta - z}\, d\zeta \\ &= \chi(z)\dot{f}(\hat{z}) - \frac{1}{2\pi i}\int_0^L \frac{d}{d\tau}\Big\{\dot{f}(\zeta(\tau))\Big\}\log_z(\zeta(\tau) - z)\, d\tau \\ &= \chi(z)\dot{f}(\hat{z}) - \frac{1}{2\pi i}\int_0^L \left(\frac{F'(\tau)}{\zeta'(\tau)}\right)' \log_z(\zeta(\tau) - z)\, d\tau.\end{aligned}$$

As in (a) above, we observe the continuity of the parametric integral on $G_0^- \cup \Gamma \cup G_0^+$, and the proof of statement (b) is complete.

q.e.d.

Remarks:

1. In order to establish Hölder-continuity for the relevant functions, one has to derive an inequality parallel to Hadamard's estimate from Theorem 4.12 in Section 4 for the parametric integral (17) with the complex logarithmic kernel.
2. On the boundary curve $\Gamma = \partial G$, we can interpret Cauchy's integral only as a Cauchy principal value

$$\hat{\Phi}(z_0) := \lim_{\varepsilon \to 0+} \left\{ \frac{1}{2\pi i} \oint\limits_{\zeta \in \partial G:\, |\zeta - z_0| \geq \varepsilon} \frac{f(\zeta)}{\zeta - z_0}\, d\zeta \right\}, \quad z_0 \in \Gamma. \tag{18}$$

A real partial integration yields

$$\hat{\Phi}(z_0) = \frac{1}{2} f(z_0) - \frac{1}{2\pi i} \int_0^L F'(\tau) \log_{z_0}(\zeta(\tau) - z_0)\, d\tau, \quad z_0 \in \Gamma. \tag{19}$$

Thus we obtain from (13) the jump relation

$$\lim_{z \to z_0,\, z \in G_0^{\mp}} \Phi(z) = \pm \frac{1}{2} f(z_0) + \hat{\Phi}(z_0), \quad z_0 \in \Gamma. \tag{20}$$

for all boundary functions $f \in C^1(\Gamma)$.

We shall utilize Cauchy's integral to solve a boundary value problem for harmonic functions on the unit disc in

Theorem 9.4. (Harmonic extension)
On the boundary $\Gamma = \partial B$ of the open unit disc $B := \{z = x+iy \in \mathbb{C} : |z| < 1\}$ we prescribe the boundary values

$$f = f(z) : \Gamma \to \mathbb{C} \in C^k(\Gamma)$$

for k=1,2. Then we have a function

$$\Psi = \Psi(z) = \Psi(x,y) : \overline{B} \to \mathbb{C} \in C^\infty(B) \cap C^{k-1}(\overline{B}), \tag{21}$$

which satisfies the Laplace equation

$$\Delta \Psi(x,y) = \Big(\Psi_{xx} + \Psi_{yy}\Big)\Big|_{(x,y)} = 4 \frac{\partial}{\partial z} \frac{\partial}{\partial \overline{z}} \Psi(z) = 0 \quad \text{for all} \quad z = x+iy \in B \tag{22}$$

and assumes continuously the boundary values

$$\lim_{(x,y) \to (x_0,y_0),\, (x,y) \in B} \Psi(x,y) = f(x_0,y_0) \quad \text{for all} \quad z_0 = (x_0,y_0) \in \Gamma. \tag{23}$$

This function has the integral representation (28) with (24)–(26) below.

Proof: We consider the holomorphic Cauchy integral

$$\Phi(z) := \frac{1}{2\pi i} \oint\limits_{\partial B} \frac{f(\zeta)}{\zeta - z}\, d\zeta, \quad z \in \mathbb{C} \setminus \Gamma. \tag{24}$$

Furthermore, the function

$$\Phi^*(z) := \Phi\Big(\frac{1}{z}\Big), \quad z \in B \setminus \{0\} \tag{25}$$

is holomorphic on its domain of definition, and for $z \to 0$, $z \neq 0$ it remains bounded. Riemann's removability theorem yields a holomorphic extension of Φ^* into the origin 0, which we shall use without renaming this function. Now

$$\Phi^{**}(z) := \Phi^*(\overline{z}) = \Phi\Big(\frac{1}{\overline{z}}\Big), \quad z \in B \tag{26}$$

gives us an antiholomorphic function due to

$$\Phi_z^{**}(z) = \Phi_w^*(\overline{z})(\overline{z})_z + \Phi_{\overline{w}}^*(\overline{z})(z)_z = \Phi_{\overline{w}}^*(\overline{z}) = 0, \quad z \in B. \tag{27}$$

Then we inherit the regularity of Φ and Φ^{**} on $\overline{B}$ from our Theorem 9.3, and we define the *harmonic function*

$$\Psi(z) := \Phi(z) - \Phi^{**}(z), \quad z \in \overline{B}. \tag{28}$$

This means that

$$\Psi_{z\overline{z}}(z) = \Phi_{z\overline{z}}(z) - \Phi_{z\overline{z}}^{**}(z) = 0\,,\ z \in B$$

holds true. Furthermore, we have the boundary behavior

$$\begin{aligned}\lim_{z \to z_0,\, z \in B} \Psi(z) &= \lim_{z \to z_0,\, z \in B} \Big(\Phi(z) - \Phi^{**}(z)\Big)\\ &= \lim_{z \to z_0,\, z \in B} \left(\Phi(z) - \Phi\Big(\frac{1}{\overline{z}}\Big)\right)\\ &= f(z_0) \quad \text{for all} \quad z_0 \in \Gamma,\end{aligned}$$

due to Proposition 9.1. q.e.d.

Remarks: When we want to prescribe complex boundary values, we can solve this problem only in the class of harmonic functions; here we leave open the question, for which boundary functions this harmonic mapping provides a diffeomorphism. A conformal mapping *searches* for its boundary representation within the given boundary contour. For boundary value problems with holomorphic functions, we can only prescribe the real part on the boundary; here we refer the reader to Theorem 2.2 (Schwarzian integral formula) in Section 2 of Chapter 9.

10 Some Historical Notices to Chapter 4

C.F. Gauß created the complex number field with his inaugural dissertation on the *Fundamental Theorem of Algebra* in 1801. Furthermore, he characterized differential-geometrically the conformal mappings – already known from the stereographic projection. It was the eminent task of Cauchy (1789–1857), Weierstraß (1815–1897), and Riemann (1826–1866) to develop the theory of holomorphic functions – via the alternative concepts of *contour integrals*, *power series*, and *complex differentiability*, respectively. Obviously, Riemann's ideas proved being most profound and constructive for the theory of partial differential equations and for the geometry in general.

Astonishingly late in the 1950s, the inhomogeneous Cauchy-Riemann equation was studied by L. Bers and I.N. Vekua – independently in New York and Moscow. This was performed in the natural desire, to extract the square root out of a second-order elliptic equation. In this context, the discovery by Carleman, in 1930, of the isolated character for the zeroes in pseudoholomorphic functions cannot be estimated highly enough!

The complete proof of Riemann's mapping theorem posed a great challenge for Dirichlet, Weierstraß, Koebe, Hilbert, Courant... The analytic boundary behavior was already solved by H.A. Schwarz via his reflection principle, the continuous boundary behavior was investigated by Carathéodory in 1913 and simplified by R. Courant (1888–1972) via his well-known lemma together with H. Lebesgue (1875–1941). However, the differentiable boundary behavior had to wait for S. Warschawski, in 1961/68, to be fully understood. The behavior of Cauchy's integral across the boundary has originally been studied by Plemelj.

Figure 1.5 PORTRAIT OF HERMANN AMANDUS SCHWARZ (1843–1921) Niedersächsische Staats- und Universitätsbibliothek Göttingen; taken from the book by *S. Hildebrandt, A. Tromba: Panoptimum – Mathematische Grundmuster des Vollkommenen*, Spektrum-Verlag Heidelberg (1986).

Chapter 5

Potential Theory and Spherical Harmonics

In this chapter we investigate solutions of the potential equation due to Laplace in the homogeneous case and due to Poisson in the inhomogeneous case. Parallel to the theory of holomorphic functions we develop the theory of harmonic functions annihilating the Laplace equation. By the ingenious Perron method we shall solve Dirichlet's problem for harmonic functions. Then we present the theory of spherical harmonics initiated by Legendre and elaborated by Herglotz to the present form. This system of functions constitutes an explicit basis for the standard Hilbert space and simultaneously provides a model for the ground states of atoms.

1 Poisson's Differential Equation in $\mathbb{R}^n$

The solutions of 2-dimensional differential equations can often be obtained via integral representations over the circle S^1. As an example we remind the reader of Cauchy's integral formula. For n-dimensional differential equations will appear integrals over the $(n-1)$-dimensional sphere

$$S^{n-1} := \Big\{\xi = (\xi_1, \ldots, \xi_n) \in \mathbb{R}^n \ : \ \xi_1^2 + \ldots + \xi_n^2 = 1\Big\}, \qquad n \geq 2. \tag{1}$$

At first, we shall determine the area of this sphere S^{n-1}. Given the function $f = f(\xi) : S^{n-1} \to \mathbb{R} \in C^0(S^{n-1}, \mathbb{R})$ we set

$$\int\limits_{S^{n-1}} f(\xi)\, d\omega_\xi = \int\limits_{|\xi|=1} f(\xi)\, d\omega_\xi := \sum_{i=1}^{N} \int\limits_{\Sigma_i} f(\xi)\, d\omega_\xi. \tag{2}$$

By the symbols $\Sigma_1, \ldots, \Sigma_N$ we denote the $N \in \mathbb{N}$ regular surface parts with their surface elements $d\omega_\xi$ satisfying

$$S^{n-1} = \bigcup_{i=1}^{N} \overline{\Sigma}_i, \qquad \overline{\Sigma}_i \cap \overline{\Sigma}_j = \partial\Sigma_i \cap \partial\Sigma_j, \quad i \neq j.$$

F. Sauvigny, *Partial Differential Equations 1*, Universitext,
DOI 10.1007/978-1-4471-2981-3_5,

We now consider a continuous function

$$f : \Big\{ x = r\xi \in \mathbb{R}^n \; : \; a < r < b, \; \xi \in S^{n-1} \Big\} \to \mathbb{R}$$

with $0 \le a < b \le +\infty$, and we define the open sets

$$\mathcal{O}_i := \Big\{ x = r\xi \; : \; \xi \in \Sigma_i, \; r \in (a,b) \Big\}, \qquad i = 1, \ldots, N.$$

We require the integrability $\int\limits_{a<|x|<b} |f(x)|\,dx < +\infty$ and set

$$\int\limits_{a<|x|<b} f(x)\,dx = \sum_{i=1}^{N} \int\limits_{\mathcal{O}_i} f(x)\,dx. \tag{3}$$

The surface parts Σ_i are parametrized as follows

$$\Sigma_i : \quad \xi = \xi(t) = \xi(t_1, \ldots, t_{n-1}) : T_i \to \Sigma_i \in C^1(T_i, \Sigma_i), \qquad i = 1, \ldots, N$$

with the parameter domains $T_i \subset \mathbb{R}^{n-1}$. By the representation

$$x = x(t,r) = x(t_1, \ldots, t_{n-1}, r) = r\xi(t_1, \ldots, t_{n-1}), \qquad t \in T_i, \quad r \in (a,b) \tag{4}$$

we obtain a parametrization of the sets $\mathcal{O}_i$ for $i = 1, \ldots, N$. The Jacobian of this mapping is evaluated as follows:

$$J_x(t,r) = \begin{vmatrix} r\xi_{t_1}(t) \\ \vdots \\ r\xi_{t_{n-1}}(t) \\ \xi(t) \end{vmatrix} = r^{n-1} \begin{vmatrix} \xi_{t_1}(t) \\ \vdots \\ \xi_{t_{n-1}}(t) \\ \xi(t) \end{vmatrix} = r^{n-1} \Big(\xi(t) \cdot \xi_{t_1} \wedge \ldots \wedge \xi_{t_{n-1}} \Big).$$

Here the symbol $\wedge$ denotes the exterior vector product in $\mathbb{R}^n$. We have

$$\xi_{t_1} \wedge \ldots \wedge \xi_{t_{n-1}} = (D_1(t), \ldots, D_n(t))$$

with

$$D_j(t) := (-1)^{n+j} \frac{\partial(\xi_1, \ldots, \xi_{j-1}, \xi_{j+1}, \ldots, \xi_n)}{\partial(t_1, \ldots, t_{n-1})}, \quad j = 1, \ldots, n.$$

We note $|\xi(t)| = 1$ and infer $\xi(t) \cdot \xi_{t_i}(t) = 0$ for all $i = 1, \ldots, n-1$. Therefore, the vectors $\xi(t)$ and $\xi_{t_1} \wedge \ldots \wedge \xi_{t_{n-1}}$ are parallel to each other and we deduce

$$J_x(t,r) = r^{n-1} \sqrt{\sum_{j=1}^{n} D_j(t)^2}. \tag{5}$$

Setting $d\omega_\xi = \sqrt{\sum\limits_{j=1}^{n} D_j(t)^2}\, dt_1 \ldots dt_{n-1}$, $t \in T_i$ we obtain

$$\begin{aligned}\int\limits_{\mathcal{O}_i} f(x)\,dx &= \int\limits_{T_i\times(a,b)} f(r\xi(t))r^{n-1}\sqrt{\sum_{j=1}^{n} D_j(t)^2}\,dt_1 \ldots dt_{n-1}\,dr \\ &= \int\limits_a^b r^{n-1}\,dr \int\limits_{\Sigma_i} f(r\xi)\,d\omega_\xi, \qquad i = 1,\ldots,N.\end{aligned}$$

Summation over $i = 1,\ldots,N$ finally yields

$$\int\limits_{a<|x|<b} f(x)\,dx = \int\limits_a^b r^{n-1}\,dr \int\limits_{S^{n-1}} f(r\xi)\,d\omega_\xi. \tag{6}$$

Especially the functions $f \in C^0(\mathbb{R}^n,\mathbb{R})$ with $\int\limits_{\mathbb{R}^n} |f(x)|dx < +\infty$ fulfill the identity

$$\int\limits_{\mathbb{R}^n} f(x)\,dx = \int\limits_0^{+\infty} r^{n-1}\,dr \int\limits_{S^{n-1}} f(r\xi)\,d\omega_\xi. \tag{7}$$

Before we continue to evaluate the area of the sphere S^{n-1}, we shall explicitly provide a calculus rule for the integral defined in (2). In this context we consider the following special parametrization of S^{n-1}:

$$\begin{aligned}\Sigma_\pm :\ \xi_i = t_i, \quad i = 1,\ldots,n-1, \qquad \xi_n = \pm\sqrt{1 - t_1^2 - \ldots - t_{n-1}^2},\\ t = (t_1,\ldots,t_{n-1}) \in T := \Big\{t \in \mathbb{R}^{n-1}\ :\ |t| < 1\Big\}.\end{aligned}$$

We calculate

$$\begin{vmatrix} \frac{\partial\xi_1}{\partial t_1} & \cdots & \frac{\partial\xi_{n-1}}{\partial t_1} & \frac{\partial\xi_n}{\partial t_1} \\ \vdots & & \vdots & \vdots \\ \frac{\partial\xi_1}{\partial t_{n-1}} & \cdots & \frac{\partial\xi_{n-1}}{\partial t_{n-1}} & \frac{\partial\xi_n}{\partial t_{n-1}} \\ \lambda_1 & \cdots & \lambda_{n-1} & \lambda_n \end{vmatrix} = \begin{vmatrix} 1 & \cdots & 0 & -\frac{\xi_1}{\xi_n} \\ \vdots & \ddots & \vdots & \vdots \\ 0 & \cdots & 1 & -\frac{\xi_{n-1}}{\xi_n} \\ \lambda_1 & \cdots & \lambda_{n-1} & \lambda_n \end{vmatrix} = \sum_{j=1}^{n-1} \lambda_j \frac{\xi_j}{\xi_n} + \lambda_n.$$

The surface element of $\Sigma_\pm$ consequently fulfills

$$d\omega_\xi = \sqrt{\sum_{j=1}^{n} D_j(t)^2}\, dt_1 \dots dt_{n-1} = \sqrt{\frac{\sum\limits_{j=1}^{n} \xi_j(t)^2}{\xi_n(t)^2}}\, dt_1 \dots dt_{n-1}$$

$$= \frac{dt_1 \dots dt_{n-1}}{\sqrt{1 - t_1^2 - \dots - t_{n-1}^2}}, \qquad t \in T.$$

Therefore, the relation (2) implies

$$\int\limits_{|\xi|=1} f(\xi)\, d\omega_\xi = \int\limits_{|t|<1} \frac{f(t_1, \dots, t_{n-1}, +\sqrt{\dots}) + f(t_1, \dots, t_{n-1}, -\sqrt{\dots})}{\sqrt{1 - t_1^2 - \dots - t_{n-1}^2}}\, dt_1 \dots dt_{n-1} \tag{8}$$

setting $\sqrt{\dots} = \sqrt{1 - t_1^2 - \dots - t_{n-1}^2}$.

We now return to evaluate the *area for the* $(n-1)$*-dimensional sphere* S^{n-1}

$$\omega_n := \int\limits_{S^{n-1}} d\omega_\xi.$$

We take a continuous function $g = g(r) : (0, +\infty) \to \mathbb{R}$, and require the function $f(x) = g(|x|)$ to fulfill

$$\int\limits_{\mathbb{R}^n} |f(x)|\, dx < +\infty.$$

Then the relation (7) yields

$$\begin{aligned} \int\limits_{\mathbb{R}^n} g(|x|)\, dx &= \Bigg(\int\limits_0^{+\infty} r^{n-1} g(r)\, dr \Bigg) \Bigg(\int\limits_{S^{n-1}} d\omega_\xi \Bigg) \\ &= \omega_n \int\limits_0^{+\infty} r^{n-1} g(r)\, dr. \end{aligned} \tag{9}$$

We insert the function $g(r) = e^{-r^2}$, $r \in (0, +\infty)$ and obtain

$$\begin{aligned} \omega_n \int\limits_0^{+\infty} r^{n-1} e^{-r^2}\, dr &= \int\limits_{\mathbb{R}^n} e^{-|x|^2}\, dx = \int\limits_{\mathbb{R}^n} e^{-x_1^2 - \dots - x_n^2}\, dx_1 \dots dx_n \\ &= \Bigg(\int\limits_{-\infty}^{+\infty} e^{-t^2}\, dt \Bigg)^n = \sqrt{\pi}^{\,n}. \end{aligned} \tag{10}$$

Here we observe

$$\int_{-\infty}^{+\infty} e^{-t^2}\,dt = \sqrt{\iint_{\mathbb{R}^2} e^{-|x|^2}\,dx\,dy} = \sqrt{2\pi \int_0^{+\infty} e^{-r^2} r\,dr}$$

$$= \sqrt{\pi}\sqrt{\Big[-e^{-r^2}\Big]_0^{+\infty}} = \sqrt{\pi}.$$

Definition 1.1. *By the symbol*

$$\Gamma(z) := \int_0^{+\infty} t^{z-1} e^{-t}\,dt, \qquad z \in \mathbb{C} \quad \textit{with} \quad \mathrm{Re}\, z > 0$$

we denote the Gamma-function.

Remark: We have

$$\Gamma(z+1) = z\Gamma(z) \qquad \text{for all} \quad z \in \mathbb{C} \quad \text{with} \quad \mathrm{Re}\, z > 0.$$

Therefore, we inductively obtain

$$\Gamma(n) = (n-1)! \qquad \text{for} \quad n = 1, 2, \ldots$$

With the aid of the substitution $t = \varrho^2$ and $dt = 2\varrho\,d\varrho$ we calculate

$$\Gamma\Big(\frac{1}{2}\Big) = \int_0^{+\infty} t^{-\frac{1}{2}} e^{-t}\,dt = \int_0^{+\infty} \frac{1}{\varrho} e^{-\varrho^2} 2\varrho\,d\varrho$$

$$= 2\int_0^{+\infty} e^{-\varrho^2}\,d\varrho = \int_{-\infty}^{+\infty} e^{-\varrho^2}\,d\varrho = \sqrt{\pi}.$$

Substituting $t = r^2$ and $dt = 2r\,dr$, we finally deduce

$$\Gamma\Big(\frac{n}{2}\Big) = \int_0^{+\infty} t^{\frac{n-2}{2}} e^{-t}\,dt = \int_0^{+\infty} r^{n-2} e^{-r^2} 2r\,dr = 2\int_0^{+\infty} r^{n-1} e^{-r^2}\,dr.$$

From the relation (10) we get the following identity for the area of the sphere S^{n-1}, namely

$$\omega_n = \frac{2\Big(\Gamma(\frac{1}{2})\Big)^n}{\Gamma(\frac{n}{2})}. \tag{11}$$

We now become acquainted with a class of functions which have similar properties as the class of holomorphic functions.

Definition 1.2. *On the open set* $\Omega \subset \mathbb{R}^n$ *with* $n \geq 2$ *we name the function* $\varphi = \varphi(x) \in C^2(\Omega, \mathbb{R})$ harmonic in Ω, *if* φ *satisfies the* Laplacian differential equation

$$\Delta\varphi(x) = \varphi_{x_1x_1}(x) + \ldots + \varphi_{x_nx_n}(x) = 0 \qquad \text{for all} \quad x \in \Omega. \tag{12}$$

At first, we shall find the radially symmetric harmonic functions in $\mathbb{R}^n \setminus \{0\}$. Here we begin with the ansatz

$$\varphi(x) = f(|x|), \qquad x \in \mathbb{R}^n \setminus \{0\}, \tag{13}$$

using the function $f = f(r) : (0, +\infty) \to \mathbb{R} \in C^2((0, +\infty), \mathbb{R})$. According to Chapter 1, Section 8 we decompose the Laplace operator with respect to n-dimensional polar coordinates $(\xi, r) \in S^{n-1} \times (0, +\infty)$ as follows:

$$\boldsymbol{\Delta} = \frac{\partial^2}{\partial r^2} + \frac{n-1}{r}\frac{\partial}{\partial r} + \frac{1}{r^2}\boldsymbol{\Lambda}. \tag{14}$$

Here the operator $\boldsymbol{\Lambda}$ is independent of the radius r . Therefore, the function φ is harmonic in $\mathbb{R}^n \setminus \{0\}$ if and only if the function f satisfies the following ordinary differential equation

$$\frac{\partial^2 f}{\partial r^2}(r) + \frac{n-1}{r}\frac{\partial f}{\partial r}(r) = 0, \qquad r \in (0, +\infty). \tag{15}$$

The linear solution space of this ordinary differential equation is 2-dimensional, and we easily verify: The general solution of (15) is given by

$$f(r) = a + b \log r, \quad r \in (0, +\infty), \quad a, b \in \mathbb{R}, \qquad \text{if} \quad n = 2,$$

$$f(r) = a + br^{2-n}, \quad r \in (0, +\infty), \quad a, b \in \mathbb{R}, \qquad \text{if} \quad n \geq 3.$$

We observe that the solutions $f \not\equiv \text{const}$ of (15) behave at the origin like

$$\lim_{r \to 0+} |f(r)| = +\infty.$$

Therefore, the radially symmetric solutions $\varphi(x) = f(|x|)$, $x \in \mathbb{R}^n \setminus \{0\}$ of the Laplacian differential equation possess a *singularity* at the point $x = 0$. This phenomenon enables us to derive an integral representation for the solutions of Poisson's differential equation. We meet with a comparable situation in Cauchy's integral.

Definition 1.3. *A domain* $G \subset \mathbb{R}^n$ *satisfying the assumptions of the Gaussian integral theorem from Chapter 1, Section 5 is named a* normal domain in $\mathbb{R}^n$.

Definition 1.4. *On the normal domain* $G \subset \mathbb{R}^n$ *we define the function*

$$\varphi(y; x) := \frac{1}{2\pi} \log |y - x| + \psi(y; x), \quad x, y \in G \quad \text{with} \quad x \neq y, \qquad n = 2, \tag{16}$$

and alternatively

$$\varphi(y;x) := \frac{1}{(2-n)\omega_n}|y-x|^{2-n}+\psi(y;x), \quad x,y \in G \quad \text{with} \quad x \neq y, \qquad n \geq 3. \tag{17}$$

Here the function $\psi(\cdot;x)$ - defined by $y \mapsto \psi(y;x)$ - is harmonic in G and belongs to the class $C^1(\overline{G})$ for each fixed $x \in G$. Furthermore, we observe the regularity property $\psi \in C^0(\overline{G} \times \overline{G})$. Then we name $\varphi(y;x)$ a fundamental solution of the Laplace equation in G.

Of central significance for the potential theory is the following

Theorem 1.5. *On the normal domain $G \subset \mathbb{R}^n$ with $n \geq 2$, we consider a solution $u = u(x) \in C^2(G) \cap C^1(\overline{G})$ of* Poisson's differential equation

$$\Delta u(x) = f(x), \qquad x \in G \tag{18}$$

prescribing the function $f = f(x) \in C^0(\overline{G})$ as its right-hand side. *Then we have the* integral representation

$$u(x) = \int\limits_{\partial G} \Big(u(y)\frac{\partial \varphi}{\partial \nu}(y;x) - \varphi(y;x)\frac{\partial u}{\partial \nu}(y)\Big)\, d\sigma(y) + \int\limits_{G} \varphi(y;x) f(y)\, dy \tag{19}$$

for all $x \in G$. Here the symbol $\nu : \partial G \to \mathbb{R}^n$ denotes the exterior unit normal for the domain ∂G, $d\sigma(y)$ means the surface element on the boundary ∂G, and $\varphi(y;x)$ indicates a fundamental solution.

Proof:

1. We present our proof only for the case $n \geq 3$. Take a fixed point $x \in G$ and choose $\varepsilon_0 > 0$ so small that the condition

$$B_\varepsilon(x) := \Big\{y \in \mathbb{R}^n : |y-x| < \varepsilon\Big\} \subset\subset G$$

is satisfied for all $0 < \varepsilon < \varepsilon_0$. We introduce the polar coordinates

$$y = x + r\xi, \qquad \xi \in \mathbb{R}^n \quad \text{with} \quad |\xi| = 1$$

about the point x, and denote the radial derivative by $\frac{\partial}{\partial r}$. On the domain $G_\varepsilon := G \setminus \overline{B_\varepsilon(x)}$ we apply Green's formula and obtain

$$\begin{aligned}
\int\limits_{G_\varepsilon} f(y)\varphi(y;x)\,dy
&= \int\limits_{G_\varepsilon} \Big(\Delta u(y)\varphi(y;x) - u(y)\Delta_y\varphi(y;x)\Big)\,dy \\
&= \int\limits_{\partial G_\varepsilon} \Big(\varphi(y;x)\frac{\partial u}{\partial \nu}(y) - u(y)\frac{\partial \varphi}{\partial \nu}(y;x)\Big)\,d\sigma(y) \\
&= \int\limits_{\partial G} \Big(\varphi(y;x)\frac{\partial u}{\partial \nu}(y) - u(y)\frac{\partial \varphi}{\partial \nu}(y;x)\Big)\,d\sigma(y) \\
&\quad - \int\limits_{\partial B_\varepsilon(x)} \Big(\varphi(y;x)\frac{\partial u}{\partial r}(y) - u(y)\frac{\partial \varphi}{\partial r}(y;x)\Big)\,d\sigma(y)
\end{aligned} \tag{20}$$

for all $\varepsilon \in (0, \varepsilon_0)$.

2. Observing (17), we now see

$$\lim_{\varepsilon\to 0+} \int\limits_{\partial B_\varepsilon(x)} \varphi(y;x)\frac{\partial u}{\partial r}(y)\,d\sigma(y) = 0. \tag{21}$$

Furthermore, we calculate

$$\begin{aligned}
\lim_{\varepsilon\to 0+} \int\limits_{\partial B_\varepsilon(x)} u(y)\frac{\partial \varphi}{\partial r}(y;x)\,d\sigma(y)
&= \lim_{\varepsilon\to 0+} \int\limits_{\partial B_\varepsilon(x)} u(y)\frac{1}{\omega_n}|y-x|^{1-n}\,d\sigma(y) \\
&\quad + \lim_{\varepsilon\to 0+} \int\limits_{\partial B_\varepsilon(x)} u(y)\frac{\partial}{\partial r}\psi(y;x)\,d\sigma(y) \\
&= \lim_{\varepsilon\to 0+} \int\limits_{\partial B_\varepsilon(x)} \Big(u(y) - u(x)\Big)\frac{1}{\omega_n}|y-x|^{1-n}\,d\sigma(y) \\
&\quad + u(x) \lim_{\varepsilon\to 0+} \int\limits_{\partial B_\varepsilon(x)} \frac{1}{\omega_n}\varepsilon^{1-n}\,d\sigma(y) \\
&= u(x).
\end{aligned} \tag{22}$$

3. From (20), (21), and (22) together with the passage to the limit $\varepsilon \to 0+$ we now infer the stated identity

$$\int\limits_G f(y)\varphi(y;x)\,dy + \int\limits_{\partial G} \Big(u(y)\frac{\partial \varphi}{\partial \nu}(y;x) - \varphi(y;x)\frac{\partial u}{\partial \nu}(y)\Big)\,d\sigma(y) = u(x)$$

for arbitrary points $x \in G$. q.e.d.

Theorem 1.6. *Given the point $\overset{\circ}{x} = (\overset{\circ}{x}_1, \ldots, \overset{\circ}{x}_n) \in \mathbb{R}^n$ and the radius $R \in (0, +\infty)$, we consider the ball $B_R(\overset{\circ}{x}) := \{x \in \mathbb{R}^n : |x - \overset{\circ}{x}| < R\}$. Let the function*

$$u = u(x_1, \ldots, x_n) \in C^2(B_R(\overset{\circ}{x})) \cap C^1(\overline{B_R(\overset{\circ}{x})})$$

solve the Laplace equation $\Delta u(x_1, \ldots, x_n) = 0$ in $B_R(\overset{\circ}{x})$. Then we have a power series

$$\mathcal{P}(x_1, \ldots, x_n) = \sum_{k_1, \ldots, k_n = 0}^{\infty} a_{k_1 \ldots k_n} x_1^{k_1} \cdot \ldots \cdot x_n^{k_n}$$

$$\text{for} \quad x_j \in \mathbb{C} \quad \text{with} \quad |x_j| \le \frac{R}{4n}, \quad j = 1, \ldots, n$$

with the real coefficients $a_{k_1 \ldots k_n} \in \mathbb{R}$ for $k_1, \ldots, k_n = 0, 1, 2, \ldots$, converging absolutely in the designated complex polycylinder such that

$$u(x) = \mathcal{P}(x_1 - \overset{\circ}{x}_1, \ldots, x_n - \overset{\circ}{x}_n) \qquad \text{for} \quad x \in \mathbb{R}^n \quad \text{with} \quad |x_j - \overset{\circ}{x}_j| \le \frac{R}{4n}. \tag{23}$$

Proof:

1. It suffices only to prove the statement above in the case $\overset{\circ}{x} = 0$ and $R = 1$, which can easily be verified with the aid of the transformation

$$Ty := \overset{\circ}{x} + Ry, y \in B_1(0) \quad \text{satisfying} \quad T : B_1(0) \to B_R(\overset{\circ}{x}).$$

Furthermore, we only consider the situation $n \ge 3$. With the function

$$\varphi(y;x) := \frac{1}{(2-n)\omega_n}|y-x|^{2-n}, \qquad y \in B := B_1(0)$$

we obtain a fundamental solution of the Laplace equation in B for each fixed $x \in B$. Theorem 1.5 yields the representation formula

$$u(x) = \int\limits_{\partial B} \Big(u(y)\frac{\partial \varphi}{\partial \nu}(y;x) - \varphi(y;x)\frac{\partial u}{\partial \nu}(y)\Big)\,d\sigma(y), \qquad x \in B. \tag{24}$$

The points $x \in B$ being fixed and $y \in \partial B$ arbitrary, we comprehend

$$\begin{aligned} \tfrac{\partial}{\partial \nu}\varphi(y;x) = y \cdot \nabla_y \varphi(y;x) &= \frac{1}{\omega_n} y \cdot \Big(|y-x|^{1-n}\nabla_y|y-x|\Big) \\ = \frac{1}{\omega_n} y \cdot \Big(|y-x|^{-n}(y-x)\Big) &= \frac{1}{\omega_n|y-x|^n}\, y \cdot (y-x). \end{aligned} \tag{25}$$

2. We take arbitrary $\lambda \in \mathbb{R}$, $y \in \partial B$ and $x = (x_1, \ldots, x_n) \in \mathbb{C}^n$ satisfying $|x_j| \le \frac{1}{4n}$ for $j = 1, \ldots, n$ and consider the composite quantity

$$|y-x|^\lambda := \Big(\sum_{j=1}^n (y_j - x_j)^2\Big)^{\frac{\lambda}{2}} = \Big(1 - 2\sum_{j=1}^n y_j x_j + \sum_{j=1}^n x_j^2\Big)^{\frac{\lambda}{2}}.$$

Abbreviating

$$\varrho := -2\sum_{j=1}^n y_j x_j + \sum_{j=1}^n x_j^2 \quad \in \mathbb{C}$$

we see

$$|y-x|^\lambda = (1+\varrho)^{\frac{\lambda}{2}} = \sum_{l=0}^\infty \binom{\frac{\lambda}{2}}{l} \varrho^l = \sum_{l=0}^\infty \binom{\frac{\lambda}{2}}{l} \Big(-2\sum_{j=1}^n y_j x_j + \sum_{j=1}^n x_j^2\Big)^l.$$

Here we observe

$$|\varrho| = \Big| -2\sum_{j=1}^n y_j x_j + \sum_{j=1}^n x_j^2 \Big| \le 2\sum_{j=1}^n |y_j|\,|x_j| + \sum_{j=1}^n |x_j|^2$$

$$\le 2\frac{1}{4n} n + \frac{1}{16n^2} n \le \frac{3}{4} < 1.$$

3. The function

$$\psi(x) := |y-x|^\lambda, \qquad x_k \in \mathbb{C} \quad \text{with} \quad |x_j| \le \frac{1}{4n}, \quad j = 1, \ldots, n$$

is consequently holomorphic for each fixed point $y \in \partial B$. On account of the relation (25), the function

$$F(x,y) := u(y)\frac{\partial \varphi}{\partial \nu}(y;x) - \varphi(y;x)\frac{\partial u}{\partial \nu}(y), \qquad |x_j| \le \frac{1}{4n}$$

is holomorphic on the given polycylinder for each fixed $y \in \partial B$ and bounded. Now Theorem 2.12 from Chapter 4, Section 2 about holomorphic parameter integrals, together with (24), now yields that the function $u(x)$ is holomorphic on the given polycylinder. Therefore, the function u can be expanded into the power series specified above. Since the function $u(x)$ is real-valued, the coefficients $a_{k_1 \ldots k_n}$ are real as well. They are namely the coefficients of the associate Taylor series.

q.e.d.

Of central interest is the following

Theorem 1.7. *Let us take the point* $\overset{\circ}{x} \in \mathbb{R}^n$, *the radius* $R \in (0, +\infty)$, *and the number* $\lambda \in \mathbb{R}$ *with* $\lambda < n$. *Furthermore, let the function* $f = f(y_1, \ldots, y_n)$ *be* *holomorphic in an open neighborhood $\mathcal{U} \subset \mathbb{C}^n$ satisfying $\mathcal{U} \supset\supset B_R(\overset{\circ}{x})$. Then the function*

$$F(x_1,\ldots,x_n) := \int\limits_{B_R(\overset{\circ}{x})} \frac{f(y)}{|y-x|^\lambda}\,dy, \qquad x \in B_R(\overset{\circ}{x}) \tag{26}$$

can be locally expanded into a convergent power series about the point $\overset{\circ}{x}$.

Proof: Applying the transformation $Ty := \overset{\circ}{x} + Ry$, $y \in B_1(0)$ we can concentrate our considerations on the case $\overset{\circ}{x} = 0$ and $R = 1$. We therefore investigate the singular integral

$$F(x_1,\ldots,x_n) := \int\limits_{|y|<1} \frac{f(y)}{|y-x|^\lambda}\,dy, \qquad x \in B := B_1(0).$$

The point $x \in B$ being fixed, we consider the transformation of variables due to E. E. Levi, namely

$$y = x + \varrho(\xi - x) = (1-\varrho)x + \varrho\xi, \qquad 0 < \varrho \le 1, \quad |\xi| = 1;$$

$$\xi_n = \xi_n(\xi_1,\ldots,\xi_{n-1}) = \pm\sqrt{1 - \sum_{i=1}^{n-1}\xi_i^2}.$$

The so-defined mapping $(\xi_1,\ldots,\xi_{n-1},\varrho) \mapsto y$ is bijective, and we have

$$\frac{\partial(y_1,\ldots,y_n)}{\partial(\xi_1,\ldots,\xi_{n-1},\varrho)} = \begin{vmatrix} \frac{\partial y_1}{\partial \xi_1} & \cdots & \frac{\partial y_n}{\partial \xi_1} \\ \vdots & & \vdots \\ \frac{\partial y_1}{\partial \xi_{n-1}} & \cdots & \frac{\partial y_n}{\partial \xi_{n-1}} \\ \frac{\partial y_1}{\partial \varrho} & \cdots & \frac{\partial y_n}{\partial \varrho} \end{vmatrix}$$

$$= \begin{vmatrix} \varrho & \cdots & 0 & -\varrho\frac{\xi_1}{\xi_n} \\ \vdots & \ddots & \vdots & \vdots \\ 0 & \cdots & \varrho & -\varrho\frac{\xi_{n-1}}{\xi_n} \\ \xi_1 - x_1 & \cdots & \xi_{n-1} - x_{n-1} & \xi_n - x_n \end{vmatrix}$$

$$= \varrho^{n-1} \begin{vmatrix} 1 & \cdots & 0 & -\frac{\xi_1}{\xi_n} \\ \vdots & \ddots & \vdots & \vdots \\ 0 & \cdots & 1 & -\frac{\xi_{n-1}}{\xi_n} \\ \xi_1 - x_1 & \cdots & \xi_{n-1} - x_{n-1} & \xi_n - x_n \end{vmatrix}$$

$$= \frac{\varrho^{n-1}}{\xi_n} \Big(\sum_{i=1}^{n-1} \xi_i(\xi_i - x_i) + \xi_n(\xi_n - x_n) \Big)$$

$$= \frac{\varrho^{n-1}}{\xi_n} \Big(1 - \sum_{i=1}^{n} \xi_i x_i \Big) \neq 0 \qquad \text{for} \quad |\xi| = 1, \quad |x| < 1.$$

The transformation formula for multiple integrals now yields

$$F(x) = \int\limits_{|y|<1} \frac{f(y)}{|y-x|^\lambda}\, dy$$

$$= \int\limits_0^1 \int\limits_{\substack{\xi_1^2+\ldots+\xi_{n-1}^2<1 \\ \xi_n(\xi_1,\ldots,\xi_{n-1})>0}} \frac{f(x+\varrho(\xi-x))}{\varrho^\lambda |\xi-x|^\lambda} \frac{\varrho^{n-1}}{|\xi_n|} \Big(1 - \sum_{k=1}^n \xi_k x_k\Big)\, d\xi_1 \ldots d\xi_{n-1}\, d\varrho$$

$$+ \int\limits_0^1 \int\limits_{\substack{\xi_1^2+\ldots+\xi_{n-1}^2<1 \\ \xi_n(\xi_1,\ldots,\xi_{n-1})<0}} \frac{f(x+\varrho(\xi-x))}{\varrho^\lambda |\xi-x|^\lambda} \frac{\varrho^{n-1}}{|\xi_n|} \Big(1 - \sum_{k=1}^n \xi_k x_k\Big)\, d\xi_1 \ldots d\xi_{n-1}\, d\varrho$$

$$= \int\limits_0^1 \varrho^{n-1-\lambda} \Bigg(\int\limits_{|\xi|=1} \frac{f(x+\varrho(\xi-x))}{|\xi-x|^\lambda} (1 - \xi\cdot x)\, d\omega_\xi \Bigg)\, d\varrho.$$

As in the proof of Theorem 1.6 we expand the function $|\xi - x|^\lambda$ into a convergent power series. With the aid of Theorem 2.12 from Chapter 4, Section 2 we infer that the function $F(x)$ can be expanded into a convergent power series in a neighborhood of the point $x = 0$.

q.e.d.

Definition 1.8. *A function $\varphi = \varphi(x_1, \ldots, x_n) : \Omega \to \mathbb{R}$ defined on the open set $\Omega \subset \mathbb{R}^n$ is named* real-analytic *in Ω if the following condition holds true: For each point $\overset{\circ}{x} = (\overset{\circ}{x}_1, \ldots, \overset{\circ}{x}_n) \in \Omega$ there exists a sufficiently small number $\varepsilon = \varepsilon(\overset{\circ}{x}) > 0$ and a convergent power series*

$$\mathcal{P}(z_1,\ldots,z_n) = \sum_{k_1,\ldots,k_n=0}^{\infty} a_{k_1\ldots k_n} z_1^{k_1}\cdot\ldots\cdot z_n^{k_n}$$

$$\text{for} \quad z_j\in\mathbb{C} \quad \text{with} \quad |z_j|\le\varepsilon, \quad j=1,\ldots,n$$

with the real coefficients

$$a_{k_1\ldots k_n}\in\mathbb{R} \quad \text{for} \quad k_1,\ldots,k_n=0,1,2,\ldots$$

such that the identity

$$\varphi(x_1,\ldots,x_n)=\mathcal{P}(x_1-\overset{\circ}{x}_1,\ldots,x_n-\overset{\circ}{x}_n), \qquad |x_j-\overset{\circ}{x}_j|\le\varepsilon, \quad j=1,\ldots,n$$

is satisfied.

Theorem 1.9. (Analyticity theorem for Poisson's equation)
The real-analytic function $f=f(x_1,\ldots,x_n):\Omega\to\mathbb{R}$ is defined on the open set $\Omega\subset\mathbb{R}^n$ with $n\ge 2$. Furthermore, let the function $u=u(x_1,\ldots,x_n)\in C^2(\Omega)$ represent a solution of Poisson's differential equation

$$\Delta u(x_1,\ldots,x_n)=f(x_1,\ldots,x_n), \qquad (x_1,\ldots,x_n)\in\Omega.$$

Then this function $u(x)$ is real-analytic in the set Ω.

Proof: Taking $\overset{\circ}{x}\in\Omega$ and $B_R(\overset{\circ}{x})\subset\subset\Omega$, Theorem 1.5 allows us to represent the solution $u(x)$ by the fundamental solution φ in the following form

$$u(x)=\int\limits_{\partial B_R(\overset{\circ}{x})}\Big(u(y)\frac{\partial\varphi}{\partial\nu}(y;x)-\varphi(y;x)\frac{\partial u}{\partial\nu}(y)\Big)\,d\sigma(y)+\int\limits_{B_R(\overset{\circ}{x})}\varphi(y;x)f(y)\,dy$$

with $x\in B_R(\overset{\circ}{x})$. According to Theorem 1.6, the first integral on the right-hand side represents a real-analytic function about the point $\overset{\circ}{x}$. From Theorem 1.7 we infer that the second integral yields a real-analytic function about the point $\overset{\circ}{x}$ as well.

q.e.d.

2 Poisson's Integral Formula with Applications

In Theorem 1.5 from Section 1 we have constructed an integral representation for the solutions of Poisson's equation in normal domains G with the aid of the fundamental solution $\varphi(y;x)$. The representation formula becomes particularly simple if the function $\varphi(.;x)$ vanishes on the boundary ∂G. This motivates the following

Definition 2.1. *On a normal domain $G \subset \mathbb{R}^n$ we have the fundamental solution $\varphi = \varphi(y;x)$ given. We call this function a* Green's function *of the domain G, if the boundary condition*

$$\varphi(y;x) = 0 \qquad \text{for all} \quad y \in \partial G \tag{1}$$

is satisfied for all $x \in G$.

Theorem 2.2. *Given the ball $B_R := \{y \in \mathbb{R}^n : |y| < R\}$ with $R \in (0, +\infty)$ and $n \geq 2$, we have the following Green's function:*

$$\varphi(y;x) = \frac{1}{2\pi} \log \left| \frac{R(y-x)}{R^2 - \overline{x}y} \right|, \qquad y \in \overline{B}_R, \quad x \in B_R, \tag{2}$$

in the case $n = 2$ and

$$\begin{aligned} \varphi(y;x) &= \frac{1}{(2-n)\omega_n} \left(\frac{1}{|y-x|^{n-2}} - \frac{\left(\frac{R}{|x|}\right)^{n-2}}{\left| y - \frac{R^2}{|x|^2} x \right|^{n-2}} \right) \\ &= \frac{1}{(2-n)\omega_n} \left(\frac{1}{|y-x|^{n-2}} - \frac{R^{n-2}}{(R^4 - 2R^2(x \cdot y) + |x|^2|y|^2)^{\frac{n-2}{2}}} \right) \end{aligned} \tag{3}$$

for $y \in \overline{B}_R$, $x \in B_R$ in the case $n \geq 3$.

Proof:

1. At first, we consider the case $n = 2$. Taking the point $x \in B_R$ as fixed, the expression

$$f(y) := \frac{R(y-x)}{R^2 - \overline{x}y} = \frac{Ry - Rx}{-\overline{x}y + R^2}, \qquad y \in \mathbb{C}$$

is a Möbius transformation with the nonsingular coefficient matrix

$$\begin{pmatrix} R & -Rx \\ -\overline{x} & R^2 \end{pmatrix}, \qquad \det \begin{pmatrix} R & -Rx \\ -\overline{x} & R^2 \end{pmatrix} = R(R^2 - |x|^2) > 0.$$

Furthermore, we have

$$|f(R)| = \left| \frac{R^2 - Rx}{-\overline{x}R + R^2} \right| = \left| \frac{R^2 - Rx}{R^2 - Rx} \right| = 1,$$

$$|f(-R)| = \left| \frac{-R^2 - Rx}{R\overline{x} + R^2} \right| = \left| \frac{R^2 + Rx}{R^2 + Rx} \right| = 1,$$

$$|f(iR)| = \left| \frac{iR^2 - Rx}{-iR\overline{x} + R^2} \right| = \left| \frac{iR^2 - Rx}{R^2 + iRx} \right| = \left| \frac{R^2 + iRx}{R^2 + iRx} \right| = 1,$$

$$f(0) = -\frac{x}{R} \quad \in B_1.$$

This implies

$$|f(y)| = 1 \qquad \text{for all} \quad y \in \partial B_R$$

and then

$$\varphi(y;x) = \frac{1}{2\pi} \log \left| \frac{R(y-x)}{R^2 - \overline{x}y} \right| = 0$$

for all $y \in \partial B_R$ and all $x \in B_R$. Finally, we note that

$$\begin{aligned}
\varphi(y;x) &= \frac{1}{2\pi} \log \left| \frac{y-x}{R - \frac{\overline{x}}{R}y} \right| = \frac{1}{2\pi} \log |y-x| - \frac{1}{2\pi} \log \left| R - \frac{\overline{x}}{R}y \right| \\
&= \frac{1}{2\pi} \log |y-x| - \frac{1}{2\pi} \log \left| -\frac{\overline{x}}{R}\left(y - \frac{R^2}{\overline{x}}\right) \right| \\
&= \frac{1}{2\pi} \log |y-x| - \frac{1}{2\pi} \log \left| y - \frac{R^2}{|x|^2}x \right| - \frac{1}{2\pi} \log \left| \frac{\overline{x}}{R} \right| \\
&=: \frac{1}{2\pi} \log |y-x| + \psi(y;x), \qquad y \in B_R, \quad x \in B_R \setminus \{0\}.
\end{aligned}$$

The function $\psi(\cdot;x)$ is harmonic in $\overline{B}_R$ as the real part of a holomorphic function.

2. We now consider the case $n \geq 3$, and begin with the following ansatz:

$$\varphi(y;x) = \frac{1}{(2-n)\omega_n} \left(\frac{1}{|y-x|^{n-2}} - \frac{K}{|y-\lambda x|^{n-2}} \right), \qquad y \in \overline{B}_R.$$

Here the point $x \in B_R$ is fixed; the constants K and λ have still to be chosen adequately. At first, we see that the function

$$\psi(y;x) := -\frac{1}{(2-n)\omega_n} \frac{K}{|y-\lambda x|^{n-2}}$$

is harmonic in $y \in \overline{B}_R$ if $\lambda x \notin \overline{B}_R$ holds true. The condition $\varphi(y;x) = 0$ for all $y \in \partial B_R$ is satisfied if and only if

$$\frac{1}{|y-x|^{n-2}} = \frac{K}{|y-\lambda x|^{n-2}}$$

or equivalently

$$K^{\frac{2}{n-2}} |y-x|^2 = |y-\lambda x|^2 \qquad \text{for all} \quad y \in \partial B_R$$

is correct. On account of $|y| = R$ we can transform this identity into

$$K^{\frac{2}{n-2}} (R^2 - 2(y \cdot x) + |x|^2) = R^2 - 2\lambda(y \cdot x) + \lambda^2 |x|^2$$

and finally into

$$R^2\left(K^{\frac{2}{n-2}}-1\right)-2(x\cdot y)\left(K^{\frac{2}{n-2}}-\lambda\right)+|x|^2\left(K^{\frac{2}{n-2}}-\lambda^2\right)=0.$$

Setting $\lambda := K^{\frac{2}{n-2}}$ we obtain

$$0=R^2(\lambda-1)+|x|^2(\lambda-\lambda^2)=(\lambda-1)\{R^2-\lambda|x|^2\}.$$

Since the case $\lambda=1$, $K=1$ and consequently $\varphi\equiv 0$ has to be excluded as the trivial one, we choose $\lambda := \left(\frac{R}{|x|}\right)^2$ and $K=\lambda^{\frac{n-2}{2}}=\left(\frac{R}{|x|}\right)^{n-2}$. Now we obtain Green's function of the domain B_R with the following expression

$$\varphi(y;x)=\frac{1}{(2-n)\omega_n}\left(\frac{1}{|y-x|^{n-2}}-\frac{\left(\frac{R}{|x|}\right)^{n-2}}{\left|y-\left(\frac{R}{|x|}\right)^2 x\right|^{n-2}}\right),\qquad y\in\overline{B}_R,$$

for $x\in B_R\setminus\{0\}$. We note

$$\frac{\frac{R}{|x|}}{\left|y-\frac{R^2}{|x|^2}x\right|}=\frac{R}{\left||x|y-R^2\frac{x}{|x|}\right|}=\left(\frac{R^2}{|x|^2|y|^2-2R^2(x\cdot y)+R^4}\right)^{\frac{1}{2}},$$

and Green's function satisfies

$$\varphi(y;x)=\frac{1}{(2-n)\omega_n}\left(\frac{1}{|y-x|^{n-2}}-\frac{R^{n-2}}{(|x|^2|y|^2-2R^2(x\cdot y)+R^4)^{\frac{n-2}{2}}}\right)$$

for all $y\in\overline{B}_R$ and $x\in B_R$. q.e.d.

Theorem 2.3. (Poisson's integral formula)
In the ball $B_R := \{y\in\mathbb{R}^n : |y|<R\}$ of radius $R\in(0,+\infty)$ in the Euclidean space $\mathbb{R}^n$ with $n\geq 2$, let the function $u=u(x)=u(x_1,\dots,x_n)\in C^2(B_R)\cap C^0(\overline{B}_R)$ solve Poisson's differential equation

$$\Delta u(x)=f(x),\qquad x\in B_R$$

for the right-hand side $f=f(x)\in C^0(\overline{B_R})$. Then we have the Poisson integral representation

$$u(x)=\frac{1}{R\omega_n}\int\limits_{|y|=R}\frac{|y|^2-|x|^2}{|y-x|^n}u(y)\,d\sigma(y)+\int\limits_{|y|\leq R}\varphi(y;x)f(y)\,dy \qquad (4)$$

for all $x\in B_R$. Here the symbol $\varphi=\varphi(y;x)$ denotes Green's function given in Theorem 2.2.

Proof:

1. At first, we assume the regularity $u \in C^2(\overline{B}_R)$. Theorem 1.5 from Section 1 yields the identity

$$u(x) = \int\limits_{|y|=R} u(y)\frac{\partial \varphi}{\partial \nu}(y;x)\,d\sigma(y) + \int\limits_{|y|\leq R} \varphi(y;x)f(y)\,dy, \qquad x \in B_R.$$

We confine ourselves to the case $n \geq 3$. According to Theorem 2.2 we have Green's function

$$\varphi(y;x) = \frac{1}{(2-n)\omega_n}\Big(|y-x|^{2-n} - K|y-\lambda x|^{2-n}\Big), \qquad y \in \overline{B}_R,\ x \in B_R,$$

$$\text{with} \quad \lambda := \left(\frac{R}{|x|}\right)^2 \quad \text{and} \quad K = \left(\frac{R}{|x|}\right)^{n-2} = \lambda^{\frac{n-2}{2}}.$$

Taking $x \in B_R$ as fixed and $y \in \partial B_R$ arbitrarily, we calculate

$$\begin{aligned}
\frac{\partial}{\partial \nu}\varphi(y;x) &= \frac{y}{R}\cdot \nabla_y \varphi(y;x) \\
&= \frac{1}{R\omega_n} y \cdot \left(|y-x|^{1-n}\frac{y-x}{|y-x|} - K|y-\lambda x|^{1-n}\frac{y-\lambda x}{|y-\lambda x|}\right) \\
&= \frac{1}{R\omega_n} y \cdot \left(\frac{y-x}{|y-x|^n} - K\frac{y-\lambda x}{|y-\lambda x|^n}\right).
\end{aligned}$$

This formula remains true for $n = 2$ as well, where $K = 1$ is fulfilled in this case. We additionally note that

$$\begin{aligned}
|y-\lambda x|^2 &= R^2 - 2\lambda(x\cdot y) + \lambda^2|x|^2 \\
&= R^2 - 2\frac{R^2}{|x|^2}(x\cdot y) + \frac{R^4}{|x|^2} \\
&= \frac{R^2}{|x|^2}\Big(|x|^2 - 2(x\cdot y) + R^2\Big) = \lambda|y-x|^2
\end{aligned}$$

and consequently

$$|y-\lambda x|^n = \lambda^{\frac{n}{2}}|y-x|^n.$$

Finally, we obtain

$$\begin{aligned}
\frac{\partial}{\partial \nu}\varphi(y;x) &= \frac{1}{R\omega_n|y-x|^n} y \cdot \Big(y - x - K\lambda^{-\frac{n}{2}}(y-\lambda x)\Big) \\
&= \frac{1}{R\omega_n|y-x|^n} y \cdot \Big((1-\lambda^{-\frac{n}{2}}K)y - (1-K\lambda^{\frac{-n+2}{2}})x\Big) \\
&= \frac{|y|^2}{R\omega_n|y-x|^n}\Big(1-\frac{1}{\lambda}\Big) \;=\; \frac{|y|^2}{R\omega_n|y-x|^n}\Big(1-\frac{|x|^2}{R^2}\Big) \\
&= \frac{|y|^2-|x|^2}{R\omega_n|y-x|^n} \qquad \text{for all} \quad y \in \partial B_R \quad \text{and} \quad x \in B_R.
\end{aligned}$$

Therefore, we get the Poisson integral representation

$$u(x) = \frac{1}{R\omega_n} \int\limits_{|y|=R} \frac{|y|^2 - |x|^2}{|y-x|^n} u(y)\,d\sigma(y) + \int\limits_{|y|\le R} \varphi(y;x) f(y)\,dy, \qquad x \in B_R.$$

2. Now assuming $u \in C^2(B_R) \cap C^0(\overline{B}_R)$, part 1 of our proof yields the following identity for all $\varrho \in (0, R)$:

$$u(x) = \frac{1}{\varrho\omega_n} \int\limits_{|y|=\varrho} \frac{|y|^2 - |x|^2}{|y-x|^n} u(y)\,d\sigma(y) + \int\limits_{|y|\le \varrho} \varphi(y;x,\varrho) f(y)\,dy.$$

Here $\varphi(y;x,\varrho)$ denotes Green's function for the ball B_ϱ. We observe the transition to the limit $\varrho \to R-$ and obtain

$$u(x) = \frac{1}{R\omega_n} \int\limits_{|y|=R} \frac{|y|^2 - |x|^2}{|y-x|^n} u(y)\,d\sigma(y) + \int\limits_{|y|\le R} \varphi(y;x,R) f(y)\,dy$$

for all $x \in B_R$. q.e.d.

Remarks:

1. In the special case $n = 2$ and $f = 0$ we obtain for $0 \le \varrho < R$ and $0 \le \vartheta < 2\pi$:

$$u(\varrho\cos\vartheta, \varrho\sin\vartheta) = \frac{1}{2\pi} \int\limits_0^{2\pi} \frac{R^2 - \varrho^2}{R^2 - 2\varrho R\cos(\lambda - \vartheta) + \varrho^2} u(R\cos\lambda, R\sin\lambda)\,d\lambda.$$

2. We name

$$P(x,y,R) := \frac{1}{R\omega_n} \frac{|y|^2 - |x|^2}{|y-x|^n}, \qquad y \in \overline{B}_R, \quad x \in B_R$$

the *Poisson kernel.*

3. Later in Chapter 9 we shall investigate the boundary behavior of Poisson's integral.

Theorem 2.4. *We consider a solution $u = u(x) \in C^2(G)$ of Poisson's differential equation $\Delta u(x) = f(x)$, $x \in G$ in the domain $G \subset \mathbb{R}^n$. For each ball $B_R(a) \subset\subset G$ we then have the identity*

$$u(a) = \frac{1}{2\pi R} \int\limits_{|x-a|=R} u(x)\,d\sigma(x) - \frac{1}{2\pi} \iint\limits_{|x-a|\le R} \log\Big(\frac{R}{|x-a|}\Big) f(x)\,dx \qquad (5)$$

in the case $n = 2$, and alternatively

$$u(a) = \frac{1}{R^{n-1}\omega_n} \int\limits_{|x-a|=R} u(x)\,d\sigma(x) - \frac{1}{(n-2)\omega_n} \int\limits_{|x-a|\le R} \Big(|x-a|^{2-n} - R^{2-n}\Big) f(x)\,dx \tag{6}$$

in the case $n \ge 3$.

Proof: Via an adequate translation we can achieve $a = 0$. We then consider Green's function

$$\varphi(y;0) = \frac{1}{2\pi}\log\left|\frac{y}{R}\right| = -\frac{1}{2\pi}\log\frac{R}{|y|}, \qquad y \in \overline{B}_R, \qquad n = 2,$$

and alternatively

$$\varphi(y;0) = -\frac{1}{(n-2)\omega_n}\left(\frac{1}{|y|^{n-2}} - \frac{1}{R^{n-2}}\right), \qquad y \in \overline{B}_R, \qquad n \ge 3.$$

Poisson's integral formula now yields

$$u(0) = \frac{1}{2\pi R} \int\limits_{|y|=R} u(y)\,d\sigma(y) - \frac{1}{2\pi} \iint\limits_{|y|\le R} \log\left(\frac{R}{|y|}\right) f(y)\,dy$$

in the case $n = 2$ and

$$u(0) = \frac{1}{R^{n-1}\omega_n} \int\limits_{|y|=R} u(y)\,d\sigma(y) - \frac{1}{(n-2)\omega_n} \int\limits_{|y|\le R} \left(\frac{1}{|y|^{n-2}} - \frac{1}{R^{n-2}}\right) f(y)\,dy$$

in the case $n \ge 3$. q.e.d.

Corollary: Harmonic functions u have the *mean value property*

$$u(a) = \frac{1}{R^{n-1}\omega_n} \int\limits_{|y-a|=R} u(y)\,d\sigma(y), \tag{7}$$

if $B_R(a) \subset\subset G$ is satisfied.

Theorem 2.5. (Harnack's inequality)
Let the function $u(x) \in C^2(B_R)$ *be harmonic in the ball* $B_R = \{y \in \mathbb{R}^n : |y| < R\}$ *of radius* $R \in (0, +\infty)$, *and we assume* $u(x) \ge 0$ *for all* $x \in B_R$. *Then we have the estimate*

$$\frac{1 - \frac{|x|}{R}}{\left(1 + \frac{|x|}{R}\right)^{n-1}}\, u(0) \le u(x) \le \frac{1 + \frac{|x|}{R}}{\left(1 - \frac{|x|}{R}\right)^{n-1}}\, u(0) \qquad \textit{for all} \quad x \in B_R. \tag{8}$$

Proof: At first we assume $u \in C^2(\overline{B}_R)$, and later we establish the inequality above for functions $u \in C^2(B_R)$ by a passage to the limit. From Theorem 2.3 we infer

$$u(x) = \int\limits_{|y|=R} P(x,y,R)u(y)\,d\sigma(y), \qquad x \in B_R.$$

For arbitrary points $y \in \mathbb{R}^n$ with $|y| = R$ and $x \in B_R$ we have the following inequality:

$$\frac{|y|^2 - |x|^2}{(R+|x|)^n} \le \frac{|y|^2 - |x|^2}{|y-x|^n} \le \frac{|y|^2 - |x|^2}{(R-|x|)^n}.$$

We multiply this inequality by $\frac{1}{R\omega_n}u(y)$ and then integrate over the boundary ∂B_R:

$$\frac{1}{R\omega_n}\frac{R^2-|x|^2}{(R+|x|)^n}\int\limits_{|y|=R} u(y)\,d\sigma(y) \le u(x) \le \frac{1}{R\omega_n}\frac{R^2-|x|^2}{(R-|x|)^n}\int\limits_{|y|=R} u(y)\,d\sigma(y).$$

Using the mean value property of harmonic functions we obtain

$$R^{n-2}\frac{R^2-|x|^2}{(R+|x|)^n}u(0) \le u(x) \le R^{n-2}\frac{R^2-|x|^2}{(R-|x|)^n}u(0)$$

and consequently

$$\frac{1-\frac{|x|^2}{R^2}}{\left(1+\frac{|x|}{R}\right)^n}u(0) \le u(x) \le \frac{1-\frac{|x|^2}{R^2}}{\left(1-\frac{|x|}{R}\right)^n}u(0), \qquad x \in B_R.$$

Finally, this implies

$$\frac{1-\frac{|x|}{R}}{\left(1+\frac{|x|}{R}\right)^{n-1}}u(0) \le u(x) \le \frac{1+\frac{|x|}{R}}{\left(1-\frac{|x|}{R}\right)^{n-1}}u(0), \qquad x \in B_R.$$

q.e.d.

Theorem 2.6. (Liouville's theorem for harmonic functions) *Let $u(x) : \mathbb{R}^n \to \mathbb{R}$ denote a harmonic function satisfying $u(x) \le M$ for all $x \in \mathbb{R}^n$, with a constant $M \in \mathbb{R}$. Then we have $u(x) \equiv const$, $x \in \mathbb{R}^n$.*

Proof: We consider the harmonic function $v(x) := M - u(x)$, $x \in \mathbb{R}^n$ and note that $v(x) \ge 0$ for all $x \in \mathbb{R}^n$. Harnack's inequality now yields

$$\frac{1-\frac{|x|}{R}}{\left(1+\frac{|x|}{R}\right)^{n-1}}v(0) \le v(x) \le \frac{1+\frac{|x|}{R}}{\left(1-\frac{|x|}{R}\right)^{n-1}}v(0), \qquad x \in B_R, \quad R > 0.$$

We observe $R \to +\infty$ and obtain $v(x) = v(0)$ for all $x \in \mathbb{R}^n$ and finally $u(x) \equiv \text{const}$, $x \in \mathbb{R}^n$.

q.e.d.

Fundamentally important in the sequel is

Definition 2.7. *Let $G \subset \mathbb{R}^n$ denote a domain and $u = u(x) = u(x_1, \dots, x_n) : G \to \mathbb{R} \in C^0(G)$ a continuous function. We name u* weakharmonic (superharmonic, subharmonic), *if*

$$u(a) \;=\; (\,\geq,\,\leq\,)\; \frac{1}{r^{n-1}\omega_n} \int\limits_{|x-a|=r} u(x)\, d\sigma(x) \;=\; \frac{1}{\omega_n} \int\limits_{|\xi|=1} u(a + r\xi)\, d\sigma(\xi)$$

for all $a \in G$ and $r \in (0, \vartheta(a))$ with a certain $\vartheta(a) \in (0, dist(a, \mathbb{R}^n \setminus G)]$ is correct.

Remarks:

1. The function $u : G \to \mathbb{R} \in C^0(G)$ is superharmonic if and only if the function $-u$ is subharmonic.
2. A function is weakharmonic if and only if this function is simultaneously superharmonic and subharmonic.
3. A weakharmonic function is characterized by the mean value property - and should be carefully distinguished from certain weak solutions of the Laplace equation in Sobolev spaces, which are not necessarily continuous functions in general.
4. If the functions $u, v : G \to \mathbb{R}$ are superharmonic and the constant $\alpha \in [0, +\infty)$ is given, then the following continuous functions

$$w_1(x) := \alpha u(x),$$

$$w_2(x) := u(x) + v(x),$$

$$w_3(x) := \min\{u(x), v(x)\}, \qquad x \in G,$$

are superharmonic as well. For w_1 and w_2 this statement is evident, and we investigate the function w_3. Taking the point $a \in G$ and the radius $r \in (0, \vartheta(a))$ we infer

$$\frac{1}{\omega_n} \int\limits_{|\xi|=1} w_3(a + r\xi)\, d\sigma(\xi) = \frac{1}{\omega_n} \int\limits_{|\xi|=1} \min\{u(a + r\xi), v(a + r\xi)\}\, d\sigma(\xi)$$

$$\leq \min\left\{ \frac{1}{\omega_n} \int\limits_{|\xi|=1} u(a + r\xi)\, d\sigma(\xi), \frac{1}{\omega_n} \int\limits_{|\xi|=1} v(a + r\xi)\, d\sigma(\xi) \right\}$$

$$\leq \min\{u(a), v(a)\} \;=\; w_3(a).$$

5. If the functions $u, v : G \to \mathbb{R}$ are subharmonic and the constant $\alpha \in [0, +\infty)$ is given, then the following functions

$$w_1(x) := \alpha u(x),$$

$$w_2(x) := u(x) + v(x),$$

$$w_3(x) := \max\{u(x), v(x)\}, \qquad x \in G,$$

are subharmonic functions in G as well.

Theorem 2.8. *Let the function $u = u(x) \in C^2(G)$ be defined on the domain $G \subset \mathbb{R}^n$. Then this twice continuously differentiable function u is weakharmonic (superharmonic, subharmonic) in G if and only if the relation*

$$\Delta u(x) = 0 \ (\leq 0, \ \geq 0) \qquad \text{for all} \quad x \in G$$

is correct.

Proof: We present our proof only in the case $n \geq 3$. We define $f(x) := \Delta u(x)$, $x \in G$ and see $f \in C^0(G)$. Theorem 2.4 yields the following identity for all points $a \in G$ and radii $r \in (0, \vartheta(a))$:

$$u(a) = \frac{1}{r^{n-1}\omega_n} \int\limits_{|x-a|=r} u(x)\, d\sigma(x) - \frac{1}{(n-2)\omega_n} \int\limits_{|x-a|\leq r} (|x-a|^{2-n} - r^{2-n}) f(x)\, dx.$$

Setting

$$\chi(a,r) := -\frac{1}{(n-2)\omega_n} \int\limits_{|x-a|\leq r} (|x-a|^{2-n} - r^{2-n}) f(x)\, dx$$

we easily see: The function u is weakharmonic (superharmonic, subharmonic) if and only if

$$\chi(a,r) = 0 \ (\geq 0, \ \leq 0) \qquad \text{for all} \quad a \in G, \quad r \in (0, \vartheta(a))$$

holds true. We finally note the inequality $|x-a|^{2-n} - r^{2-n} \geq 0$ for all $x \in G$ with $|x-a| \leq r$, and we obtain the statement above.

q.e.d.

Theorem 2.9. (Maximum and minimum principle)
The superharmonic (subharmonic) function $u = u(x) : G \to \mathbb{R}$ - defined on the domain $G \subset \mathbb{R}^n$ - may attain its global minimum (maximum) at a point $\overset{\circ}{x} \in G$; this means

$$u(x) \geq u(\overset{\circ}{x}) \quad \Big(u(x) \leq u(\overset{\circ}{x})\Big) \qquad \text{for all} \quad x \in G.$$

Then we have

$$u(x) \equiv const \qquad \text{in} \quad G.$$

Proof: Since the reflection $u \to -u$ transfers subharmonic functions into superharmonic ones, the statement has only to be shown for superharmonic functions. Now the superharmonic function $u : G \to \mathbb{R} \in C^0(G)$ may attain its global minimum at the point $\overset{\circ}{x} \in G$. We then consider the nonvoid set

$$G^* := \Big\{ x \in G : \ u(x) = \inf_{y \in G} u(y) = u(\overset{\circ}{x}) \Big\}$$

which is closed in the domain G. We now show that this set G^* is open as well. If namely $a \in G^*$ is an arbitrary point, we observe

$$\inf_{y \in G} u(y) = u(a) \geq \frac{1}{\omega_n} \int\limits_{|\xi|=1} u(a + r\xi)\, d\sigma(\xi) \qquad \text{for all} \quad r \in (0, \vartheta(a)). \tag{9}$$

This implies $u(x) = u(a)$ for all points $x \in \mathbb{R}^n$ with $|x - a| < \vartheta(a)$. Consequently, the set G^* is open. Since G is a domain and especially connected, we easily see by continuation along paths: $u(x) \equiv u(\overset{\circ}{x})$ for all $x \in G$. We finally obtain $u(x) \equiv const$, $x \in G$.

q.e.d.

Theorem 2.10. *Let the function $u : G \to \mathbb{R} \in C^0(G)$ be superharmonic (subharmonic) in the bounded domain $G \subset \mathbb{R}^n$. Furthermore, all sequences of points $\{x^{(k)}\}_{k=1,2,\ldots} \subset G$ satisfying $\lim_{k \to \infty} x^{(k)} = x \in \partial G$ have the property*

$$\liminf_{k \to \infty} u(x^{(k)}) \geq M \qquad \Big(\limsup_{k \to \infty} u(x^{(k)}) \leq M \Big)$$

with a constant $M \in \mathbb{R}$. Then we have the behavior

$$u(x) \geq M \quad \Big(u(x) \leq M \Big) \qquad \text{for all} \quad x \in G.$$

Proof: It suffices to consider superharmonic functions $u : G \to \mathbb{R}$. If the statement $u(x) \geq M$ for all $x \in G$ were false, we have a point $\xi \in G$ with $\mu := u(\xi) < M$. We now construct a sequence of connected compact subsets of G exhausting the set G; this means $\Theta_j \uparrow G$ for $j \to \infty$ satisfying

$$\xi \in \Theta_1 \subset \Theta_2 \subset \ldots .$$

Due to Theorem 2.9, the superharmonic function u attains its minimum at a boundary point $y^{(j)} \in \partial \Theta_j$ of each compact set Θ_j. Therefore, we have the inequalities

$$u(y^{(j)}) \leq u(\xi) = \mu \qquad \text{for} \quad j = 1, 2, \ldots$$

From the sequence $\{y^{(j)}\}_{j=1,2,\ldots} \subset \overline{G}$ we now select a convergent subsequence $\{x^{(k)}\}_{k=1,2,\ldots} \subset \{y^{(j)}\}_{j=1,2,\ldots}$. We then obtain a sequence $\{x^{(k)}\}_{k=1,2,\ldots} \subset G$ satisfying

$$\lim_{k\to\infty} x^{(k)} = x \in \partial G \qquad \text{and} \qquad \liminf_{k\to\infty} u(x^{(k)}) \le \mu < M.$$

However, this contradicts the assumption

$$\liminf_{k\to\infty} u(x^{(k)}) \ge M \qquad \text{for all} \quad \{x^{(k)}\}_{k=1,2,\ldots} \subset G \quad \text{with} \quad \lim_{k\to\infty} x^{(k)} \in \partial G.$$

q.e.d.

Theorem 2.11. *Let $G \subset \mathbb{R}^n$ denote a bounded domain. Furthermore, we consider two functions $u = u(x)$, $v = v(x) : \overline{G} \to \mathbb{R} \in C^0(\overline{G})$, which are weakharmonic in G. Then we have the estimate*

$$\sup_{x\in\overline{G}} |u(x) - v(x)| \le \sup_{x\in\partial G} |u(x) - v(x)|.$$

Proof: The function $w(x) := u(x) - v(x)$, $x \in \overline{G}$ is continuous in $\overline{G}$ and weakharmonic in G. Setting $M := \sup\limits_{x\in\partial G} |u(x)-v(x)|$, Theorem 2.10 yields the inequality

$$-M \le w(x) \le M \qquad \text{for all} \quad x \in G.$$

This implies the stated estimate. q.e.d.

Theorem 2.12. *Let $G \subset \mathbb{R}^n$ denote a bounded domain. Then the Green function $\varphi_G(y;x)$ for this domain is uniquely determined, and we have*

$$\varphi_G(y;x) < 0 \qquad \textit{for all} \quad y \in G \quad \textit{and fixed} \quad x \in G. \tag{10}$$

Proof: (Only for $n \ge 3$.)

1. Let the two Green functions

$$\varphi_j(y;x) = \frac{1}{(2-n)\omega_n}|y-x|^{2-n} + \psi_j(y;x), \qquad y \in \overline{G}, \quad x \in G; \quad j = 1,2$$

   be given. Then we infer $0 = \varphi_1(y;x) = \varphi_2(y;x)$ for $y \in \partial G$, $x \in G$ and therefore

$$\psi_1(y;x) = \psi_2(y;x), \qquad y \in \partial G, \quad x \in G.$$

   Theorem 2.11 now implies $\psi_1(y;x) \equiv \psi_2(y;x)$, and finally

$$\varphi_1 \equiv \varphi_2, \qquad y \in G, \quad x \in G.$$

2. We take the point $x \in G$ as fixed and consider Green's function

$$\varphi_G(y;x) = \frac{1}{(2-n)\omega_n}|y-x|^{2-n} + \psi(y;x), \qquad y \in \overline{G}$$

   for the domain G. Then the function $\chi(y) := \varphi(y;x) : G \setminus \{x\} \to \mathbb{R}$ is harmonic. Arbitrary sequences $\{y^{(k)}\}_{k=1,2,\ldots} \subset G' := G \setminus \{x\}$ with $\lim\limits_{k\to\infty} y^{(k)} \in \partial G' = \partial G \cup \{x\}$ now satisfy

$$\limsup_{k\to\infty} \chi(y^{(k)}) \le 0.$$

Therefore, Theorem 2.10 yields $\chi(y) \le 0$ for all $y \in G'$ and Theorem 2.9 implies the inequality (10).

q.e.d.

Remark: The existence question for Green's function on *Dirichlet domains G* will be answered affirmatively in the next section.

3 Dirichlet's Problem for the Laplace Equation in $\mathbb{R}^n$

In this paragraph the symbol $G \subset \mathbb{R}^n$ always means a bounded domain, and $f = f(x) : \partial G \to \mathbb{R} \in C^0(\partial G)$ denotes a continuous function on its boundary ∂G. Our interest is devoted to the following *Dirichlet's boundary value problem for the Laplace equation*

$$\begin{aligned} &u = u(x) \in C^2(G) \cap C^0(\overline{G}), \\ &\Delta u(x) = 0 \qquad \text{for all} \quad x \in G, \\ &u(x) = f(x) \qquad \text{for all} \quad x \in \partial G. \end{aligned} \tag{1}$$

Theorem 3.1. (Uniqueness theorem)
Consider two solutions $u(x)$, $v(x)$ of the Dirichlet problem (1) for the data G and f. Then we have

$$u(x) \equiv v(x) \qquad \text{in} \quad \overline{G}.$$

Proof: The function $w(x) := v(x) - u(x)$, $x \in \overline{G}$ belonging to the class $C^2(G) \cap C^0(\overline{G})$ is especially weakharmonic in G and has the boundary values

$$\begin{aligned} w(x) &= v(x) - u(x) \\ &= f(x) - f(x) = 0 \qquad \text{for all} \quad x \in \partial G. \end{aligned}$$

Theorem 2.11 from Section 2 implies $w(x) \equiv 0$ in $\overline{G}$ and therefore

$$v(x) \equiv u(x), \qquad x \in \overline{G}.$$

q.e.d.

With the aid of Poisson's integral formula we can explicitly solve the Dirichlet problem on balls.

Theorem 3.2. *On the ball $B_R(a) := \{y \in \mathbb{R}^n : |y - a| < R\}$ with the center $a \in \mathbb{R}^n$ and the radius $R \in (0, +\infty)$ we consider Poisson's integral*

$$u(x) := \frac{1}{R\omega_n} \int\limits_{|y-a|=R} \frac{|y-a|^2 - |x-a|^2}{|y-x|^n} f(y)\, d\sigma(y), \qquad x \in B_R(a). \tag{2}$$

Then the function u belongs to the regularity class $C^2(B_R(a)) \cap C^0(\overline{B_R(a)})$ and is harmonic in $B_R(a)$. Furthermore, we have the boundary behavior

$$\lim_{\substack{x \to \mathring{x} \\ x \in B_R(a)}} u(x) = f(\mathring{x}) \qquad \textit{for all} \quad \mathring{x} \in \partial B_R(a). \tag{3}$$

Consequently, the given function u solves Dirichlet's problem (1) on the ball $G = B_R(a)$ for the continuous boundary function $f : \partial B_R(a) \to \mathbb{R}$ being prescribed.

Proof:

1. At first, we consider the situation $a = 0$, $R = 1$ and set $B := B_1(0) \subset \mathbb{R}^n$. Then we obtain the function

$$u(x) = \frac{1}{\omega_n} \int\limits_{|y|=1} \frac{|y|^2 - |x|^2}{|y-x|^n} f(y)\, d\sigma(y) = \int\limits_{|y|=1} P(y;x) f(y)\, d\sigma(y), \quad x \in B \tag{4}$$

with Poisson's kernel

$$P(y;x) := \frac{1}{\omega_n} \frac{|y|^2 - |x|^2}{|y-x|^n}, \qquad y \in \partial B, \quad x \in B.$$

2. Formula (4) immediately implies the regularity $u \in C^2(B)$. According to part 1 in the proof of Theorem 2.3 from Section 2 the following identity is satisfied:

$$\begin{aligned} P(y;x) &= \frac{1}{\omega_n} \frac{|y|^2 - |x|^2}{|y-x|^n} = \frac{\partial}{\partial \nu} \varphi(y;x) \\ &= y \cdot \nabla_y \varphi(y;x), \qquad y \in \partial B, \quad x \in B. \end{aligned} \tag{5}$$

Here the symbol $\varphi(y;x)$ denotes Green's function for the unit ball B described in Section 2, Theorem 2.2. We note that φ is symmetric, more precisely

$$\varphi(x;y) = \varphi(y;x) \qquad \text{for all} \quad x, y \in B \quad \text{with} \quad x \neq y. \tag{6}$$

Furthermore, we have

$$\Delta_x P(y;x) = y \cdot \nabla_y \Big(\Delta_x \varphi(y;x) \Big) = 0, \qquad x \in B, \quad y \in \partial B. \tag{7}$$

Consequently, we obtain

$$\Delta u(x) = \int\limits_{|y|=1} \Delta_x P(y;x) f(y)\, d\sigma(y) = 0 \qquad \text{for all} \quad x \in B. \tag{8}$$

3. Applying Theorem 2.3 from Section 2 to the harmonic function $v(x) \equiv 1$, $x \in \overline{B}$ we deduce
$$1 = \frac{1}{\omega_n} \int\limits_{|y|=1} \frac{|y|^2 - |x|^2}{|y-x|^n} 1 \, d\sigma(y) = \int\limits_{|y|=1} P(y;x) \, d\sigma(y) \quad \text{for all } x \in B. \tag{9}$$
Furthermore, $P(y;x) > 0$ for all $y \in \partial B$ and all $x \in B$ is satisfied.
4. We now show that the relation
$$\lim_{\substack{x \to \overset{\circ}{x} \\ x \in B}} u(x) = f(\overset{\circ}{x})$$
is correct for all boundary points $\overset{\circ}{x} \in \partial B$. We take an arbitrary point $x \in B$ and see
$$\begin{aligned} u(x) - f(\overset{\circ}{x}) &= \frac{1}{\omega_n} \int\limits_{|y|=1} \frac{|y|^2 - |x|^2}{|y-x|^n} \Big(f(y) - f(\overset{\circ}{x}) \Big) \, d\sigma(y) \\ &= \frac{1}{\omega_n} \int\limits_{\substack{y \in \partial B \\ |y - \overset{\circ}{x}| \geq 2\delta}} \frac{|y|^2 - |x|^2}{|y-x|^n} \Big(f(y) - f(\overset{\circ}{x}) \Big) \, d\sigma(y) \\ &\quad + \frac{1}{\omega_n} \int\limits_{\substack{y \in \partial B \\ |y - \overset{\circ}{x}| \leq 2\delta}} \frac{|y|^2 - |x|^2}{|y-x|^n} \Big(f(y) - f(\overset{\circ}{x}) \Big) \, d\sigma(y). \end{aligned} \tag{10}$$
The function f is continuous at the point $\overset{\circ}{x}$. Given the quantity $\varepsilon > 0$ we therefore have a number $\delta = \delta(\varepsilon) > 0$ such that $|f(y) - f(\overset{\circ}{x})| \leq \varepsilon$ holds true for all points $y \in \partial B$ with $|y - \overset{\circ}{x}| \leq 2\delta$. This implies
$$\begin{aligned} &\left| \frac{1}{\omega_n} \int\limits_{\substack{y \in \partial B \\ |y - \overset{\circ}{x}| \leq 2\delta}} \frac{|y|^2 - |x|^2}{|y-x|^n} \Big(f(y) - f(\overset{\circ}{x}) \Big) \, d\sigma(y) \right| \\ &\qquad \leq \frac{1}{\omega_n} \int\limits_{\substack{y \in \partial B \\ |y - \overset{\circ}{x}| \leq 2\delta}} \frac{|y|^2 - |x|^2}{|y-x|^n} \Big| f(y) - f(\overset{\circ}{x}) \Big| \, d\sigma(y) \\ &\qquad \leq \varepsilon \qquad \text{for all} \quad x \in B. \end{aligned} \tag{11}$$
Choosing a point $x \in B$ with $|x - \overset{\circ}{x}| \leq \delta$ we infer the following estimate for all $y \in \partial B$ with $|y - \overset{\circ}{x}| \geq 2\delta$, namely
$$|y - x| \geq |y - \overset{\circ}{x}| - |\overset{\circ}{x} - x| \geq 2\delta - \delta = \delta.$$

Consequently, for all $y \in \partial B$ with $|y - \overset{\circ}{x}| \geq 2\delta$ and $x \in B$ with $|x - \overset{\circ}{x}| \leq \eta < \delta$ we have

$$\begin{aligned}\frac{|y|^2 - |x|^2}{|y-x|^n} &\leq \frac{(|y|+|x|)(|y|-|x|)}{\delta^n} \\ &\leq \frac{2}{\delta^n}(|\overset{\circ}{x}| - |x|) \leq \frac{2}{\delta^n}|\overset{\circ}{x} - x| \\ &\leq \frac{2\eta}{\delta^n}.\end{aligned}$$

Setting $M := \sup_{y \in \partial B} |f(y)|$ we now can estimate as follows:

$$\begin{aligned}&\left| \frac{1}{\omega_n} \int\limits_{\substack{y \in \partial B \\ |y - \overset{\circ}{x}| \geq 2\delta}} \frac{|y|^2 - |x|^2}{|y-x|^n} \Big(f(y) - f(\overset{\circ}{x}) \Big)\, d\sigma(y) \right| \\ &\qquad \leq \frac{1}{\omega_n} \int\limits_{\substack{y \in \partial B \\ |y - \overset{\circ}{x}| \geq 2\delta}} \frac{|y|^2 - |x|^2}{|y-x|^n} \Big| f(y) - f(\overset{\circ}{x}) \Big|\, d\sigma(y) \\ &\qquad \leq \frac{2M}{\omega_n} \int\limits_{\substack{y \in \partial B \\ |y - \overset{\circ}{x}| \geq 2\delta}} \frac{|y|^2 - |x|^2}{|y-x|^n}\, d\sigma(y) \\ &\qquad \leq \frac{2M}{\omega_n \delta^n} 2\eta\omega_n \;\leq\; \varepsilon,\end{aligned} \tag{12}$$

if we choose $\eta \in (0, \delta)$ sufficiently small. With the aid of (10), (11), and (12) we deduce

$$|u(x) - f(\overset{\circ}{x})| \leq 2\varepsilon \qquad \text{for all} \quad x \in B \quad \text{with} \quad |x - \overset{\circ}{x}| \leq \eta. \tag{13}$$

This implies

$$\lim_{\substack{x \to \overset{\circ}{x} \\ x \in B}} u(x) = f(\overset{\circ}{x}) \qquad \text{for all} \quad \overset{\circ}{x} \in \partial B.$$

5. The function

$$u(x) := \frac{1}{\omega_n} \int\limits_{|y|=1} \frac{|y|^2 - |x|^2}{|y-x|^n} f(y)\, d\sigma(y), \qquad x \in B$$

solves Dirichlet's problem on the unit ball B. We now utilize the transformation

$$x = T\xi = \frac{1}{R}(\xi - a), \qquad \xi \in \overline{B_R(a)}.$$

Then the function $v(\xi) := u(T\xi)$, $\xi \in \overline{B_R(a)}$ gives us a solution of Dirichlet's problem

$$\begin{aligned} &v = v(\xi) \in C^2(B_R(a)) \cap C^0(\overline{B_R(a)}), \\ &\Delta v(\xi) = 0 \qquad \text{for all} \quad \xi \in B_R(a), \\ &v(\xi) = g(\xi) \qquad \text{for all} \quad \xi \in \partial B_R(a), \end{aligned} \tag{14}$$

where we have set $g(\xi) := f(T\xi)$, $\xi \in \partial B_R(a)$. Taking

$$\eta := T^{-1}y = Ry + a, \qquad y \in \partial B$$

we see $\eta \in \partial B_R(a)$ and $d\sigma(\eta) = R^{n-1}\,d\sigma(y)$. On this basis we calculate

$$\begin{aligned} v(\xi) = u(T\xi) &= \frac{1}{\omega_n} \int\limits_{|y|=1} \frac{|y|^2 - |T\xi|^2}{|y - T\xi|^n} f(y)\,d\sigma(y) \\ &= \frac{1}{\omega_n} \int\limits_{|\eta-a|=R} \frac{|T\eta|^2 - |T\xi|^2}{|T\eta - T\xi|^n} f(T\eta) \frac{1}{R^{n-1}}\,d\sigma(\eta) \\ &= \frac{1}{R^{n-1}\omega_n} \int\limits_{|\eta-a|=R} \frac{\frac{1}{R^2}\Big(|\eta - a|^2 - |\xi - a|^2\Big)}{\frac{1}{R^n}|\eta - \xi|^n} g(\eta)\,d\sigma(\eta) \\ &= \frac{1}{R\omega_n} \int\limits_{|\eta-a|=R} \frac{|\eta - a|^2 - |\xi - a|^2}{|\eta - \xi|^n} g(\eta)\,d\sigma(\eta), \qquad \xi \in B_R(a). \end{aligned}$$

q.e.d.

Theorem 3.3. (Regularity theorem for weakharmonic functions) *Let the weakharmonic function $u = u(x) : G \to \mathbb{R} \in C^0(G)$ be given on the domain $G \subset \mathbb{R}^n$. Then the function u is real-analytic in G and satisfies the Laplace equation $\Delta u(x) = 0$ for all $x \in G$.*

Proof: Let the point $a \in G$ be chosen arbitrarily. For a suitable radius $R \in (0, +\infty)$ we then consider the ball $B_R(a) \subset\subset G$, where we solve Dirichlet's problem with the aid of Theorem 3.2, namely

$$\begin{aligned} &v = v(x) \in C^2(B_R(a)) \cap C^0(\overline{B_R(a)}), \\ &\Delta v(x) = 0 \qquad \text{for all} \quad x \in B_R(a), \\ &v(x) = u(x) \qquad \text{for all} \quad x \in \partial B_R(a). \end{aligned} \tag{15}$$

Theorem 2.11 from Section 2 now yields $u(x) \equiv v(x)$ in $\overline{B_R(a)}$. Consequently, we have $u \in C^2(G)$ and $\Delta u(x) = 0$ for all $x \in G$. According to Theorem 1.9 in Section 1, the function u is real-analytic in G.

q.e.d.

We now intend to solve Dirichlet's problem (1) for a large class of domains G. In this context we use an ingenious *method* proposed *by O. Perron.*

Definition 3.4. *Let $G \subset \mathbb{R}^n$ denote a bounded domain on which the continuous function $u = u(x) : G \to \mathbb{R} \in C^0(G)$ is given. Then we define the* harmonically modified function

$$v(x) := [u]_{a,R}(x)$$
$$:= \begin{cases} u(x), & x \in G \text{ with } |x-a| \geq R \\ \dfrac{1}{R\omega_n} \displaystyle\int\limits_{|y-a|=R} \frac{|y-a|^2 - |x-a|^2}{|y-x|^n} u(y)\, d\sigma(y), & x \in G \text{ with } |x-a| < R \end{cases}$$

for all $a \in G$ and $R \in (0, dist(a, \mathbb{R}^n \setminus G))$.

Remark: The function $v = v(x) : G \to \mathbb{R} \in C^0(G)$ is harmonic in $B_R(a)$ and coincides with the original function on the complement of this ball $G \setminus B_R(a)$.

In the sequel we need the important

Proposition 3.5. *Let the point $a \in G$ and the radius $R \in (0, dist(a, \mathbb{R}^n \setminus G))$ be chosen as fixed, whereas $u = u(x)$ denotes a superharmonic function in G. Then the harmonically modified function*

$$v(x) := [u]_{a,R}(x), \qquad x \in G$$

is superharmonic in G as well, and we have

$$v(x) \leq u(x) \qquad \textit{for all} \quad x \in G.$$

Proof:

1. At first, we show the inequality $v(x) \leq u(x)$ for all $x \in G$. In this context we only have to verify $v(x) \leq u(x)$ for all $x \in \overline{B_R(a)}$. The function

$$w(x) := u(x) - v(x), \qquad x \in \overline{B_R(a)}$$

is superharmonic in the ball $B_R(a)$. Each sequence of points

$$\{x^{(k)}\}_{k=1,2,\ldots} \subset B_R(a)$$

with $\lim\limits_{k\to\infty} x^{(k)} = \overset{\circ}{x} \in \partial B_R(a)$ satisfies

$$\liminf_{k\to\infty} w(x^{(k)}) = w(\overset{\circ}{x}) = 0.$$

From Section 2, Theorem 2.10 we infer $w(x) \geq 0$, $x \in B_R(a)$ and consequently

$$v(x) \leq u(x) \qquad \text{for all} \quad x \in B_R(a).$$

2. We now show that v is superharmonic in G. Choose an arbitrary point $\xi \in \partial B_R(a)$ and a quantity $\vartheta(\xi) \in (0, \operatorname{dist}(\xi, \mathbb{R}^n \setminus G)]$. Using part 1 of our proof, we then obtain

$$\frac{1}{\varrho^{n-1}\omega_n} \int\limits_{|x-\xi|=\varrho} v(x)\,d\sigma(x) \leq \frac{1}{\varrho^{n-1}\omega_n} \int\limits_{|x-\xi|=\varrho} u(x)\,d\sigma(x) \leq u(\xi) = v(\xi)$$

for all $\varrho \in (0, \vartheta(\xi))$. Consequently, the function v is superharmonic in G: In the ball $B_R(a)$ the function v is harmonic anyway, and in $G \setminus \overline{B_R(a)}$ this function v is superharmonic.

q.e.d.

We additionally need the following

Proposition 3.6. (Harnack's lemma)
We consider a sequence $w_k(x) : G \to \mathbb{R}$, $k = 1, 2, \ldots$ of harmonic functions in G, which are descending in the following way:

$$w_1(x) \geq w_2(x) \geq w_3(x) \geq \ldots \qquad \text{for all} \quad x \in G.$$

Furthermore, let the sequence converge at one point $\overset{\circ}{x} \in G$ which means

$$\lim_{k\to\infty} w_k(\overset{\circ}{x}) > -\infty.$$

Then the sequence of functions $\{w_k(x)\}_{k=1,2,}$ uniformly converges in each compact set $\Theta \subset G$ towards a function harmonic in G, namely

$$w(x) := \lim_{k\to\infty} w_k(x), \qquad x \in G.$$

Proof: Without loss of generality we assume $\overset{\circ}{x}= 0$ and for the ball the inclusion $B_R \subset G$ with a radius $R \in (0, +\infty)$. For the indices $k, l \in \mathbb{N}$ with $k \leq l$ we define the nonnegative functions $v_{kl}(x) := w_k(x) - w_l(x) \geq 0$, $x \in B_R$. We apply Harnack's inequality and obtain

$$0 \leq v_{kl}(x) \leq \frac{1 + \frac{|x|}{R}}{\left(1 - \frac{|x|}{R}\right)^{n-1}} v_{kl}(0) \leq \frac{1 + \frac{1}{2}}{\left(1 - \frac{1}{2}\right)^{n-1}} v_{kl}(0), \qquad x \in \overline{B_{\frac{R}{2}}}.$$

Setting $K := \frac{3}{2} \cdot (\frac{1}{2})^{1-n} = 3 \cdot 2^{n-2}$ we infer

$$\begin{aligned} &|w_k(x) - w_l(x)| \leq K|w_k(0) - w_l(0)| \\ &\text{for all} \quad x \in \overline{B_{\frac{R}{2}}} \quad \text{and all} \quad k, l \in \mathbb{N}. \end{aligned} \tag{16}$$

Since the limit $\lim\limits_{k\to\infty} w_k(0)$ exists, the sequence $\{w_k(x)\}_{k=1,2,\ldots}$ converges uniformly in $\overline{B_{\frac{R}{2}}}$ towards the function $w(x)$. When we cover a compact set

$\Theta \subset G$ by finitely many balls we comprehend that the sequence of functions $\{w_k(x)\}_{k=1,2,\ldots}$ converges uniformly in Θ towards the function $w(x)$. The transition to the limit in Poisson's integral formula shows that the limit function $w(x)$ is harmonic in G.

q.e.d.

In order to solve Dirichlet's problem we utilize the following set of admissible functions

$$\mathcal{M} := \Big\{ v : G \to \mathbb{R} \in C^0(G) \; : \; v \text{ is in } G \text{ superharmonic, and}$$
$$\text{for all sequences } \{x^{(k)}\}_{k=1,2,\ldots} \subset G \text{ with } \lim_{k\to\infty} x^{(k)} = x^* \in \partial G$$
$$\text{we have } \liminf_{k\to\infty} v(x^{(k)}) \geq f(x^*) \Big\}.$$

Here the symbol $f : \partial G \to \mathbb{R}$ denotes a continuous boundary function. Since

$$v(x) := M := \max_{x \in \partial G} f(x) \quad \in \mathcal{M}$$

holds true, we have $\mathcal{M} \neq \emptyset$.

Proposition 3.7. *Let us define the function*

$$u(x) := \inf_{v \in \mathcal{M}} v(x), \qquad x \in G.$$

Then u is harmonic in G and we have

$$m \leq u(x) \leq M \qquad \text{for all} \quad x \in G.$$

Here we abbreviate $m := \inf_{x \in \partial G} f(x)$ *and* $M := \sup_{x \in \partial G} f(x)$.

Proof:

1. We take a sequence of points $\{x^i\}_{i=1,2,3,\ldots} \subset G$ which are dense in G. For each index $i \in \mathbb{N}$, there exists a sequence of functions $\{v_{ij}\}_{j=1,2,\ldots} \subset \mathcal{M}$ satisfying

$$\lim_{j\to\infty} v_{ij}(x^i) = u(x^i).$$

The minimum principle implies the estimate $v_{ij}(x) \geq m$ for all $x \in G$ and all $i, j \in \mathbb{N}$. We now define the functions

$$v_k(x) := \min_{1 \leq i,j \leq k} v_{ij}(x), \qquad x \in G$$

for each index $k \in \mathbb{N}$. Evidently, we have $v_k(x) \geq v_{k+1}(x)$, $x \in G$ for all $k \in \mathbb{N}$. The minimum of finitely many superharmonic functions is superharmonic again according to a previous remark, and we infer

$$v_k \in \mathcal{M}, \qquad k = 1, 2, \ldots$$

We observe $u(x^i) \leq v_k(x^i) \leq v_{ik}(x^i)$ for $1 \leq i \leq k$, and we obtain

$$\lim_{k\to\infty} v_k(x^i) = u(x^i) \qquad \text{for all} \quad i = 1, 2, \ldots$$

2. In the disc $B_R(a) \subset\subset G$ we harmonically modify the function v_k to the following function
$$w_k(x) := [v_k]_{a,R}(x), \qquad x \in G.$$
With the aid of Proposition 3.5 we see $\{w_k\}_{k=1,2,\dots} \subset \mathcal{M}$. Furthermore, we have $w_k(x) \geq w_{k+1}(x)$ in $B_R(a)$ for all $k \in \mathbb{N}$ and
$$u(x^i) \leq w_k(x^i) \leq v_k(x^i) \qquad \text{for all} \quad i, k \in \mathbb{N}.$$
Therefore, we obtain
$$\lim_{k\to\infty} w_k(x^i) = u(x^i) \qquad \text{for all} \quad i \in \mathbb{N}.$$
According to Harnack's lemma the sequence $\{w_k(x)\}_{k=1,2,\dots}$ converges uniformly in $B_R(a)$ towards a harmonic function $w(x)$, and we comprehend
$$w(x^i) = u(x^i) \qquad \text{for all} \quad x^i \in B_R(a), \quad i = 1, 2, \dots$$
Since w and u are continuous functions, we infer the identity $u(x) = w(x)$, $x \in \overline{B_R(a)}$. Consequently, the function u has to be harmonic in G, because the ball $B_R(a) \subset\subset G$ has been chosen arbitrarily.
3. The inclusion $M \in \mathcal{M}$ implies the estimate $u(x) \leq M$ for all $x \in G$. Since the inequality $v_{ij}(x) \geq m$ for all $x \in G$ and all $i, j \in \mathbb{N}$ holds true and consequently $v_k(x) \geq m$ in G for all $k \in \mathbb{N}$ is valid, we finally obtain
$$u(x) = \lim_{k\to\infty} v_k(x) \geq m \qquad \text{for all} \quad x \in G.$$

q.e.d.

Definition 3.8. *Let us consider the bounded domain $G \subset \mathbb{R}^n$. We name a* boundary point $x \in \partial G$ regular *if we have a superharmonic function*
$$\Phi(y) = \Phi(y; x) : G \to \mathbb{R} \quad \textit{with} \quad \lim_{\substack{y\to x\\ y\in G}} \Phi(y) = 0$$
and
$$\varrho(\varepsilon) := \inf_{\substack{y\in G\\ |y-x|\geq\varepsilon}} \Phi(y) > 0 \qquad \textit{for all} \quad \varepsilon > 0.$$
If each boundary point of the domain G is regular, we speak of a Dirichlet domain.

Remark: A point $x \in \partial G$ is regular if and only if we have a number $r > 0$ and a superharmonic function $\Psi = \Psi(y) : G \cap B_r(x) \to \mathbb{R}$ satisfying
$$\lim_{\substack{y\to x\\ y\in G\cap B_r(x)}} \Psi(y) = 0 \qquad \text{and} \qquad \inf_{\substack{r>|y-x|\geq\varepsilon\\ y\in G}} \Psi(y) > 0, \quad 0 < \varepsilon < r.$$

Here we set $m := \inf\limits_{\substack{r>|y-x|\geq\frac{1}{2}r \\ y\in G}} \Psi(y) > 0$ and consider the following function

$$\Phi(y) := \begin{cases} \min\left(1, \frac{2\Psi(y)}{m}\right), & y \in G \cap B_r(x) \\ 1, & y \in G \setminus B_r(x) \end{cases}$$

which is superharmonic in G.

Theorem 3.9. (Dirichlet problem for the Laplacian)
Let $G \subset \mathbb{R}^n$ denote a bounded domain with $n \geq 2$. Then the Dirichlet problem

$$\begin{aligned} & u = u(x) \in C^2(G) \cap C^0(\overline{G}), \\ & \Delta u(x) = 0 \qquad in \quad G, \\ & u(x) = f(x) \qquad on \quad \partial G \end{aligned} \tag{17}$$

can be solved for all continuous boundary functions $f : \partial G \to \mathbb{R}$ if and only if G is a Dirichlet domain in the sense of Definition 3.8.

Proof:

'$\Longrightarrow$' Let the Dirichlet problem be solvable for all continuous boundary functions $f : \partial G \to \mathbb{R}$. Taking an arbitrary point $\xi \in \partial G$ we define the function $f(y) := |y - \xi|$, $y \in \partial G$, and we solve Dirichlet's problem (17) for these boundary values. We apply the minimum principle to the harmonic function $u = u(x) : \overline{G} \to \mathbb{R}$ and obtain

$$u(x) > 0 \qquad \text{for all} \quad x \in \overline{G} \setminus \{\xi\}.$$

Therefore, the boundary point ξ is regular.

'$\Longleftarrow$' Let G be a Dirichlet domain and $x \in \partial G$ an arbitrary regular boundary point. Then we have an associate superharmonic function $\Phi(y) = \Phi(y; x) : G \to \mathbb{R}$ due to Definition 3.8. Since the function $f : \partial G \to \mathbb{R}$ is continuous, we can prescribe $\varepsilon > 0$ and obtain a quantity $\delta = \delta(\varepsilon) > 0$ satisfying

$$|f(y) - f(x)| \leq \varepsilon \quad \text{for all} \quad y \in \partial G \quad \text{with} \quad |y - x| \leq \delta.$$

We now define

$$\eta(\varepsilon) := \inf_{\substack{y\in G \\ |y-x|\geq\delta(\varepsilon)}} \Phi(y) > 0.$$

1. Let the *upper barrier function*

$$v^+(y) := f(x) + \varepsilon + (M - m)\frac{\Phi(y)}{\eta(\varepsilon)}, \qquad y \in G$$

be given. Evidently, the function v^+ is superharmonic in G. Furthermore, an arbitrary sequence $\{y^{(k)}\}_{k=1,2,\ldots} \subset G$ with $y^{(k)} \to y^+ \in \partial G$ for $k \to \infty$ satisfies

$$\liminf_{k\to\infty} v^+(y^{(k)}) \geq f(y^+).$$

Consequently, $v^+ \in \mathcal{M}$ holds true.

2. Now we consider the *lower barrier function*

$$v^-(y) := f(x) - \varepsilon - (M-m)\frac{\Phi(y)}{\eta(\varepsilon)}, \qquad y \in G.$$

We choose $v \in \mathcal{M}$ arbitrarily. Considering a sequence $\{y^{(k)}\}_{k=1,2,\ldots} \subset G$ with $y^{(k)} \to y^- \in \partial G$ for $k \to \infty$, we can estimate

$$\begin{aligned} &\liminf_{k\to\infty} \Big(v(y^{(k)}) - v^-(y^{(k)})\Big) \\ &\qquad \geq \liminf_{k\to\infty} \Big(v(y^{(k)}) - f(y^-)\Big) + \liminf_{k\to\infty} \Big(f(y^-) - v^-(y^{(k)})\Big) \\ &\qquad \geq 0. \end{aligned}$$

Furthermore, the function $v - v^-$ is superharmonic in G, and Theorem 2.10 from Section 2 yields $v - v^- \geq 0$ in G. This implies

$$v(y) \geq v^-(y), \qquad y \in G \quad \text{for all} \quad v \in \mathcal{M}.$$

3. The harmonic function

$$u(y) := \inf_{v \in \mathcal{M}} v(y), \qquad y \in G$$

constructed in Proposition 3.7 now attains the prescribed boundary values f continuously. On account of 1. and 2. the estimate

$$v^-(y) \leq u(y) \leq v^+(y) \qquad \text{for all} \quad y \in G$$

is fulfilled, which means

$$f(x) - \varepsilon - (M-m)\frac{\Phi(y)}{\eta(\varepsilon)} \leq u(y) \leq f(x) + \varepsilon + (M-m)\frac{\Phi(y)}{\eta(\varepsilon)}, \qquad y \in G.$$

Using the relation $\lim\limits_{\substack{y\in G\\ y\to x}} \Phi(y) = 0$ we obtain

$$|f(x) - u(y)| \leq \varepsilon + (M-m)\frac{\Phi(y)}{\eta(\varepsilon)} \leq 2\varepsilon$$

for all $y \in G$ with $|y - x| \leq \delta^*(\varepsilon)$. This implies

$$\lim_{\substack{y\in G\\ y\to x}} u(y) = f(x).$$

Therefore, the function u solves Dirichlet's problem (17) for the boundary values f. q.e.d.

Figure 1.6 POINCARÉ'S CONDITION OF EXTERIOR SUPPORT BALLS

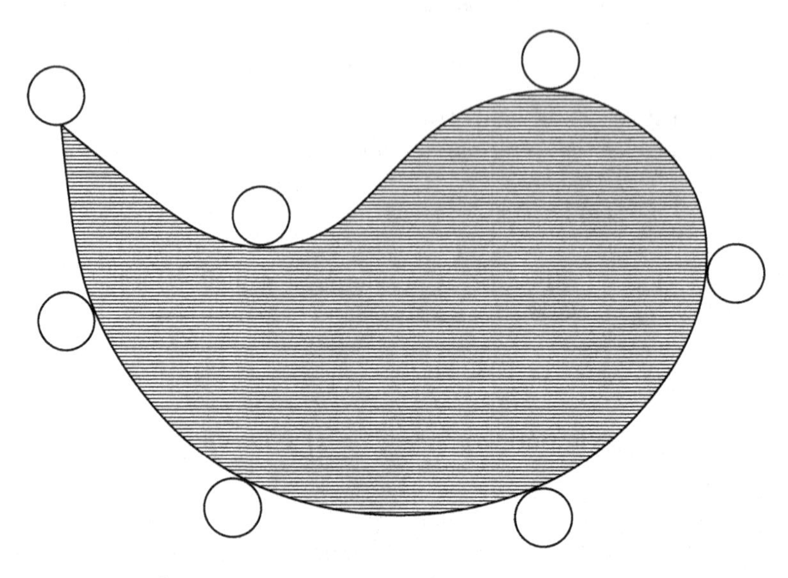

Theorem 3.10. (Poincaré's condition)
A boundary point $x \in \partial G$ is regular, if we have a ball $B_r(a)$ with the center $a \in \mathbb{R}^n$ and the radius $r \in (0, +\infty)$ satisfying $\overline{G} \cap \overline{B_r(a)} = \{x\}$. Especially, bounded domains with a regular C^2-boundary are Dirichlet domains.

Proof: For $n = 2$ we consider in G the harmonic function

$$\Phi(y) := \log\left(\frac{|y-a|}{r}\right), \qquad y \in G,$$

and for $n \geq 3$ we consider the harmonic function

$$\Phi(y) := r^{2-n} - |y-a|^{2-n}, \qquad y \in G.$$

Then we immediately obtain the statements above. q.e.d.

Theorem 3.11. *Let $B_R := \{x \in \mathbb{R}^n : |x| < R\}$ denote the ball about the origin of radius $R > 0$ and consider the pointed ball $\dot{B}_R := B_R \setminus \{0\}$. The function $u = u(x) \in C^2(\dot{B}_R) \cap C^0(\overline{B_R})$ is assumed to be harmonic in $\dot{B}_R$. Then the function u is harmonic in B_R.*

Proof: We restrict our considerations to the case $n \geq 3$ and set

$$v(x) := \frac{1}{R\omega_n} \int\limits_{|y|=R} \frac{R^2 - |x|^2}{|y-x|^n} u(y)\, d\sigma(y), \qquad x \in B_R.$$

This function v is harmonic in B_R and continuous in $\overline{B_R}$ with the boundary values

$$v(x) = u(x), \qquad x \in \partial B_R.$$

Since the functions u and v are continuous in $\overline{B_R}$, we have a constant $M > 0$ such that

$$\sup_{x \in B_R} |u(x) - v(x)| \le M$$

holds true. Given the quantity $\varepsilon > 0$, we now can choose a sufficiently small number $\delta = \delta(\varepsilon) \in (0, R)$ such that

$$M \le \varepsilon\Big(|x|^{2-n} - R^{2-n}\Big) \qquad \text{for all} \quad x \in \mathbb{R}^n \quad \text{with} \quad |x| = \delta(\varepsilon).$$

We consider the spherical shell $K_\varepsilon := \{x \in \mathbb{R}^n \, : \, \delta(\varepsilon) \le |x| \le R\}$ and see

$$|u(x) - v(x)| \le \varepsilon\Big(|x|^{2-n} - R^{2-n}\Big) \qquad \text{for all} \quad x \in \partial K_\varepsilon.$$

The maximum principle for harmonic functions now yields

$$|u(x) - v(x)| \le \varepsilon\Big(|x|^{2-n} - R^{2-n}\Big) \qquad \text{for all} \quad x \in K_\varepsilon.$$

Since the number $\varepsilon > 0$ has been chosen arbitrarily and the behavior $\delta(\varepsilon) \downarrow 0$ for $\varepsilon \downarrow 0$ can be achieved, we obtain

$$u(x) \equiv v(x), \qquad x \in \dot{B}_R.$$

Now the functions u and v are continuous in $\overline{B_R}$, and we infer

$$u(x) \equiv v(x), \qquad x \in \overline{B_R}.$$

Therefore, the function u is harmonic in B_R. q.e.d.

Remarks:

1. When we consider the Riemannian theorem on removable singularities for holomorphic functions, it suffices to assume the boundedness of the functions in the neighborhood of a singular point in order to continue them holomorphically into this point.
2. There are bounded domains, where the Dirichlet problem cannot be solved for arbitrary boundary values. For example, we consider the domain

$$G := \dot{B}_R, \qquad \partial G = \partial B_R \cup \{0\}.$$

On account of Theorem 3.11, there does not exist a harmonic function for the boundary values $f(x) = 1$, $|x| = R$ and $f(0) = 0$.

4 Theory of Spherical Harmonics in 2 Variables: Fourier Series

The theory of spherical harmonics has been founded by Laplace and Legendre and is applied in quantum mechanics to the investigation of the spectrum for the hydrogen atom. We owe the theory in arbitrary spatial dimensions $n \geq 2$ to G.Herglotz. In the next two paragraphs we utilize Banach and Hilbert spaces introduced in Chapter 2, Section 6. At first, we consider the case $n = 2$.

On the unit circle line $S^1 := \{x \in \mathbb{R}^2 : |x| = 1\}$ we consider the functions $u = u(x) \in C^0(S^1, \mathbb{R})$. They are identified with the 2π-periodic continuous functions

$$C^0_{2\pi}(\mathbb{R}, \mathbb{R}) := \left\{ v : \mathbb{R} \to \mathbb{R} \in C^0(\mathbb{R}, \mathbb{R}) : \begin{array}{c} v(\varphi + 2\pi k) = v(\varphi) \\ \text{for all } \varphi \in \mathbb{R},\ k \in \mathbb{Z} \end{array} \right\}$$

via $\hat{u}(\varphi) := u(e^{i\varphi})$, $0 \leq \varphi \leq 2\pi$. We endow the space $C^0(S^1, \mathbb{R})$ with the norm

$$\|u\|_0 := \max_{x \in S^1} |u(x)|, \qquad u \in C^0(S^1, \mathbb{R}) \tag{1}$$

and get a Banach space with the topology of uniform convergence. By the inner product

$$(u, v) := \int_0^{2\pi} u(e^{i\varphi}) v(e^{i\varphi})\, d\varphi, \qquad u, v \in C^0(S^1, \mathbb{R}) \tag{2}$$

the set $C^0(S^1, \mathbb{R})$ becomes a pre-Hilbert-space. We complete this space with respect to the L^2-norm induced by the inner product (2), namely

$$\|u\| := \sqrt{(u, u)}, \qquad u \in C^0(S^1, \mathbb{R}), \tag{3}$$

and obtain the Lebesgue space $L^2(S^1, \mathbb{R})$ of the square integrable, measurable functions on S^1. Furthermore, we note the inequality

$$\|u\| \leq \sqrt{2\pi} \|u\|_0 \qquad \text{for all} \quad u \in C^0(S^1, \mathbb{R}). \tag{4}$$

If a sequence converges with respect to the Banach-space-norm $\|\cdot\|_0$, this is as well the case with respect to the Hilbert-space-norm $\|\cdot\|$. However, the opposite direction is not true, since the Hilbert space $L^2(S^1, \mathbb{R})$ also contains discontinuous functions.

Theorem 4.1. (Fourier series)
The system of functions

$$\frac{1}{\sqrt{2\pi}}, \quad \frac{1}{\sqrt{\pi}} \cos k\varphi, \quad \frac{1}{\sqrt{\pi}} \sin k\varphi, \qquad \varphi \in [0, 2\pi], \quad k = 1, 2, \ldots$$

represents a complete orthonormal system - briefly c.o.n.s. - in the pre-Hilbert-space $\mathcal{H} := C^0(S^1, \mathbb{R})$ *endowed with the inner product from (2).*

Proof:

1. We easily verify that the system of functions $\mathcal{S}$ given is orthonormal, which means $\|u\| = 1$ for all $u \in \mathcal{S}$ and $(u,v) = 0$ for all $u, v \in \mathcal{S}$ with $u \neq v$. It remains for us to comprehend that this orthonormal system of functions is complete in the pre-Hilbert-space $\mathcal{H}$. According to Theorem 6.19 from Chapter 2, Section 6 we have to show that the Fourier series for each element $u \in \mathcal{H}$ approximates this element with respect to the Hilbert-space-norm $\|\cdot\|$ from (3).
2. Let the function

$$u = u(x) \in \mathcal{H} = C^0(S^1, \mathbb{R})$$

be given arbitrarily. We then continue u harmonically onto the disc

$$B = \{x \in \mathbb{R}^2 : |x| < 1\}$$

via

$$u(z) = \frac{1}{2\pi} \int_0^{2\pi} \frac{1 - r^2}{|e^{i\varphi} - z|^2} u(e^{i\varphi})\, d\varphi, \qquad |z| < 1; \tag{5}$$

here we have set $z = re^{i\vartheta}$. We now expand Poisson's kernel as follows:

$$\begin{aligned}
\frac{1-r^2}{|e^{i\varphi} - z|^2} &= \frac{1-r^2}{|e^{i\varphi} - re^{i\vartheta}|^2} \\
&= \frac{1-r^2}{|1 - re^{i(\vartheta-\varphi)}|^2} \\
&= \frac{1-r^2}{(1 - re^{i(\vartheta-\varphi)})(1 - re^{i(\varphi-\vartheta)})} \\
&= -1 + \frac{1}{1 - re^{i(\varphi-\vartheta)}} + \frac{1}{1 - re^{-i(\varphi-\vartheta)}} \\
&= -1 + \sum_{k=0}^{\infty} r^k e^{ik(\varphi-\vartheta)} + \sum_{k=0}^{\infty} r^k e^{-ik(\varphi-\vartheta)} \\
&= 1 + 2\sum_{k=1}^{\infty} r^k \cos k(\varphi - \vartheta).
\end{aligned} \tag{6}$$

Here the series converges locally uniformly for $0 \leq r < 1$ and $\varphi, \vartheta \in \mathbb{R}$. Now we have

$$\cos k(\varphi - \vartheta) = \cos k\varphi \cos k\vartheta + \sin k\varphi \sin k\vartheta,$$

and we obtain the following identity with $g(\varphi) := u(e^{i\varphi})$, $\varphi \in [0, 2\pi)$:

$$
\begin{aligned}
u(re^{i\vartheta}) &= \frac{1}{2\pi}\int_0^{2\pi}\left\{1+2\sum_{k=1}^{\infty} r^k\Big(\cos k\varphi\cos k\vartheta+\sin k\varphi\sin k\vartheta\Big)\right\}g(\varphi)\,d\varphi \\
&= \frac{1}{2\pi}\int_0^{2\pi} g(\varphi)\,d\varphi+\sum_{k=1}^{\infty}\left\{\left(\frac{1}{\pi}\int_0^{2\pi} g(\varphi)\cos k\varphi\,d\varphi\right)r^k\cos k\vartheta\right. \\
&\quad \left.+\left(\frac{1}{\pi}\int_0^{2\pi} g(\varphi)\sin k\varphi\,d\varphi\right)r^k\sin k\vartheta\right\}.
\end{aligned}
$$

Finally, we set

$$
a_k := \frac{1}{\pi}\int_0^{2\pi} g(\varphi)\cos k\varphi\,d\varphi, \qquad k=0,1,2,\ldots \tag{7}
$$

and

$$
b_k := \frac{1}{\pi}\int_0^{2\pi} g(\varphi)\sin k\varphi\,d\varphi, \qquad k=1,2,\ldots. \tag{8}
$$

With the representation

$$
u(re^{i\vartheta}) = \frac{1}{2}a_0+\sum_{k=1}^{\infty}\Big(a_k\cos k\vartheta+b_k\sin k\vartheta\Big)r^k,\ 0\le r<1,\ 0\le\vartheta<2\pi \tag{9}
$$

we obtain the *Fourier expansion of a harmonic function within the unit disc.*

3. Since the function $u(z)$ is continuous in $\overline{B}$, we find a radius $r\in(0,1)$ to each given $\varepsilon>0$, such that

$$
|u(re^{i\vartheta})-g(\vartheta)|\le\varepsilon \qquad \text{for all}\quad \vartheta\in[0,2\pi). \tag{10}
$$

Furthermore, we can choose an integer $N=N(\varepsilon)\in\mathbb{N}$ so large that

$$
\left|\frac{a_0}{2}+\sum_{k=1}^{N} r^k\Big(a_k\cos k\vartheta+b_k\sin k\vartheta\Big)-u(re^{i\vartheta})\right|\le\varepsilon \quad \text{for all}\quad \vartheta\in[0,2\pi) \tag{11}
$$

is satisfied. For the quantity $\varepsilon>0$ given, we therefore find real coefficients $A_0,\ldots,A_N$ and $B_1,\ldots,B_N$, such that the *trigonometric polynomial*

$$
F_\varepsilon(\vartheta) := A_0+\sum_{k=1}^{N}\Big(A_k\sin k\vartheta+B_k\cos k\vartheta\Big), \qquad 0\le\vartheta<2\pi
$$

fulfills the following inequality

$$|F_\varepsilon(\vartheta) - g(\vartheta)| \leq 2\varepsilon \qquad \text{for all} \quad \vartheta \in [0, 2\pi). \tag{12}$$

From the relation (4) we infer

$$\|F_\varepsilon - g\| \leq 2\sqrt{2\pi}\,\varepsilon. \tag{13}$$

On account of the minimal property for the Fourier coefficients due to Chapter 2, Section 6, Proposition 6.17, the Fourier series belonging to the system of functions above approximates the given function with respect to the Hilbert-space-norm. From Theorem 6.19 in Chapter 2, Section 6 we infer that this system of functions represents a complete orthonormal system in $\mathcal{H}$.

q.e.d.

Remark: We leave the following question unanswered: Which functions $g = g(\vartheta)$ satisfy the identity (9) pointwise even for the radius $r = 1$, which concerns the validity of the pointwise equation

$$u(e^{i\vartheta}) = \frac{1}{2}a_0 + \sum_{k=1}^{\infty}\Big(a_k \cos k\vartheta + b_k \sin k\vartheta\Big), \qquad 0 \leq \vartheta < 2\pi.$$

We have shown only the convergence in the square mean. For continuous functions the identity above is *not* satisfied, in general. The investigations on the convergence of Fourier series gave an important motivation for the development of the analysis.

We now present the relationship of trigonometric functions to the Laplace operator. At first, we remind the reader of the decomposition for the Laplacian in polar coordinates:

$$\Delta = \frac{\partial^2}{\partial r^2} + \frac{1}{r}\frac{\partial}{\partial r} + \frac{1}{r^2}\frac{\partial^2}{\partial \varphi^2}. \tag{14}$$

For an arbitrary C^2-function $f = f(r)$ we therefore have the identity

$$\Delta\Big(f(r)\begin{matrix}\cos k\varphi \\ \sin k\varphi\end{matrix}\Big) = \Big(f''(r) + \frac{1}{r}f'(r) - \frac{k^2}{r^2}f(r)\Big)\begin{matrix}\cos k\varphi \\ \sin k\varphi\end{matrix} = \Big(L_k f(r)\Big)\begin{matrix}\cos k\varphi \\ \sin k\varphi\end{matrix}.$$

Here we abbreviate

$$L_k f(r) := f''(r) + \frac{1}{r}f'(r) - \frac{k^2}{r^2}f(r), \qquad r > 0.$$

We note that

$$L_k(r^k) = k(k-1)r^{k-2} + kr^{k-2} - k^2 r^{k-2} = 0, \qquad k = 0, 1, 2, \ldots$$

and obtain

$$\Delta(r^k \cos k\varphi) = 0 = \Delta(r^k \sin k\varphi), \qquad k = 0, 1, 2, \ldots \tag{15}$$

Proposition 4.2. *Let the function* $u = u(x_1, x_2) \in C^2(B_R)$ *be given on the disc* $B_R := \{(x_1, x_2) \in \mathbb{R}^2 : x_1^2 + x_2^2 < R^2\}$. *By the symbols*

$$a_k(r) = \frac{1}{\pi}\int_0^{2\pi} u(re^{i\varphi})\cos k\varphi\, d\varphi, \qquad b_k(r) = \frac{1}{\pi}\int_0^{2\pi} u(re^{i\varphi})\sin k\varphi\, d\varphi \tag{16}$$

we denote the Fourier coefficients of the function u *and by*

$$\tilde{a}_k(r) = \frac{1}{\pi}\int_0^{2\pi} \Delta u(re^{i\varphi})\cos k\varphi\, d\varphi, \qquad \tilde{b}_k(r) = \frac{1}{\pi}\int_0^{2\pi} \Delta u(re^{i\varphi})\sin k\varphi\, d\varphi \tag{17}$$

we mean the Fourier coefficients of the function Δu *for* $0 < r < R$. *Now we have the equation*

$$\tilde{a}_k(r) = L_k a_k(r), \qquad \tilde{b}_k(r) = L_k b_k(r), \qquad 0 < r < R. \tag{18}$$

Remark: The Fourier coefficients of Δu are consequently obtained by formal differentiation of the Fourier series

$$u(re^{i\vartheta}) = \frac{1}{2}a_0(r) + \sum_{k=1}^{\infty}\Big(a_k(r)\cos k\vartheta + b_k(r)\sin k\vartheta\Big).$$

Proof of Proposition 4.2: We evaluate as follows:

$$\begin{aligned}
\tilde{a}_k(r) &= \frac{1}{\pi}\int_0^{2\pi} \Delta u(re^{i\varphi})\cos k\varphi\, d\varphi, \\
&= \frac{1}{\pi}\int_0^{2\pi}\left\{\left(\frac{\partial^2}{\partial r^2} + \frac{1}{r}\frac{\partial}{\partial r} + \frac{1}{r^2}\frac{\partial^2}{\partial \varphi^2}\right)u(re^{i\varphi})\right\}\cos k\varphi\, d\varphi \\
&= \left(\frac{\partial^2}{\partial r^2} + \frac{1}{r}\frac{\partial}{\partial r}\right)\left\{\frac{1}{\pi}\int_0^{2\pi} u(re^{i\varphi})\cos k\varphi\, d\varphi\right\} - \frac{k^2}{\pi r^2}\int_0^{2\pi} u(re^{i\varphi})\cos k\varphi\, d\varphi \\
&= L_k a_k(r), \qquad 0 < r < R, \quad k = 0, 1, 2, \ldots
\end{aligned}$$

Similarly we show the relation (18) for the functions $b_k(r)$. q.e.d.

Theorem 4.3. *We choose* $k \in \mathbb{R}$ *and define* $\dot{\mathbb{R}}^2 := \mathbb{R}^2 \setminus \{0\}$. *Furthermore, the symbol* $H_k = H_k(\xi) : S^1 \to \mathbb{R}$ *denotes a function defined on the unit circle* S^1 *with the properties*

$$|x|^k H_k\left(\frac{x}{|x|}\right) \in C^2(\dot{\mathbb{R}}^2) \quad \text{and} \quad \Delta\left\{|x|^k H_k\left(\frac{x}{|x|}\right)\right\} = 0,\ x \in \dot{\mathbb{R}}^2.$$

Then we infer $k \in \mathbb{Z}$, and we have the identity

$$H_k(e^{i\vartheta}) = A_k \cos k\vartheta + B_k \sin k\vartheta$$

with the real constants A_k, B_k.

Proof: At first, we calculate

$$\begin{aligned} 0 &= \Delta\left\{|x|^k H_k\left(\frac{x}{|x|}\right)\right\} \\ &= \left(\frac{\partial^2}{\partial r^2} + \frac{1}{r}\frac{\partial}{\partial r} + \frac{1}{r^2}\frac{\partial^2}{\partial \varphi^2}\right)\left[r^k H_k(e^{i\varphi})\right] \\ &= \left[k(k-1)r^{k-2} + kr^{k-2}\right] H_k(e^{i\varphi}) + r^{k-2}\frac{\partial^2}{\partial\varphi^2} H_k(e^{i\varphi}). \end{aligned}$$

Therefore, the functions $H_k(e^{i\varphi})$ satisfy the linear ordinary differential equation

$$\frac{d^2}{d\varphi^2} H_k(e^{i\varphi}) + k^2 H_k(e^{i\varphi}) = 0, \qquad 0 \le \varphi \le 2\pi.$$

This means that

$$H_k(e^{i\varphi}) = A_k \cos k\varphi + B_k \sin k\varphi, \qquad A_k, B_k \in \mathbb{R}$$

holds true if $k \neq 0$ is correct. Since the function H_k is periodic in $[0, 2\pi]$, we infer $k \in \mathbb{Z}$. In the case $k = 0$ we obtain the solution

$$H_0(e^{i\varphi}) = A_0 + B_0\varphi, \qquad A_0, B_0 \in \mathbb{R}.$$

Therefore, $B_0 = 0$ holds true, and the theorem is proved. q.e.d.

5 Theory of Spherical Harmonics in n Variables

Theorem 4.3 from Section 4 suggests the following definition of the spherical harmonics in $\mathbb{R}^n$:

Definition 5.1. *Let $H_k = H_k(x_1, \ldots, x_n) \in C^2(\dot{\mathbb{R}}^n)$ denote a harmonic function on the set $\dot{\mathbb{R}}^n := \mathbb{R}^n \setminus \{0\}$ which is homogeneous of degree k, more precisely*

$$H_k(tx_1, \ldots, tx_n) = t^k H(x_1, \ldots, x_n) \qquad \text{for all} \quad x \in \dot{\mathbb{R}}^n, \quad t \in (0, +\infty).$$

Then we name

$$H_k = H_k(\xi_1, \ldots, \xi_n) : S^{n-1} \to \mathbb{R}$$

*an n-*dimensional spherical harmonic (or spherically harmonic function) of degree *k; here the symbol*

$$S^{n-1} := \{\xi = (\xi_1, \ldots, \xi_n) \in \mathbb{R}^n \;:\; \xi_1^2 + \ldots + \xi_n^2 = 1\}$$

denotes the $(n-1)$-dimensional unit sphere in the Euclidean space $\mathbb{R}^n$.

In this paragraph we answer the following questions for $n \geq 2$:

1. Are there spherical harmonics in all spatial dimensions, and for which degrees of homogeneity k do they exist?
2. Is the system of spherically harmonic functions complete?
3. In which relationship do the spherical harmonics appear with respect to the Laplace operator?

In Chapter 1, Section 8 we have represented the Laplace operator in $\mathbb{R}^n$ with respect to spherical coordinates. We utilize $r \in (0, +\infty)$ and $\xi = (\xi_1, \ldots, \xi_n) \in S^{n-1}$, and the function $u = u(r\xi)$ satisfies the identity

$$\boldsymbol{\Delta} u(r\xi) = \frac{\partial^2}{\partial r^2} u(r\xi) + \frac{n-1}{r}\frac{\partial}{\partial r} u(r\xi) + \frac{1}{r^2} \boldsymbol{\Lambda} u(r\xi); \tag{1}$$

here the symbol $\boldsymbol{\Lambda}$ denotes the invariant Laplace-Beltrami operator on the sphere S^{n-1}. We now endow the space of functions $C^0(S^{n-1}, \mathbb{R})$ with the inner product

$$(u, v) := \int_{S^{n-1}} u(\xi) v(\xi)\, d\sigma(\xi), \qquad u, v \in C^0(S^{n-1}, \mathbb{R}) \tag{2}$$

and we obtain a pre-Hilbert-space $\mathcal{H} = C^0(S^{n-1}, \mathbb{R})$. Setting

$$\|u\| := \sqrt{(u, u)}$$

the set $\mathcal{H}$ becomes a normed space.

Theorem 5.2. *The function*

$$H_k = H_k(\xi_1, \ldots, \xi_n) : S^{n-1} \to \mathbb{R}$$

is an n-dimensional spherical harmonic of the degree $k \in \mathbb{R}$ if and only if the following differential equation

$$\boldsymbol{\Lambda} H_k(\xi) + k\Big\{k + (n-2)\Big\} H_k(\xi) = 0, \qquad \xi \in S^{n-1} \tag{3}$$

is satisfied. If H_k and H_l are two spherical harmonics with different degrees $k \neq l$ satisfying $k + l \neq 2 - n$, we then have the orthogonality relation

$$(H_k, H_l) = 0. \tag{4}$$

Proof:

1. On account of (1) we have the identity

$$\begin{aligned} 0 = \boldsymbol{\Delta} H_k(r\xi) &= \boldsymbol{\Delta}\Big\{r^k H_k(\xi)\Big\} \\ &= \Big\{k(k-1)r^{k-2} + k(n-1)r^{k-2}\Big\} H_k(\xi) + r^{k-2} \boldsymbol{\Lambda} H_k(\xi) \end{aligned}$$

and equivalently

$$\boldsymbol{\Lambda} H_k(\xi) + \Big\{k^2 + (n-2)k\Big\} H_k(\xi) = 0, \qquad \xi \in S^{n-1}.$$

2. The symmetry of the operator $\boldsymbol{\Lambda}$ from Theorem 8.7 in Chapter 1, Section 8 yields

$$\begin{aligned}\Big\{k^2 + (n-2)k\Big\} \int\limits_{S^{n-1}} & H_k(\xi) H_l(\xi)\, d\sigma(\xi) \\ &= - \int\limits_{S^{n-1}} \Big(\boldsymbol{\Lambda} H_k(\xi)\Big) H_l(\xi)\, d\sigma(\xi) \\ &= - \int\limits_{S^{n-1}} H_k(\xi) \Big(\boldsymbol{\Lambda} H_l(\xi)\Big)\, d\sigma(\xi) \\ &= \Big\{l^2 + (n-2)l\Big\} \int\limits_{S^{n-1}} H_k(\xi) H_l(\xi)\, d\sigma(\xi).\end{aligned}$$

This implies that

$$0 = \Big\{k^2 - l^2 + (n-2)(k-l)\Big\}(H_k, H_l) = \{k-l\}\{k+l+n-2\}(H_k, H_l)$$

and therefore $(H_k, H_l) = 0$ if $k \neq l$ and $k + l \neq 2 - n$ is fulfilled. q.e.d.

Remarks: The spherical harmonics of the degree k are consequently eigenfunctions of the Laplace-Beltrami operator $\boldsymbol{\Lambda}$ on the sphere S^{n-1} to the eigenvalue $-k\{k + (n-2)\}$. The orthogonality condition (4) is especially satisfied in the case $k \geq 0$, $l \geq 0$ and $k \neq l$.

At this moment we do not yet know for which degrees $k \in \mathbb{R}$ (nonvanishing) spherical harmonics of the degree k exist. This will be investigated now: Given the continuous boundary function, we shall construct a harmonic function with the aid of Poisson's integral and shall decompose this function into homogeneous harmonic functions of the degrees $k = 0, 1, 2, \ldots$. Here we have to expand Poisson's kernel suitably with the aid of power series.

We take $\nu > 0$ as fixed and choose $h = \cos\vartheta \in [-1, +1]$ with $\vartheta \in [0, \pi]$; then we consider the following expression in $t \in (-1, +1)$:

$$\begin{aligned}(1-2ht+t^2)^{-\nu} &= (1-2(\cos\vartheta)t+t^2)^{-\nu}\\ &= (1-e^{i\vartheta}t)^{-\nu}(1-e^{-i\vartheta}t)^{-\nu}\\ &= \left\{\sum_{m=0}^{\infty}\binom{-\nu}{m}(-e^{i\vartheta}t)^m\right\}\left\{\sum_{m=0}^{\infty}\binom{-\nu}{m}(-e^{-i\vartheta}t)^m\right\}\\ &= \left\{\sum_{m=0}^{\infty}\begin{bmatrix}\nu\\ m\end{bmatrix}e^{im\vartheta}t^m\right\}\left\{\sum_{m=0}^{\infty}\begin{bmatrix}\nu\\ m\end{bmatrix}e^{-im\vartheta}t^m\right\}.\end{aligned}$$

Here we set

$$\begin{aligned}\begin{bmatrix}\nu\\ m\end{bmatrix} := \binom{-\nu}{m}(-1)^m &= \frac{-\nu(-\nu-1)(-\nu-2)\dots(-\nu-m+1)}{m!}(-1)^m\\ &= \frac{\nu(\nu+1)(\nu+2)\dots(\nu+m-1)}{m!}, \qquad m\in\mathbb{N},\end{aligned}$$

$$\begin{bmatrix}\nu\\ 0\end{bmatrix} := 1.$$

Defining the real coefficients

$$\begin{aligned}c_m^{(\nu)}(h) &:= \sum_{k=0}^{m}\begin{bmatrix}\nu\\ k\end{bmatrix}\begin{bmatrix}\nu\\ m-k\end{bmatrix}e^{ik\vartheta}e^{-i(m-k)\vartheta}\\ &= \sum_{k=0}^{m}\begin{bmatrix}\nu\\ k\end{bmatrix}\begin{bmatrix}\nu\\ m-k\end{bmatrix}e^{-i(m-2k)\vartheta}\\ &= \frac{1}{2}\sum_{k=0}^{m}\begin{bmatrix}\nu\\ k\end{bmatrix}\begin{bmatrix}\nu\\ m-k\end{bmatrix}\left\{e^{i(m-2k)\vartheta}+e^{-i(m-2k)\vartheta}\right\}\\ &= \sum_{k=0}^{m}\begin{bmatrix}\nu\\ k\end{bmatrix}\begin{bmatrix}\nu\\ m-k\end{bmatrix}\cos(m-2k)\vartheta,\end{aligned}$$

we obtain the following identity for $t\in(-1,+1)$:

$$(1-2ht+t^2)^{-\nu} = \sum_{m=0}^{\infty}c_m^{(\nu)}(h)t^m, \qquad t\in(-1,+1). \tag{5}$$

On account of the Binomial Theorem, we have the following expansion for $p\in\mathbb{Z}$:

$$\begin{aligned}\cos p\vartheta &= \frac{1}{2}\left(e^{ip\vartheta} + e^{-ip\vartheta}\right) = \frac{1}{2}\left\{(e^{i\vartheta})^p + (e^{-i\vartheta})^p\right\} \\ &= \frac{1}{2}\left\{(\cos\vartheta + i\sin\vartheta)^p + (\cos\vartheta - i\sin\vartheta)^p\right\} \\ &= (\cos\vartheta)^p - \binom{p}{2}(\cos\vartheta)^{p-2}(\sin\vartheta)^2 + \binom{p}{4}(\cos\vartheta)^{p-4}(\sin\vartheta)^4 - \ldots\end{aligned}$$

Due to the formula $\sin^2\vartheta = 1 - \cos^2\vartheta$, *Gegenbaur's polynomials* $c_m^{(\nu)}(h)$ are polynomials in $h = \cos\vartheta$ of the degree m. Furthermore, we utilize the relation

$$\sum_{m=0}^{\infty} c_m^{(\nu)}(-h)(-t)^m = (1 - 2ht + t^2)^{-\nu} = \sum_{m=0}^{\infty} c_m^{(\nu)}(h)t^m,$$

and comparison of the coefficients yields

$$c_m^{(\nu)}(-h) = (-1)^m c_m^{(\nu)}(h), \qquad m = 0, 1, 2, \ldots \tag{6}$$

Therefore, Gegenbaur's polynomials can be represented in the form

$$c_m^{(\nu)}(h) = \gamma_m^{(\nu)} h^m + \gamma_{m-2}^{(\nu)} h^{m-2} + \ldots \tag{7}$$

with the real constants $\gamma_m^{(\nu)}, \gamma_{m-2}^{(\nu)}, \ldots$. Furthermore, we have the estimate

$$\left|c_m^{(\nu)}(h)\right| \le \sum_{k=0}^{m} \begin{bmatrix}\nu \\ k\end{bmatrix}\begin{bmatrix}\nu \\ m-k\end{bmatrix} = c_m^{(\nu)}(1) \qquad \text{for all} \quad h \in [-1, +1]. \tag{8}$$

With $\nu = \frac{1}{2}$ we obtain the *Legendre polynomials* by $c_m^{(\frac{1}{2})}(h)$. We now choose $n \in \mathbb{N} \setminus \{1\}$. With the aid of (5) we expand as follows for $t \in (-1, +1)$ and $h \in [-1, +1]$:

$$\frac{1 - t^2}{(1 - 2ht + t^2)^{\frac{n}{2}}} = \sum_{m=0}^{\infty} c_m^{(\frac{n}{2})}(h)(1 - t^2)t^m =: \sum_{m=0}^{\infty} P_m(h; n)t^m. \tag{9}$$

For the case $n = 2$ we have derived the following expansion in the proof of Theorem 4.1 from Section 4 (compare the formula (6)):

$$\frac{1 - t^2}{1 - 2ht + t^2} = 1 + 2\sum_{m=1}^{\infty}(\cos m\vartheta)t^m, \qquad t \in (-1, +1). \tag{10}$$

Therefore, we have $P_0(h; 2) = 1$ and $P_m(h; 2) = 2\cos m\vartheta$, $m = 1, 2, \ldots$. For the case $n \ge 3$ we calculate

$$\begin{aligned}\left(1 + \frac{2t}{n-2}\frac{\partial}{\partial t}\right)\frac{1}{(1 - 2ht + t^2)^{\frac{n}{2}-1}} &= \frac{1 - 2ht + t^2 + \frac{2-n}{2}\frac{2t}{n-2}(-2h + 2t)}{(1 - 2ht + t^2)^{\frac{n}{2}}} \\ &= \frac{1 - t^2}{(1 - 2ht + t^2)^{\frac{n}{2}}}.\end{aligned}$$

Therefore, we have the identity

$$\frac{1-t^2}{(1-2ht+t^2)^{\frac{n}{2}}} = \left(1+\frac{2t}{n-2}\frac{\partial}{\partial t}\right)\frac{1}{(1-2ht+t^2)^{\frac{n}{2}-1}}, \qquad t\in(-1,+1). \quad (11)$$

Together with (9) we infer

$$\sum_{m=0}^{\infty} P_m(h;n)t^m = \frac{1-t^2}{(1-2ht+t^2)^{\frac{n}{2}}} = \left(1+\frac{2t}{n-2}\frac{\partial}{\partial t}\right)\sum_{m=0}^{\infty} c_m^{(\frac{n}{2}-1)}(h)t^m,$$

and comparision of the coefficients yields the formula

$$P_m(h;n) = c_m^{(\frac{n}{2}-1)}(h)\Big(\frac{2m}{n-2}+1\Big), \qquad m=0,1,2,\ldots \quad (12)$$

The relations (8) and (12) imply the estimate

$$|P_m(h;n)| \le P_m(1;n), \qquad h\in[-1,+1], \quad m\in\{0,1,2,\ldots\}. \quad (13)$$

This inequality holds true for $n=2,3,\ldots$

We now can expand the Poisson kernel: We choose $\eta\in S^{n-1}$ as fixed and $x=r\xi$ with $r\in[0,1)$ and $\xi\in S^{n-1}$ to be variable. We utilize the parameter of homogeneity $\tau\in\mathbb{R}$ with $|\tau r|<1$, and obtain the following relation with the aid of the expansion (9):

$$\begin{aligned}\frac{|\eta|^2-|\tau x|^2}{|\eta-\tau x|^n} &= \frac{1-(\tau r)^2}{\Big\{|\eta-(\tau r)\xi|^2\Big\}^{\frac{n}{2}}} \\ &= \frac{1-(\tau r)^2}{\Big\{1-2(\tau r)(\xi,\eta)+(\tau r)^2\Big\}^{\frac{n}{2}}} \\ &= \sum_{m=0}^{\infty}\Big\{P_m\Big((\xi,\eta);n\Big)r^m\Big\}\tau^m.\end{aligned} \quad (14)$$

For each $x\in\mathbb{R}^n$ with $|x|<1$ and each $\tau\in\mathbb{R}$ with $|\tau x|<1$ we have the identity

$$0=\Delta_x\left\{\frac{|\eta|^2-|\tau x|^2}{|\eta-\tau x|^n}\right\} = \sum_{m=0}^{\infty}\Delta_x\Big\{P_m\Big((\xi,\eta);n\Big)r^m\Big\}\tau^m.$$

Taking $\eta\in S^{n-1}$ fixed, the comparison of coefficients yields

$$\Delta_x\Big\{P_m\Big((\xi,\eta);n\Big)r^m\Big\}=0, \qquad |x|<1, \quad m=0,1,2,\ldots. \quad (15)$$

On account of (7) and (12) we have the representation

$$P_m\Big((\xi,\eta);n\Big)r^m = \Big(\pi_m^{(m)}(\xi,\eta)^m + \pi_{m-2}^{(m)}(\xi,\eta)^{m-2} + \ldots\Big)r^m$$

$$= \pi_m^{(m)}(x,\eta)^m + \pi_{m-2}^{(m)}(x,\eta)^{m-2}|x|^2 + \ldots$$

with the real constants $\pi_m^{(m)}, \pi_{m-2}^{(m)}, \ldots$. Therefore, $P_m((\xi,\eta);n)r^m$ is a homogeneous polynomial of the degree m in the variables $x_1,\ldots,x_n$. On account of (15), we obtain an n-dimensional spherical harmonic of the degree $m \in \{0,1,2,\ldots\}$ with $P_m((\xi,\eta);n)$ for each fixed $\eta \in S^{n-1}$. Given the function $f = f(\eta) : S^{n-1} \to \mathbb{R} \in C^0(S^{n-1},\mathbb{R})$, then the integral

$$\tilde{f}(\xi) := \frac{1}{\omega_n} \int\limits_{|\eta|=1} P_m\Big((\xi,\eta);n\Big) f(\eta)\,d\sigma(\eta), \qquad \xi \in S^{n-1}$$

represents an n-dimensional spherical harmonic of the degree m. Here $\tilde{f}(\xi)r^m$ means a homogeneous polynomial in the variables $x_1,\ldots,x_n$.

Theorem 5.3. *Let the function $f = f(x) : S^{n-1} \to \mathbb{R} \in C^0(S^{n-1},\mathbb{R})$ be prescribed, and the function $u = u(x) : B := \{x \in \mathbb{R}^n : |x| < 1\} \to \mathbb{R}$ of the class $C^2(B) \cap C^0(\overline{B})$ solves the Dirichlet problem*

$$\Delta u(x) = 0 \qquad \text{for all} \quad x \in B,$$
$$u(x) = f(x) \qquad \text{for all} \quad x \in \partial B = S^{n-1}.$$

For each $R \in (0,1)$ we then have the representation

$$u(x) = \sum_{m=0}^{\infty} \left\{ \frac{1}{\omega_n} \int\limits_{|\eta|=1} P_m\Big(\xi_1\eta_1 + \ldots + \xi_n\eta_n;n\Big) f(\eta)\,d\sigma(\eta) \right\} r^m \tag{16}$$

with $x = r\xi$, $\xi \in S^{n-1}$ and $0 \le r \le R$. The series on the right-hand side converges uniformly.

Proof: The unique solution of the Dirichlet problem above is given by Poisson's integral. With the aid of the expansion (14) for $\tau = 1$ we infer

$$u(x) = \frac{1}{\omega_n} \int\limits_{|\eta|=1} \frac{|\eta|^2 - |x|^2}{|\eta - x|^n} f(\eta)\,d\sigma(\eta)$$

$$= \frac{1}{\omega_n} \int\limits_{|\eta|=1} \left\{ \sum_{m=0}^{\infty} P_m\Big((\xi,\eta);n\Big) r^m \right\} f(\eta)\,d\sigma(\eta), \qquad x \in B.$$

For all $\xi,\eta \in S^{n-1}$ and $0 \le r \le R < 1$ we obtain the inequality

$$\left|\sum_{m=0}^{\infty} P_m\Big((\xi,\eta);n\Big)r^m\right| \le \sum_{m=0}^{\infty}\left|P_m\Big((\xi,\eta);n\Big)\right|r^m \le \sum_{m=0}^{\infty} P_m(1;n)R^m$$

$$= \frac{1-R^2}{(1-2R+R^2)^{\frac{n}{2}}} = \frac{1+R}{(1-R)^{n-1}}$$

respecting (9) and (13). Due to the Weierstraß majorant test, the following series

$$\sum_{m=0}^{\infty} P_m\Big((\xi,\eta);n\Big)r^m$$

converges uniformly on $S^{n-1}\times S^{n-1}\times[0,R]$ for all $R\in(0,1)$. This implies

$$u(x) = \sum_{m=0}^{\infty}\left\{\frac{1}{\omega_n}\int\limits_{|\eta|=1} P_m\Big(\xi_1\eta_1+\ldots+\xi_n\eta_n;n\Big)f(\eta)\,d\sigma(\eta)\right\}r^m, \qquad |x|\le R,$$

where the given series converges uniformly for all $R\in(0,1)$.

q.e.d.

We choose $k=0,1,2,\ldots$ and denote by

$$\mathcal{M}_k := \Big\{f: S^{n-1}\to\mathbb{R}: f \text{ is } n\text{-dimensional spherical harmonic of degree } k\Big\}$$

the *linear space of the n-dimensional spherical harmonics of the order k*. We already know $\dim\mathcal{M}_k\ge 1$ for $k=0,1,2,\ldots$ and intend to show $\dim\mathcal{M}_k<+\infty$ in the sequel. For the function $f=f(\eta)\in\mathcal{H}=C^0(S^{n-1},\mathbb{R})$ we define the *projector on* $\mathcal{M}_k$ by

$$\boldsymbol{P}_k f(\xi) = \hat{f}(\xi) := \frac{1}{\omega_n}\int\limits_{|\eta|=1} P_k\Big(\xi_1\eta_1+\ldots+\xi_n\eta_n;n\Big)f(\eta)\,d\sigma(\eta).$$

Theorem 5.4. *For each integer $k=0,1,2,\ldots$ the linear operator $\boldsymbol{P}_k:\mathcal{H}\to\mathcal{H}$ has the following properties:*

a) $(\boldsymbol{P}_k f,g)=(f,\boldsymbol{P}_k g)$ *for all* $f,g\in\mathcal{H}$;
b) $\boldsymbol{P}_k(\mathcal{H})=\mathcal{M}_k$;
c) $\boldsymbol{P}_k\circ\boldsymbol{P}_k=\boldsymbol{P}_k$.

Proof:

a) Let the functions $f,g\in\mathcal{H}$ be chosen arbitrarily. Then we have

$$\begin{aligned}(\boldsymbol{P}_k f, g) &= \int\limits_{|\xi|=1} \boldsymbol{P}_k f(\xi) g(\xi)\, d\sigma(\xi) \\ &= \int\limits_{|\xi|=1}\int\limits_{|\eta|=1} P_k(\xi_1\eta_1 + \ldots + \xi_n\eta_n) f(\eta) g(\xi)\, d\sigma(\eta)\, d\sigma(\xi) \\ &= (f, \boldsymbol{P}_k g).\end{aligned}$$

b) and c) In our considerations preceding Theorem 5.3 we already have seen that

$$\hat{f}(\xi) = \boldsymbol{P}_k f(\xi) \quad \in \mathcal{M}_k \qquad \text{for all} \quad f \in \mathcal{H}.$$

Therefore, we have $\boldsymbol{P}_k(\mathcal{H}) \subset \mathcal{M}_k$. Choosing $f \in \mathcal{M}_k$ arbitrarily we infer $\Delta_x(f(\xi)r^k) = 0$ in $\dot{\mathbb{R}}^n$ with $x = r\xi$. Now our Theorem 5.3 yields the representation

$$f(\xi)r^k = \sum_{m=0}^{\infty} \Big(\boldsymbol{P}_m f(\xi)\Big) r^m, \quad \xi \in S^{n-1}, \quad r \in [0,1).$$

Comparison of the coefficients implies

$$f(\xi) = \boldsymbol{P}_k f(\xi), \qquad \xi \in S^{n-1}.$$

Consequently, we obtain $\mathcal{M}_k \subset \boldsymbol{P}_k(\mathcal{H})$ and $\boldsymbol{P}_k \circ \boldsymbol{P}_k = \boldsymbol{P}_k$. q.e.d.

We now show that $\dim \mathcal{M}_k \in \mathbb{N}$ for $k = 0, 1, 2, \ldots$ is correct. For a fixed index $k \in \{0, 1, 2, \ldots\}$ we choose an orthonormal system $\{\varphi_\alpha\}_{\alpha=1,\ldots,N}$ of dimension $N \in \mathbb{N}$ in the linear subspace $\mathcal{M}_k \subset \mathcal{H}$. Then we have

$$(\varphi_\alpha, \varphi_\beta) = \delta_{\alpha\beta} \qquad \text{for all} \quad \alpha, \beta \in \{1, \ldots, N\}$$

and

$$\boldsymbol{P}_k \varphi_\alpha(\xi) = \varphi_\alpha(\xi), \qquad \alpha = 1, \ldots, N.$$

For each $\xi \in S^{n-1}$ we infer

$$\int\limits_{|\eta|=1} \frac{1}{\omega_n} P_k\Big((\xi,\eta); n\Big) \varphi_\alpha(\eta)\, d\sigma(\eta) = \varphi_\alpha(\xi), \qquad \alpha = 1, \ldots, N.$$

Bessel's inequality now yields

$$\begin{aligned}\sum_{\alpha=1}^{N} \varphi_\alpha^2(\xi) &= \sum_{\alpha=1}^{N} \Bigg\{ \int\limits_{|\eta|=1} \frac{1}{\omega_n} P_k\Big((\xi,\eta); n\Big) \varphi_\alpha(\eta)\, d\sigma(\eta) \Bigg\}^2 \\ &\le \int\limits_{|\eta|=1} \Bigg\{ \frac{1}{\omega_n} P_k\Big((\xi,\eta); n\Big) \Bigg\}^2 d\sigma(\eta) \qquad \text{for all} \quad \xi \in S^{n-1}.\end{aligned}$$

Therefore, we have

$$N = \int\limits_{|\xi|=1} \sum_{\alpha=1}^{N} \varphi_\alpha^2(\xi)\, d\sigma(\xi)$$
$$\leq \int\limits_{|\xi|=1} \int\limits_{|\eta|=1} \left\{ \frac{1}{\omega_n} P_k\Big((\xi,\eta);n\Big) \right\}^2 d\sigma(\eta)\, d\sigma(\xi).$$

Consequently, we get the following estimate for the dimension of $\mathcal{M}_k$, namely

$$\dim \mathcal{M}_k \leq \int\limits_{|\xi|=1} \int\limits_{|\eta|=1} \left\{ \frac{1}{\omega_n} P_k\Big((\xi,\eta);n\Big) \right\}^2 d\sigma(\eta)\, d\sigma(\xi) < +\infty, \quad k = 0,1,2,\ldots \tag{17}$$

We now set $N = N(k,n) := \dim \mathcal{M}_k$ and choose N orthonormal functions $H_{k1}(\xi), \ldots, H_{kN}(\xi)$ in $\mathcal{M}_k$ spanning the vector space $\mathcal{M}_k$. Each element $f \in \mathcal{M}_k$ can be represented in the form

$$f(\xi) = c_1 H_{k1}(\xi) + \ldots + c_N H_{kN}(\xi), \qquad \xi \in S^{n-1},$$

with the real coefficients $c_j = c_j[f]$ for $j = 1, \ldots, N$. More generally, taking $f = f(\xi) \in \mathcal{H}$ we have the identity

$$\frac{1}{\omega_n} \int\limits_{|\eta|=1} P_k\Big((\xi,\eta);n\Big) f(\eta)\, d\sigma(\eta) = c_1[f] H_{k1}(\xi) + \ldots + c_N[f] H_{kN}(\xi)$$

with the real constants $c_1[f], \ldots, c_N[f]$. This implies

$$c_l[f] = \int\limits_{|\xi|=1} H_{kl}(\xi) \left\{ \frac{1}{\omega_n} \int\limits_{|\eta|=1} P_k\Big((\xi,\eta);n\Big) f(\eta)\, d\sigma(\eta) \right\} d\sigma(\xi)$$
$$= \int\limits_{|\eta|=1} f(\eta) \left\{ \frac{1}{\omega_n} \int\limits_{|\xi|=1} P_k\Big((\xi,\eta);n\Big) H_{kl}(\xi)\, d\sigma(\xi) \right\} d\sigma(\eta)$$
$$= \int\limits_{|\eta|=1} f(\eta) H_{kl}(\eta)\, d\sigma(\eta).$$

Therefore, we obtain

$$\frac{1}{\omega_n} \int\limits_{|\eta|=1} P_k\Big((\xi,\eta);n\Big) f(\eta)\, d\sigma(\eta) = \int\limits_{|\eta|=1} \left\{ \sum_{l=1}^{N(k,n)} H_{kl}(\xi) H_{kl}(\eta) \right\} f(\eta)\, d\sigma(\eta)$$

and consequently

$$\int\limits_{|\eta|=1} \left\{ \frac{1}{\omega_n} P_k\Big((\xi,\eta);n\Big) - \sum_{l=1}^{N(k,n)} H_{kl}(\xi)H_{kl}(\eta) \right\} f(\eta)\,d\sigma(\eta) = 0$$

for all $\xi \in S^{n-1}$ and each $f = f(\eta) \in \mathcal{H}$. Since the functions $P_k((\xi,\eta);n)$ and $H_{kl}(\xi)$ are continuous, we get the *addition theorem for the n-dimensional spherical harmonics*

$$\sum_{l=1}^{N(k,n)} H_{kl}(\xi)H_{kl}(\eta) = \frac{1}{\omega_n} P_k\Big(\xi_1\eta_1 + \ldots + \xi_n\eta_n; n\Big), \qquad \xi,\eta \in S^{n-1} \tag{18}$$

for $k = 0,1,2,\ldots$ and $n = 2,3,\ldots$. We insert $\xi = \eta$ into (18) and integrate over the unit sphere S^{n-1}. Then we obtain

$$N(k,n) = \int\limits_{|\xi|=1} \sum_{l=1}^{N(k,n)} \Big(H_{kl}(\xi)\Big)^2 d\sigma(\xi) = P_k(1;n).$$

On account of (9), we finally deduce the expansion

$$\sum_{k=0}^{\infty} N(k,n)t^k = \sum_{k=0}^{\infty} P_k(1;n)t^k = \frac{1-t^2}{(1-t)^n} = \frac{1+t}{(1-t)^{n-1}}, \qquad |t| < 1.$$

We summarize our results as follows:

Theorem 5.5. *I. The cardinality $N(k,n)$ of all linear independent spherical harmonics in $\mathbb{R}^n$ of the order k is finite. The number $N(k,n) = \dim \mathcal{M}_k$ is determined by the equation*

$$\frac{1+t}{(1-t)^{n-1}} = \sum_{k=0}^{\infty} N(k,n)t^k, \quad |t| < 1. \tag{19}$$

II. Let $H_{k1}(\xi),\ldots,H_{kN}(\xi)$ represent the $N = N(k,n)$ orthonormal spherical harmonics of the order k, which means

$$\int\limits_{|\xi|=1} H_{kl}(\xi)H_{kl'}(\xi)\,d\sigma(\xi) = \delta_{ll'} \qquad \text{for} \quad l,l' \in \{1,\ldots,N\} \tag{20}$$

is satisfied. Then we have the representation

$$\sum_{l=1}^{N(k,n)} H_{kl}(\xi)H_{kl}(\eta) = \frac{1}{\omega_n} P_k\Big(\xi_1\eta_1 + \ldots + \xi_n\eta_n; n\Big) \tag{21}$$

for all $\xi,\eta \in S^{n-1}$. Here the functions $P_k(h;n)$ are defined by the equation

$$\frac{1-t^2}{(1-2ht+t^2)^{\frac{n}{2}}} = \sum_{k=0}^{\infty} P_k(h;n)t^k, \qquad -1 < t < +1, \quad -1 \le h \le +1. \tag{22}$$

III. Each solution $u = u(x) \in C^2(B) \cap C^0(\overline{B})$ *of Dirichlet's problem*

$$\Delta u(x) = 0 \qquad in \quad B,$$
$$u(x) = f(x) \qquad on \quad \partial B = S^{n-1}$$

possesses the representation as uniformly convergent series

$$u(x) = \sum_{k=0}^{\infty} \left\{ \sum_{l=1}^{N(k,n)} \left(\int\limits_{|\eta|=1} f(\eta) H_{kl}(\eta)\, d\sigma(\eta) \right) H_{kl}(\xi) \right\} r^k \tag{23}$$

with $x = r\xi$, $\xi \in S^{n-1}$ *and* $0 \le r \le R$*; here* $R \in (0,1)$ *can be chosen arbitrarily.*

Proof: Statement III immediately follows from (18) together with Theorem 5.3. q.e.d.

Analogously to Theorem 4.1 from Section 4, we obtain the following result for arbitrary dimensions $n \ge 2$:

Theorem 5.6. (Completeness of spherical harmonics)
The n-dimensional spherical harmonics $\{H_{kl}(\xi)\}_{k=0,1,2,\ldots;\ l=1,\ldots,N(k,n)}$ *constitute a complete orthonormal system of functions in* $\mathcal{H}$. *More precisely,*

$$(H_{kl}, H_{k'l'}) = \delta_{kk'}\delta_{ll'}, \qquad k,k' = 0,1,2,\ldots, \qquad l,l' = 1,\ldots,N(k,n)$$

holds true, and for each element $f \in \mathcal{H}$ *we have the relation*

$$\lim_{M\to\infty} \left\| f(\xi) - \sum_{k=0}^{M} \sum_{l=1}^{N(k,n)} f_{kl} H_{kl}(\xi) \right\| = 0$$

or equivalently

$$\|f\|^2 = \sum_{k=0}^{\infty} \sum_{l=1}^{N(k,n)} f_{kl}^2.$$

Here we have used the following abbreviations

$$f_{kl} := (f, H_{kl}), \qquad k = 0,1,2,\ldots, \qquad l = 1,\ldots,N(k,n)$$

for the Fourier coefficients.

Proof: We have only to show the completeness for the system of the n-dimensional spherical harmonics. To each element $f \in \mathcal{H}$ we have a function $u = u(x)$ with the following properties:

1. the function u is harmonic for all $|x| < 1$;
2. the function u is continuous for $|x| \le 1$ and satisfies the boundary condition

$$u(x) = f(x) \quad \text{for all} \quad |x| = 1.$$

According to Theorem 5.5, Statement III we see: For each $\varepsilon > 0$ there exists a radius $r \in (0,1)$ and an index $M = M(\varepsilon) \in \mathbb{N}$, such that

$$\left| f(\xi) - \sum_{k=0}^{M(\varepsilon)} r^k \sum_{l=1}^{N(k,n)} f_{kl} H_{kl}(\xi) \right| \le \varepsilon \qquad \text{for all} \quad \xi \in S^{n-1}.$$

This implies

$$\left\| f(\xi) - \sum_{k=0}^{M(\varepsilon)} r^k \sum_{l=1}^{N(k,n)} f_{kl} H_{kl}(\xi) \right\| \le \sqrt{\omega_n}\, \varepsilon,$$

and the minimal property of the Fourier coefficients yields

$$\left\| f(\xi) - \sum_{k=0}^{M(\varepsilon)} \sum_{l=1}^{N(k,n)} f_{kl} H_{kl}(\xi) \right\| \le \sqrt{\omega_n}\, \varepsilon.$$

From this relation we immediately infer the statement. q.e.d.

Corollaries from Theorem 5.6:

1. With $f(\xi)$ and $g(\xi)$ we consider two real, continuous functions on S^{n-1}, and then *Parseval's equation*

$$\int_{|\xi|=1} f(\xi) g(\xi)\, d\sigma(\xi) = \sum_{k=0}^{\infty} \sum_{l=1}^{N(k,n)} f_{kl} g_{kl}$$

   holds true with

$$f_{kl} = \int_{|\xi|=1} f(\xi) H_{kl}(\xi)\, d\sigma(\xi), \qquad g_{kl} = \int_{|\xi|=1} g(\xi) H_{kl}(\xi)\, d\sigma(\xi).$$

2. Nontrivial spherical harmonics H_j of the order $j \neq 0, \pm 1, \pm 2, \ldots$ do not exist. Due to Theorem 5.2 such a function would satisfy the orthogonality relations $(H_j, H_{kl}) = 0$. The system of functions $\{H_{kl}\}_{k=0,1,2,\ldots;\ l=1,\ldots,N(k,n)}$ being complete in $\mathcal{H}$, we infer $H_j = 0$ for all $j \neq 0, \pm 1, \pm 2, \ldots$

At the end of this paragraph we shall investigate the relationship of the spherical harmonics to the Laplace operator in $\mathbb{R}^n$. From (1) we infer the decomposition

$$\boldsymbol{\Delta} = \frac{\partial^2}{\partial r^2} + \frac{n-1}{r} \frac{\partial}{\partial r} + \frac{1}{r^2} \boldsymbol{\Lambda} \qquad \text{in} \quad \mathbb{R}^n.$$

We note (3) and obtain the following identity for arbitrary C^2-functions $f = f(r)$:

$$\begin{aligned} \Delta\Big\{ f(r) H_{kl}(\xi) \Big\} &= \Big\{ f''(r) + \frac{n-1}{r} f'(r) - \frac{k(k+(n-2))}{r^2} f(r) \Big\} H_{kl}(\xi) \\ &= \Big(L_{k,n} f(r) \Big) H_{kl}(\xi), \qquad l = 1, \ldots, N(k,n) \end{aligned} \tag{24}$$

with the operator

$$L_{k,n}f(r) := \left(\frac{\partial^2}{\partial r^2} + \frac{n-1}{r}\frac{\partial}{\partial r} - \frac{k(k+(n-2))}{r^2}\right)f(r).$$

Evidently, we have $L_{k,2} = L_k$ with the operator L_k from Section 4.

Let the function $u = u(x_1, \ldots, x_n) \in C^2(B_R)$ with $B_R := \{x \in \mathbb{R}^n : |x| < R\}$ be chosen arbitrarily. We now expand u in $\mathcal{H}$ with respect to the spherical harmonics

$$u = u(r\xi) = \sum_{k=0}^{\infty} \sum_{l=1}^{N(k,n)} f_{kl}(r) H_{kl}(\xi), \qquad 0 \le r < R, \quad \xi \in S^{n-1}. \tag{25}$$

Here we utilize the n-dimensional Fourier coefficients

$$f_{kl}(r) := \int\limits_{|\eta|=1} u(r\eta) H_{kl}(\eta)\, d\sigma(\eta), \qquad k = 0, 1, 2, \ldots, \quad l = 1, \ldots, N(k,n). \tag{26}$$

We then expand the function $\tilde{u}(x) = \Delta u(x)$, $x \in B_R$ in $\mathcal{H}$ with respect to spherical harmonics as well, and we obtain the n-dimensional Fourier series

$$\Delta u(x) = \Delta u(r\xi) = \sum_{k=0}^{\infty} \sum_{l=1}^{N(k,n)} \tilde{f}_{kl}(r) H_{kl}(\xi), \qquad 0 \le r < R, \quad \xi \in S^{n-1}, \tag{27}$$

with the Fourier coefficients $\tilde{f}_{kl}(r) = L_{k,n} f_{kl}(r)$. We consequently obtain the series for Δu in $\mathcal{H}$ by formal differentiation of the series for u. This is the content of the following

Proposition 5.7. *Let the function $u = u(x) \in C^2(B_R)$ be given, and its Fourier coefficients $f_{kl}(r)$ are defined due to the formula (26). Then the Fourier coefficients $\tilde{f}_{kl}(r)$ of Δu, namely*

$$\tilde{f}_{kl}(r) := \int\limits_{|\eta|=1} \Delta u(r\eta) H_{kl}(\eta)\, d\sigma(\eta), \qquad k = 0, 1, 2, \ldots, \quad l = 1, \ldots, N(k,n),$$

satisfy the identity

$$\tilde{f}_{kl}(r) = L_{k,n} f_{kl}(r), \qquad k = 0, 1, 2, \ldots, \quad l = 1, \ldots, N(k,n), \tag{28}$$

with $0 \le r < R$.

Proof: We choose $0 \le r < R$, and calculate with the aid of (3) as follows:

$$\begin{aligned}\tilde{f}_{kl}(r) &= \int\limits_{|\xi|=1} \Delta u(r\xi) H_{kl}(\xi)\, d\sigma(\xi) \\ &= \int\limits_{|\xi|=1} \left\{ \left(\frac{\partial^2}{\partial r^2} + \frac{n-1}{r}\frac{\partial}{\partial r} + \frac{1}{r^2}\boldsymbol{\Lambda} \right) u(r\xi) \right\} H_{kl}(\xi)\, d\sigma(\xi) \\ &= \left(\frac{\partial^2}{\partial r^2} + \frac{n-1}{r}\frac{\partial}{\partial r} \right) \int\limits_{|\xi|=1} u(r\xi) H_{kl}(\xi)\, d\sigma(\xi) \\ &\quad + \frac{1}{r^2} \int\limits_{|\xi|=1} u(r\xi) \boldsymbol{\Lambda} H_{kl}(\xi)\, d\sigma(\xi) \\ &= \left(\frac{\partial^2}{\partial r^2} + \frac{n-1}{r}\frac{\partial}{\partial r} - \frac{k(k+(n-2))}{r^2} \right) \int\limits_{|\xi|=1} u(r\xi) H_{kl}(\xi)\, d\sigma(\xi) \\ &= L_{k,n} f_{kl}(r) \qquad \text{for} \quad k = 0,1,2,\ldots, \quad l = 1,\ldots,N(k,n). \end{aligned}$$

q.e.d.

Remark: The most important partial differential equation of the second order in quantum mechanics, namely the Schrödinger equation, contains the Laplacian as its principal part. Therefore, the investigation of eigenvalues of this operator is of central interest. This will be presented in Chapter 8.

Figure 1.7 PORTRAIT OF JOSEPH A. F. PLATEAU (1801–1883) Universitätsbibliothek der Rheinischen Friedrich-Wilhelms-Univerität Bonn; taken from the book by *S. Hildebrandt and A. Tromba: Panoptimum – Mathematische Grundmuster des Vollkommenen*, Spektrum-Verlag Heidelberg (1986).

Chapter 6

Linear Partial Differential Equations in $\mathbb{R}^n$

In this chapter we become familiar with the different types of partial differential equations in $\mathbb{R}^n$. We treat the maximum principle for elliptic differential equations and prove the uniqueness of the mixed boundary value problem for quasilinear elliptic differential equations. Then we consider the initial value problem of the parabolic heat equation. Finally, we solve the Cauchy initial value problem for the hyperbolic wave equation in $\mathbb{R}^n$ and show its invariance under Lorentz transformations. The differential equations presented are situated in the center of mathematical physics.

1 The Maximum Principle for Elliptic Differential Equations

We shall consider a class of differential operators and equations, which contains the Laplace operator and equation as its characteristic representative.

Definition 1.1. *Let $\Omega \subset \mathbb{R}^n$ be a domain with $n \in \mathbb{N}$, where the continuous coefficient functions $a_{ij}(x), b_i(x), c(x) : \Omega \to \mathbb{R} \in C^0(\Omega)$ for $i, j = 1, \ldots, n$ are defined. Furthermore, let the matrix $(a_{ij}(x))_{i,j=1,\ldots,n}$ be symmetric for all $x \in \Omega$. The linear partial* differential operator *of the second order*

$$\mathcal{L} : C^2(\Omega) \to C^0(\Omega) \quad \textit{defined by}$$

$$\mathcal{L}u(x) := \sum_{i,j=1}^n a_{ij}(x) \frac{\partial^2}{\partial x_i \partial x_j} u(x) + \sum_{i=1}^n b_i(x) \frac{\partial}{\partial x_i} u(x) + c(x)u(x), \; x \in \Omega, \quad (1)$$

is named elliptic *(or alternatively* degenerate elliptic*), if and only if*

$$\sum_{i,j=1}^n a_{ij}(x)\xi_i\xi_j > 0 \qquad \left(\textit{or alternatively} \; \sum_{i,j=1}^n a_{ij}(x)\xi_i\xi_j \geq 0 \right)$$

F. Sauvigny, *Partial Differential Equations 1*, Universitext,
DOI 10.1007/978-1-4471-2981-3_6,

for all $\xi = (\xi_1, \ldots, \xi_n) \in \mathbb{R}^n \setminus \{0\}$ *and all* $x \in \Omega$ *is satisfied. When we have the* ellipticity constants $0 < m \leq M < +\infty$ *such that*

$$m|\xi|^2 \leq \sum_{i,j=1}^{n} a_{ij}(x)\xi_i\xi_j \leq M|\xi|^2$$

for all $\xi = (\xi_1, \ldots, \xi_n) \in \mathbb{R}^n$ *and all* $x \in \Omega$ *holds true, the operator* $\mathcal{L}$ *is called* uniformly elliptic. *In the case* $c(x) \equiv 0$, $x \in \Omega$, *we use the notation* $\mathcal{M}u(x) := \mathcal{L}u(x)$, $x \in \Omega$ *for the* reduced differential operator .

Remark: A uniformly elliptic differential operator is elliptic, and an elliptic differential operator is degenerate elliptic. The Laplace operator appears for $a_{ij}(x) \equiv \delta_{ij}$, $b_i(x) \equiv 0$, $c(x) \equiv 0$ with $i, j = 1, \ldots, n$ and is consequently uniformly elliptic with $m = M = 1$.

Proposition 1.2. *Let* $\mathcal{M} = \mathcal{M}u$, $u \in C^2(\Omega)$, *be a reduced, degenerate elliptic differential operator on the domain* $\Omega \subset \mathbb{R}^n$. *The function* u *attains its maximum at the point* $z \in \Omega$, *that means*

$$u(x) \leq u(z) \qquad \text{for all} \quad x \in \Omega.$$

Then we have $\{\mathcal{M}u(x)\}_{x=z} \leq 0$.

Proof: Since $u(x)$ attains its maximum at the point $z \in \Omega$, we infer $u_{x_i}(z) = 0$ for $i = 1, \ldots, n$ and consequently

$$\mathcal{M}u(z) = \sum_{i,j=1}^{n} a_{ij}(z)u_{x_ix_j}(z) + \sum_{i=1}^{n} b_i(z)u_{x_i}(z) = \sum_{i,j=1}^{n} u_{x_ix_j}(z)a_{ij}(z).$$

Now the $n \times n$-matrix $A := (a_{ij}(z))_{i,j=1,\ldots,n}$ is symmetric and positive-semidefinite. Therefore, we have an orthogonal matrix $S = (s_{ij})_{i,j=1,\ldots,n}$ and a diagonal matrix

$$\Lambda := \begin{pmatrix} \lambda_1 & & 0 \\ & \ddots & \\ 0 & & \lambda_n \end{pmatrix}$$

with the entries $\lambda_j \geq 0$ for $j = 1, \ldots, n$, such that

$$A = S^* \circ \Lambda \circ S \tag{2}$$

holds true (Theorem on the principal axes transformation). Now we set

$$\Lambda^{\frac{1}{2}} := \begin{pmatrix} \sqrt{\lambda_1} & & 0 \\ & \ddots & \\ 0 & & \sqrt{\lambda_n} \end{pmatrix}$$

and see

$$A = S^* \circ \Lambda \circ S = S^* \circ (\Lambda^{\frac{1}{2}})^* \circ \Lambda^{\frac{1}{2}} \circ S$$
$$= (\Lambda^{\frac{1}{2}} \circ S)^* \circ \Lambda^{\frac{1}{2}} \circ S = T^* \circ T \tag{3}$$

with $T := \Lambda^{\frac{1}{2}} \circ S =: (t_{ij})_{i,j=1,\ldots,n}$. Consequently, we obtain

$$A = T^* \circ T = \left(\sum_{k=1}^{n} t_{ki} t_{kj}\right)_{i,j=1,\ldots,n}. \tag{4}$$

Since the Hessian $(u_{x_i x_j}(z))_{i,j=1,\ldots,n}$ is negative-semidefinite, we conclude

$$\begin{aligned}\mathcal{M}u(z) &= \sum_{i,j=1}^{n} u_{x_i x_j}(z) a_{ij}(z) \\ &= \sum_{i,j,k=1}^{n} u_{x_i x_j}(z) t_{ki} t_{kj} \\ &= \sum_{k=1}^{n} \left(\sum_{i,j=1}^{n} u_{x_i x_j}(z) t_{ki} t_{kj} \right) \leq 0.\end{aligned}$$

q.e.d.

Theorem 1.3. (Uniqueness and stability)

I. Let $\mathcal{L}$ define a degenerate elliptic differential operator on the bounded domain $\Omega \subset \mathbb{R}^n$ with the coefficient function $c(x) \leq 0$, $x \in \Omega$.

II. We have the constants $0 < m \leq M < +\infty$, such that

$$m \leq a_{11}(x) \leq M, \quad |b_1(x)| \leq M, \quad |c(x)| \leq M \qquad \text{for all} \quad x \in \Omega;$$
$$\Omega \subset B_M := \Big\{ x \in \mathbb{R}^n \; : \; |x| < M \Big\} \tag{5}$$

is satisfied.

III. Finally, let $u = u(x) \in C^2(\Omega) \cap C^0(\overline{\Omega})$ be a solution of the Dirichlet problem

$$\mathcal{L}u(x) = f(x) \quad \text{in} \;\; \Omega, \qquad u(x) = g(x) \quad \text{auf} \;\; \partial\Omega \tag{6}$$

with the given functions $f = f(x) \in C^0(\Omega) \cap L^\infty(\Omega)$ and $g = g(x) \in C^0(\partial\Omega)$.

Statement: *Then we have a constant $\gamma = \gamma(m, M) \in [0, +\infty)$, such that*

$$|u(x)| \leq \max_{y \in \partial\Omega} |g(y)| + \gamma(m, M) \sup_{y \in \Omega} |f(y)|, \qquad x \in \overline{\Omega}. \tag{7}$$

Proof:

1. We consider the auxiliary function $v(x) := e^{\beta x_1}$, $x \in \overline{\Omega}$, with the arbitrary parameter $\beta > 0$. Then we calculate

$$\begin{aligned}\mathcal{L}v(x) &= a_{11}(x)\beta^2 e^{\beta x_1} + b_1(x)\beta e^{\beta x_1} + c(x)e^{\beta x_1}\\ &\geq e^{\beta x_1}\Big(m\beta^2 - M\beta - M\Big)\\ &\geq e^{-\beta(m,M)M}, \qquad x \in \Omega,\end{aligned}$$

choosing $\beta = \beta(m, M)$ so large that $m\beta^2 - M\beta - M \geq 1$ is satisfied.

2. The quantity $\varrho > 0$ still to be fixed, we define the auxiliary function

$$w(x) := \pm u(x) + \varrho\Big(v(x) - e^{\beta M}\Big) - \max_{y\in\partial\Omega} |g(y)|, \qquad x \in \overline{\Omega}.$$

On account of $c(x) \leq 0$ in Ω, we can estimate as follows:

$$\begin{aligned}\mathcal{L}w(x) &= \pm\mathcal{L}u(x) + \varrho\mathcal{L}v(x) - c(x)\Big(\varrho e^{\beta M} + \max_{y\in\partial\Omega}|g(y)|\Big)\\ &\geq \pm f(x) + \varrho e^{-\beta M} \qquad\qquad (8)\\ &\geq -\sup_{y\in\Omega}|f(y)| + \varrho e^{-\beta M}, \qquad x \in \Omega.\end{aligned}$$

Choosing $\varrho = e^{\beta(m,M)M}(\sup_{y\in\Omega}|f(y)| + \varepsilon)$ with a fixed number $\varepsilon > 0$, we obtain

$$\mathcal{L}w(x) \geq \varepsilon > 0 \qquad \text{for all} \quad x \in \Omega. \tag{9}$$

3. We calculate

$$\begin{aligned}w(x) &= \pm u(x) + \varrho(v(x) - e^{\beta M}) - \max_{y\in\partial\Omega}|g(y)|\\ &\leq \pm g(x) - \max_{y\in\partial\Omega}|g(y)| \;\leq\; 0\end{aligned}$$

for $x \in \partial\Omega$. Now $w(x) \leq 0$ even holds true for all $x \in \overline{\Omega}$. If this were violated, there would exist a point $z \in \Omega$ with $w(x) \leq w(z)$ for all $x \in \Omega$. Proposition 1.2 yields

$$\mathcal{L}w(z) = \mathcal{M}w(z) + c(z)w(z) \leq 0$$

in contradiction to (9). This implies

$$\pm u(x) \leq \max_{y\in\partial\Omega}|g(y)| + \varrho e^{\beta M} = \max_{y\in\partial\Omega}|g(y)| + e^{2\beta M}\Big(\sup_{y\in\Omega}|f(y)| + \varepsilon\Big)$$

for all $x \in \overline{\Omega}$ and all $\varepsilon > 0$. Passing to the limit $\varepsilon \downarrow 0$, we finally obtain

$$|u(x)| \leq \max_{y\in\partial\Omega}|g(y)| + \gamma(m, M)\sup_{y\in\Omega}|f(y)|, \qquad x \in \overline{\Omega},$$

with $\gamma(m, M) := e^{2\beta(m,M)M}$. q.e.d.

Remarks to Theorem 1.3:

1. The estimate (7) is already interesting for ordinary differential equations ($n=1$). This inequality is valid for uniformly elliptic differential operators in $\mathbb{R}^n$ with $n=2,3,\ldots$, and additionally for parabolic differential operators as
$$\Delta_x - \frac{\partial}{\partial t}, \quad (x,t) \in \mathbb{R}^n \times [0,+\infty)$$
appearing in the heat equation (compare Section 3).
2. We cannot omit the assumption $c(x) \le 0$, $x \in \Omega$ in Theorem 1.3, which is illustrated by the following example: For the function
$$u = u(x) = \sin x_1 \cdot \ldots \cdot \sin x_n, \qquad x = (x_1, \ldots, x_n) \in \Omega := (0,\pi)^n \subset \mathbb{R}^n,$$
we calculate
$$\Delta u(x) = \sum_{i=1}^{n} u_{x_i x_i}(x) = -nu(x), \qquad x \in \Omega.$$
Therefore, u satisfies the homogeneous Dirichlet problem
$$\Delta u(x) + nu(x) = 0 \quad \text{in} \quad \Omega, \qquad u(x) = 0 \quad \text{on} \quad \partial\Omega.$$
An estimate of the form (7) evidently does not hold here.
3. Let $u_j(x) \in C^2(\Omega) \cap C^0(\overline{\Omega})$ be two solutions of the problems
$$\mathcal{L}u_j(x) = f_j(x) \quad \text{in} \quad \Omega, \qquad u_j(x) = g_j(x) \quad \text{on} \quad \partial\Omega, \qquad j = 1,2.$$
Applied on the function $u(x) := u_1(x) - u_2(x)$, Theorem 1.3 yields the following estimate
$$|u_1(x) - u_2(x)| \le \max_{y \in \partial\Omega} |g_1(y) - g_2(y)| + \gamma(m,M) \sup_{y \in \Omega} |f_1(y) - f_2(y)| \quad (10)$$
for all $x \in \overline{\Omega}$. This implies the unique solvability of the Dirichlet problem (6) and the continuous dependence of the solution from the boundary values and the right-hand side of the differential equation.
4. The question of existence for a solution $u = u(x) \in C^2(\Omega) \cap C^0(\overline{\Omega})$ of the Dirichlet problem (6) can be answered in the affirmative for uniformly elliptic differential operators $\mathcal{L}$ with $c(x) \le 0$, $x \in \Omega$ under the following assumptions: The functions $a_{ij}(x)$, $b_i(x)$, $c(x)$, $f(x)$ are Hölder continuous in $\overline{\Omega}$ and the boundary $\partial\Omega$ of the bounded domain $\Omega \subset \mathbb{R}^n$ can locally be represented as the zero-set of a nondegenerate C^2-function $\varphi = \varphi(x)$ with Hölder continuous second derivatives, and $g : \partial\Omega \to \mathbb{R}$ has to be continuous. We shall establish this existence theorem in Chapter 9, departing from the Poisson equation $\Delta u(x) = f(x)$, $x \in \Omega$ and extending the result to the class of uniformly elliptic differential operators by the *continuity method.* This has been discovered by J. Leray and P. Schauder, and can as well be studied in the monograph [GT], Chapter 4 and Chapter 6.

Proposition 1.4. (Boundary point lemma of E. Hopf)

I. The coefficient functions $a_{ij}(x), b_i(x) \in C^0(\overline{G})$ *are given on the ball*

$$G := B_r(\xi) := \{x \in \mathbb{R}^n \, : \, |x - \xi| < r\}$$

in such a way that the reduced partial differential operator

$$\mathcal{M}u(x) := \sum_{i,j=1}^{n} a_{ij}(x) \frac{\partial^2}{\partial x_i \partial x_j} u(x) + \sum_{i=1}^{n} b_i(x) \frac{\partial}{\partial x_i} u(x), \qquad x \in G,$$

is uniformly elliptic in G *with the ellipticity constants* $0 < m \le M < +\infty$.

II. Let a solution $u = u(x) \in C^2(G) \cap C^0(\overline{G})$ *of the differential inequality*

$$\mathcal{M}u(x) \ge 0 \qquad \text{for all} \quad x \in G$$

be given, and for a fixed point $z \in \partial G$ *we have*

$$u(x) \le u(z) \quad \text{for all} \;\; x \in \overline{G} \qquad \text{and} \qquad u(\xi) < u(z). \tag{11}$$

III. Finally the derivative of u in direction of the exterior normal

$$\nu = \nu(z) := |z - \xi|^{-1}(z - \xi) \in S^{n-1}$$

may exist at the point $z \in \partial G$, *namely*

$$\frac{\partial u}{\partial \nu}(z) := \lim_{t \to 0-} \frac{d}{dt} u(z + t\nu(z)) = \lim_{t \to 0-} \frac{u(z) - u(z + t\nu(z))}{-t}.$$

Statement: *Then we have*

$$\frac{\partial u}{\partial \nu}(z) > 0. \tag{12}$$

Proof:

1. It is sufficient to prove the theorem for the case $G = B := B_1(0)$ and $u(z) = 0$. Given a function $u = u(x)$ with the properties I, II, III, we consider the composition

$$v(y) := u(\xi + ry) - u(\xi + r\eta), \qquad y \in B.$$

If we show (12) for $v(y)$ at the point $\eta \in \partial B$, then we see (12) for $u(x)$ at the point $z = \xi + r\eta \in \partial B_r(\xi)$.

2. Now let the function $u = u(x)$, $x \in \overline{B}$, with the properties I, II, III be given and $u(z) = 0$ hold true. For a parameter $\alpha > 0$ still to be fixed, we consider the auxiliary function

$$\varphi(x) := e^{-\alpha|x|^2} - e^{-\alpha} = e^{-\alpha(x_1^2 + \ldots + x_n^2)} - e^{-\alpha}, \qquad x = (x_1, \ldots, x_n) \in \overline{B}.$$

We remark $\varphi(x) = 0$ for all $x \in \partial B$ and calculate

$$\varphi_{x_i}(x) = -2\alpha x_i e^{-\alpha(x_1^2+\ldots+x_n^2)},$$

$$\varphi_{x_i x_j}(x) = \Big(4\alpha^2 x_i x_j - 2\alpha\delta_{ij}\Big)e^{-\alpha(x_1^2+\ldots x_n^2)}, \qquad x \in B.$$

Consequently, we obtain

$$\begin{aligned}\mathcal{M}\varphi(x) &= \Big\{4\alpha^2 \sum_{i,j=1}^{n} a_{ij}(x)x_i x_j - 2\alpha\sum_{i=1}^{n} a_{ii}(x) - 2\alpha\sum_{i=1}^{n} x_i b_i(x)\Big\}e^{-\alpha|x|^2} \\ &\geq 4\alpha^2 e^{-\alpha|x|^2}\Big\{m|x|^2 - \frac{1}{2\alpha}\sum_{i=1}^{n}\Big(a_{ii}(x) + x_i b_i(x)\Big)\Big\}, \qquad x \in \overline{B}.\end{aligned} \tag{13}$$

3. Now we determine numbers $r_1 \in (0,1)$ and $k_1 \in (-\infty, 0)$, such that

$$u(x) \leq k_1 \qquad \text{for all} \quad x \in \partial B_{r_1}(0) \tag{14}$$

is valid. On account of (13), we can choose $\alpha \in (0, +\infty)$ so large that the inequality

$$\mathcal{M}\varphi(x) > 0 \qquad \text{for all} \quad x \in \Omega := \Big\{x \in \mathbb{R}^n \,:\, r_1 < |x| < 1\Big\} \tag{15}$$

is satisfied. Then we define the auxiliary function

$$v(x) := u(x) + \varepsilon\varphi(x), \qquad x \in \Omega.$$

Here we choose $\varepsilon > 0$ so small that the inequality

$$v(x) \leq 0 \qquad \text{for all} \quad x \in \partial\Omega \tag{16}$$

holds true on account of (14). Furthermore, (15) and the assumption II yield

$$\mathcal{M}v(x) = \mathcal{M}u(x) + \varepsilon\mathcal{M}\varphi(x) > 0 \qquad \text{for all} \quad x \in \Omega. \tag{17}$$

Due to Proposition 1.2, the function $v(x)$ attains its maximum on $\partial\Omega$. Therefore, (16) implies $v(x) \leq 0$, $x \in \overline{\Omega}$ and

$$u(x) \leq -\varepsilon\varphi(x) = \varepsilon\Big(e^{-\alpha} - e^{-\alpha|x|^2}\Big), \qquad x \in \overline{\Omega}. \tag{18}$$

Now we define the functions

$$\tilde{u}(r) := u(rz), \quad \tilde{v}(r) := -\varepsilon\varphi(rz), \qquad r_1 \leq r \leq 1.$$

Since $\tilde{u}(r) \leq \tilde{v}(r)$ for $r_1 \leq r \leq 1$ and $\tilde{u}(1) = \tilde{v}(1) = 0$ hold true, we obtain

$$\frac{d}{dr}\tilde{u}(r)\Big|_{r=1} \geq \frac{d}{dr}\tilde{v}(r)\Big|_{r=1} = \frac{d}{dr}\Big\{\varepsilon(e^{-\alpha} - e^{-\alpha r^2})\Big\}_{r=1} = 2\alpha\varepsilon e^{-\alpha} > 0.$$

This implies the inequality (12) stated above. q.e.d.

Theorem 1.5. (Maximum principle of E. Hopf)

I. Let $\mathcal{M} = \mathcal{M}u$, $u \in C^2(\Omega)$, denote a reduced elliptic differential operator on the domain $\Omega \subset \mathbb{R}^n$, $n \in \mathbb{N}$.

II. The function $u = u(x) \in C^2(\Omega)$ satisfies the differential inequality

$$\mathcal{M}u(x) \geq 0, \qquad x \in \Omega,$$

and attains its maximum at a point $z \in \Omega$, that means

$$u(z) \geq u(x) \qquad \text{for all} \quad x \in \Omega.$$

Statement: *Then we have $u(x) \equiv u(z)$ for all $x \in \Omega$.*

Proof: We consider the following nonvoid set which is closed in Ω, namely

$$\Theta := \Big\{x \in \Omega \,:\, u(x) = \sup_{y \in \Omega} u(y) =: s\Big\} \neq \emptyset.$$

Then we show this set being open. Since Ω is a domain, the continuation along a path yields the identity $\Theta = \Omega$ and consequently

$$u(x) \equiv s = u(z) \qquad \text{for all} \quad x \in \Omega.$$

We choose $\xi \in \Theta$ arbitrarily. For a given $\eta \in \Omega$ with

$$|\eta - \xi| < \frac{1}{2}\,\mathrm{dist}(\xi, \mathbb{R}^n \setminus \Omega)$$

we consider the ball $G := B_\varrho(\eta)$ of radius $\varrho := |\eta - \xi|$ about the center η. Obviously, we have $G \subset\subset \Omega$ und $\xi \in \partial G$. Therefore, we find ellipticity constants $0 < m \leq M < +\infty$ such that $\mathcal{M}u$, $u \in C^2(G)$, is uniformly elliptic. If the inequality $u(\eta) < s = u(\xi)$ were fulfilled, Proposition 1.4 would yield

$$\frac{\partial u}{\partial \nu}(\xi) = \nabla u(\xi) \cdot \nu > 0$$

in contradiction to $\nabla u(\xi) = 0$. This implies $u(\eta) = s$. Since this is correct for arbitrary $\eta \in \Omega$ with $|\eta - \xi| < \frac{1}{2}\,\mathrm{dist}(\xi, \mathbb{R}^n \setminus \Omega)$, we obtain $B_r(\xi) \subset \Theta$ with a radius $0 < r < \frac{1}{2}\,\mathrm{dist}(\xi, \mathbb{R}^n \setminus \Omega)$. Therefore, the set Θ is open. q.e.d.

Theorem 1.6. (Strong maximum principle)

I. Let $\Omega \subset \mathbb{R}^n$ be a domain and $z \in \partial\Omega$ a boundary point of Ω with the following property: There exists a ball $B_\varrho(z)$ and a function

$$\varphi = \varphi(x) \in C^2(B_\varrho(z)) \quad \text{with} \quad \nabla\varphi(z) \neq 0 \quad \text{and} \quad \varphi(z) = 0,$$

such that we have

$$\Omega \cap B_\varrho(z) = \Big\{x \in B_\varrho(z) \,:\, \varphi(x) < 0\Big\}.$$

II. The coefficient functions $a_{ij}(x), b_i(x) \in C^0(\overline{\Omega})$, $i,j = 1,\ldots,n$, are given in such a way that the reduced partial differential operator

$$\mathcal{M}u(x) = \sum_{i,j=1}^{n} a_{ij}(x)\frac{\partial^2}{\partial x_i \partial x_j}u(x) + \sum_{i=1}^{n} b_i(x)\frac{\partial}{\partial x_i}u(x), \qquad x \in \Omega,$$

is uniformly elliptic on Ω.

III. The function $u = u(x) \in C^2(\Omega) \cap C^0(\overline{\Omega})$ may satisfy the following differential inequality

$$\mathcal{M}u(x) \geq 0 \qquad \textit{for all} \quad x \in \Omega.$$

IV. The function u attains its maximum at the point z, namely

$$u(x) \leq u(z) \qquad \textit{for all} \quad x \in \Omega.$$

Finally its derivative in the direction of the exterior normal ν to $\partial\Omega$ exists there and satisfies

$$\frac{\partial u}{\partial \nu}(z) = 0.$$

Statement: *Then we have $u(x) \equiv u(z)$ for all $x \in \overline{\Omega}$.*

Proof: On account of the assumption I, we can find a ball $G = B_r(\xi)$ with the center $\xi \in \Omega$ and the radius $r > 0$ such that

$$G \subset \Omega, \qquad \overline{G} \cap \partial\Omega = \{z\}, \qquad \nu(z) = |z-\xi|^{-1}(z-\xi)$$

is valid. If the inequality $u(\xi) < u(z)$ would be fulfilled, the Hopf boundary point lemma implies $\frac{\partial u}{\partial \nu}(z) > 0$ in contradiction to the assumption IV. Consequently, the function u attains its maximum at an interior point $\xi \in \Omega$. Theorem 1.5 gives us the identity $u(x) \equiv u(z)$ for all $x \in \Omega$.

q.e.d.

Example 1.7. For $n = 2, 3, \ldots$ we consider the sector

$$S := \Big\{x + iy = re^{i\varphi} \;:\; r > 0,\; \varphi \in \Big(-\frac{\pi}{2n}, \frac{\pi}{2n}\Big)\Big\}$$

and the function $v = v(x,y) : \overline{S} \to \mathbb{R}$ defined by

$$v(x,y) := -\mathrm{Re}((x+iy)^n) = -r^n \cos n\varphi, \qquad x + iy = re^{i\varphi} \in \overline{S}.$$

Obviously, we have:

$$\begin{aligned} &v \in C^2(\overline{S}), \qquad \Delta v(x,y) = 0 \quad \text{in} \quad S, \qquad v(x,y) < 0 \quad \text{in} \quad S, \\ &v(x,y) = 0 \quad \text{on} \quad \partial S, \qquad v(0,0) = 0, \qquad \nabla v(0,0) = 0. \end{aligned} \tag{19}$$

The harmonic, nonconstant function v takes on its maximum at a boundary point with vanishing gradient. Consequently, the assumption I in Theorem 1.6 cannot be deleted.

The applicability of the maximum principle for linear elliptic differential operators $\mathcal{L}$ depends decisively on the sign-condition $c(x) \leq 0$, $x \in \Omega$.

Definition 1.8. *We denote the linear elliptic* differential operator $\mathcal{L}$ *being* stable, *if there exists a function* $v(x) : \Omega \to (0, +\infty) \in C^2(\Omega)$ *satisfying*

$$\mathcal{L}v(x) \leq 0 \qquad \textit{for all} \quad x \in \Omega.$$

If the sign-condition above is fulfilled, then the operator $\mathcal{L}$ is stable with the function $v(x) \equiv 1$, $x \in \Omega$. In the general situation of a stable differential operator $\mathcal{L}$, we apply the fundamental *product device* as follows

$$u(x) = w(x)v(x), \quad x \in \Omega, \quad \text{or equivalently} \quad w(x) = \frac{u(x)}{v(x)}, \quad x \in \Omega.$$

We then calculate

$$\begin{aligned}
\mathcal{L}u(x) &= \sum_{i,j=1}^{n} a_{ij}(x)[w(x)v(x)]_{x_i x_j} + \sum_{i=1}^{n} b_i(x)[w(x)v(x)]_{x_i} + c(x)w(x)v(x) \\
&= \sum_{i,j=1}^{n} a_{ij}(x) w_{x_i x_j} v + \sum_{i,j=1}^{n} a_{ij}(x)[w_{x_i} v_{x_j} + w_{x_j} v_{x_i}] + \sum_{i,j=1}^{n} a_{ij}(x) w v_{x_i x_j} \\
&\quad + \sum_{i=1}^{n} b_i(x) w_{x_i}(x) v(x) + \sum_{i=1}^{n} b_i(x) w(x) v_{x_i}(x) + c(x)w(x)v(x) \\
&= \sum_{i,j=1}^{n} \Big\{ v(x) a_{ij}(x) \Big\} w_{x_i x_j} \\
&\quad + \sum_{i=1}^{n} \Big\{ v(x) b_i(x) + \sum_{j=1}^{n} [a_{ij}(x) + a_{ji}(x)] v_{x_j}(x) \Big\} w_{x_i} + \Big\{ \mathcal{L}v(x) \Big\} w \\
&=: \sum_{i,j=1}^{n} \tilde{a}_{ij}(x) w_{x_i x_j}(x) + \sum_{i=1}^{n} \tilde{b}_i(x) w_{x_i}(x) + \tilde{c}(x) w(x) \quad =: \quad \widetilde{\mathcal{L}}v(x).
\end{aligned}$$

Therefore, we obtain an elliptic differential operator $\widetilde{\mathcal{L}}$ for $w(x)$ satisfying a sign-condition for the coefficient function $\tilde{c}(x) := \mathcal{L}v(x)$. The differential operator $\widetilde{\mathcal{L}}$ is subject to the maximum principle, and we conclude

Theorem 1.9. *Let* $\mathcal{L}$ *be a stable elliptic differential operator defined on the bounded domain* $\Omega \subset \mathbb{R}^n$ *and the function* $u = u(x) \in C^2(\Omega) \cap C^0(\overline{\Omega})$ *may solve the homogeneous Dirichlet problem*

$$\mathcal{L}u(x) = 0 \quad \textit{in} \ \ \Omega, \qquad u(x) = 0 \quad \textit{on} \ \ \partial\Omega.$$

Then we have $u(x) \equiv 0$ *in* $\overline{\Omega}$.

2 Quasilinear Elliptic Differential Equations

Now we investigate a class of elliptic differential equations, which contains the linear ones as a special case, namely the quasilinear differential equations. Let $\Omega \subset \mathbb{R}^n$ be a bounded domain. We consider the coefficient functions

$$A^{ij} = A^{ij}(x,p) = A^{ij}(x_1,\ldots,x_n;p_1,\ldots,p_n) \in C^0(\overline{\Omega} \times \mathbb{R}^n, \mathbb{R})$$

for $i,j = 1,\ldots,n$. The matrix $(A^{ij}(x,p))_{i,j=1,\ldots,n}$ is symmetric and positive-definite for each $(x,p) \in \overline{\Omega} \times \mathbb{R}^n$. The partial derivatives

$$A^{ij}_{p_k}(x,p) := \frac{\partial}{\partial p_k} A^{ij}(x,p) \in C^0(\overline{\Omega} \times \mathbb{R}^n, \mathbb{R}), \qquad i,j,k = 1\ldots,n$$

should exist. Furthermore, we choose a function

$$B = B(x,z,p) = B(x_1,\ldots,x_n;z;p_1,\ldots,p_n) : \overline{\Omega} \times \mathbb{R} \times \mathbb{R}^n \to \mathbb{R}$$

of the regularity class $C^0(\overline{\Omega} \times \mathbb{R}^{1+n})$, whose partial derivatives $B_z, B_{p_1}, \ldots, B_{p_n}$ exist and belong to the class $C^0(\overline{\Omega} \times \mathbb{R}^{1+n})$. Finally we abbreviate

$$B_p := (B_{p_1},\ldots,B_{p_n}) \qquad \text{and} \qquad A^{ij}_p := (A^{ij}_{p_1},\ldots,A^{ij}_{p_n}), \quad i,j = 1,\ldots,n.$$

We now consider the following differential operator acting on the functions $u = u(x) \in C^2(\Omega)$, namely

$$\mathcal{Q}u(x) := \sum_{i,j=1}^{n} A^{ij}(x, \nabla u(x)) \frac{\partial^2}{\partial x_i \partial x_j} u(x) + B(x, u(x), \nabla u(x)), \qquad x \in \Omega. \tag{1}$$

We denote by

$$\sum_{i,j=1}^{n} A^{ij}(x, \nabla u(x)) \frac{\partial^2}{\partial x_i \partial x_j} u(x)$$

the *principal part* of the operator $\mathcal{Q}$. The term $B(x, u(x), \nabla u(x))$ of lower order is called *subordinate part* of $\mathcal{Q}$.

Remark: Here we investigate quasilinear operators $\mathcal{Q}$ whose principal parts only have coefficient functions A^{ij} independent of u.

Example 2.1. We consider the case $n = 2$ and abbreviate $(x_1, x_2) =: (x,y)$. The function

$$H = H(x,y,z) : \overline{\Omega} \times \mathbb{R} \to \mathbb{R}$$

is assumed to be continuous and may possess the continuous partial derivative $H_z(x,y,z) : \overline{\Omega} \times \mathbb{R} \to \mathbb{R}$. We consider a solution $z = \zeta(x,y) : \Omega \to \mathbb{R} \in C^2(\Omega)$ of the *nonparametric equation of prescribed mean curvature*

$$\begin{aligned}\mathcal{M}\zeta(x,y) &:= \Big(1+\zeta_y^2(x,y)\Big)\zeta_{xx} - 2\zeta_x\zeta_y\zeta_{xy} + \Big(1+\zeta_x^2(x,y)\Big)\zeta_{yy}(x,y) \\ &= 2H(x,y,\zeta(x,y))\Big(1+|\nabla\zeta(x,y)|^2\Big)^{\frac{3}{2}}, \qquad (x,y)\in\Omega.\end{aligned} \tag{2}$$

The surface

$$X(x,y) := (x,y,\zeta(x,y)), \qquad (x,y)\in\Omega$$

represents a graph over the x,y-plane with the prescribed mean curvature $H(X(x,y))$ at each point $X(x,y)$. We name $\mathcal{M}$ the *minimal surface operator* and

$$\mathcal{M}\zeta(x,y) = 0, \qquad (x,y)\in\Omega,$$

is the intensively studied *minimal surface equation.* The operator $\mathcal{M}\zeta$ represents the principal part of a quasilinear elliptic differential operator with the coefficients

$$\begin{pmatrix} A^{11}(p) & A^{12}(p) \\ A^{21}(p) & A^{22}(p)\end{pmatrix} := \begin{pmatrix} 1+p_2^2 & -p_1p_2 \\ -p_1p_2 & 1+p_1^2\end{pmatrix}. \tag{3}$$

Setting

$$B(x,y,z,p) := -2H(x,y,z)(1+p_1^2+p_2^2)^{\frac{3}{2}}, \tag{4}$$

the equation (2) appears as the quasilinear elliptic differential equation

$$\mathcal{Q}\zeta(x,y) = 0, \qquad (x,y)\in\Omega.$$

Now we consider two solutions $u = u(x)\in C^2(\Omega)$ and $v = v(x)\in C^2(\Omega)$ of the general quasilinear elliptic differential equation

$$\mathcal{Q}u(x) := \sum_{i,j=1}^{n} A^{ij}(x,\nabla u(x))\frac{\partial^2}{\partial x_i\partial x_j}u(x) + B(x,u(x),\nabla u(x)) = 0, \qquad x\in\Omega \tag{5}$$

and

$$\mathcal{Q}v(x) := \sum_{i,j=1}^{n} A^{ij}(x,\nabla v(x))\frac{\partial^2}{\partial x_i\partial x_j}v(x) + B(x,v(x),\nabla v(x)) = 0, \qquad x\in\Omega, \tag{6}$$

respectively. For the difference function

$$w(x) := u(x) - v(x) \in C^2(\Omega,\mathbb{R})$$

we derive a linear elliptic differential equation. The relations (5) and (6) imply

$$
\begin{aligned}
0 &= \mathcal{Q}u(x) - \mathcal{Q}v(x) \\
&= \sum_{i,j=1}^{n} A^{ij}(x, \nabla u(x)) \frac{\partial^2}{\partial x_i \partial x_j} w(x) \\
&\quad + \sum_{i,j=1}^{n} \Big\{ A^{ij}(x, \nabla u(x)) - A^{ij}(x, \nabla v(x)) \Big\} \frac{\partial^2}{\partial x_i \partial x_j} v(x) \\
&\quad + \Big\{ B(x, u(x), \nabla u(x)) - B(x, u(x), \nabla v(x)) \Big\} \\
&\quad + \Big\{ B(x, u(x), \nabla v(x)) - B(x, v(x), \nabla v(x)) \Big\}, \qquad x \in \Omega.
\end{aligned} \tag{7}
$$

We set

$$
a_{ij} = a_{ij}(x) := A^{ij}(x, \nabla u(x)), \quad x \in \Omega \quad \text{for} \quad i, j = 1, \ldots, n \tag{8}
$$

and see $a_{ij} \in C^0(\Omega, \mathbb{R})$. Furthermore, we calculate

$$
\begin{aligned}
B(x, u(x), \nabla v(x)) &- B(x, v(x), \nabla v(x)) \\
&= \int_0^1 \frac{d}{dt} B(x, v(x) + t w(x), \nabla v(x))\, dt \\
&= w(x) \int_0^1 B_z(x, v(x) + t w(x), \nabla v(x))\, dt
\end{aligned} \tag{9}
$$

and define the continuous function

$$
c(x) := \int_0^1 B_z(x, v(x) + t w(x), \nabla v(x))\, dt, \qquad x \in \Omega. \tag{10}
$$

Finally we note that

$$
\begin{aligned}
B(x, u(x), \nabla u(x)) &- B(x, u(x), \nabla v(x)) \\
&= \int_0^1 \frac{d}{dt} B(x, u(x), \nabla v(x) + t \nabla w(x))\, dt \\
&= \nabla w(x) \cdot \int_0^1 B_p(x, u(x), \nabla v(x) + t \nabla w(x))\, dt
\end{aligned} \tag{11}
$$

and

$$A^{ij}(x, \nabla u(x)) - A^{ij}(x, \nabla v(x)) = \int_0^1 \frac{d}{dt} A^{ij}(x, \nabla v(x) + t\nabla w(x))\, dt$$
$$= \nabla w(x) \cdot \int_0^1 A_p^{ij}(x, \nabla v(x) + t\nabla w(x))\, dt.$$

We define the coefficient functions

$$(b_1(x), \ldots, b_n(x)) = b(x) := \sum_{i,j=1}^{n} v_{x_i x_j}(x) \int_0^1 A_p^{ij}(x, \nabla v(x) + t\nabla w(x))\, dt + \int_0^1 B_p(x, u(x), \nabla v(x) + t\nabla w(x))\, dt, \qquad x \in \Omega, \tag{12}$$

and observe $b_i = b_i(x) \in C^0(\Omega)$ for $i = 1, \ldots, n$. Altogether we obtain the following linear elliptic differential equation for $w = w(x)$, namely

$$\sum_{i,j=1}^{n} a_{ij}(x) \frac{\partial^2}{\partial x_i \partial x_j} w(x) + \sum_{i=1}^{n} b_i(x) \frac{\partial}{\partial x_i} w(x) + c(x) w(x) = 0, \qquad x \in \Omega, \tag{13}$$

with the coefficient functions from (8), (10), and (12).

Example 2.2. On the domain $\Omega \subset \mathbb{R}^2$ we consider two solutions

$$u = u(x, y) \in C^2(\Omega) \quad \text{and} \quad v = v(x, y) \in C^2(\Omega)$$

of the minimal surface equation

$$\mathcal{M}u(x, y) = 0 = \mathcal{M}v(x, y), \qquad (x, y) \in \Omega.$$

Then the difference function $w(x, y) := u(x, y) - v(x, y)$, $(x, y) \in \Omega$, satisfies the linear elliptic differential equation

$$a(x, y) w_{xx}(x, y) + 2b(x, y) w_{xy} + c(x, y) w_{yy}(x, y) + d(x, y) w_x(x, y) + e(x, y) w_y(x, y) = 0 \quad \text{in} \quad \Omega \tag{14}$$

with the coefficient functions

$$a(x, y) = 1 + u_y^2, \qquad b(x, y) = -u_x u_y, \qquad c(x, y) = 1 + u_x^2$$

and

$$d(x, y) = -(u_y + v_y) v_{xy} + (u_x + v_x) v_{yy}, \qquad e(x, y) = (u_y + v_y) v_{xx} - (u_x + v_x) v_{xy}.$$

Theorem 2.3. (Uniqueness of the mixed boundary value problem)

I. Let $\Omega \subset \mathbb{R}^2$ be a bounded domain, whose boundary $\partial\Omega$ may contain an - eventually void - subset $\Gamma \subsetneqq \partial\Omega$ with the following properties:
a) The set $\partial\Omega \setminus \Gamma$ is closed.
b) For all $\xi \in \Gamma$ we have a number $\varrho = \varrho(\xi) \in (0, +\infty)$ and a function $\varphi = \varphi(x) \in C^2(B_\varrho(\xi))$ with $\varphi(\xi) = 0$ and $\nabla\varphi(\xi) \neq 0$, such that

$$\Omega \cap B_\varrho(\xi) = \Big\{ y \in B_\varrho(\xi) \,:\, \varphi(y) < 0 \Big\}.$$

II. The continuous functions $f = f(x) : \partial\Omega \setminus \Gamma \to \mathbb{R}$ and $g = g(x) : \Gamma \to \mathbb{R}$ are given.

III. The two functions $u = u(x) : \overline{\Omega} \to \mathbb{R}$ and $v = v(x) : \overline{\Omega} \to \mathbb{R}$ of the regularity class $C^2(\Omega) \cap C^0(\overline{\Omega}) \cap C^1(\Omega \cup \Gamma)$ are solutions of the mixed quasilinear elliptic boundary value problem

$$\sum_{i,j=1}^{n} A^{ij}(x, \nabla u(x)) \frac{\partial^2}{\partial x_i \partial x_j} u(x) + B(x, u(x), \nabla u(x)) = 0, \qquad x \in \Omega, \quad (15)$$

$$u(x) = f(x), \qquad x \in \partial\Omega \setminus \Gamma, \quad (16)$$

$$\frac{\partial}{\partial\nu} u(x) = g(x), \qquad x \in \Gamma. \quad (17)$$

Here $\nu = \nu(x) : \Gamma \to S^{n-1}$ denotes the exterior normal on Γ to the domain Ω.

IV. Finally, we require

$$B_z(x, z, p) \leq 0 \qquad \text{for all} \quad (x, z, p) \in \Omega \times \mathbb{R}^{1+n}.$$

Statement: *Then we have $u(x) \equiv v(x)$ for all $x \in \overline{\Omega}$.*

Remark: The boundary condition (16) is called the *Dirichlet boundary condition*, whereas in the equation (17) the *Neumann boundary condition* appears.

Proof of Theorem 2.3: The function

$$w(x) := u(x) - v(x) \in C^2(\Omega) \cap C^0(\overline{\Omega}) \cap C^1(\Omega \cup \Gamma)$$

satisfies the linear elliptic differential equation (13), which is even uniformly elliptic in a neighborhood of Γ. Furthermore, w fulfills the homogeneous boundary conditions

$$w(x) = 0, \quad x \in \partial\Omega \setminus \Gamma, \qquad \text{and} \qquad \frac{\partial}{\partial\nu} w(x) = 0, \quad x \in \Gamma. \quad (18)$$

The coefficient (10) has the correct sign $c(x) \leq 0$ for all $x \in \Omega$, due to the assumption IV. From Theorem 1.5 and Theorem 1.6 of Section 1 we infer that $w(x)$ can neither in Ω nor on Γ attain its global maximum and global minimum. This implies $w(x) \equiv 0$ and therefore

$$u(x) \equiv v(x) \qquad \text{in} \quad \Omega.$$

q.e.d.

Example 2.4. The Dirichlet problem for the nonparametric equation of prescribed mean curvature

$$\begin{aligned}
&\zeta = \zeta(x,y) \in C^2(\Omega) \cap C^0(\overline{\Omega}), \\
&\mathcal{M}\zeta(x,y) = 2H(x,y,\zeta(x,y))\Big(1 + |\nabla\zeta(x,y)|^2\Big)^{\frac{3}{2}} \qquad \text{in} \quad \Omega, \qquad (19) \\
&\zeta(x,y) = f(x,y) \qquad \text{on} \quad \partial\Omega
\end{aligned}$$

has at most one solution, if $H_z \geq 0$ in $\Omega \times \mathbb{R}$ is assumed.

Remarks:

a) The existence question for a solution of the mixed boundary value problem in Theorem 2.3 is very difficult. Already for the minimal surface equation (this means $H \equiv 0$ in (19)) the Dirichlet problem (19) can only be solved on convex domains for arbitrary continuous boundary values $f : \partial\Omega \to \mathbb{R}$. For a direct parametric approach of the Dirichlet problem (19) we refer the reader to the paper by
F. Sauvigny: *Flächen vorgeschriebener mittlerer Krümmung mit eineindeutiger Projektion auf eine Ebene.* Mathematische Zeitschrift, Bd. 180 (1982), S. 41-67.
b) A general theory for quasilinear elliptic differential equations is developed in the book of D. Gilbarg and N. Trudinger [GT], Part 2 (especially in the Chapters 14-16).
c) Finally, we emphasize that C^0-stability with respect to the boundary values for quasilinear elliptic differential equations can only be achieved by controlling the first derivatives up to the boundary.

3 The Heat Equation

We set $\mathbb{R}_+ := (0, +\infty)$ and denote the constant *heat conductivity coefficient* by $\kappa \in \mathbb{R}_+$. We consider functions

$$u = u(x,t) = u(x_1, \ldots, x_n, t) : \mathbb{R}^n \times \mathbb{R}_+ \to \mathbb{R} \in C^2(\mathbb{R}^n \times \mathbb{R}_+) \qquad (1)$$

satisfying the *heat equation*

$$\frac{\partial}{\partial t} u(x,t) = \kappa \Delta_x u(x,t), \qquad (x,t) \in \mathbb{R}^n \times \mathbb{R}_+. \qquad (2)$$

For $n = 1$ the solution of (2) models the distribution of temperature in an insulating wire, and for $n = 3$ we obtain the temperature distribution in a heat conducting medium. The transition from the solution $u = u(x,t)$ of (1) and (2) to the function

$$v = v(x,t) := u(\sqrt{\kappa}x_1, \dots, \sqrt{\kappa}x_n, t), \qquad (x,t) \in \mathbb{R}^n \times \mathbb{R}_+$$

yields the following differential equation

$$\frac{\partial}{\partial t} v(x,t) = \Delta_x v(x,t) \qquad \text{in} \quad \mathbb{R}^n \times \mathbb{R}_+. \tag{3}$$

The equations (2) and (3) are not invariant with respect to reflections in the time: $t \to (-t)$. Therefore, the heat equation describes an irreversible process distinguishing between the past and the future. However, the heat equation is invariant with respect to linear substitutions

$$\xi = ax, \quad x \in \mathbb{R}^n; \qquad \tau = a^2 t, \quad t \in \mathbb{R}_+ \tag{4}$$

choosing the number $a \in \mathbb{R} \setminus \{0\}$. The quantity $|x|^2/t$ is invariant under the transformation (4) as well and often appears in connection with this differential equation.

We are looking for a solution of (3) by the ansatz

$$v = v(x,t) = \exp i(\lambda t + \xi \cdot x), \qquad (x,t) \in \mathbb{R}^n \times \mathbb{R}_+,$$

with $\lambda \in \mathbb{C}$ and $\xi = (\xi_1, \dots, \xi_n) \in \mathbb{R}^n$. Then the relation (3) implies

$$\begin{aligned} 0 &= \frac{\partial}{\partial t} v(x,t) - \Delta_x v(x,t) \\ &= e^{i(\lambda t + \xi \cdot x)} \Big(i\lambda + |\xi|^2 \Big), \qquad (x,t) \in \mathbb{R}^n \times \mathbb{R}_+. \end{aligned}$$

Inserting $i\lambda = -|\xi|^2$ we obtain a solution of the heat equation (3) as follows:

$$v(x,t) = e^{-|\xi|^2 t} e^{i\xi \cdot x}, \qquad (x,t) \in \mathbb{R}^n \times \mathbb{R}_+. \tag{5}$$

For each fixed $t \in \mathbb{R}_+$, the function $v(\cdot, t)$ describes a plane wave which is constant on the planes $\xi \cdot x = const$. The phase plane has the unit normal vector $|\xi|^{-1}\xi$, and the length of the wave is given by $L = 2\pi|\xi|^{-1}$. More precisely, we have

$$v\Big(x + \frac{2\pi l}{|\xi|} \frac{\xi}{|\xi|}, t\Big) = v(x,t) \qquad \text{for all} \quad (x,t) \in \mathbb{R}^n \times \mathbb{R}_+ \quad \text{and all} \quad l \in \mathbb{Z}. \tag{6}$$

The amplitude of the wave is determined by

$$|v(x,t)| = e^{-|\xi|^2 t} = \exp\Big(- \frac{4\pi^2 t}{L^2} \Big),$$

and consequently the solutions decay exponentially.

For a given function $g = g(\xi) \in C_0^\infty(\mathbb{R}^n, \mathbb{R})$ we consider the integral

$$\begin{aligned} u(x,t) &:= (2\pi)^{-\frac{n}{2}} \int_{\mathbb{R}^n} e^{(i\xi\cdot x - |\xi|^2 t)} g(\xi)\, d\xi \\ &= (2\pi)^{-\frac{n}{2}} \int_{-\infty}^{+\infty} \cdots \int_{-\infty}^{+\infty} e^{i(\xi_1 x_1 + \ldots \xi_n x_n)} e^{-|\xi|^2 t} g(\xi)\, d\xi_1 \ldots d\xi_n. \end{aligned} \tag{7}$$

Now we calculate

$$u_t(x,t) = (2\pi)^{-\frac{n}{2}} \int_{\mathbb{R}^n} e^{i\xi\cdot x} e^{-|\xi|^2 t} (-|\xi|^2) g(\xi)\, d\xi$$

and

$$\Delta_x u(x,t) = (2\pi)^{-\frac{n}{2}} \int_{\mathbb{R}^n} e^{i\xi\cdot x} e^{-|\xi|^2 t} (-|\xi|^2) g(\xi)\, d\xi$$

for all $(\xi, t) \in \mathbb{R}^n \times \mathbb{R}_+$. Therefore, follows

$$\Delta_x u(x,t) - u_t(x,t) = 0, \qquad (x,t) \in \mathbb{R}^n \times \mathbb{R}_+. \tag{8}$$

Furthermore, the function $u(x,t)$ satisfies the initial condition

$$u(x,0) = (2\pi)^{-\frac{n}{2}} \int_{\mathbb{R}^n} e^{i\xi\cdot x} g(\xi)\, d\xi, \qquad x \in \mathbb{R}^n. \tag{9}$$

Now the question arises for which functions $f(x) : \mathbb{R}^n \to \mathbb{R}$ the initial value problem

$$u(x,0) = f(x), \quad x \in \mathbb{R}^n; \qquad u \in C^0(\mathbb{R}^n \times [0+\infty)) \tag{10}$$

of the heat equation (8) can be solved. In this context we need the following

Theorem 3.1. (Fourier-Plancherel)
The linear operator

$$\tilde{g}(x) := \mathbf{F}^{-1}(g)\Big|_x := (2\pi)^{-\frac{n}{2}} \int_{\mathbb{R}^n} e^{i\xi\cdot x} g(\xi)\, d\xi, \qquad g \in C_0^\infty(\mathbb{R}^n) \tag{11}$$

has a continuous extension on the Hilbert space

$$\mathcal{H} := L^2(\mathbb{R}^n) := \left\{ \varphi : \mathbb{R}^n \to \mathbb{C} : \begin{array}{l} \varphi \text{ is Lebesgue-measurable and} \\ \text{we have } \displaystyle\int_{\mathbb{R}^n} |\varphi(\xi)|^2\, d\xi < +\infty \end{array} \right\}$$

with the inner product

$$(\varphi, \psi) := \int_{\mathbb{R}^n} \varphi(\xi)\overline{\psi}(\xi)\, d\xi, \qquad \varphi, \psi \in \mathcal{H}.$$

The mapping $\mathbf{F}^{-1} : \mathcal{H} \to \mathcal{H}$ *has the inverse*

$$\hat{f}(\xi) := \mathbf{F}(f)\Big|_{\xi} := (2\pi)^{-\frac{n}{2}} \int_{\mathbb{R}^n} e^{-i\xi\cdot x} f(x)\, dx, \qquad f \in C_0^\infty(\mathbb{R}^n) \tag{12}$$

which can be extended continuously on $\mathcal{H}$ *as well. Furthermore, the operators* $\mathbf{F}$ *and* $\mathbf{F}^{-1}$ *are isometric on the Hilbert space* $\mathcal{H}$*, more precisely*

$$(\mathbf{F}\varphi, \mathbf{F}\psi) = (\varphi, \psi) = (\mathbf{F}^{-1}\varphi, \mathbf{F}^{-1}\psi) \qquad \textit{for all} \quad \varphi, \psi \in \mathcal{H},$$

and we have

$$(\mathbf{F}\varphi, \psi) = (\varphi, \mathbf{F}^{-1}\psi) \qquad \textit{for all} \quad \varphi, \psi \in \mathcal{H}.$$

Proof: This Theorem 5.11 will be proved in Chapter 8, Section 5.

Definition 3.2. *We name the operator* $\mathbf{F} : \mathcal{H} \to \mathcal{H}$ *the* Fourier transformation *and* $\mathbf{F}^{-1}$ *the* inverse Fourier transformation.

In (7) we choose the function

$$g(\xi) = \mathbf{F}(f)\Big|_{\xi} = \hat{f}(\xi) = (2\pi)^{-\frac{n}{2}} \int_{\mathbb{R}^n} e^{-i\xi\cdot x} f(x)\, dx, \qquad f \in \mathcal{H}$$

such that $g(\xi) \in C_0^\infty(\mathbb{R}^n)$ holds true. Now (9) implies

$$\begin{aligned} u(x,0) &= (2\pi)^{-\frac{n}{2}} \int_{\mathbb{R}^n} e^{i\xi\cdot x} \hat{f}(\xi)\, d\xi \\ &= \mathbf{F}^{-1} \circ \mathbf{F}(f)\Big|_{x} = f(x), \qquad x \in \mathbb{R}^n. \end{aligned} \tag{13}$$

Furthermore, we calculate via Fubini's tbeorem

$$\begin{aligned} u(x,t) &= (2\pi)^{-\frac{n}{2}} \int_{\mathbb{R}^n} e^{i\xi\cdot x} e^{-|\xi|^2 t} \hat{f}(\xi)\, d\xi \\ &= (2\pi)^{-n} \int_{\mathbb{R}^n}\int_{\mathbb{R}^n} e^{i\xi\cdot x} e^{-|\xi|^2 t} e^{-i\xi\cdot y} f(y)\, dy\, d\xi \\ &= (2\pi)^{-n} \int_{\mathbb{R}^n} \left(\int_{\mathbb{R}^n} e^{i\xi\cdot(x-y) - |\xi|^2 t}\, d\xi \right) f(y)\, dy. \end{aligned} \tag{14}$$

With the aid of the substitution

$$\xi = \frac{i(x-y)}{2t} + \frac{1}{\sqrt{t}}\eta, \quad d\xi = t^{-\frac{n}{2}}\, d\eta, \qquad \eta \in \mathbb{R}^n$$

we determine

$$\begin{aligned}
\int\limits_{\mathbb{R}^n} e^{i\xi\cdot(x-y)-|\xi|^2 t}\, d\xi &= \int\limits_{\mathbb{R}^n} e^{\left(-\frac{|x-y|^2}{2t} + \frac{i}{\sqrt{t}}\eta\cdot(x-y) + \frac{|x-y|^2}{4t} - \frac{i}{\sqrt{t}}\eta\cdot(x-y) - |\eta|^2\right)} t^{-\frac{n}{2}}\, d\eta \\
&= \int\limits_{\mathbb{R}^n} e^{-\frac{|x-y|^2}{4t}} e^{-|\eta|^2} t^{-\frac{n}{2}}\, d\eta \\
&= t^{-\frac{n}{2}} e^{-\frac{|x-y|^2}{4t}} \int\limits_{\mathbb{R}^n} e^{-|\eta|^2}\, d\eta \\
&= t^{-\frac{n}{2}} e^{-\frac{|x-y|^2}{4t}} \left(\int\limits_{-\infty}^{+\infty} e^{-\varrho^2}\, d\varrho \right)^n,
\end{aligned}$$

and consequently

$$\int\limits_{\mathbb{R}^n} e^{i\xi\cdot(x-y)-|\xi|^2 t}\, d\xi = \pi^{\frac{n}{2}} t^{-\frac{n}{2}} e^{-\frac{|x-y|^2}{4t}}. \tag{15}$$

Inserting (15) into (14) we obtain

$$u(x,t) = (4\pi t)^{-\frac{n}{2}} \int\limits_{\mathbb{R}^n} e^{-\frac{|x-y|^2}{4t}} f(y)\, dy, \qquad (x,t) \in \mathbb{R}^n \times (0,+\infty). \tag{16}$$

With the integral (16) we have found a solution of the initial value problem (10) for the heat equation (8).

Definition 3.3. *The function*

$$K(x,y,t) := (4\pi t)^{-\frac{n}{2}} \exp\Big\{ -\frac{|x-y|^2}{4t} \Big\}, \qquad x \in \mathbb{R}^n, \quad y \in \mathbb{R}^n, \quad t \in \mathbb{R}_+$$

represents the kernel function of the heat equation.

Proposition 3.4. *We have the following statements for the kernel function $K(x,y,t) : \mathbb{R}^n \times \mathbb{R}^n \times \mathbb{R}_+ \to \mathbb{R}_+$ of the heat equation:*

(i) $K \in C^2(\mathbb{R}^n \times \mathbb{R}^n \times \mathbb{R}_+)$ and

$$\Big(\frac{\partial}{\partial t} - \Delta_x\Big) K(x,y,t) = 0 \qquad \text{in} \quad \mathbb{R}^n \times \mathbb{R}^n \times \mathbb{R}_+.$$

(ii) For all $(x,t) \in \mathbb{R}^n \times \mathbb{R}_+$ *we have*

$$\int\limits_{\mathbb{R}^n} K(x,y,t)\,dy = 1.$$

(iii) For each $\delta > 0$ *the following integral converges uniformly*

$$\int\limits_{y:|y-x|>\delta} K(x,y,t)\,dy \to 0 \ (t \to 0+) \qquad \text{for all} \quad x \in \mathbb{R}^n.$$

Proof:

(i) Formula (15) gives us

$$K(x,y,t) = (4\pi t)^{-\frac{n}{2}} \exp\Big\{-\frac{|x-y|^2}{4t}\Big\} = (2\pi)^{-n} \int\limits_{\mathbb{R}^n} e^{i(x-y)\cdot\xi - |\xi|^2 t}\,d\xi.$$

The heat equation

$$\Big(\frac{\partial}{\partial t} - \Delta_x\Big)\Big\{e^{i(x-y)\cdot\xi-|\xi|^2 t}\Big\} = 0 \qquad \text{in} \quad \mathbb{R}^n \times \mathbb{R}^n \times \mathbb{R}_+$$

holds true, and the given integral remains absolutely convergent while differentiating with respect to t and x. Therefore, we comprehend

$$\Big(\frac{\partial}{\partial t} - \Delta_x\Big)K(x,y,t) = (2\pi)^{-n} \int\limits_{\mathbb{R}^n} \Big(\frac{\partial}{\partial t} - \Delta_x\Big)\Big\{e^{i(x-y)\cdot\xi-|\xi|^2 t}\Big\}\,d\xi = 0$$

for all $(x,y,t) \in \mathbb{R}^n \times \mathbb{R}^n \times \mathbb{R}_+$.

(ii) We calculate for $\delta \geq 0$ with the aid of the substitution $y = x + \sqrt{4t}\eta$, $dy = (4t)^{\frac{n}{2}}\,d\eta$ as follows:

$$\begin{aligned} \int\limits_{y:|y-x|>\delta} K(x,y,t)\,dy &= (4\pi t)^{-\frac{n}{2}} \int\limits_{y:|y-x|>\delta} \exp\Big\{-\frac{|x-y|^2}{4t}\Big\}\,dy \\ &= \pi^{-\frac{n}{2}} \int\limits_{\eta:|\eta|>\frac{\delta}{\sqrt{4t}}} e^{-|\eta|^2}\,d\eta \qquad \text{for all} \quad (x,t) \in \mathbb{R}^n \times \mathbb{R}_+. \end{aligned} \tag{17}$$

Inserting $\delta = 0$ into (17), we obtain

$$\int\limits_{\mathbb{R}^n} K(x,y,t)\,dy = \pi^{-\frac{n}{2}} \int\limits_{\mathbb{R}^n} \exp(-|\eta|^2)\,d\eta = \pi^{-\frac{n}{2}}\pi^{\frac{n}{2}} = 1.$$

(iii) In the case $\delta > 0$ the formula (17) implies

$$\int\limits_{y:|y-x|>\delta} K(x,y,t)\,dy = \pi^{-\frac{n}{2}} \int\limits_{\eta:|\eta|>\frac{\delta}{\sqrt{4t}}} e^{-|\eta|^2}\,d\eta \quad \to 0\,(t \to 0+)$$

uniformly for all $x \in \mathbb{R}^n$. q.e.d.

Independently of the Fourier-Plancherel integral theorem, we now prove the following

Theorem 3.5. *Let $f = f(x) : \mathbb{R}^n \to \mathbb{R} \in C^0(\mathbb{R}^n)$ be a continuous, bounded function and we define*

$$\begin{aligned} u(x,t) &:= \int\limits_{\mathbb{R}^n} K(x,y,t) f(y)\, dy \\ &= (4\pi t)^{-\frac{n}{2}} \int\limits_{\mathbb{R}^n} e^{-\frac{|x-y|^2}{4t}} f(y)\, dy, \qquad (x,t) \in \mathbb{R}^n \times \mathbb{R}_+. \end{aligned} \tag{18}$$

Then the following statements hold true:

(i) Setting $\mathbb{C}_+ := \{t = \sigma + i\tau \in \mathbb{C} : \sigma > 0\}$ there exists a holomorphic function $U : \mathbb{C}^n \times \mathbb{C}_+ \to \mathbb{C}$, such that

$$u(x,t) = U(x,t) \qquad \text{for all} \quad (x,t) \in \mathbb{R}^n \times \mathbb{R}_+$$

is correct. This especially implies $u \in C^\infty(\mathbb{R}^n \times \mathbb{R}_+)$.

(ii) The function u satisfies the heat equation

$$\Delta_x u(x,t) - \frac{\partial}{\partial t} u(x,t) = 0 \qquad \text{in} \quad \mathbb{R}^n \times \mathbb{R}_+.$$

(iii) We have $u \in C^0(\mathbb{R}^n \times [0,+\infty))$ and u fulfills the initial condition

$$u(x,0) = f(x) \qquad \text{for all} \quad x \in \mathbb{R}^n.$$

(iv) Finally, we have the inequality

$$\inf_{y \in \mathbb{R}^n} f(y) \le u(x,t) \le \sup_{y \in \mathbb{R}^n} f(y) \qquad \text{for all} \quad (x,t) \in \mathbb{R}^n \times \mathbb{R}_+, \tag{19}$$

where only for constant functions $f : \mathbb{R}^n \to \mathbb{R}$ equality is attained.

Proof:

(i) At first, we extend the kernel $K : \mathbb{R}^n \times \mathbb{R}^n \times \mathbb{R}_+ \to \mathbb{R}_+$ to the domain $\mathbb{C}^n \times \mathbb{R}^n \times \mathbb{C}_+$ as follows: With the notation $x = \xi + i\eta \in \mathbb{C}^n$, $y \in \mathbb{R}^n$ and $t = \sigma + i\tau \in \mathbb{C}_+$ we define

$$K(x,y,t) := (4\pi)^{-\frac{n}{2}} (t^2)^{-\frac{n}{4}} \exp\Big\{ -\frac{(x-y)\cdot(x-y)}{4t} \Big\}$$

for $(x,y,t) \in \mathbb{C}^n \times \mathbb{R}^n \times \mathbb{C}_+$. To each fixed $y \in \mathbb{R}^n$ the function $K(x,y,t) : \mathbb{C}^n \times \mathbb{C}_+ \to \mathbb{C}$ is holomorphic, and the kernel $K : \mathbb{C}^n \times \mathbb{R}^n \times \mathbb{C}_+ \to \mathbb{C}$ is continuous. Furthermore, we have

$$
\begin{aligned}
|K(x,y,t)| &= (4\pi)^{-\frac{n}{2}}(|t|^2)^{-\frac{n}{4}} \exp\Big\{-\operatorname{Re}\frac{(x-y)\cdot(x-y)}{4t}\Big\} \\
&= (4\pi)^{-\frac{n}{2}}(\sigma^2+\tau^2)^{-\frac{n}{4}} \exp\Big\{-\frac{1}{4}\operatorname{Re}\frac{(\xi-y+i\eta)\cdot(\xi-y+i\eta)}{\sigma+i\tau}\Big\}.
\end{aligned}
\tag{20}
$$

Now we calculate

$$
\begin{aligned}
&\operatorname{Re}\frac{(\xi-y+i\eta)\cdot(\xi-y+i\eta)}{\sigma+i\tau} \\
&\quad= \frac{1}{\sigma^2+\tau^2}\operatorname{Re}\Big\{(\sigma-i\tau)\Big[(\xi-y)^2-\eta^2+2i\eta\cdot(\xi-y)\Big]\Big\} \\
&\quad= \frac{1}{\sigma^2+\tau^2}\Big\{\sigma(\xi-y)^2-\sigma\eta^2+2\tau\eta\cdot(\xi-y)\Big\} \\
&\quad= \frac{1}{\sigma(\sigma^2+\tau^2)}\Big\{|\sigma(\xi-y)+\tau\eta|^2-\tau^2|\eta|^2-\sigma^2|\eta|^2\Big\} \\
&\quad= \frac{1}{\sigma(\sigma^2+\tau^2)}|\sigma(\xi-y)+\tau\eta|^2-\frac{1}{\sigma}|\eta|^2.
\end{aligned}
\tag{21}
$$

From (20) and (21) we derive the relation

$$
\begin{aligned}
&|K(x,y,t)| \\
&\quad= (4\pi)^{-\frac{n}{2}}(\sigma^2+\tau^2)^{-\frac{n}{4}} \exp\Big\{\frac{1}{4\sigma}|\eta|^2-\frac{1}{4\sigma(\sigma^2+\tau^2)}|\sigma(\xi-y)+\tau\eta|^2\Big\} \\
&\quad= \Big(1+\frac{\tau^2}{\sigma^2}\Big)^{-\frac{n}{4}} \exp\Big\{\frac{|\eta|^2}{4\sigma}\Big\}\sigma^{-\frac{n}{2}}(4\pi)^{-\frac{n}{2}} \exp\Big\{-\frac{|\sigma(\xi-y)+\tau\eta|^2}{4\sigma(\sigma^2+\tau^2)}\Big\} \\
&\quad= \Big(1+\frac{\tau^2}{\sigma^2}\Big)^{+\frac{n}{4}} \exp\Big\{\frac{|\eta|^2}{4\sigma}\Big\}K\Big(\xi+\frac{\tau}{\sigma}\eta, y, \sigma+\frac{\tau^2}{\sigma}\Big) \\
&\quad=: \Theta_{x,t}(y) \qquad \text{for all} \quad (x,y,t)\in\mathbb{C}^n\times\mathbb{R}^n\times\mathbb{C}_+.
\end{aligned}
$$

Consequently, the parameter integral

$$
U(x,t) := \int_{\mathbb{R}^n} K(x,y,t)f(y)\,dy, \qquad (x,t)\in\mathbb{C}^n\times\mathbb{C}_+
$$

has an integrable majorant. Due to Theorem 2.12 in Chapter 4, Section 2 the function $U:\mathbb{C}^n\times\mathbb{C}_+\to\mathbb{C}$ is holomorphic.

(ii) The heat equation for $u(x,t)$ can immediately be derived from (18) and Proposition 3.4, (i).

(iii) Now we show that the initial values are continuously attained: For given $\xi\in\mathbb{R}^n$ and $\varepsilon>0$ we have a number $\delta=\delta(\xi,\varepsilon)>0$ such that

$$
|f(y)-f(\xi)|<\varepsilon \qquad \text{for all} \quad |y-\xi|<2\delta
$$

holds true. We set $M := \sup\{|f(y)| \ : \ y \in \mathbb{R}^n\} < +\infty$ and obtain the following inequality for all $(x,t) \in \mathbb{R}^n \times \mathbb{R}_+$ with $|x-\xi| < \delta$ and $0 < t < \vartheta$:

$$\begin{aligned}
|u(x,t) - f(\xi)| &= \left| \int\limits_{\mathbb{R}^n} K(x,y,t)(f(y) - f(\xi))\, dy \right| \\
&\leq \int\limits_{y:|y-x|\leq\delta} K(x,y,t)|f(y) - f(\xi)|\, dy \\
&\quad + \int\limits_{y:|y-x|\geq\delta} K(x,y,t)|f(y) - f(\xi)|\, dy \\
&\leq \int\limits_{y:|y-\xi|\leq 2\delta} K(x,y,t)|f(y) - f(\xi)|\, dy \\
&\quad + 2M \int\limits_{y:|y-x|\geq\delta} K(x,y,t)\, dy \\
&\leq \varepsilon + 2M\varepsilon.
\end{aligned} \tag{22}$$

Here we have chosen $\vartheta > 0$ sufficiently small and used Proposition 3.4, (ii) and (iii) in the last inequality. From the estimate (22) we obtain the desired relation

$$\lim_{t\to 0+} u(x,t) = f(x) \qquad \text{for all} \quad x \in \mathbb{R}^n.$$

(iv) The statement (19) follows directly from the integral representation (18) combined with Proposition 3.4, (ii).

q.e.d.

Remarks:

1. Bounded continuous functions $f : \mathbb{R}^n \to \mathbb{R}$ being given, we obtain a bounded solution of the initial value problem for the heat equation by the function $u(x,t)$ defined in (18). However, there are further (unbounded) solutions of the same problem; see the monograph of F. John [J], Chapter 7, Section 1. Later in this section, we shall prove that the initial value problem for the heat equation is uniquely determined in the class of bounded solutions.
2. With the aid of (18) we can also construct solutions of the problem for initial values $f : \mathbb{R}^n \to \mathbb{R}$ subject to the growth condition

$$|f(x)| \leq M e^{a|x|^2}, \qquad x \in \mathbb{R}^n.$$

However, then the solution (18) only exists for the times $0 \leq t < \frac{1}{4a}$.

Now we are going to prove a maximum principle for parabolic differential equations: Over the bounded domain $\Omega \subset \mathbb{R}^n$, where $n \in \mathbb{N}$ is given, we consider the *parabolic cylinder*

$$\Omega_T := \Big\{(x,t) \in \mathbb{R}^n \times \mathbb{R}_+ \,:\, x \in \Omega,\ t \in (0,T]\Big\}$$

with the given height $T \in \mathbb{R}_+$. The *parabolic boundary* denotes the following set

$$\Delta\Omega_T := \Big\{(x,t) \in \mathbb{R}^n \times [0,+\infty) \,:\, (x,t) \in (\partial\Omega \times [0,T]) \cup (\Omega \times \{0\})\Big\}.$$

Proposition 3.6. *Let function $u = u(x,t) \in C^2(\Omega_T)$ satisfy the differential inequality*

$$\Delta_x u(x,t) - \frac{\partial}{\partial t} u(x,t) > 0, \qquad (x,t) \in \Omega_T.$$

Then u cannot attain its maximum at any point of Ω_T.

Proof: We assume that u would attain its maximum at a point $(\xi,\tau) \in \Omega_T$. If $(\xi,\tau) \in \overset{\circ}{\Omega}_T$ is correct, Proposition 1.2 from Section 1 yields the inequality

$$\Big(\Delta_x - \frac{\partial}{\partial t}\Big) u(\xi,\tau) \le 0$$

in contradiction to the assumption. Consequently, we have $(\xi,\tau) \in \Omega_T \setminus \overset{\circ}{\Omega}_T$ and especially $\tau = T$ holds true. Furthermore, the differential inequality implies

$$\Delta_x u(\xi,T) > \frac{\partial}{\partial t} u(\xi,T) \ge 0. \tag{23}$$

Now also the function $\tilde{u}(x) := u(x,T)$, $x \in \Omega$ takes on its maximum at the point $\xi \in \Omega$. From Proposition 1.2 in Section 1 we infer $\Delta\tilde{u}(\xi) \le 0$ contradicting (23).

q.e.d.

Proposition 3.7. *Let the function $u = u(x,t) \in C^2(\Omega_T) \cap C^0(\Omega_T \cup \Delta\Omega_T)$ be a solution of the differential inequality*

$$\Delta_x u(x,t) - \frac{\partial}{\partial t} u(x,t) \ge 0, \qquad (x,t) \in \Omega_T,$$

and fulfill the boundary condition

$$u(x,t) \le 0, \qquad (x,t) \in \Delta\Omega_T.$$

Then we have $u(x,t) \le 0$ in $\Omega_T \cup \Delta\Omega_T$.

Proof: For the given $\varepsilon > 0$ we consider the following auxiliary function $w(x,t) := u(x,t) - \varepsilon t$ and observe

$$\Big(\Delta_x - \frac{\partial}{\partial t}\Big) w(x,t) = \Big(\Delta_x - \frac{\partial}{\partial t}\Big) u(x,t) + \varepsilon > 0 \qquad \text{in} \quad \Omega_T.$$

Then we determine the boundary condition

$$w(x,t) = u(x,t) - \varepsilon t \leq 0 \qquad \text{on} \quad \Delta\Omega_T.$$

Proposition 3.6 implies $w(x,t) \leq 0$ and consequently $u(x,t) \leq \varepsilon t$ in Ω_T. The transition to the limit $\varepsilon \downarrow 0$ yields

$$u(x,t) \leq 0 \qquad \text{in} \quad \Omega_T \cup \Delta\Omega_T.$$

q.e.d.

Theorem 3.8. (Parabolic maximum-minimum principle)
Let $u = u(x,y) \in C^2(\Omega_T) \cap C^0(\Omega_T \cup \Delta\Omega_T)$ be a solution of the heat equation

$$\Delta_x u(x,t) - \frac{\partial}{\partial t} u(x,t) = 0, \qquad (x,t) \in \Omega_T.$$

Then we have

$$\min_{(\xi,\tau)\in\Delta\Omega_T} u(\xi,\tau) =: m \leq u(x,t) \leq M := \max_{(\xi,\tau)\in\Delta\Omega_T} u(\xi,\tau), \qquad (x,t) \in \Omega_T.$$

Proof: Applying Proposition 3.7 to the auxiliary function

$$u(x,t) - M \quad \text{and} \quad m - u(x,t), \qquad (x,t) \in \Omega_T \cup \Delta\Omega_T,$$

we obtain the statement immediately. q.e.d.

Theorem 3.9. (Uniqueness for the initial value problem of the heat equation)
The bounded, continuous function $f = f(x) : \mathbb{R}^n \to \mathbb{R} \in C^0(\mathbb{R}^n)$ is given. Then there is exactly one bounded solution u of the initial value problem for the heat equation, attributed to this function f, with the following properties:

$$\begin{aligned}
& u = u(x,t) \in C^2(\mathbb{R}^n \times \mathbb{R}_+, \mathbb{R}) \cap C^0(\mathbb{R}^n \times [0,+\infty), \mathbb{R}), \\
& \Delta_x u(x,t) - \frac{\partial}{\partial t} u(x,t) = 0 \qquad \textit{in} \quad \mathbb{R}^n \times \mathbb{R}_+, \\
& u(x,0) = f(x), \qquad x \in \mathbb{R}^n, \\
& \sup_{(x,t)\in\mathbb{R}^n\times\mathbb{R}_+} |u(x,t)| < +\infty.
\end{aligned} \tag{24}$$

Proof: Let $u = u(x,t)$ and $v = v(x,t)$ be two solutions of (24), and we set

$$M := \sup_{\mathbb{R}^n \times \mathbb{R}_+} |u(x,t)| + \sup_{\mathbb{R}^n \times \mathbb{R}_+} |v(x,t)| \in [0, +\infty).$$

For the function

$$w(x,t) := u(x,t) - v(x,t) \in C^2(\mathbb{R}^n \times \mathbb{R}_+, \mathbb{R}) \cap C^0(\mathbb{R}^n \times [0,+\infty), \mathbb{R})$$

we have

$$\begin{aligned} &\Delta_x w(x,t) - \frac{\partial}{\partial t} w(x,t) = 0 \qquad \text{in} \quad \mathbb{R}^n \times \mathbb{R}_+, \\ &u(x,0) = 0, \qquad x \in \mathbb{R}^n, \\ &|w(x,t)| \le M \qquad \text{for all} \quad (x,t) \in \mathbb{R}^n \times [0,+\infty). \end{aligned} \tag{25}$$

Now we choose numbers $T \in \mathbb{R}_+$, $R \in \mathbb{R}_+$, define the ball $B_R := \{x \in \mathbb{R}^n \, : \, |x| < R\}$, and attribute the parabolic cylinder

$$B_{R,T} := \Big\{(x,t) \in \mathbb{R}^n \times \mathbb{R}_+ \, : \, x \in B_R, \; t \in (0,T]\Big\}$$

with the parabolic boundary

$$\Delta B_{R,T} = \Big\{(x,t) \in \overline{B_R} \times [0,T] \, : \, x \in \partial B_R \quad \text{or} \quad t = 0\Big\}.$$

On the domain $B_{R,T}$ we consider both the solution $w(x,t)$ of the problem (25) and the function

$$W(x,t) := \frac{2nM}{R^2}\Big(\frac{|x|^2}{2n} + t\Big). \tag{26}$$

Now the comparison function W satisfies the differential equation

$$\Big(\Delta_x - \frac{\partial}{\partial t}\Big) W(x,t) = \frac{2nM}{R^2}(1-1) = 0, \qquad (x,t) \in B_{R,T},$$

and the following inequality holds true on the parabolic boundary

$$|w(x,t)| \le W(x,t), \qquad (x,t) \in \Delta B_{R,T}.$$

Application of the parabolic maximum-minimum principle yields

$$|w(x,t)| \le W(x,t) = \frac{2nM}{R^2}\Big(\frac{|x|^2}{2n} + t\Big), \qquad (x,t) \in B_{R,T}. \tag{27}$$

We observe the transition $R \to +\infty$ in formula (27) and obtain

$$w(x,t) = 0, \qquad x \in \mathbb{R}^n, \quad t \in (0,T],$$

with an arbitrary $T \in \mathbb{R}_+$. Consequently, we have $w(x,t) \equiv 0$ and finally $u(x,t) \equiv v(x,t)$ in $\mathbb{R}^n \times \mathbb{R}_+$.

q.e.d.

Let us consider the following

Example 3.10. (An initial-boundary-value problem for the one-dimensional heat equation)
We are looking for a solution $v = v(x,t)$, $0 < x < L$, $t > 0$ of the *one-dimensional heat equation*

$$v_{xx}(x,t) - v_t(x,t) = 0, \qquad x \in (0,L), \quad t \in (0,+\infty), \tag{28}$$

with the *boundary conditions*

$$v(0,t) = 0 = v(L,t), \qquad t \in [0,+\infty), \tag{29}$$

and the *initial condition*

$$v(x,0) = f(x), \qquad x \in (0,L). \tag{30}$$

Here $f = f(x) : [0,L] \to \mathbb{R}$ is a continuous function satisfying $f(0) = 0 = f(L)$.

The problem (28)-(30) models a temperature distribution in an insulating wire with fixed temperature at the boundary. We shall construct a solution of (28)-(30) by reflection methods. Therefore, we apply an uneven reflection to the function f at the points $x = 0$ and $x = L$ such that

$$f(-x) = -f(x), \quad f(L + (L-x)) = -f(x), \qquad x \in \mathbb{R} \tag{31}$$

is satisfied. Setting

$$\varphi(x) := \begin{cases} f(x), \; 0 \le x \le L \\ 0, \qquad \text{otherwise} \end{cases},$$

the continued function f appears in the form

$$f(x) = \sum_{n=-\infty}^{+\infty} \Big\{ \varphi(2nL + x) - \varphi(2nL - x) \Big\}, \qquad x \in \mathbb{R}. \tag{32}$$

For this continuous and bounded initial distribution $f : \mathbb{R} \to \mathbb{R}$ we globally solve the heat equation.With the aid of the substitutions

$$\xi = 2nL \pm y, \quad d\xi = \pm dy, \qquad n = 0, \pm 1, \pm 2, \dots$$

we obtain

$$
\begin{aligned}
u(x,t) &= \int_{-\infty}^{+\infty} K(x,y,t) f(y)\, dy \\
&= \int_{-\infty}^{+\infty} K(x,y,t) \sum_{n=-\infty}^{+\infty} \Big\{ \varphi(2nL+y) - \varphi(2nL-y) \Big\}\, dy \\
&= \int_{-\infty}^{+\infty} \varphi(\xi) \sum_{n=-\infty}^{+\infty} \Big\{ K(x,\xi-2nL,t) - K(x,2nL-\xi,t) \Big\}\, d\xi \\
&= \int_{0}^{L} G(x,\xi,t) f(\xi)\, d\xi.
\end{aligned}
\tag{33}
$$

Here we define

$$
\begin{aligned}
G(x,\xi,t) &:= \sum_{n=-\infty}^{+\infty} \Big\{ K(x,\xi-2nL,t) - K(x,2nL-\xi,t) \Big\} \\
&= \frac{1}{\sqrt{4\pi t}} \sum_{n=-\infty}^{+\infty} \Big\{ e^{-\frac{1}{4t}(x-\xi+2nL)^2} - e^{-\frac{1}{4t}(x+\xi-2nL)^2} \Big\} \\
&= \frac{1}{2L} \Big\{ \vartheta\Big(\frac{x-\xi}{2L}, \frac{i\pi t}{L^2}\Big) - \vartheta\Big(\frac{x+\xi}{2L}, \frac{i\pi t}{L^2}\Big) \Big\}
\end{aligned}
\tag{34}
$$

denoting by

$$
\vartheta(z,\tau) := \frac{1}{\sqrt{-i\tau}} \sum_{n=-\infty}^{+\infty} \exp\Big(-i\pi \frac{(z+n)^2}{\tau} \Big) \tag{35}
$$

the *Theta function.*

The functions $u(x,t) + u(-x,t)$ and $u(x,t) + u(2L-x,t)$ are solutions of the heat equation (28) in $\mathbb{R} \times \mathbb{R}_+$ with homogeneous initial conditions. Theorem 3.9 now implies

$$
u(x,t) + u(-x,t) \equiv 0 \equiv u(x,t) + u(2L-x,t), \qquad (x,t) \in \mathbb{R} \times [0,+\infty). \tag{36}
$$

Consequently, $v(x,t) := u(x,t)$, $x \in [0,L]$, $t \in [0,+\infty)$ solves the initial-boundary-value problem (28)-(30) for the one-dimensional heat equation.

Finally, we refer the reader to the book [GuLe], Chapters 5 and 9 with further results on the heat equation.

4 Characteristic Surfaces and an Energy Estimate

On a domain $\Omega \subset \mathbb{R}^{n+1}$ with $n \in \mathbb{N}$ and for the functions $u \in C^2(\Omega)$ we consider the *linear partial differential equation of second order*

$$\mathcal{L}u(y) := \sum_{j,k=1}^{n+1} a_{jk}(y) \frac{\partial^2}{\partial y_j \partial y_k} u(y) + \sum_{j=1}^{n+1} b_j(y) \frac{\partial}{\partial y_j} u(y) + c(y)u(y) = h(y) \quad (1)$$

for $y \in \Omega$. The coefficient functions $a_{jk}(y)$, $b_j(y)$ and $c(y)$ for $j,k = 1,\ldots,n+1$ as well as the right-hand side $h(y)$ belong to the regularity class $C^0(\Omega,\mathbb{R})$. Furthermore, the matrix $(a_{jk}(y))_{j,k=1,\ldots,n+1}$ is symmetric for all $y \in \Omega$.

Definition 4.1. *Let $\varphi = \varphi(y_1,\ldots,y_{n+1}) : \Omega \to \mathbb{R} \in C^2(\Omega)$ be a nonconstant function which defines the following nonvoid set*

$$\mathcal{F} := \Big\{ y \in \Omega \; : \; \varphi(y) = 0 \Big\}.$$

We name $\mathcal{F}$ a characteristic surface *for the differential equation (1), if the adjoint quadratic form*

$$Q[\varphi](y) := \sum_{j,k=1}^{n+1} a_{jk}(y) \frac{\partial \varphi}{\partial y_j}(y) \frac{\partial \varphi}{\partial y_k}(y), \qquad y \in \Omega, \tag{2}$$

fulfills the condition

$$Q[\varphi](y) = 0 \qquad \textit{for all} \quad y \in \mathcal{F}.$$

Otherwise $\mathcal{F}$ is called a noncharacteristic surface, *if namely*

$$Q[\varphi](y) \neq 0 \qquad \textit{for all} \quad y \in \mathcal{F}.$$

In the case $n = 1$ we speak of characteristic *and alternatively of* noncharacteristic curves.

Remark: Since we do not assume $\nabla\varphi(y) \neq 0$ on $\mathcal{F}$, the set $\mathcal{F}$ may have singular points. Therefore, $\mathcal{F} \subset \mathbb{R}^{n+1}$ is not a hypersurface in general.

Example 4.2. If $\mathcal{L}$ is elliptic in Ω, where the matrix $(a_{jk}(x))_{j,k=1,\ldots,n+1}$ is positive-definite for all $x \in \Omega$, then characteristic surfaces do not exist.

Example 4.3. We consider the differential operator of the heat equation

$$\mathcal{L} := \Delta_x - \frac{\partial}{\partial t} \qquad \text{in} \quad \mathbb{R}^n \times \mathbb{R}_+.$$

Then we obtain the quadratic form

$$Q[\varphi](x,t) = \sum_{j=1}^{n} (\varphi_{x_j}(x,t))^2, \qquad (x,t) \in \mathbb{R}^n \times \mathbb{R}_+.$$

For fixed $\tau \in \mathbb{R}$ we choose the function $\varphi(x,t) := t - \tau$ and see that the surface

$$\mathcal{F} := \Big\{(x,t) \in \mathbb{R}^n \times \mathbb{R} \ : \ \varphi(x,t) = 0\Big\} = \mathbb{R}^n \times \{\tau\}$$

is characteristic. Especially, the plane $\mathbb{R}^n \times \{0\}$ is a characteristic surface for the heat equation.

Definition 4.4. *For the given domain $\Omega \subset \mathbb{R}^n$ and the numbers $-\infty \le t_1 < t_2 \le +\infty$ we construct the box*

$$\Omega_{t_1,t_2} := \Big\{(x,t) \in \mathbb{R}^n \times \mathbb{R} \ : \ x \in \Omega, \ t \in (t_1,t_2)\Big\}.$$

We define the d'Alembert operator $\Box : C^2(\Omega_{t_1,t_2}) \to C^0(\Omega_{t_1,t_2})$ *setting*

$$\Box u(x_1,\ldots,x_n,t) := \frac{\partial^2}{\partial t^2} u(x_1,\ldots,x_n,t) - c^2 \Delta_x u(x_1,\ldots,x_n,t) \tag{3}$$

for $(x_1,\ldots,x_n,t) \in \Omega \times (t_1,t_2)$. Here $c > 0$ denotes a fixed positive constant which represents the velocity of light in physics.

Example 4.5. For the *homogeneous wave equation*

$$\Box u(x_1,\ldots,x_n,t) = 0 \qquad \text{in} \quad \mathbb{R}^n \times \mathbb{R}$$

we obtain the associate quadratic form

$$Q[\varphi](x,t) = (\varphi_t(x,t))^2 - c^2 |\nabla_x \varphi(x,t)|^2, \qquad (x,t) \in \mathbb{R}^n \times \mathbb{R}.$$

Given the vector $(\xi,\tau) = (\xi_1,\ldots,\xi_n,\tau) \in \mathbb{R}^n \times \mathbb{R}$, we consider the function

$$\varphi(x,t) := \frac{c^2}{2}(t-\tau)^2 - \frac{1}{2}|x-\xi|^2, \qquad (x,t) \in \mathbb{R}^n \times \mathbb{R}, \tag{4}$$

and calculate

$$\begin{aligned} Q[\varphi](x,t) &= c^4 (t-\tau)^2 - c^2 |x-\xi|^2 \\ &= 2c^2 \Big\{\frac{c^2}{2}(t-\tau)^2 - \frac{1}{2}|x-\xi|^2\Big\} \\ &= 2c^2 \varphi(x,t), \qquad (x,t) \in \mathbb{R}^n \times \mathbb{R}. \end{aligned}$$

The set

$$\begin{aligned} \mathcal{F}(\xi,\tau) &:= \Big\{(x,t) \in \mathbb{R}^{n+1} \ : \ \varphi(x,t) = 0\Big\} \\ &= \Big\{(x,t) \in \mathbb{R}^{n+1} \ : \ |x-\xi| = c|t-\tau|\Big\} \end{aligned} \tag{5}$$

represents a characteristic surface of the wave equation for each $(\xi,\tau) \in \mathbb{R}^{n+1}$. These are conical surfaces with the singular tip (ξ,τ) and the opening angle $\alpha = \arctan c$.

We now present a fundamental result for the wave equation.

Theorem 4.6. (Energy estimate for the wave equation)
Let the point $(\xi, \tau) = (\xi_1, \ldots, \xi_n, \tau) \in \mathbb{R}^n \times \mathbb{R}_+$ *with the associate cone*

$$K = K(\xi, \tau) := \Big\{ (x,t) \in \mathbb{R}^n \times \mathbb{R}_+ \,:\, t \in (0, \tau),\ |x - \xi| < c(\tau - t) \Big\}$$

be given. Furthermore, we have a solution $u = u(x,t) \in C^2(K) \cap C^1(\overline{K})$ *of the homogeneous wave equation*

$$\Box u(x,t) + q(x,t) \frac{\partial}{\partial t} u(x,t) = 0 \qquad \text{in} \quad K. \tag{6}$$

Here $q = q(x,t) \in C^0(K, [0, +\infty))$ *represents a nonnegative continuous potential on* K.
Then the following energy inequality *holds true for all* $s \in (0, \tau)$, *namely*

$$\begin{aligned} &\int\limits_{x:|x-\xi|<c(\tau-s)} \Big\{ c^2 |\nabla_x u(x,s)|^2 + |\frac{\partial}{\partial t} u(x,s)|^2 \Big\}\, dx \\ &\qquad\qquad \leq \int\limits_{x:|x-\xi|<c\tau} \Big\{ c^2 |\nabla_x u(x,0)|^2 + |\frac{\partial}{\partial t} u(x,0)|^2 \Big\}\, dx. \end{aligned} \tag{7}$$

Proof:

Figure 1.8 Illustration of the Energy Estimate

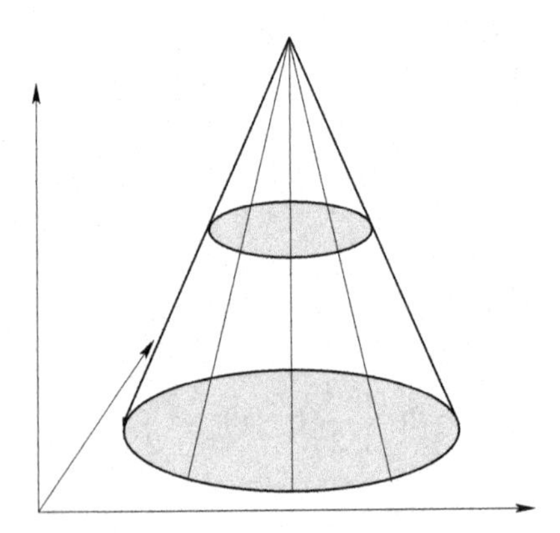

1. With the aid of the transformation $(x,t) \mapsto (c(x+\xi), t)$ we can confine ourselves to the case $\xi = 0$ and $c = 1$. The coefficient matrix of the d'Alembert operator takes on the form

$$(a_{jk})_{j,k=1,\ldots,n+1} = \begin{pmatrix} -1 & & & 0 \\ & \ddots & & \\ & & -1 & \\ 0 & & & +1 \end{pmatrix}. \tag{8}$$

For $s \in (0,\tau)$ we consider the box

$$D = D(s) := \Big\{(x,t) \in \mathbb{R}^n \times \mathbb{R}_+ \,:\, |x| < \tau - t,\ t \in (0,s)\Big\},$$

whose boundary $\partial D = \mathcal{F}_0 \cup \mathcal{F}_s \cup \mathcal{F}$ consists of the hypersurfaces $\mathcal{F}_0$, $\mathcal{F}_s$ and $\mathcal{F}$. Here $\mathcal{F} = \partial D \cap \partial K(0,\tau)$ represents a characteristic surface of the differential equation (6) with the exterior normal

$$\begin{aligned} \nu = \nu(x,t) &= (\nu_1(x,t),\dots,\nu_n(x,t),\nu_{n+1}(x,t)) \\ = (\tilde{\nu}(x,t),\nu_{n+1}(x,t)) &= \Big(\frac{1}{\sqrt{2}}\frac{x}{|x|},\frac{1}{\sqrt{2}}\Big), \qquad (x,t) \in \mathcal{F}. \end{aligned} \tag{9}$$

The surfaces

$$\mathcal{F}_0 := \Big\{(x,t) \in \partial D \setminus \partial K(0,\tau) \,:\, t = 0\Big\}$$

and

$$\mathcal{F}_s := \Big\{(x,t) \in \partial D \setminus \partial K(0,\tau) \,:\, t = s\Big\}$$

possess the exterior normals

$$\begin{aligned} \nu = \nu(x,0) &= (0,\dots,0,-1), \qquad (x,0) \in \mathcal{F}_0 \quad \text{and} \\ \nu = \nu(x,s) &= (0,\dots,0,+1), \qquad (x,s) \in \mathcal{F}_s, \end{aligned}$$

respectively.

2. We multiply (6) by $2u_t(x,t)$ and calculate for all $(x,t) \in D$ the identity

$$\begin{aligned} 0 &= 2u_t\Big(u_{tt} - \Delta_x u(x,t)\Big) + 2q(x,t)\Big(u_t(x,t)\Big)^2 \\ &= \frac{\partial}{\partial t}\Big[(u_t)^2\Big] - 2\operatorname{div}_x(u_t\nabla_x u) + 2\nabla_x u_t \cdot \nabla_x u + 2q(u_t)^2 \\ &= \frac{\partial}{\partial t}\Big[|\nabla_x u(x,t)|^2 + |\frac{\partial}{\partial t}u(x,t)|^2\Big] + \operatorname{div}_x(-2u_t\nabla_x u) + 2q(u_t)^2. \end{aligned} \tag{10}$$

Integration of (10) via the Gaussian integral theorem over the box $D = D(s)$ yields

$$\begin{aligned} 0 &= 2\int\limits_D q(x,t)(u_t(x,t))^2\,dx\,dt + \int\limits_{\mathcal{F}_s}\Big\{|\nabla_x u(x,s)|^2 + |\frac{\partial}{\partial t}u(x,s)|^2\Big\}\,dx \\ &\quad - \int\limits_{\mathcal{F}_0}\Big\{|\nabla_x u(x,0)|^2 + |\frac{\partial}{\partial t}u(x,0)|^2\Big\}\,dx \\ &\quad + \int\limits_{\mathcal{F}}\Big\{-2u_t\nabla_x u\cdot\tilde{\nu} + \frac{1}{\sqrt{2}}\Big(|\nabla_x u|^2 + |u_t|^2\Big)\Big\}\,d\sigma(x,t) \\ &\geq \int\limits_{\mathcal{F}_s}\Big\{|\nabla_x u(x,s)|^2 + |u_t(x,s)|^2\Big\}\,dx - \int\limits_{\mathcal{F}_0}\Big\{|\nabla_x u(x,0)|^2 + |u_t(x,0)|^2\Big\}\,dx. \end{aligned}$$

Here we observe that the term $q(u_t)^2$ is nonnegative, and formula (9) implies

$$|2u_t\nabla_x u\cdot\tilde{\nu}|\le 2|u_t||\nabla_x u||\tilde{\nu}|=\frac{2}{\sqrt{2}}|u_t||\nabla_x u|\le\frac{1}{\sqrt{2}}\Big(|\nabla_x u|^2+|u_t|^2\Big)\text{ on }\mathcal{F}.$$

We conclude

$$\int\limits_{\mathcal{F}_s}\Big\{|\nabla_x u(x,s)|^2+|\frac{\partial}{\partial t}u(x,s)|^2\Big\}\,dx\le\int\limits_{\mathcal{F}_0}\Big\{|\nabla_x u(x,0)|^2+|\frac{\partial}{\partial t}u(x,0)|^2\Big\}\,dx.$$

q.e.d.

As a corollary of Theorem 4.6 we obtain

Theorem 4.7. (Uniqueness of Cauchy's initial value problem for the wave equation)
Let the assumptions of Theorem 4.6 be fulfilled, and additionally $u=u(x,t)$ may satisfy the homogeneous Cauchy initial conditions

$$u(x,0)=0=u_t(x,0)\qquad\text{for all}\quad x\in\mathbb{R}^n\quad\text{with}\quad|x-\xi|<c\tau. \tag{11}$$

Then we have $u(x,t)\equiv 0$ on $K=K(\xi,\tau)$.

Proof: From the initial conditions (11) we obtain

$$c^2|\nabla_x u(x,0)|^2+|\frac{\partial}{\partial t}u(x,0)|^2=0,\qquad|x-\xi|<c\tau,$$

and the energy estimate in Theorem 4.6 yields

$$\int\limits_{x:|x-\xi|<c(\tau-s)}\Big\{c^2|\nabla_x u(x,s)|^2+|\frac{\partial}{\partial t}u(x,s)|^2\Big\}\,dx=0\qquad\text{for all}\quad s\in(0,\tau).$$

This implies $\nabla_x u(x,t)\equiv 0\equiv u_t(x,t)$ on K and consequently $u(x,t)\equiv$ *const*. Again from (11) we obtain

$$u(x,t)\equiv 0\qquad\text{in}\quad K.$$

q.e.d.

In the sections Section 5 and Section 6, we shall explicitly solve Cauchy's initial value problem for the wave equation, with the aid of integral formulas: At first, in the odd dimensions $n=1,3,5,\ldots$ and later for the even dimensions $n=2,4,6,\ldots$.

We now return to the general differential equation (1): For an ordinary differential equation of second order we prescribe the value of the function and its first derivative at one point as initial values. Here we shall treat the higher-dimensional analogue, namely *Cauchy's initial value problem for the partial differential equation (1).*

In the domain $\Omega \subset \mathbb{R}^{n+1}$ the partial differential equation

$$\mathcal{L}u(y) = h(y), \quad y \in \Omega \quad \text{with} \quad u = u(y) \in C^2(\Omega) \tag{12}$$

may be given due to (1). Furthermore, let the function

$$\varphi = \varphi(y) : \Omega \to \mathbb{R} \in C^3(\Omega) \quad \text{with} \quad \nabla\varphi(y) \neq 0,\, y \in \Omega$$

represent a hypersurface in $\mathbb{R}^{n+1}$, namely

$$\emptyset \neq \mathcal{F} := \Big\{ y \in \Omega : \; \varphi(y) = 0 \Big\} \subset \Omega.$$

On $\mathcal{F}$ we prescribe the function $f = f(y) : \mathcal{F} \to \mathbb{R} \in C^2(\mathcal{F})$, and we require the following initial condition of order 0 in

$$u(y) = f(y), \qquad y \in \mathcal{F}. \tag{13}$$

The derivatives of u tangential to the surface $\mathcal{F}$ are already prescribed by this initial condition. Denoting the normal to the surface $\mathcal{F}$ by

$$\nu(y) := |\nabla\varphi(y)|^{-1}\nabla\varphi(y), \quad y \in \mathcal{F}$$

we additionally prescribe a function

$$g = g(y) : \mathcal{F} \to \mathbb{R} \in C^1(\mathcal{F})$$

and require the following initial condition of the first order:

$$\frac{\partial}{\partial\nu}u(y) = g(y), \quad y \in \mathcal{F}. \tag{14}$$

The function $\nabla u(y)$ on $\mathcal{F}$ is already determined by (13) and (14).

We now recommend to introduce new coordinates as follows: Take the domain $\Gamma \subset \mathbb{R}^n$ and the regular parameter representation

$$\gamma = \gamma(x_1, \ldots, x_n) : \Gamma \to \mathcal{F} \in C^2(\Gamma)$$

of the surface $\mathcal{F}$. We denote the parametric normal to the surface $\mathcal{F}$ by

$$\nu(x_1, \ldots, x_n) := \nu(\gamma(x_1, \ldots, x_n)), \qquad x = (x_1, \ldots, x_n) \in \Gamma.$$

Then we consider the *parameter transformation*

$$\theta = \theta(x_1, \ldots, x_n, t) : \Gamma_\varepsilon \to \Omega \in C^2(\Gamma_\varepsilon, \mathbb{R}^{n+1})$$

on the box $\Gamma_\varepsilon := \Gamma \times (-\varepsilon, \varepsilon) \subset \mathbb{R}^{n+1}$ defined as follows:

$$\begin{aligned} &\theta(x_1, \ldots, x_n, t) := \gamma(x_1, \ldots, x_n) + t\nu(x_1, \ldots, x_n), \\ &x = (x_1, \ldots, x_n) \in \Gamma, \quad t \in (-\varepsilon, \varepsilon). \end{aligned} \tag{15}$$

Here $\varepsilon > 0$ is chosen sufficiently small, and we set $\Omega := \theta(\Gamma_\varepsilon)$. Now

$$\partial\theta = (\theta_{x_1}, \ldots, \theta_{x_n}, \theta_t) : \Gamma_\varepsilon \to \mathbb{R}^{(n+1)\times(n+1)}$$

denotes the functional matrix of θ, and we see

$$\partial\theta(x,0) = (\gamma_{x_1}, \ldots, \gamma_{x_n}, \nu)\Big|_x . \tag{16}$$

The symmetric matrix function

$$B^{-1}(x,t) := \partial\theta(x,t) \circ (\partial\theta(x,t))^*, \qquad (x,t) \in \Gamma_\varepsilon \tag{17}$$

satisfies

$$B^{-1}(x,0) \circ \nu(x) = (\gamma_{x_1}, \ldots, \gamma_{x_n}, \nu) \circ \begin{pmatrix} \gamma_{x_1}^* \\ \vdots \\ \gamma_{x_n}^* \\ \nu^* \end{pmatrix} \circ \nu(x) = \nu(x), \qquad x \in \Gamma. \tag{18}$$

Therefore, $\nu(x)$ is an eigenvector of the matrix $B^{-1}(x,0)$ to the eigenvalue 1, and the same is correct for the matrix $B(x,0)$. We now consider the function

$$\tilde{u}(x_1, \ldots, x_n, t) := u \circ \theta(x_1, \ldots, x_n, t) : \Gamma_\varepsilon \to \mathbb{R}. \tag{19}$$

Setting

$$\tilde{f}(x) := f \circ \gamma(x), \quad \tilde{g}(x) := g \circ \gamma(x), \qquad x \in \Gamma, \tag{20}$$

we obtain the initial condition of order 0 equivalent to (13), namely

$$\tilde{u}(x,0) = \tilde{f}(x), \qquad x \in \Gamma, \tag{21}$$

and the initial condition of order 1 equivalent to (14), namely

$$\frac{\partial}{\partial t}\tilde{u}(x,0) = \tilde{g}(x), \qquad x \in \Gamma. \tag{22}$$

We now prove the following

Theorem 4.8. *Let $\mathcal{F}$ be a characteristic surface for the differential operator $\mathcal{L}$ in (1), and the function $u = u(y) \in C^2(\Omega)$ may satisfy the Cauchy initial conditions (13) and (14) on $\mathcal{F}$. Then the expression $\mathcal{L}u(y)$ for all $y \in \mathcal{F}$ is already determined by the initial values $f \in C^2(\mathcal{F})$ and $g \in C^1(\mathcal{F})$.*

Proof:

1. With the aid of the parameter transformation θ we make the transition to the function $\tilde{u}(x,t)$,$(x,t) \in \Gamma_\varepsilon$ - due to (19) - satisfying the initial conditions (21) and (22) on Γ. By differentiation to x_j and x_k we obtain

$$\tilde{u}_{x_j}(x,0) = \tilde{f}_{x_j}(x), \quad \tilde{u}_{x_j x_k}(x,0) = \tilde{f}_{x_j x_k}(x), \quad \tilde{u}_{x_j t}(x,0) = \tilde{g}_{x_j}(x)$$

for all $x \in \Gamma$ and $j,k = 1,\dots,n$. The relation (19) implies

$$\nabla_{(x,t)}\tilde{u}(x,t) = \nabla_y u(\theta(x,t)) \circ \partial\theta(x,t), \qquad (x,t) \in \Gamma_\varepsilon. \tag{23}$$

We set

$$\partial u(y) := (\nabla_y u(y))^*,\ y \in \Omega \quad \text{and} \quad \partial\tilde{u}(x,t) := (\nabla_{(x,t)}\tilde{u}(x,t))^*,\ (x,t) \in \Gamma_\varepsilon,$$

and rewrite (23) into the form

$$\partial\tilde{u}(x,t) = (\partial\theta(x,t))^* \circ \partial u(\theta(x,t)), \quad (x,t) \in \Gamma_\varepsilon.$$

Multiplying this identity by $\partial\theta(x,t)$ from the left, we see

$$\partial\theta(x,t) \circ \partial\tilde{u}(x,t) = \partial\theta(x,t) \circ (\partial\theta(x,t))^* \circ \partial u(\theta(x,t)).$$

Noticing $B^{-1} = \partial\theta \circ (\partial\theta)^*$, we conclude

$$\partial u(\theta(x,t)) = B(x,t) \circ \partial\theta(x,t) \circ \partial\tilde{u}(x,t), \qquad (x,t) \in \Gamma_\varepsilon. \tag{24}$$

2. The identity (24) for $t = 0$ implies that $\nabla_y u(y)$ is determined for all $y \in \mathcal{F}$. Now we note for all $y \in \Omega$ that

$$\begin{aligned} \mathcal{L}u(y) &= \sum_{j,k=1}^{n+1} a_{jk}(y)\frac{\partial^2}{\partial y_j \partial y_k}u(y) + \sum_{j=1}^{n+1} b_j(y)\frac{\partial}{\partial y_j}u(y) + c(y)u(y) \\ &= \mathcal{M}u(y) + \sum_{k=1}^{n+1}\left\{b_k(y) - \sum_{j=1}^{n+1}\frac{\partial a_{jk}}{\partial y_j}(y)\right\}\frac{\partial}{\partial y_k}u(y) + c(y)u(y) \end{aligned} \tag{25}$$

holds true, setting

$$\mathcal{M}u(y) := \sum_{j=1}^{n+1}\frac{\partial}{\partial y_j}\left\{\sum_{k=1}^{n+1} a_{jk}(y)\frac{\partial}{\partial y_k}u(y)\right\}, \qquad y \in \Omega.$$

Therefore, we only have to show that $\mathcal{M}u(y)$ is determined by the initial data for all $y \in \mathcal{F}$. In this context we utilize the so called *weak differential equation* as follows: Let $\chi = \chi(y_1,\dots,y_{n+1}) \in C_0^\infty(\Omega)$ be an arbitrary test function and

$$\tilde{\chi} = \tilde{\chi}(x,t) = \chi \circ \theta(x,t) : \Gamma_\varepsilon \to \mathbb{R} \in C_0^2(\Gamma_\varepsilon)$$

denote the transformed test function. Parallel to (24) we obtain the relation

$$\partial\chi(\theta(x,t)) = B(x,t) \circ \partial\theta(x,t) \circ \partial\tilde{\chi}(x,t), \qquad (x,t) \in \Gamma_\varepsilon. \tag{26}$$

The test function $\chi \in C_0^\infty(\Omega)$ being given arbitrarily, the Gaussian integral theorem yields the identity

$$\begin{aligned}\int_\Omega \chi(y)\mathcal{M}u(y)\,dy &= \int_\Omega \left\{\chi(y)\sum_{j=1}^{n+1}\frac{\partial}{\partial y_j}\left(\sum_{k=1}^{n+1} a_{jk}(y)\frac{\partial}{\partial y_k}u(y)\right)\right\}dy\\ &= -\int_\Omega \left\{\sum_{j,k=1}^{n+1} a_{jk}(y)\frac{\partial}{\partial y_j}\chi(y)\frac{\partial}{\partial y_k}u(y)\right\}dy \qquad (27)\\ &= -\int_\Omega \left\{(\partial\chi(y))^* \circ A(y)\circ \partial u(y)\right\}dy\end{aligned}$$

with the symmetric matrix $A(y) := (a_{jk}(y))_{j,k=1,\ldots,n+1}$, $y \in \Omega$.

3. We now apply the transformation formula for multiple integrals: For the mapping $y = \theta(x,t)$, $(x,t) \in \Gamma_\varepsilon$ we denote the modulus of the functional determinant by

$$J_\theta(x,t) := |\det \partial\theta(x,t)|, \qquad (x,t) \in \Gamma_\varepsilon.$$

Taking the identities (24) and (26) into account, the relation (27) implies

$$\begin{aligned}\int_\Omega \chi(y)\mathcal{M}u(y)\,dy &= -\int_\Omega \left\{(\partial\chi(y))^* \circ A(y)\circ \partial u(y)\right\}dy\\ &= -\int_{\Gamma_\varepsilon} (\partial\chi(\theta(x,t)))^* \circ A(\theta(x,t))\circ \partial u(\theta(x,t))J_\theta(x,t)\,dx\,dt\\ &= -\int_{\Gamma_\varepsilon} (\partial\tilde\chi(x,t))^* \circ C(x,t)\circ \partial\tilde u(x,t)J_\theta(x,t)\,dx\,dt\end{aligned}$$

with

$$\begin{aligned}(c_{jk}(x,t))_{j,k=1,\ldots,n+1} &= C(x,t)\\ &:= (\partial\theta(x,t))^* \circ B(x,t)\circ A(\theta(x,t))\circ B(x,t)\circ \partial\theta(x,t).\end{aligned}$$

Then we calculate

$$\begin{aligned}c_{n+1,n+1}(x,0) &= \nu(x)^* \circ B(x,0)\circ A(\gamma(x))\circ B(x,0)\circ \nu(x)\\ &= (B(x,0)\circ\nu(x))^* \circ A(\gamma(x))\circ B(x,0)\circ \nu(x)\\ &= \nu(x)^* \circ A(\gamma(x))\circ \nu(x) \;=\; 0, \qquad x \in \Gamma,\end{aligned}$$

since $\mathcal{F}$ is a characteristic surface for $\mathcal{L}$. Consequently, we have

$$c_{n+1,n+1}(x,0) = 0 \qquad \text{for all} \quad x \in \Gamma. \tag{28}$$

Additionally, we observe

$$\begin{aligned}
\int_{\Omega} \chi(y)\mathcal{M}u(y)\,dy = &-\int_{\Gamma_\varepsilon} \Big\{ \sum_{j,k=1}^{n} c_{jk}(x,t)\tilde{\chi}_{x_j}\tilde{u}_{x_k} \Big\} J_\theta(x,t)\,dx\,dt \\
&-\int_{\Gamma_\varepsilon} \Big\{ \sum_{j=1}^{n} c_{j,n+1}(x,t)\tilde{\chi}_{x_j}\tilde{u}_t \Big\} J_\theta(x,t)\,dx\,dt \\
&-\int_{\Gamma_\varepsilon} \Big\{ \sum_{k=1}^{n} c_{n+1,k}(x,t)\tilde{\chi}_t\tilde{u}_{x_k} \Big\} J_\theta(x,t)\,dx\,dt \\
&-\int_{\Gamma_\varepsilon} c_{n+1,n+1}(x,t)\tilde{\chi}_t\tilde{u}_t J_\theta(x,t)\,dx\,dt
\end{aligned}$$

and find

$$\begin{aligned}
\int_{\Omega} \chi(y)\mathcal{M}u(y)\,dy = &\sum_{j,k=1}^{n} \int_{\Gamma_\varepsilon} \Big(c_{jk}(x,t)\tilde{u}_{x_k} J_\theta\Big)_{x_j} \tilde{\chi}(x,t)\,dx\,dt \\
&+\sum_{j=1}^{n} \int_{\Gamma_\varepsilon} \Big(c_{j,n+1}(x,t)\tilde{u}_t J_\theta\Big)_{x_j} \tilde{\chi}(x,t)\,dx\,dt \\
&+\sum_{k=1}^{n} \int_{\Gamma_\varepsilon} \Big(c_{n+1,k}(x,t)\tilde{u}_{x_k} J_\theta\Big)_{t} \tilde{\chi}(x,t)\,dx\,dt \qquad (29) \\
&+\int_{\Gamma_\varepsilon} \Big(c_{n+1,n+1}(x,t) J_\theta\Big)_{t} \tilde{u}_t\tilde{\chi}(x,t)\,dx\,dt \\
&+\int_{\Gamma_\varepsilon} c_{n+1,n+1}(x,t)\tilde{u}_{tt} J_\theta\tilde{\chi}(x,t)\,dx\,dt.
\end{aligned}$$

4. On the ball

$$B := \{z \in \mathbb{R}^{n+1} \;:\; |z| < 1\}$$

let us take the test function

$$\psi = \psi(z) \in C_0^\infty(B, [0,+\infty)) \quad \text{satisfying} \quad \int_B \psi(z)\,dz = 1.$$

The point $\eta \in \mathcal{F}$ being fixed, we consider the sequence of test functions

$$\chi_l(y) := l^{n+1}\psi(l(y-\eta)),\; y \in \mathbb{R}^{n+1} \quad \text{for} \quad l = 1,2,\ldots$$

Now the statement

$$\chi_l(y) = 0 \quad \text{for all} \quad |y - \eta| \geq l^{-1}$$

is correct, and we see

$$\int\limits_{\mathbb{R}^{n+1}} \chi_l(y)\, dy = 1 \qquad \text{for} \quad l = 1, 2, \ldots$$

With the notation

$$\tilde{\chi}_l(x,t) := \chi_l \circ \theta(x,t) \quad \text{for} \quad l = 1, 2, \ldots$$

the formula (28) implies

$$\lim_{l \to \infty} \int\limits_{\Gamma_\varepsilon} c_{n+1,n+1}(x,t)\tilde{u}_{tt}(x,t) J_\theta(x,t)\tilde{\chi}_l(x,t)\, dx\, dt = 0. \tag{30}$$

Inserting the sequence of test functions $\{\chi_l(y)\}_{l=1,2,\ldots}$ into the identity (29), the last term on the right-hand side disappears in the limit. Since the other terms are uniquely determined by $\mathcal{F}$, $f(y)$ and $g(y)$, $y \in \mathcal{F}$ - due to part 1 of the proof - and the left-hand side satisfies

$$\mathcal{M}u(\eta) = \lim_{l \to \infty} \int\limits_{\Omega} \chi_l(y)\mathcal{M}u(y)\, dy, \qquad \eta \in \mathcal{F}$$

in the limit, the proof of the theorem is established. q.e.d.

Remarks:

1. The Cauchy initial value problem (12), (13), (14) cannot be solved for arbitrary right-hand sides h, if $\mathcal{F}$ represents a characteristic surface. Therefore, one should start with noncharacteristic initial surfaces $\mathcal{F}$ in order to solve the Cauchy initial value problem. For the wave equation we shall choose the noncharacteristic basic plane of the cone from above as the initial plane.
2. When we consider the Cauchy initial value problem (12), (13),(14) on a noncharacteristic surface $\mathcal{F}$, the condition

$$c_{n+1,n+1}(x,0) = \nu(x)^* \circ A(\gamma(x)) \circ \nu(x) \neq 0, \qquad x \in \Gamma,$$

is valid. Localizing the equation (29) as in part 4 of the proof above, we can determine $\tilde{u}_{tt}(x,0)$, $x \in \Gamma$ by the differential equation (12) and the initial data (13), (14). Consequently, the second derivatives

$$(u_{y_j y_k}(y))_{j,k=1,\ldots,n+1}, \quad y \in \mathcal{F}$$

are already prescribed for the Cauchy initial value problem (12), (13), (14) on noncharacteristic surfaces. Similar statements can be established for the higher derivatives in case they exist.

5 The Wave Equation in $\mathbb{R}^n$ for $n = 1, 3, 2$

We take sufficiently regular functions $f = f(x_1, \ldots, x_n), g = g(x_1, \ldots, x_n) : \mathbb{R}^n \to \mathbb{R}$ and intend to solve the *Cauchy initial value problem for the n-dimensional wave equation:*

$$u = u(x,t) = u(x_1, \ldots, x_n, t) \in C^2(\mathbb{R}^n \times [0, +\infty), \mathbb{R}), \tag{1}$$

$$\Box u(x,t) = \frac{\partial^2}{\partial t^2} u(x,t) - c^2 \Delta_x u(x,t) = 0, \qquad (x,t) \in \mathbb{R}^n \times \mathbb{R}_+, \tag{2}$$

$$u(x,0) = f(x), \qquad \frac{\partial}{\partial t} u(x,0) = g(x), \qquad x \in \mathbb{R}^n. \tag{3}$$

Here $c > 0$ denotes a positive constant. We combine (1), (2),(3) to the problem $\mathcal{P}(f,g,n)$ or briefly $\mathcal{P}(n)$. At first, we consider the **case n=1** of the one-dimensional wave equation

$$u_{tt}(x,t) - c^2 u_{xx}(x,t) = 0, \qquad (x,t) \in \mathbb{R} \times \mathbb{R}. \tag{4}$$

In physics the function $u(x,t)$ describes the vertical deviation of a swinging string from the resting position $x \in \mathbb{R}$ in dependence of the time $t \in \mathbb{R}$.
Due to Section 4, Example 4.5, we obtain the characteristic lines of the one-dimensional wave equation

$$x = \alpha \pm ct, \qquad t \in \mathbb{R}, \tag{5}$$

with arbitrary $\alpha \in \mathbb{R}$. We now introduce these characteristics according to

$$\xi = x + ct, \qquad \eta = x - ct \tag{6}$$

as *characteristic parameters* into the differential equation (4). We observe (6) and deduce

$$x = \frac{1}{2}(\xi + \eta), \qquad t = \frac{1}{2c}(\xi - \eta). \tag{7}$$

Then we consider the function

$$U(\xi, \eta) := u\Big(\frac{1}{2}(\xi + \eta), \frac{1}{2c}(\xi - \eta)\Big), \qquad (\xi, \eta) \in \mathbb{R}^2. \tag{8}$$

On account of

$$U_\xi = u_x\Big(\frac{1}{2}(\xi + \eta), \frac{1}{2c}(\xi - \eta)\Big)\frac{1}{2} + u_t\Big(\frac{1}{2}(\xi + \eta), \frac{1}{2c}(\xi - \eta)\Big)\frac{1}{2c}$$

we obtain

$$\begin{aligned} U_{\xi\eta} &= \frac{1}{4} u_{xx}\Big(\frac{1}{2}(\xi + \eta), \frac{1}{2c}(\xi - \eta)\Big) - \frac{1}{4c} u_{xt}(\ldots) + \frac{1}{4c} u_{tx}(\ldots) - \frac{1}{4c^2} u_{tt}(\ldots) \\ &= -\frac{1}{4c^2}\Big\{u_{tt}\Big(\frac{1}{2}(\xi + \eta), \frac{1}{2c}(\xi - \eta)\Big) - c^2 u_{xx}\Big(\frac{1}{2}(\xi + \eta), \frac{1}{2c}(\xi - \eta)\Big)\Big\} \\ &= -\frac{1}{4c^2} \Box u\Big(\frac{1}{2}(\xi + \eta), \frac{1}{2c}(\xi - \eta)\Big). \end{aligned}$$

Therefore, the wave equation (4) appears with characteristic parameters in the form

$$\frac{\partial^2}{\partial\xi\partial\eta}U(\xi,\eta)=0, \qquad (\xi,\eta)\in\mathbb{R}^2. \tag{9}$$

On account of $\frac{\partial}{\partial\eta}U_\xi(\xi,\eta)=0$ the function $U_\xi=F'(\xi)$ is independent of η and we see

$$U(\xi,\eta)=F(\xi)+G(\eta).$$

Returning to the parameters (x,t) we obtain

$$u(x,t)=F(x+ct)+G(x-ct), \qquad (x,t)\in\mathbb{R}^2. \tag{10}$$

The solution belongs to the class $C^2(\mathbb{R}^2)$ if and only if $F,G\in C^2(\mathbb{R})$ is valid. Now the functions $v(x,t):=F(x+ct)$ and $w(x,t):=G(x-ct)$ satisfy the equations

$$v_t-cv_x=0 \quad \text{and} \quad w_t+cw_x=0 \qquad \text{in} \quad \mathbb{R}^2,$$

respectively. Then we obtain a solution of (4) by superposition with $C^2(\mathbb{R}^2)$-solutions for these equations. Here we remark that the one-dimensional d'Alembert operator can be decomposed as follows:

$$\Box=\frac{\partial^2}{\partial t^2}-c^2\frac{\partial^2}{\partial x^2}=\Big(\frac{\partial}{\partial t}+c\frac{\partial}{\partial x}\Big)\Big(\frac{\partial}{\partial t}-c\frac{\partial}{\partial x}\Big). \tag{11}$$

In the physical interpretation, the solution (10) of (4) consists of an incoming and an outgoing wave, each moving with the same absolute velocity into opposite directions.

We are now going to solve the initial value problem $\mathcal{P}(f,g,1)$: We require

$$f=f(x)\in C^2(\mathbb{R},\mathbb{R}), g=g(x)\in C^1(\mathbb{R},\mathbb{R})$$

and deduce the following relation for the function $u(x,y)$ given in (10):

$$\begin{aligned}
&u(x,0)=F(x)+G(x)=f(x),\\
&u_x(x,0)=F'(x)+G'(x)=f'(x),\\
&u_t(x,0)=cF'(x)-cG'(x)=g(x), \qquad x\in\mathbb{R}.
\end{aligned}$$

The last two equations yield

$$\begin{aligned}
F'(x)&=\frac{1}{2c}\{cf'(x)+g(x)\}=\frac{1}{2}f'(x)+\frac{1}{2c}g(x),\\
G'(x)&=\frac{1}{2c}\{cf'(x)-g(x)\}=\frac{1}{2}f'(x)-\frac{1}{2c}g(x), \qquad x\in\mathbb{R},
\end{aligned}$$

and integration from 0 to x gives

$$F(x) = \frac{1}{2} f(x) + \frac{1}{2c} \int_0^x g(\xi)\, d\xi + c_1,$$

$$G(x) = \frac{1}{2} f(x) - \frac{1}{2c} \int_0^x g(\xi)\, d\xi + c_2$$

with the two constants $c_1, c_2 \in \mathbb{R}$. The equation $F(x) + G(x) = f(x)$ implies $c_1 + c_2 = 0$, and the solution of Cauchy's initial value problem $\mathcal{P}(f, g, 1)$ has the representation

$$u(x,t) = F(x+ct) + G(x-ct) = \frac{1}{2}\Big\{ f(x+ct) + f(x-ct) \Big\} + \frac{1}{2c} \int_{x-ct}^{x+ct} g(\xi)\, d\xi$$

for $x \in \mathbb{R}$, $t \in \mathbb{R}$. Utilizing Theorem 4.7 in Section 4, we finally obtain the following

Theorem 5.1. (d'Alembert)
For the given functions $f = f(x) \in C^2(\mathbb{R})$ and $g = g(x) \in C^1(\mathbb{R})$ the expression

$$u(x,t) = \frac{1}{2}\Big\{ f(x+ct) + f(x-ct) \Big\} + \frac{1}{2c} \int_{x-ct}^{x+ct} g(\xi)\, d\xi, \qquad (x,t) \in \mathbb{R}^2, \quad (12)$$

represents the uniquely determined solution of Cauchy's initial value problem for the one-dimensional wave equation $\mathcal{P}(f, g, 1)$.

Remark: (Domain of dependence for the one-dimensional wave equation)
The value of the function $u(x,t)$ depends only on the initial values in the interval $[x - ct, x + ct]$, which means only on data within the characteristic cone with the tip (x, t). This is in accordance with the statement of Section 4, Theorem 4.7. On the other hand, the initial value at the point ξ can only become effective within the double cone

$$\Big\{ (x,t) \in \mathbb{R}^2 \; : \; |x - \xi| = c|t| \Big\}.$$

Therefore, the signals can propagate with the velocity c at most.

We now consider Cauchy's initial value problem for the wave equation in $\mathbb{R}^n$ and arbitrary $n \in \mathbb{N}$. In d'Alembert's solution formula (12) already appears a *spherical mean value*, which will enable us in higher dimensions $n \in \mathbb{N}$ as well, to solve the problems $\mathcal{P}(f, g, n)$ explicitly.

Definition 5.2. *For $f = f(x) \in C^2(\mathbb{R}^n)$ the associate function*

$$v = v(x,r) = M(x,r;f) := \frac{1}{\omega_n} \int\limits_{|\xi|=1} f(x + r\xi)\, d\sigma(\xi),\ (x,r) \in \mathbb{R}^n \times \mathbb{R} \quad (13)$$

is denoted as the spherical integral-mean-value of f *over the sphere*

$$\partial B_{|r|}(x) := \Big\{ y \in \mathbb{R}^n \ : \ |y - x| = |r| \Big\}.$$

Theorem 5.3. (F. John)
For $f = f(x) \in C^k(\mathbb{R}^n)$ given with $k \geq 2$, the function

$$v = v(x,r) = M(x,r;f) : \mathbb{R}^n \times \mathbb{R} \to \mathbb{R}$$

belongs to the regularity class $C^k(\mathbb{R}^n \times \mathbb{R})$, and the following statements hold true:

a) $v(x,0) = f(x)$ *for all* $x \in \mathbb{R}^n$ *(Initial value),*
b) $v(x,-r) = v(x,r)$ *for all* $x \in \mathbb{R}^n$, $r \in \mathbb{R}$ *(Even radial symmetry),*
c) $\frac{\partial}{\partial r} v(x,0) = 0$ *for all* $x \in \mathbb{R}^n$ *(Radial orthogonality),*
d) $\frac{\partial^2}{\partial r^2} v(x,r) + \frac{n-1}{r}\frac{\partial}{\partial r} v(x,r) - \Delta_x v(x,r) = 0$ *in* $\mathbb{R}^n \times (\mathbb{R} \setminus \{0\})$

*(***Darboux's differential equation***).*

Proof:

a) From (13) we infer $v \in C^k(\mathbb{R}^n \times \mathbb{R})$ and

$$v(x,0) = \frac{1}{\omega_n} \int\limits_{|\xi|=1} f(x)\, d\sigma(\xi) = f(x) \qquad \text{for all} \quad x \in \mathbb{R}^n.$$

b) and c) Once more we refer to (13) and see $v(x,-r) = v(x,r)$ immediately, while differentiation yields $-v_r(x,0) = v_r(x,0)$ for all $x \in \mathbb{R}^n$.

d) We introduce polar coordinates on the sphere

$$S^{n-1}(x) := \{y \in \mathbb{R}^n \ : \ |y - x| = 1\}$$

as follows:

$$y = x + r\xi, \qquad \xi \in S^{n-1}, \quad r > 0.$$

We remind the reader of Section 8 in Chapter 1, and the Laplace operator with these coordinates appears in the form

$$\Delta = \frac{\partial^2}{\partial r^2} + \frac{n-1}{r}\frac{\partial}{\partial r} + \frac{1}{r^2}\mathbf{\Lambda},$$

denoting the Laplace-Beltrami operator on the sphere S^{n-1} by $\mathbf{\Lambda}$. In Theorem 8.7 of Section 8 from Chapter 1, we have proved the symmetry of $\mathbf{\Lambda}$

on S^{n-1}. Consequently, we obtain the following equation for all $x \in \mathbb{R}^n$ and $r > 0$:

$$\begin{aligned}
\Delta_x v(x,r) &= \frac{1}{\omega_n} \int\limits_{|\xi|=1} \Delta_x f(x + r\xi)\, d\sigma(\xi) \\
&= \frac{1}{\omega_n} \int\limits_{|\xi|=1} \left\{ \frac{\partial^2}{\partial r^2} + \frac{n-1}{r}\frac{\partial}{\partial r} + \frac{1}{r^2}\mathbf{\Lambda} \right\} f(x + r\xi)\, d\sigma(\xi) \\
&= \left\{ \frac{\partial^2}{\partial r^2} + \frac{n-1}{r}\frac{\partial}{\partial r} \right\} v(x,r) + \frac{1}{r^2\omega_n} \int\limits_{|\xi|=1} 1 \cdot \mathbf{\Lambda} f(x + r\xi)\, d\sigma(\xi) \\
&= \left\{ \frac{\partial^2}{\partial r^2} + \frac{n-1}{r}\frac{\partial}{\partial r} \right\} v(x,r) + \frac{1}{r^2\omega_n} \int\limits_{|\xi|=1} (\mathbf{\Lambda} 1) \cdot f(x + r\xi)\, d\sigma(\xi) \\
&= \left\{ \frac{\partial^2}{\partial r^2} + \frac{n-1}{r}\frac{\partial}{\partial r} \right\} v(x,r)
\end{aligned}$$

utilizing $\mathbf{\Lambda} 1 = 0$. Therefore, Darboux's differential equation is satified for all $x \in \mathbb{R}^n$ and $r > 0$. The invariance with respect to the reflection $r \mapsto -r$ implies that the Darboux equation remains valid for all $x \in \mathbb{R}^n$ and $r < 0$.

q.e.d.

We now consider the **case n=3** of the three-dimensional wave equation. In physics their solutions represent waves from acoustics and optics. We prove the following

Theorem 5.4. (Kirchhoff)
Let the functions $f = f(x) \in C^3(\mathbb{R}^3)$ and $g = g(x) \in C^2(\mathbb{R}^3)$ be given. Then Cauchy's initial value problem $\mathcal{P}(f, g, 3)$ for the three-dimensional wave equation has the unique solution

$$\begin{aligned}
u(x,t) &= \frac{\partial}{\partial t}\Big\{ tM(x, ct; f) \Big\} + tM(x, ct; g) \\
&= \frac{1}{4\pi c^2 t^2} \iint\limits_{|y-x|=ct} \Big\{ tg(y) + f(y) + \nabla f(y) \cdot (y - x) \Big\}\, d\sigma(y)
\end{aligned} \tag{14}$$

for $(x,t) \in \mathbb{R}^3 \times \mathbb{R}_+$.

Proof:

1. We specialize Theorem 5.3 to the case $n = 3$, and the function

$$v(x,r) = M(x, r; g), (x,r) \in \mathbb{R}^3 \times (\mathbb{R} \setminus \{0\})$$

satisfies Darboux's differential equation

$$0 = v_{rr}(x,r) + \frac{2}{r}v_r(x,r) - \Delta_x v(x,r) = \frac{1}{r}\{rv(x,r)\}_{rr} - \Delta_x v(x,r).$$

The multiplication with r yields

$$0 = \frac{\partial^2}{\partial r^2}\{rv(x,r)\} - \Delta_x\{rv(x,r)\}, \qquad (x,r) \in \mathbb{R}^3 \times \mathbb{R}.$$

We now consider the function

$$\psi(x,t) := \frac{1}{c}\Big\{ctv(x,ct)\Big\} = tv(x,ct) = t\frac{1}{4\pi}\iint\limits_{|\xi|=1} g(x+ct\xi)\,d\sigma(\xi)$$

with $(x,t) \in \mathbb{R}^3 \times \mathbb{R}$. This function fulfills the wave equation

$$\Box\psi(x,t) = \frac{\partial^2}{\partial t^2}\psi(x,t) - c^2\Delta_x\psi(x,t) = 0 \qquad \text{in} \quad \mathbb{R}^3 \times \mathbb{R} \tag{15}$$

and is subject to the initial conditions

$$\psi(x,0) = 0, \quad \frac{\partial}{\partial t}\psi(x,0) = v(x,0) = g(x) \qquad \text{for all} \quad x \in \mathbb{R}^3. \tag{16}$$

2. Parallel to part 1 of the proof we see that the function

$$\chi(x,t) := tM(x,ct;f) = \frac{t}{4\pi}\iint\limits_{|\xi|=1} f(x+ct\xi)\,d\sigma(\xi), \qquad (x,t) \in \mathbb{R}^3 \times \mathbb{R},$$

satifies the wave equation

$$\Box\chi(x,t) = 0 \quad \text{in} \quad \mathbb{R}^3 \times \mathbb{R}.$$

Furthermore, we have $\chi \in C^3(\mathbb{R}^3 \times \mathbb{R})$. Then we consider the function

$$\begin{aligned}
\varphi(x,t) := \frac{\partial}{\partial t}\chi(x,t) &= \frac{\partial}{\partial t}\{tM(x,ct;f)\} \\
&= M(x,ct;f) + t\frac{\partial}{\partial t}M(x,ct;f) \\
&= \frac{1}{4\pi}\iint\limits_{|\xi|=1} f(x+ct\xi)\,d\sigma(\xi) + \frac{t}{4\pi}\frac{\partial}{\partial t}\Big\{\iint\limits_{|\xi|=1} f(x+ct\xi)\,d\sigma(\xi)\Big\} \\
&= \frac{1}{4\pi}\iint\limits_{|\xi|=1}\Big\{f(x+ct\xi) + ct\nabla f(x+ct\xi)\cdot\xi\Big\}\,d\sigma(\xi).
\end{aligned}$$

As well the function φ fulfills the wave equation, and we have the initial conditions

$$\begin{aligned}
&\varphi(x,0) = M(x,0;f) = f(x), \\
&\frac{\partial}{\partial t}\varphi(x,0) = \frac{\partial^2}{\partial t^2}\chi(x,0) = c^2 \Delta_x \chi(x,0) \\
&= c^2 \Big\{ t\Delta_x M(x,ct;f) \Big\}_{t=0} = 0 \quad \text{for all} \quad x \in \mathbb{R}^3.
\end{aligned} \tag{17}$$

3. With the expression

$$\begin{aligned}
u(x,t) := \varphi(x,t) + \psi(x,t) \quad &= \quad \frac{\partial}{\partial t}\Big\{ tM(x,ct;f) \Big\} + tM(x,ct;g) \\
&= \frac{1}{4\pi} \iint\limits_{|\xi|=1} \Big\{ f(x+ct\xi) + ct\nabla f(x+ct\xi)\cdot\xi + tg(x+ct\xi) \Big\}\, d\sigma(\xi)
\end{aligned}$$

for $(x,t) \in \mathbb{R}^3 \times \mathbb{R}$ we obtain a solution of the wave equation. The relations (16) and (17) imply the initial conditions

$$u(x,0) = f(x), \quad \frac{\partial}{\partial t}u(x,0) = g(x), \qquad x \in \mathbb{R}^3.$$

With the aid of the substitution

$$y = x + ct\xi, d\sigma(y) = c^2 t^2 d\sigma(\xi)$$

we deduce the following identity for all $(x,t) \in \mathbb{R}^3 \times \mathbb{R}_+$:

$$u(x,t) = \frac{1}{4\pi c^2 t^2} \iint\limits_{|y-x|=ct} \Big\{ tg(y) + f(y) + \nabla f(y)\cdot(y-x) \Big\}\, d\sigma(y).$$

q.e.d.

We now treat the **case n=2** with the aid of *Hadamard's method of descent.* Solutions of the two-dimensional wave equation model the movements of surfaces, for instance those of water waves.

Theorem 5.5. *Given the functions* $f = f(y) = f(y_1,y_2) \in C^3(\mathbb{R}^2)$ *and* $g = g(y) = g(y_1,y_2) \in C^2(\mathbb{R}^2)$, *the unique solution of Cauchy's initial value problem* $\mathcal{P}(f,g,2)$ *appears in the form*

$$\begin{aligned}
u(x,t) &= u(x_1,x_2,t) \\
&= \frac{1}{2\pi ct} \iint\limits_{y:|y-x|<ct} \Big\{ tg(y) + f(y) + \nabla f(y)\cdot(y-x) \Big\} \frac{1}{\sqrt{c^2t^2 - r^2}}\, dy_1\, dy_2
\end{aligned}$$

for $(x,t) \in \mathbb{R}^2 \times \mathbb{R}_+$ *with* $r := |y-x| = \sqrt{(y_1-x_1)^2 + (y_2-x_2)^2}$.

Proof: For $x \in \mathbb{R}^2 = \mathbb{R}^2 \times \{0\}$ and $t \in \mathbb{R}_+$ we consider the function

$$\begin{aligned} u(x,t) &= \frac{1}{4\pi c^2 t^2} \iint\limits_{|\eta - x| = ct} \Big\{ tg(\eta) + f(\eta) + \nabla f(\eta) \cdot (\eta - x) \Big\}\, d\sigma(\eta) \\ &= \frac{1}{2\pi c^2 t^2} \iint\limits_{\substack{|\eta - x| = ct \\ z>0}} \Big\{ tg(y_1,y_2) + f(y_1,y_2) + \nabla f(y_1,y_2) \cdot (y - x) \Big\}\, d\sigma(\eta) \end{aligned} \tag{18}$$

with $\eta = (y_1, y_2, z) \in \mathbb{R}^3$ and $f(\eta) := f(y_1, y_2)$, $g(\eta) := g(y_2, y_2)$. We parametrize the upper hemisphere $|\eta - x| = ct$, $z > 0$ as follows:

$$c^2t^2 = |\eta - x|^2 = (y_1 - x_1)^2 + (y_2 - x_2)^2 + z^2,$$

$$z = z(y) = z(y_1, y_2) = \sqrt{c^2t^2 - (y_1 - x_1)^2 - (y_2 - x_2)^2} = \sqrt{c^2t^2 - r^2}$$

using the parameter domain $\{y = (y_1, y_2) \in \mathbb{R}^2 \ : \ |y - x| < ct\}$. On account of

$$\sqrt{1 + \Big(\frac{\partial}{\partial r} z(r)\Big)^2} = \sqrt{1 + \Big(\frac{-r}{\sqrt{c^2t^2 - r^2}}\Big)^2} = \frac{ct}{\sqrt{c^2t^2 - r^2}}, \qquad r < ct,$$

the surface element of the upper hemisphere is determined by

$$d\sigma(y_1, y_2) = \sqrt{1 + \Big(\frac{\partial}{\partial r} z(r)\Big)^2}\, r\, dr\, d\varphi = \frac{ct}{\sqrt{c^2t^2 - r^2}}\, dy_1\, dy_2. \tag{19}$$

Inserting (19) into (18), we deduce

$$u(x,t) = \frac{1}{2\pi ct} \iint\limits_{y:|y-x|<ct} \Big\{ tg(y) + f(y) + \nabla f(y) \cdot (y - x) \Big\} \frac{1}{\sqrt{c^2t^2 - r^2}}\, dy_1\, dy_2.$$

q.e.d.

Remarks to Theorem 5.4 and Theorem 5.5:

1. In the case $n = 1$ the initial regularity $f \in C^2(\mathbb{R})$, $g \in C^1(\mathbb{R})$ being sufficient, we have to assume the initial regularity $f \in C^3(\mathbb{R}^n)$, $g \in C^2(\mathbb{R}^n)$ for $n = 2, 3$. Consequently, the wave equation implies a loss of regularity in the case $n = 2, 3$. This phenomenon is even strengthened in higher dimensions (compare Section 6).
2. From Theorem 5.4 we infer that the value of the solution u in the three-dimensional wave equation at the point (x, t) depends only on the initial values f and g on the sphere

$$\partial B_{ct}(x) = \Big\{ y \in \mathbb{R}^3 \ : \ |y - x| = ct \Big\},$$

that means $\partial B_{ct}(x)$ is the domain of dependence for $u(x, t)$.

On the other hand, the initial values f and g near a point y at the time $t = 0$ have an influence only on the points (x, t) near the conical surface $|x - y| = ct$ with the tip $x \in \mathbb{R}^3$. The signals in the ball

$$B_\varrho := \{x \in \mathbb{R}^3 \; : \; |x - y| < \varrho\}$$

have an effect only on $u(x, t)$ in the domain

$$\Omega := \bigcup_{x \in B_\varrho} M(x) \quad \text{where} \quad M(x) := \Big\{(z, t) \in \mathbb{R}^3 \times \mathbb{R}_+ \; : \; |z - x| = ct\Big\}.$$

Only the signals in $\mathbb{R}^3$ can be *sharply* transmitted. This is possible, since the domain of dependence of $u(x, t)$ is a sphere and not an open set in $\mathbb{R}^3$. This is known as *Huygens's principle in the sharp form.* Already for the wave equation in $\mathbb{R}^2$ (and many other hyperbolic equations) this principle is violated. According to Theorem 5.5, the solution $u(x, t)$ depends on the initial values in an open disc for the two-dimensional wave equation. Therefore, the perturbations especially of water waves propagate infinitely.

3. Assuming $f \in C_0^3(B_\varrho)$ and $g \in C_0^2(B_\varrho)$ with $\varrho > 0$, Theorem 3 yields a constant $C \in (0, +\infty)$ such that

$$|u(x, t)| \leq \frac{C}{t}, \qquad (x, t) \in \mathbb{R}^3 \times \mathbb{R}_+.$$

Therefore, the waves in $\mathbb{R}^3$, and in $\mathbb{R}^2$ as well (compare the proof of Theorem 5.5), have an amplitude with the asymptotic behavior $\frac{1}{t}$ for $t \to +\infty$.

6 The Wave Equation in $\mathbb{R}^n$ for $n \geq 2$

We now continue our considerations from Section 5, however, we fix the constant in the wave equation to $c = 1$ – and leave the translation of the results for arbitrary $c > 0$ to the reader. We begin with

Theorem 6.1. (Mean value theorem of Asgeirsson)
The following two statements are equivalent for a function $u = u(x, y) = u(x_1, \ldots, x_n, y_1, \ldots, y_n) \in C^2(\mathbb{R}^n \times \mathbb{R}^n)$ with $n \geq 2$:

I. We have the ultrahyperbolic differential equation

$$\left(\sum_{i=1}^n \frac{\partial^2}{\partial x_i^2} - \sum_{i=1}^n \frac{\partial^2}{\partial y_i^2}\right) u(x_1, \ldots, x_n, y_1, \ldots, y_n) = 0 \qquad in \quad \mathbb{R}^n \times \mathbb{R}^n. \quad (1)$$

II. For all $(x, y, r) \in \mathbb{R}^n \times \mathbb{R}^n \times \mathbb{R}_+$ we have the identity

$$\frac{1}{\omega_n} \int\limits_{|\xi|=1} u(x + r\xi, y)\, d\sigma(\xi) = \frac{1}{\omega_n} \int\limits_{|\xi|=1} u(x, y + r\xi)\, d\sigma(\xi). \qquad (2)$$

Proof:

I $\Rightarrow$ *II:* Since the relation (2) is invariant with respect to the reflection $r \to -r$, we can assume $r \geq 0$ without loss of generality. We consider the functions

$$\mu = \mu(x, y, r) := \frac{1}{\omega_n} \int\limits_{|\xi|=1} u(x + r\xi, y)\, d\sigma(\xi)$$

and

$$\nu = \nu(x, y, r) := \frac{1}{\omega_n} \int\limits_{|\xi|=1} u(x, y + r\xi)\, d\sigma(\xi)$$

for $(x, y, r) \in \mathbb{R}^n \times \mathbb{R}^n \times [0, +\infty)$. At first, we note that

$$\mu(x, y, 0) = u(x, y) = \nu(x, y, 0) \qquad \text{for all} \quad (x, y) \in \mathbb{R}^n \times \mathbb{R}^n \tag{3}$$

and

$$\mu_r(x, y, 0) = 0 = \nu_r(x, y, 0) \qquad \text{for all} \quad (x, y) \in \mathbb{R}^n \times \mathbb{R}^n. \tag{4}$$

According to Theorem 5.3 in Section 5, the functions μ and ν fulfill the Darboux differential equations

$$\begin{aligned} \mu_{rr}(x, y, r) + \frac{n-1}{r}\mu_r(x, y, r) - \Delta_x \mu(x, y, r) &= 0, \\ \nu_{rr}(x, y, r) + \frac{n-1}{r}\nu_r(x, y, r) - \Delta_y \nu(x, y, r) &= 0 \end{aligned} \tag{5}$$

for $(x, y, r) \in \mathbb{R}^n \times \mathbb{R}^n \times \mathbb{R}_+$. Furthermore, the ultrahyperbolic differential equation (1) yields the identity

$$\begin{aligned} \Delta_y \nu(x, y, r) &= \Delta_y \Big\{ \frac{1}{\omega_n} \int\limits_{|\xi|=1} u(x, y + r\xi)\, d\sigma(\xi) \Big\} \\ &= \frac{1}{\omega_n} \int\limits_{|\xi|=1} \Delta_y u(x, y + r\xi)\, d\sigma(\xi) \\ &= \frac{1}{\omega_n} \int\limits_{|\xi|=1} \Delta_x u(x, y + r\xi)\, d\sigma(\xi) \\ &= \Delta_x \Big\{ \frac{1}{\omega_n} \int\limits_{|\xi|=1} u(x, y + r\xi)\, d\sigma(\xi) \Big\} \\ &= \Delta_x \nu(x, y, r) \qquad \text{for all} \quad (x, y, r) \in \mathbb{R}^n \times \mathbb{R}^n \times \mathbb{R}_+. \end{aligned}$$

On account of (3), (4), and (5) the function $\varphi(x, y, r) := \mu(x, y, r) - \nu(x, y, r)$ solves the boundary value problem

$$\begin{aligned}
&\varphi_{rr}(x,y,r) + \frac{n-1}{r}\varphi_r(x,y,r) - \Delta_x\varphi(x,y,r) = 0 \text{ in } \mathbb{R}^n \times \mathbb{R}^n \times \mathbb{R}_+, \\
&\varphi(x,y,0) = 0, \quad \varphi_r(x,y,0) = 0 \text{ in } \mathbb{R}^n \times \mathbb{R}^n.
\end{aligned} \tag{6}$$

Theorem 4.7 from Section 4 implies $\varphi(x,y,r) = 0$ in each compact subset of the half-space $\mathbb{R}^n \times \mathbb{R}^n \times \mathbb{R}_+$ and $\mu(x,y,r) = \nu(x,y,r)$ in $\mathbb{R}^n \times \mathbb{R}^n \times \mathbb{R}_+$ or (2), respectively.

II ⇒ I: Let $(x_0, y_0) \in \mathbb{R}^n \times \mathbb{R}^n$ be a point, where the ultrahyperbolic differential equation (1) is not valid, which means

$$\Delta_x(x_0, y_0) \neq \Delta_y(x_0, y_0).$$

Respecting $u \in C^2(\mathbb{R}^n \times \mathbb{R}^n)$, we find a number $\varrho > 0$ such that

$$\Delta_x u(x', y') \neq \Delta_y(x'', y'') \qquad \text{for all} \quad (x', y'), (x'', y'') \in Z_\varrho(x_0, y_0). \tag{7}$$

Here we have set $Z_\varrho(x_0, y_0) := \{(x,y) \in \mathbb{R}^n \times \mathbb{R}^n : |x - x_0| + |y - y_0| \leq \varrho\}$. We now differentiate (2) with respect to r, and the Gaussian integral theorem combined with the mean value theorem of integral calculus yield the following relation at the point $(x,y,r) = (x_0, y_0, \varrho)$:

$$\begin{aligned}
0 &= \int\limits_{|\xi|=1} \nabla_x u(x_0 + \varrho\xi, y_0) \cdot \xi \, d\sigma(\xi) - \int\limits_{|\xi|=1} \nabla_y u(x_0, y_0 + \varrho\xi) \cdot \xi \, d\sigma(\xi) \\
&= \frac{1}{\varrho^{n-1}} \int\limits_{|x-x_0|\leq\varrho} \Delta_x u(x, y_0) \, dx - \frac{1}{\varrho^{n-1}} \int\limits_{|y-y_0|\leq\varrho} \Delta_y(x_0, y) \, dy \\
&= \Big(\Delta_x u(\tilde{x}, y_0) - \Delta_y u(x_0, \tilde{y})\Big)|B|\varrho.
\end{aligned} \tag{8}$$

Here we have chosen $\tilde{x} \in \mathbb{R}^n$ with $|\tilde{x} - x_0| \leq \varrho$ as well as $\tilde{y} \in \mathbb{R}^n$ with $|\tilde{y} - y_0| \leq \varrho$ suitably, and $|B|$ denotes the volume of the n-dimensional unit ball. With (8) a contradiction to (7) now appears. Consequently, the differential equation (1) is satisfied at all points $(x,y) \in \mathbb{R}^n \times \mathbb{R}^n$. q.e.d.

We now utilize calculations and arguments presented at the beginning of the n-dimensional potential theory in Section 1 of Chapter 5. Furthermore, we supply the following statement:

Proposition 6.2. *For each number $n \in \mathbb{N}$ with $n \geq 3$ and each continuous function $h = h(t) : [-1,1] \to \mathbb{R} \in C^0([-1,1])$ we have the identity*

$$\int\limits_{\xi \in \mathbb{R}^n : \, |\xi|=1} h(\xi_n) \, d\sigma(\xi) = \omega_{n-1} \int\limits_{-1}^{1} h(s)(1 - s^2)^{\frac{n-3}{2}} \, ds.$$

Proof: At first, we parametrize

$$\int\limits_{|\xi|=1} h(\xi_n)\,d\sigma(\xi)$$

$$= \int\limits_{t\in\mathbb{R}^{n-1}:\ |t|<1} \frac{h(\sqrt{1-t_1^2-\ldots-t_{n-1}^2})+h(-\sqrt{1-t_1^2-\ldots-t_{n-1}^2})}{\sqrt{1-t_1^2-\ldots-t_{n-1}^2}}\,dt_1\ldots dt_{n-1}.$$

Setting $t=\varrho\tau$ with $\varrho\in(0,1)$ and $\tau\in\mathbb{R}^{n-1}$, $|\tau|=1$, the formula (1) from Chapter 5, Section 1 implies

$$\int\limits_{|\xi|=1} h(\xi_n)\,d\sigma(\xi)=\omega_{n-1}\int\limits_0^1 \frac{h(\sqrt{1-\varrho^2})+h(-\sqrt{1-\varrho^2})}{\sqrt{1-\varrho^2}}\,\varrho^{n-2}\,d\varrho.$$

Finally, we obtain the following identity with the aid of the transformation $s=\sqrt{1-\varrho^2}$, $d\varrho=-\dfrac{s}{\sqrt{1-s^2}}\,ds$ as follows:

$$\begin{aligned}
\int\limits_{|\xi|=1} h(\xi_n)\,d\sigma(\xi) &= \omega_{n-1}\int\limits_0^1 (h(s)+h(-s))\frac{1}{s}(1-s^2)^{\frac{n-2}{2}}\frac{s}{\sqrt{1-s^2}}\,ds\\
&= \omega_{n-1}\int\limits_0^1 (h(s)+h(-s))(1-s^2)^{\frac{n-3}{2}}\,ds\\
&= \omega_{n-1}\int\limits_{-1}^1 h(s)(1-s^2)^{\frac{n-3}{2}}\,ds.
\end{aligned}$$

q.e.d.

Proposition 6.3. (Integral equation of Abel)
For the given function $f=f(x)\in C^2(\mathbb{R}^n)$ with $n\geq 3$, let $u=u(x,t)\in C^2(\mathbb{R}^n,\mathbb{R})$ be a solution of the problem

$$\begin{aligned}
&\Box u(x,t)=\Big(\frac{\partial^2}{\partial t^2}-\Delta_x\Big)u(x,t)=0 \qquad \text{in}\quad \mathbb{R}^n\times\mathbb{R},\\
&u(x,0)=f(x),\quad u_t(x,0)=0 \qquad \text{in}\quad \mathbb{R}^n.
\end{aligned} \tag{9}$$

Then the function u is symmetric with respect to reflections at the plane $t=0$, which means

$$u(x,-t)=u(x,t) \qquad \text{for all}\quad (x,t)\in\mathbb{R}^n\times\mathbb{R},$$

and satisfies the integral equation

$$\int\limits_{-r}^{r} u(x,\varrho)(r^2-\varrho^2)^{\frac{n-3}{2}}\,d\varrho=\frac{\omega_n}{\omega_{n-1}}r^{n-2}M(x,r;f), \qquad (x,r)\in\mathbb{R}^n\times\mathbb{R}_+. \tag{10}$$

Proof: At first, we note that u has to be symmetric with respect to reflections at the plane $t = 0$. More precisely, if $u = u(x,t)$, $(x,t) \in \mathbb{R}^n \times [0, +\infty)$ is the given solution of $\Box u = 0$ in $\mathbb{R}^n \times [0, +\infty)$, the function $\tilde{u}(x,t) := u(x,-t)$ satisfies this equation in $\mathbb{R}^n \times (-\infty, 0]$. The composition

$$w(x,t) := \begin{cases} u(x,t), & x \in \mathbb{R}^n, \ t \geq 0 \\ \tilde{u}(x,t), & x \in \mathbb{R}^n, \ t \leq 0 \end{cases}$$

fulfills $w(x,0) = f(x)$, $w_t(x,0) = 0$ for all $x \in \mathbb{R}^n$. Theorem 4.7 in Section 4 implies $w = u$ in $\mathbb{R}^n \times \mathbb{R}$, and we finally obtain the symmetry property stated.

We now consider the function $v = v(x,y) := u(x_1, \ldots, x_n, y_n)$, which satisfies the ultrahyperbolic differential equation

$$\Delta_x v(x,y) = \Delta_y v(x,y) \qquad \text{in} \quad \mathbb{R}^n \times \mathbb{R}^n$$

according to (9). Furthermore, we have $v(x,0) = f(x)$, $x \in \mathbb{R}^n$ and Theorem 6.1 yields

$$\begin{aligned} \int\limits_{|\xi|=1} v(x, r\xi)\, d\sigma(\xi) &= \int\limits_{|\xi|=1} v(x + r\xi, 0)\, d\sigma(\xi) \\ &= \int\limits_{|\xi|=1} f(x + r\xi)\, d\sigma(\xi) \\ &= \omega_n M(x, r; f), \qquad (x,r) \in \mathbb{R}^n \times \mathbb{R}_+. \end{aligned}$$

Taking Proposition 6.2 into account and applying the transformation $\varrho = rs$ with $ds = \dfrac{d\varrho}{r}$, we conclude

$$\begin{aligned} \omega_n M(x,r;f) = \int\limits_{|\xi|=1} v(x, r\xi)\, d\sigma(\xi) &= \int\limits_{|\xi|=1} u(x, r\xi_n)\, d\sigma(\xi) \\ &= \omega_{n-1} \int\limits_{-1}^{1} u(x, rs)(1 - s^2)^{\frac{n-3}{2}}\, ds \\ &= \omega_{n-1} \int\limits_{-r}^{r} u(x, \varrho)\Big(1 - \frac{\varrho^2}{r^2}\Big)^{\frac{n-3}{2}} \frac{1}{r}\, d\varrho \\ &= \frac{\omega_{n-1}}{r^{n-2}} \int\limits_{-r}^{r} u(x, \varrho)(r^2 - \varrho^2)^{\frac{n-3}{2}}\, d\varrho, \end{aligned}$$

which is equivalent to (10). q.e.d.

We are going to solve the Abel integral equation (10) for odd dimensions $n \geq 3$, at first. In this context we set $m = \frac{n-3}{2} \in \{0, 1, 2, \ldots\}$ and define the functions

$$\varphi(r) := \frac{\omega_n}{\omega_{n-1}} r^{n-2} M(x, r; f), \quad \psi(r) := u(x, r), \qquad r \in \mathbb{R}_+$$

for $x \in \mathbb{R}^n$ being fixed. Then we rewrite (10) into the form

$$\int_{-r}^{r} \psi(\varrho)(r^2 - \varrho^2)^m \, d\varrho = \varphi(r), \qquad r \in \mathbb{R}_+. \tag{11}$$

We now assume $f \in C^{m+3}(\mathbb{R}^n) = C^{\frac{n+3}{2}}(\mathbb{R}^n)$. Both sides of the identity (11) tending to zero when $r \to 0+$, the relation (11) is equivalent to

$$\int_{-r}^{r} \psi(\varrho)(r^2 - \varrho^2)^{m-1} \, d\varrho = \frac{1}{2mr} \frac{d}{dr} \varphi(r), \qquad r \in \mathbb{R}_+.$$

Again both sides tend to zero for $r \to 0+$, and repeating this differentiation m times we comprehend that (11) is equivalent to the identity

$$\int_{-r}^{r} \psi(\varrho) \, d\varrho = \frac{1}{2^m m!} \Big(\frac{1}{r} \frac{d}{dr}\Big)^m \varphi(r), \qquad r \in \mathbb{R}_+. \tag{12}$$

An additional differentiation reveals the following relation equivalent to (11), namely

$$\psi(r) + \psi(-r) = \frac{1}{2^m m!} \frac{d}{dr} \Big(\frac{1}{r} \frac{d}{dr}\Big)^m \varphi(r), \qquad r \in \mathbb{R}_+. \tag{13}$$

We now set $n = 2k + 1$ with $k \in \mathbb{N}$ and obtain $m = k - 1 \in 0, 1, 2, \ldots$. Due to the formula (11) from Chapter 5, Section 1 we have the equation

$$\omega_n = \frac{2(\Gamma(\frac{1}{2}))^n}{\Gamma(\frac{n}{2})}. \tag{14}$$

Therefore, we determine

$$\begin{aligned}
\frac{\omega_n}{\omega_{n-1}} &= \frac{\omega_{2k+1}}{\omega_{2k}} = \frac{2(\Gamma(\frac{1}{2}))^{2k+1}}{\Gamma(k + \frac{1}{2})} : \frac{2(\Gamma(\frac{1}{2}))^{2k}}{\Gamma(k)} \\
&= \frac{\Gamma(\frac{1}{2})\Gamma(k)}{\Gamma(k + \frac{1}{2})} = \frac{\Gamma(\frac{1}{2})(k-1)!}{\frac{1}{2}(\frac{1}{2} + 1) \ldots (\frac{1}{2} + (k-1))\Gamma(\frac{1}{2})} \\
&= \frac{2^k (k-1)!}{1 \cdot 3 \cdot \ldots \cdot (2k-1)}.
\end{aligned}$$

This implies

$$\frac{1}{2^m m!}\frac{\omega_n}{\omega_{n-1}} = \frac{1}{2^{k-1}(k-1)!}\frac{(k-1)!2^k}{1\cdot 3\cdot\ldots\cdot(2k-1)} = \frac{2}{1\cdot 3\cdot\ldots\cdot(n-2)}. \tag{15}$$

Restricting ourselves to solutions symmetric with respect to reflections at the plane $t = 0$, more precisely

$$u(x,-t) = u(x,t) \qquad \text{for all} \quad (x,t) \in \mathbb{R}^n \times \mathbb{R}, \tag{16}$$

we infer the following solution of Abel's integral equation (10) from (13) and (15):

$$u(x,t) = \frac{1}{1\cdot 3\cdot\ldots\cdot(n-2)}\frac{\partial}{\partial t}\Big(\frac{1}{t}\frac{\partial}{\partial t}\Big)^{\frac{n-3}{2}}\Big\{t^{n-2}M(x,t;f)\Big\} \tag{17}$$

for $(x,t) \in \mathbb{R}^n \times \mathbb{R}_+$ and $n = 3,5,7,\ldots$ We now prove

Proposition 6.4. *Let $f \in C^{\frac{n+3}{2}}(\mathbb{R}^n)$ for $n = 3,5,7,\ldots$ be given, the function $u = u(x,t)$, $(x,t) \in \mathbb{R}^n \times \mathbb{R}$ defined in (17) and reflected due to (16) then belongs to the regularity class $C^2(\mathbb{R}^n \times \mathbb{R})$ and represents the uniquely determined solution of Cauchy's initial value problem (9).*

Proof: The function $\chi(x,t) := M(x,t;f)$ belongs to the regularity class $C^{\frac{n+3}{2}}(\mathbb{R}^n \times \mathbb{R})$. The differential operator $\dfrac{1}{t}\dfrac{\partial}{\partial t}$ diminishes the order of differentiation by 1. We reflect the function defined in (17) according to (16) and obtain

$$u = u(x,t) \in C^{\frac{n+3}{2}-\frac{n-3}{2}-1}(\mathbb{R}^n \times \mathbb{R}) = C^2(\mathbb{R}^n \times \mathbb{R}).$$

We note that

$$\Big(\frac{1}{t}\frac{d}{dt}\Big)t^k = kt^{k-2}, \qquad k \in \mathbb{Z}$$

and calculate

$$\begin{aligned}
&1\cdot 3\cdot\ldots\cdot(n-2)\,u(x,t)\\
&= \frac{\partial}{\partial t}\Big(\frac{1}{t}\frac{\partial}{\partial t}\Big)^{\frac{n-3}{2}}\Big\{t^{n-2}\chi(x,t)\Big\}\\
&= \frac{\partial}{\partial t}\Big(\frac{1}{t}\frac{\partial}{\partial t}\Big)^{\frac{n-3}{2}-1}\Big\{(n-2)t^{n-4}\chi + t^{n-3}\chi_t\Big\}\\
&= \frac{\partial}{\partial t}\Big(\frac{1}{t}\frac{\partial}{\partial t}\Big)^{\frac{n-3}{2}-2}\Big\{(n-2)(n-4)t^{n-6}\chi + ct^{n-5}\chi_t + t^{n-4}\mu\Big\}\\
&= \ldots = \frac{\partial}{\partial t}\Big\{(n-2)(n-4)\ldots\cdot 1\,t\chi + ct^2\chi_t + t^3\mu\Big\}\\
&= 1\cdot 3\cdot\ldots\cdot(n-2)\Big[\chi(x,t) + ct\chi_t(x,t) + t^2\mu(x,t)\Big]
\end{aligned}$$

with the constants $c \in \mathbb{R}$ and the functions $\mu = \mu(x,t)$. Comparing with Section 5, Theorem 5.3 we see

$$u(x,0) = \chi(x,0) = f(x), \quad u_t(x,0) = c\chi_t(x,0) = 0, \qquad x \in \mathbb{R}^n.$$

We now show that u satisfies the wave equation: With the aid of Darboux's differential equation we deduce

$$\begin{aligned} &1 \cdot 3 \cdot \ldots \cdot (n-2)\Big\{u_{tt}(x,t) - \Delta_x u(x,t)\Big\} \\ &= \Big(\frac{\partial}{\partial t}\Big)^3 \Big(\frac{1}{t}\frac{\partial}{\partial t}\Big)^{\frac{n-3}{2}} \Big\{t^{n-2}\chi(x,t)\Big\} - \frac{\partial}{\partial t}\Big(\frac{1}{t}\frac{\partial}{\partial t}\Big)^{\frac{n-3}{2}} \Big\{t^{n-2}\Delta_x \chi(x,t)\Big\} \\ &= \Big(\frac{\partial}{\partial t}\Big)^3 \Big(\frac{1}{t}\frac{\partial}{\partial t}\Big)^{\frac{n-3}{2}} \Big\{t^{n-2}\chi\Big\} - \frac{\partial}{\partial t}\Big(\frac{1}{t}\frac{\partial}{\partial t}\Big)^{\frac{n-3}{2}} \Big\{t^{n-2}\chi_{tt} + (n-1)t^{n-3}\chi_t\Big\}. \end{aligned}$$

We consider the ordinary linear differential operator $L : C^{\frac{n+3}{2}}(\mathbb{R}) \to C^0(\mathbb{R})$ defined by

$$L\varphi := \Big(\frac{d}{dt}\Big)^3 \Big(\frac{1}{t}\frac{d}{dt}\Big)^{\frac{n-3}{2}} \Big\{t^{n-2}\varphi\Big\} - \frac{d}{dt}\Big(\frac{1}{t}\frac{d}{dt}\Big)^{\frac{n-3}{2}} \Big\{t^{n-2}\frac{d^2}{dt^2}\varphi + (n-1)t^{n-3}\frac{d}{dt}\varphi\Big\}$$

for $\varphi = \varphi(t) \in C^{\frac{n+3}{2}}(\mathbb{R})$. We show the claim $L : C^{\frac{n+3}{2}}(\mathbb{R}) \to \mathcal{O}$ with $\mathcal{O}(t) \equiv 0$. This relation is proved on the dense space of polynomials, and the Weierstraß approximation theorem gives the complete statement, which implies $\Box u = 0$ in $\mathbb{R}^n \times \mathbb{R}$.

We take $\varphi(t) = t^k$ with $k \in \mathbb{N} \cup \{0\}$ and calculate

$$\begin{aligned} L\varphi &= \Big(\frac{d}{dt}\Big)^3 \Big(\frac{1}{t}\frac{d}{dt}\Big)^{\frac{n-3}{2}} \Big\{t^{n+k-2}\Big\} \\ &\quad - \frac{d}{dt}\Big(\frac{1}{t}\frac{d}{dt}\Big)^{\frac{n-3}{2}} \Big\{k(k-1)t^{k+n-4} + k(n-1)t^{k+n-4}\Big\} \\ &= \Big(\frac{d}{dt}\Big)^3 \Big(\frac{1}{t}\frac{d}{dt}\Big)^{\frac{n-3}{2}} \Big\{t^{n+k-2}\Big\} - \frac{d}{dt}\Big(\frac{1}{t}\frac{d}{dt}\Big)^{\frac{n-3}{2}} \Big\{k(k+n-2)t^{k+n-4}\Big\} \\ &= \frac{d}{dt}\Big\{\Big(\frac{d}{dt}\Big)^2 \Big[(n+k-2) \cdot \ldots \cdot (k+3)t^{k+1}\Big] \\ &\quad -(n+k-2) \cdot \ldots \cdot (k+1)kt^{k-1}\Big\} \\ &= \frac{d}{dt}\Big\{(n+k-2) \cdot \ldots \cdot (k+3)(k+1)kt^{k-1} \\ &\quad -(n+k-2) \cdot \ldots \cdot (k+1)kt^{k-1}\Big\} = 0. \end{aligned}$$

From the linearity of L we obtain the statement for arbitrary polynomials, and consequently the Proposition is proved. q.e.d.

With the aid of Proposition 6.4 we show

Theorem 6.5. *Let the functions $f = f(x), g = g(x) \in C^{\frac{n+3}{2}}(\mathbb{R}^n)$ with odd $n \geq 3$ be given. Then Cauchy's initial value problem $\mathcal{P}(f, g, n)$ for the n-dimensional wave equation is uniquely solved by the following function*

$$\psi(x,t) = \frac{1}{1 \cdot 3 \cdot \ldots \cdot (n-2)} \left\{ \frac{\partial}{\partial t} \left(\frac{1}{t} \frac{\partial}{\partial t} \right)^{\frac{n-3}{2}} \left(t^{n-2} M(x,t;f) \right) + \left(\frac{1}{t} \frac{\partial}{\partial t} \right)^{\frac{n-3}{2}} \left(t^{n-2} M(x,t;g) \right) \right\}, \qquad (x,t) \in \mathbb{R}^n \times \mathbb{R}_+. \tag{18}$$

Here we abbreviate

$$M(x,t;f) := \frac{1}{\omega_n} \int\limits_{|\xi|=1} f(x + t\xi) \, d\sigma(\xi), \qquad (x,t) \in \mathbb{R}^n \times \mathbb{R}_+.$$

Proof: According to Proposition 6.4, we consider the following function reflected at the plane $t = 0$ with the aid of (16), namely

$$u(x,t) := \frac{1}{1 \cdot 3 \cdot \ldots \cdot (n-2)} \frac{\partial}{\partial t} \left(\frac{1}{t} \frac{\partial}{\partial t} \right)^{\frac{n-3}{2}} \left\{ t^{n-2} M(x,t;f) \right\}, \quad (x,t) \in \mathbb{R}^n \times \mathbb{R}_+.$$

Now $u(x,t)$ solves the Cauchy initial value problem

$$\begin{aligned} \Box u(x,t) &= 0 \quad \text{in} \quad \mathbb{R}^n \times \mathbb{R}, \\ u(x,0) &= f(x), \quad u_t(x,0) = 0 \quad \text{in} \quad \mathbb{R}^n. \end{aligned} \tag{19}$$

Analogously, the function

$$v(x,t) := \frac{1}{1 \cdot 3 \cdot \ldots \cdot (n-2)} \frac{\partial}{\partial t} \left(\frac{1}{t} \frac{\partial}{\partial t} \right)^{\frac{n-3}{2}} \left\{ t^{n-2} M(x,t;g) \right\}, \quad (x,t) \in \mathbb{R}^n \times \mathbb{R}_+$$

- reflected due to (16) - solves the Cauchy problem

$$\begin{aligned} \Box v(x,t) &= 0 \quad \text{in} \quad \mathbb{R}^n \times \mathbb{R}, \\ v(x,0) &= g(x), \quad v_t(x,0) = 0 \quad \text{in} \quad \mathbb{R}^n. \end{aligned} \tag{20}$$

We now define the function

$$\begin{aligned} w(x,t) &:= \int\limits_0^t v(x,\tau) \, d\tau \\ &= \frac{1}{1 \cdot 3 \cdot \ldots \cdot (n-2)} \int\limits_0^t \frac{\partial}{\partial \tau} \left(\frac{1}{\tau} \frac{\partial}{\partial \tau} \right)^{\frac{n-3}{2}} \left\{ \tau^{n-2} M(x,\tau;g) \right\} d\tau \\ &= \frac{1}{1 \cdot 3 \cdot \ldots \cdot (n-2)} \left(\frac{1}{t} \frac{\partial}{\partial t} \right)^{\frac{n-3}{2}} \left\{ t^{n-2} M(x,t;g) \right\}, \quad (x,t) \in \mathbb{R}^n \times \mathbb{R}_+, \end{aligned}$$

and easily see

$$w(x,0)=0, \quad w_t(x,0)=v(x,0)=g(x) \qquad \text{in} \quad \mathbb{R}^n. \tag{21}$$

Observing (20) we deduce

$$\begin{aligned} w_{tt}(x,t) &= \frac{\partial}{\partial t}v(x,t) = \int_0^t v_{\tau\tau}(x,\tau)\,d\tau \\ &= \int_0^t \Delta_x v(x,\tau)\,d\tau = \Delta_x \int_0^t v(x,\tau)\,d\tau \\ &= \Delta_x w(x,t), \end{aligned}$$

and consequently

$$\Box w(x,t) = 0 \qquad \text{in} \quad \mathbb{R}^n \times \mathbb{R}_+. \tag{22}$$

On account of (19), (21), and (22) the composition

$$\psi(x,t) := u(x,t) + w(x,t), \quad (x,t) \in \mathbb{R}^n \times \mathbb{R}_+$$

gives us the solution of $\mathcal{P}(f,g,n)$ defined in (18). From Theorem 4.7 in Section 4 we infer the uniqueness of the solution. q.e.d.

With the aid of Hadamard's method of descent we want to solve $\mathcal{P}(f,g,n)$ for even dimensions $n \geq 2$:

Theorem 6.6. *Let the even, positive integer $n \geq 2$ and the two functions $f = f(x), g = g(x) \in C^{\frac{n+4}{2}}(\mathbb{R}^n)$ be given. Then Cauchy's initial value problem $\mathcal{P}(f,g,n)$ for the n-dimensional wave equation is uniquely solved by the following function*

$$\begin{aligned} \psi(x,t) = \alpha_n \Bigg\{ &\frac{\partial}{\partial t}\Big(\frac{1}{t}\frac{\partial}{\partial t}\Big)^{\frac{n-2}{2}} \Bigg[\int_0^t \frac{s^{n-1}}{\sqrt{t^2-s^2}} M(x,s;f)\,ds\Bigg] \\ &+\Big(\frac{1}{t}\frac{\partial}{\partial t}\Big)^{\frac{n-2}{2}} \Bigg[\int_0^t \frac{s^{n-1}}{\sqrt{t^2-s^2}} M(x,s;g)\,ds\Bigg]\Bigg\}, \qquad (x,t) \in \mathbb{R}^n \times \mathbb{R}_+. \end{aligned}$$

Here we abbreviate $\alpha_2 = 1$ and

$$\alpha_n = \frac{1}{2 \cdot 4 \cdot \ldots \cdot (n-2)} \qquad \textit{for} \quad n = 4, 6, \ldots$$

Proof:

1. We extend the initial values onto the whole space $\mathbb{R}^{n+1}$ as follows:

$$f^*(x_1, \ldots, x_n, x_{n+1}) := f(x_1, \ldots, x_n),$$
$$g^*(x_1, \ldots, x_n, x_{n+1}) := g(x_1, \ldots, x_n)$$

for $(x_1, \ldots, x_{n+1}) =: y \in \mathbb{R}^{n+1}$. Observing $f^*, g^* \in C^{\frac{(n+1)+3}{2}}(\mathbb{R}^{n+1})$, we can explicitly determine the unique solution of $\mathcal{P}(f^*, g^*, n+1)$ with the aid of Theorem 6.5 as follows:

$$\psi(y,t) = \frac{1}{1 \cdot 3 \cdot \ldots \cdot (n-1)} \Big\{ \frac{\partial}{\partial t} \Big(\frac{1}{t} \frac{\partial}{\partial t} \Big)^{\frac{n-2}{2}} \Big(t^{n-1} M(y,t; f^*) \Big)$$
$$+ \Big(\frac{1}{t} \frac{\partial}{\partial t} \Big)^{\frac{n-2}{2}} \Big(t^{n-1} M(y,t; g^*) \Big) \Big\}, \qquad (y,t) \in \mathbb{R}^{n+1} \times \mathbb{R}_+. \tag{23}$$

2. We evaluate the integral-mean-value

$$M(y,t; f^*) = M(x,t; f^*) = \frac{1}{\omega_{n+1}} \int\limits_{\xi \in \mathbb{R}^{n+1},\ |\xi|=1} f(x_1 + t\xi_1, \ldots, x_n + t\xi_n)\, d\sigma(\xi).$$

In this context we choose the parametrization

$$\xi_{n+1} = \pm\sqrt{1 - \xi_1^2 - \ldots - \xi_n^2}, \quad d\sigma(\xi) = \frac{d\xi_1 \ldots d\xi_n}{\sqrt{1 - \xi_1^2 - \ldots - \xi_n^2}}$$

for $\xi_1^2 + \ldots \xi_n^2 < 1$ and obtain

$$M(x,t; f^*) = \frac{2}{\omega_{n+1}} \int\limits_{\xi_1^2 + \ldots \xi_n^2 < 1} \frac{f(x_1 + t\xi_1, \ldots, x_n + t\xi_n)}{\sqrt{1 - \xi_1^2 - \ldots - \xi_n^2}}\, d\xi_1 \ldots d\xi_n.$$

We now introduce polar coordinates in $\mathbb{R}^n$ as follows: $\xi_i = \varrho\eta_i$ for $i = 1, \ldots, n$ with $\varrho = |\xi|$ and $\eta = (\eta_1, \ldots, \eta_n)$ satisfying $|\eta| = 1$. Then we obtain

$$\begin{aligned} M(x,t; f^*) &= \frac{2}{\omega_{n+1}} \int_0^1 \Big\{ \frac{\varrho^{n-1}}{\sqrt{1-\varrho^2}} \int\limits_{|\eta|=1} f(x + t\varrho\eta)\, d\sigma(\eta) \Big\}\, d\varrho \\ &= \frac{2\omega_n}{\omega_{n+1}} \int_0^1 \frac{\varrho^{n-1}}{\sqrt{1-\varrho^2}} M(x, t\varrho; f)\, d\varrho \\ &\overset{s=t\varrho}{=} \frac{2\omega_n}{\omega_{n+1}} \int_0^t \frac{s^{n-1}}{t^n \sqrt{1 - (\frac{s}{t})^2}} M(x,s;f)\, ds \\ &= \frac{2\omega_n}{\omega_{n+1}} \frac{1}{t^{n-1}} \int_0^t \frac{s^{n-1}}{\sqrt{t^2 - s^2}} M(x,s;f)\, ds. \end{aligned}$$

3. In the case $n = 2$ we have $\frac{2\omega_n}{\omega_{n+1}} = 1$. In the case $n = 4, 6, \ldots$ we utilize (14) and calculate

$$\frac{2\omega_n}{\omega_{n+1}} = 2\frac{2(\Gamma(\frac{1}{2}))^n}{\Gamma(\frac{n}{2})} : \frac{2(\Gamma(\frac{1}{2}))^{n+1}}{\Gamma(\frac{n+1}{2})} = \frac{2}{\Gamma(\frac{1}{2})}\frac{\Gamma(\frac{n+1}{2})}{\Gamma(\frac{n}{2})}$$

$$= \frac{2}{\Gamma(\frac{1}{2})}\frac{\Gamma(\frac{1}{2}) \cdot \frac{1}{2}(\frac{1}{2}+1)\ldots(\frac{1}{2}+(\frac{n}{2}-1))}{1 \cdot 2 \cdot \ldots \cdot (\frac{n}{2}-1)}$$

$$= \frac{1 \cdot 3 \cdot \ldots \cdot (n-1)}{2^{\frac{n}{2}-1} \cdot 1 \cdot 2 \cdot \ldots \cdot (\frac{n}{2}-1)} = \frac{1 \cdot 3 \cdot \ldots \cdot (n-1)}{2 \cdot 4 \cdot \ldots \cdot (n-2)}.$$

4. The function ψ given in (23) does not depend on x_{n+1}. According to the results from 2. and 3. the solution $\psi(x,t) = \psi(x, x_{n+1}, t)$ of the problem $\mathcal{P}(f, g, n)$ can be represented in the form

$$\psi(x,t) = \frac{1}{2 \cdot 4 \cdot \ldots \cdot (n-2)}\left\{\frac{\partial}{\partial t}\Big(\frac{1}{t}\frac{\partial}{\partial t}\Big)^{\frac{n-2}{2}}\left[\int_0^t \frac{s^{n-1}}{\sqrt{t^2-s^2}}M(x,s;f)\,ds\right]\right.$$

$$\left. +\Big(\frac{1}{t}\frac{\partial}{\partial t}\Big)^{\frac{n-2}{2}}\left[\int_0^t \frac{s^{n-1}}{\sqrt{t^2-s^2}}M(x,s;g)\,ds\right]\right\}, \qquad (x,t) \in \mathbb{R}^n \times \mathbb{R}_+,$$

for $n = 4, 6, 8, \ldots$ In the case $n = 2$ we obtain

$$\psi(x,t) = \frac{\partial}{\partial t}\left[\int_0^t \frac{s}{\sqrt{t^2-s^2}}M(x,s;f)\,ds\right] + \int_0^t \frac{s}{\sqrt{t^2-s^2}}M(x,s;g)\,ds$$

for $(x,t) \in \mathbb{R}^2 \times \mathbb{R}_+$. q.e.d.

7 The Inhomogeneous Wave Equation and an Initial-boundary-value Problem

We prescribe the function $h = h(x,t) \in C^2(\mathbb{R}^n \times [0,+\infty), \mathbb{R})$ and consider Cauchy's initial value problem $\mathcal{P}(f, g, h, n)$ for the *inhomogeneous wave equation*, namely

$$\begin{aligned} &u = u(x,t) = u(x_1, \ldots, x_n, t) \in C^2(\mathbb{R}^n \times [0,+\infty), \mathbb{R}),\\ &\Box u(x,t) = h(x,t) \quad \text{in} \quad \mathbb{R}^n \times \mathbb{R}_+,\\ &u(x,0) = f(x), \quad u_t(x,0) = g(x) \quad \text{in} \quad \mathbb{R}^n. \end{aligned} \tag{1}$$

When we are able to solve the inhomogeneous wave equation for the initial values $f(x) \equiv 0$, $g(x) \equiv 0$ in $\mathbb{R}^n$, the superposition with a solution of the initial value problem considered in Section 5 and Section 6 for the homogeneous wave equation yields a solution of the problem $\mathcal{P}(f, g, h, n)$. Consequently, we assume

$$f(x) \equiv 0, \quad g(x) \equiv 0, \qquad x \in \mathbb{R}^n \tag{2}$$

in the sequel. We now construct a solution $u(x,t)$ of $\mathcal{P}(0, 0, h, n)$ with the following *ansatz of Duhamel*:

$$u(x,t) = \int_0^t U(x,t,s)\, ds, \qquad (x,t) \in \mathbb{R}^n \times \mathbb{R}_+. \tag{3}$$

Here the functions $U = U(x,t,s)$ are solutions of the wave equation for each fixed $s \in [0,t]$

$$\Box U(x,t,s) = \frac{\partial^2}{\partial t^2} U(x,t,s) - c^2 \Delta_x U(x,t,s) \equiv 0 \qquad \text{in} \quad \mathbb{R}^n \times \mathbb{R}_+, \tag{4}$$

satisfying the initial conditions

$$U(x,s,s) = 0, \quad U_t(x,s,s) = h(x,s), \qquad x \in \mathbb{R}^n. \tag{5}$$

Then the function u from (3) solves the problem $\mathcal{P}(0, 0, h, n)$. We evidently have $u(x,0) = 0$, and the equation

$$u_t(x,t) = U(x,t,t) + \int_0^t U_t(x,t,s)\, ds = \int_0^t U_t(x,t,s)\, ds$$

implies $u_t(x,0) = 0$. Furthermore, we calculate

$$\begin{aligned} u_{tt}(x,t) &= U_t(x,t,t) + \int_0^t U_{tt}(x,t,s)\, ds \\ &= h(x,t) + c^2 \int_0^t \Delta_x U(x,t,s)\, ds \\ &= h(x,t) + c^2 \Delta_x u(x,t), \qquad (x,t) \in \mathbb{R}^n \times \mathbb{R}_+. \end{aligned}$$

Alternatively to the problem (4) and (5), we recommend the transition to the function

$$V(x,t,s) = U(x,t+s,s), \tag{6}$$

satisfying the problem

$$\begin{aligned} &\Box V(x,t,s) \equiv 0, \qquad (x,t) \in \mathbb{R}^n \times \mathbb{R}_+, \\ &V(x,0,s) = 0, \quad V_t(x,0,s) = h(x,s), \qquad x \in \mathbb{R}^n, \end{aligned} \tag{7}$$

for each fixed $s \in [0,t]$. Now we can explicitly solve problem (7) with the aid of integral formulas from Section 5 and Section 6. We confine ourselves to the important case $n = 3$ in physics, and obtain from Section 5, Theorem 5.4 the formula

$$V(x,t,s) = \frac{1}{4\pi c^2 t} \iint_{|y-x|=ct} h(y,s)\, d\sigma(y),$$

for $h = h(x,t) \in C^2(\mathbb{R}^3 \times [0,+\infty))$. Inserting

$$U(x,t,s) = V(x,t-s,s) = \frac{1}{4\pi c^2 (t-s)} \iint_{|y-x|=c(t-s)} h(y,s)\, d\sigma(y), \qquad s \in [0,t]$$

into Duhamel's formula (3), we have proved the following

Theorem 7.1. *Let $h = h(x,t) \in C^2(\mathbb{R}^3 \times [0,+\infty))$ be given. Then Cauchy's initial value problem $\mathcal{P}(0,0,h,3)$ for the inhomogeneous wave equation is uniquely solved by the function*

$$u = u(x,t) = \frac{1}{4\pi c^2} \int_0^t \left\{ \frac{1}{t-s} \iint_{|y-x|=c(t-s)} h(y,s)\, d\sigma(y) \right\} ds, \quad (x,t) \in \mathbb{R}^3 \times \mathbb{R}_+. \tag{8}$$

Remark: The solution $u(x,t)$ only depends on the values of h restricted to the backward characteristic cone

$$\Big\{ (y,s) \in \mathbb{R}^3 \times \mathbb{R} : \; |y-x| = c(t-s), \; 0 < s < t \Big\}$$

with the tip $(x,t) \in \mathbb{R}^3 \times \mathbb{R}_+$ and the basic surface in the plane $t = 0$.

So far, we only have considered solutions of the wave equation extending onto the whole space $\mathbb{R}^n$, $n \in \mathbb{N}$. We now choose a bounded open set $\Omega \subset \mathbb{R}^n$ with smooth regular C^2-boundary, and investigate the following *initial-boundary-value problem $\mathcal{P}_0(f,g,\Omega)$ for the n-dimensional wave equation*: We look for a function $u = u(x,t) : \overline{\Omega} \to \mathbb{R}$ in the class

$$\mathcal{F} := \left\{ v(x,t) \in C^2(\Omega \times [0,+\infty)) : \begin{array}{l} v(\cdot,t), v_t(\cdot,t), v_{tt}(\cdot,t) \in C^0(\overline{\Omega}) \\ \text{for all } t \in [0,+\infty) \end{array} \right\},$$

satisfying

$$\begin{aligned} &\Box u(x,t) = u_{tt}(x,t) - c^2 \Delta_x u(x,t) \equiv 0 \quad \text{in} \quad \Omega \times (0,+\infty), \\ &u(x,0) = f(x) \quad \text{in} \quad \Omega, \\ &u_t(x,0) = g(x) \quad \text{in} \quad \Omega, \\ &u(x,t) = 0 \quad \text{in} \quad \partial\Omega \times [0,+\infty). \end{aligned} \tag{9}$$

Here $f, g : \overline{\Omega} \to \mathbb{R}$ are initial data of the regularity class $C^1(\overline{\Omega})$. The *energy of u at the time t* is determined by

$$E(t) := \frac{1}{2} \int_{\Omega} \Big\{ |u_t(x,t)|^2 + c^2 |\nabla_x u(x,t)|^2 \Big\}\, dx, \qquad 0 \le t < +\infty. \tag{10}$$

A solution $u \in \mathcal{F}$ of (9) possesses finite energy. This can be seen by partial integration, where we utilize Proposition 9.1 (Giesecke, E. Heinz) from Chapter 8, Section 9. Let $u \in \mathcal{F}$ be a solution of the problem $\mathcal{P}_0(f, g, \Omega)$. We now deduce

$$\begin{aligned} \frac{d}{dt} E(t) &= \int_{\Omega} \Big\{ u_t u_{tt} + c^2 \nabla_x u \cdot \nabla_x u_t \Big\}\, dx \\ &= \int_{\Omega} \Big\{ u_t u_{tt} - c^2 u_t \Delta_x u \Big\}\, dx \\ &= \int_{\Omega} u_t \Box u \, dx \;=\; 0, \qquad t \in [0, +\infty). \end{aligned}$$

This implies

$$E(t) = const, \qquad t \in [0, +\infty) \tag{11}$$

for the solutions $u = u(x,t) \in \mathcal{F}$ of $\mathcal{P}_0(f, g, \Omega)$. For initial data

$$f(x) \equiv 0, \quad g(x) \equiv 0, \qquad x \in \Omega$$

we obtain

$$E(0) = \frac{1}{2} \int_{\Omega} \Big\{ (g(x))^2 + c^2 |\nabla f(x)|^2 \Big\}\, dx.$$

Consequently, the condition $E(t) \equiv 0$, $t \in [0, +\infty)$ holds true which implies

$$u_t \equiv 0, \quad \nabla_x u \equiv 0 \qquad \text{in} \quad \Omega \times [0, +\infty).$$

We get $u(x,t) \equiv 0$ for homogeneous initial values $f \equiv 0 \equiv g$. Since (9) is a linear problem, we have established the following

Theorem 7.2. *The problem $\mathcal{P}_0(f, g, \Omega)$ admits at most one solution.*

With the aid of the spectral theory (compare Chapter 8, Section 9) we are going to construct a solution of $\mathcal{P}_0(f, g, \Omega)$. There we prove the *spectral theorem of H. Weyl*: For the domain Ω given as above, there exists a sequence of eigenfunctions $v_k = v_k(x) : \overline{\Omega} \to \mathbb{R} \in C^2(\Omega) \cap C^0(\overline{\Omega})$ with

$$v_k(x) = 0, \quad x \in \partial\Omega, \qquad \text{and} \qquad \int_{\Omega} (v_k(x))^2\, dx = 1$$

belonging to the eigenvalues $0 < \lambda_1 \le \lambda_2 \le \lambda_3 \le \ldots \to +\infty$ such that

$$\Delta v_k(x) + \lambda_k v_k(x) = 0, \qquad x \in \Omega, \tag{12}$$

for $k = 1, 2, \ldots$ is fulfilled. These functions $\{v_k(x)\}_{k=1,2,\ldots}$ represent a complete orthonormal system in $L^2(\Omega)$.

Example 7.3. In the case $n = 1$ and $\Omega = [0, \pi]$, we find the eigenfunctions and solutions of (12) as follows

$$v_k(x) = \sqrt{\frac{2}{\pi}} \sin(kx), \qquad x \in [0, \pi], \qquad k = 1, 2, \ldots$$

We now construct a solution of $\mathcal{P}_0(f, g, \Omega)$ with the ansatz

$$u(x,t) = \sum_{k=1}^{\infty} a_k(t) v_k(x), \qquad (x,t) \in \Omega \times [0, +\infty). \tag{13}$$

Evidently, $u(x,t) = 0$ on $\partial\Omega \times [0, +\infty)$ holds true. Furthermore, the wave equation

$$0 = \Box u(x,t) = \sum_{k=1}^{\infty} \Big(a_k''(t) + c^2 \lambda_k a_k(t) \Big) v_k(x)$$

turns out being equivalent to the ordinary differential equations

$$a_k''(t) + c^2 \lambda_k a_k(t) = 0, \qquad k = 1, 2, \ldots \tag{14}$$

Additionally, we use the initial conditions

$$f(x) = u(x,0) = \sum_{k=1}^{\infty} a_k(0) v_k(x), \qquad x \in \Omega, \tag{15}$$

and

$$g(x) = u_t(x,0) = \sum_{k=1}^{\infty} a_k'(0) v_k(x), \qquad x \in \Omega. \tag{16}$$

These are equivalent to

$$a_k(0) = \int\limits_{\Omega} f(x) v_k(x)\, dx, \quad a_k'(0) = \int\limits_{\Omega} g(x) v_k(x)\, dx, \qquad k = 1, 2, \ldots \tag{17}$$

We find the uniquely determined coefficient functions from (14) and (17) as follows:

$$a_k(t) = \int\limits_{\Omega} \Big\{ f(x) \cos(c\sqrt{\lambda_k} t) + g(x) \frac{\sin(c\sqrt{\lambda_k} t)}{c\sqrt{\lambda_k}} \Big\} v_k(x)\, dx, \quad k = 1, 2, \ldots \tag{18}$$

Finally, we obtain

Theorem 7.4. *The uniquely determined solution of $\mathcal{P}_0(f, g, \Omega)$ is given by the series of functions (13) with the coefficients (18).*

8 Classification, Transformation and Reduction of Partial Differential Equations

We consider the following linear partial differential equation of second order on the domain $\Omega \subset \mathbb{R}^n$ with $n \geq 2$:

$$\mathcal{L}u(x) := \sum_{i,j=1}^{n} a_{ij}(x)\frac{\partial^2}{\partial x_i \partial x_j}u(x) + \sum_{i=1}^{n} b_i(x)\frac{\partial}{\partial x_i}u(x) + c(x)u(x) = d(x) \quad (1)$$

with $x \in \Omega$ and $u = u(x) \in C^2(\Omega)$. The coefficient functions $a_{ij}(x)$, $b_i(x)$, $c(x)$ for $i, j = 1, \ldots, n$ and the right-hand side $d(x)$ belong to the regularity class $C^0(\Omega)$, and the matrix $(a_{ij}(x))_{i,j=1,\ldots,n}$ is symmetric for all $x \in \Omega$. The domain $\Theta \subset \mathbb{R}^n$ being given, we consider the diffeomorphism

$$\xi = \xi(x) = (\xi_1(x_1, \ldots, x_n), \ldots, \xi_n(x_1, \ldots, x_n)) : \Omega \to \Theta \in C^2(\Omega, \mathbb{R}^n) \quad (2)$$

with the inverse mapping

$$x = x(\xi) = (x_1(\xi_1 \ldots, \xi_n), \ldots, x_n(\xi_1, \ldots, \xi_n)) : \Theta \to \Omega \in C^2(\Theta, \mathbb{R}^n). \quad (3)$$

We define the function

$$v(\xi) := u(x_1(\xi_1, \ldots, \xi_n), \ldots, x_n(\xi_1, \ldots, \xi_n)) : \Theta \to \mathbb{R} \in C^2(\Theta).$$

Then we calculate

$$\frac{\partial u}{\partial x_i} = \sum_{k=1}^{n} \frac{\partial v}{\partial \xi_k}\frac{\partial \xi_k}{\partial x_i}, \qquad i = 1, \ldots, n \quad (4)$$

and

$$\frac{\partial^2 u}{\partial x_i \partial x_j} = \sum_{k,l=1}^{n} \frac{\partial^2 v}{\partial \xi_k \partial \xi_l}\frac{\partial \xi_k}{\partial x_i}\frac{\partial \xi_l}{\partial x_j} + \ldots, \qquad i, j = 1, \ldots, n. \quad (5)$$

Here the points $\ldots$ denote terms in $1 = v^0$, v and $\frac{\partial v}{\partial \xi_k}$ for $k = 1, \ldots, n$. Now the differential equation below is derived from (1), (4) and (5) using the convention above:

$$\sum_{k,l=1}^{n} A_{kl}(\xi)\frac{\partial^2}{\partial \xi_k \partial \xi_l}v(\xi) + \ldots = 0 \qquad \text{in} \quad \Theta, \quad (6)$$

with the coefficients

$$A_{kl}(\xi) := \sum_{i,j=1}^{n} a_{ij}(x(\xi))\frac{\partial \xi_k}{\partial x_i}\frac{\partial \xi_l}{\partial x_j}, \qquad \xi \in \Theta, \qquad k, l = 1, \ldots, n. \quad (7)$$

We denote the Jacobi matrix by $\partial\xi := (\frac{\partial \xi_i}{\partial x_j})_{i,j=1,\ldots,n}$. For each $\xi \in \Theta$ the expression

$$(A_{kl}(\xi))_{k,l=1,\ldots,n} = \partial\xi(x(\xi)) \circ (a_{ij}(x(\xi))_{i,j=1,\ldots,n} \circ (\partial\xi(x(\xi)))^*$$

is a real, symmetric $n \times n$-matrix with $\frac{n}{2}(n+1)$ independent coefficients. We now intend to choose the parameter transformation (2) in such a way that the leading coefficient matrix $(A_{kl})(\xi))_{k,l=1,\ldots,n}$ appears as simple as possible. Here we have n functions $\xi_1(x), \ldots, \xi_n(x)$ at our disposal. Dividing by one factor $A_{kl}(\xi)$ in the homogeneous differential equation (6), we can achieve one coefficient being normed to 1. Therefore, we can at most fulfill $(n+1)$ conditions. We distinguish between

The case n=2: We have $\frac{n}{2}(n+1) = n+1$, and the parameter transformation (2) can be chosen such that $(A_{kl}(\xi))_{k,l=1,2}$ appears in the neighborhood of each point x with $(a_{ij}(x))_{i,j=1,2} \neq 0$ in one of the following *normal forms*:

$$\begin{pmatrix} 1 & 0 \\ 0 & 1 \end{pmatrix}, \quad \begin{pmatrix} 1 & 0 \\ 0 & -1 \end{pmatrix}, \quad \begin{pmatrix} 1 & 0 \\ 0 & 0 \end{pmatrix}. \tag{8}$$

In principle, this transformation has been already established by C. F. Gauß. As we shall see in Chapter 11 and 12, this possibility to locally – and sometimes even globally – reduce the equation into the normal form distinguishes the two-dimensional theory of partial differential equations from the higher-dimensional situation.

The case n=3: We can use the three transformation functions $\xi_1(x)$, $\xi_2(x)$, $\xi_3(x)$ to render the coefficients in (6) being zero outside the diagonal. Thus we achieve

$$A_{12}(\xi) \equiv 0, \quad A_{13}(\xi) \equiv 0, \quad A_{23}(\xi) \equiv 0 \qquad \text{in} \quad \Theta.$$

If $(a_{ij}(x))_{i,j=1,2,3} \neq 0$ holds true, we still normalize one of the diagonal-elements to 1 on account of the homogeneity in the equation (6); for instance $A_{11}(\xi) \equiv 1$ in Θ. The other two diagonal-elements (in our case $A_{22}(\xi)$ and $A_{33}(\xi)$) remain undetermined. Therefore, a transformation into one of the forms

$$\begin{pmatrix} 1 & 0 & 0 \\ 0 & 1 & 0 \\ 0 & 0 & 1 \end{pmatrix}, \quad \begin{pmatrix} 1 & 0 & 0 \\ 0 & -1 & 0 \\ 0 & 0 & -1 \end{pmatrix}, \quad \begin{pmatrix} 1 & 0 & 0 \\ 0 & 1 & 0 \\ 0 & 0 & 0 \end{pmatrix}$$

is impossible, in general. The matrices above correspond to the Laplace equation in $\mathbb{R}^3$, to the wave equation in $\mathbb{R}^2$, and to the heat equation in $\mathbb{R}^2$, respectively.

The cases n=4,5,...: In the case $n = 4$ we have six matrix-elements outside the diagonal, which cannot be transformed into zero by the four parameter functions $\xi_1(x), \ldots, \xi_4(x)$, in general. We remark that all time-dependent partial differential equations in $\mathbb{R}^3$ (as the wave and heat equation) are differential equations in the space $\mathbb{R}^4$. In higher dimensions these problems are even increasing.

However, it is possible in the dimensions $n = 2, 3, \ldots$ to transform the differential equation (1) into the normal form at a fixed point $x^0 \in \Omega$. For the sake of simplicity, we choose $x^0 = 0 \in \Omega$ which can always be achieved by a translation in $\mathbb{R}^n$. We consider the homogeneously linear transformation

$$\xi_i = \sum_{j=1}^{n} f_{ij} x_j, \quad i = 1, \ldots, n, \qquad \xi = \xi(x) = F \circ x, \tag{9}$$

with the real coefficient matrix $F = (f_{ij})_{i,j=1,\ldots,n} \in \mathbb{R}^{n \times n}$. We expand

$$a_{ij}(x) = \alpha_{ij} + o(1), \quad x \to 0, \qquad i, j = 1, \ldots,$$

set $A := (\alpha_{ij})_{i,j=1,\ldots,n} \in \mathbb{R}^{n \times n}$, and transform the coefficient matrix

$$(a_{ij}(x))_{i,j=1,\ldots,n}$$

into

$$(A_{kl}(\xi))_{k,l=1,\ldots,n} = F \circ A \circ F^* + o(1), \qquad \xi \to 0. \tag{10}$$

Due to the symmetry of A, we find an orthogonal matrix F such that

$$\Lambda = \begin{pmatrix} \lambda_1 & & 0 \\ & \ddots & \\ 0 & & \lambda_n \end{pmatrix} := F \circ A \circ F^* \tag{11}$$

becomes a diagonal matrix. Choosing

$$G := \begin{pmatrix} \mu_1 & & 0 \\ & \ddots & \\ 0 & & \mu_n \end{pmatrix} \qquad \text{with} \quad \mu_k = \begin{cases} 1, & \text{if } \lambda_k = 0 \\ \dfrac{1}{\sqrt{|\lambda_k|}}, & \text{if } \lambda_k \neq 0 \end{cases}, \qquad k = 1, \ldots, n,$$

we obtain

$$(G \circ F) \circ A \circ (G \circ F)^* = G \circ \Lambda \circ G^* = \begin{pmatrix} \varepsilon_1 & & 0 \\ & \ddots & \\ 0 & & \varepsilon_n \end{pmatrix} \tag{12}$$

with $\varepsilon_k \in \{-1, 0, 1\}$ for $k = 1, \ldots, n$. Setting $M := G \circ F$, we have proved the following

Theorem 8.1. *For each fixed point $x^0 \in \Omega$, we have an affine-linear transformation $\xi = \xi(x) = M \circ (x - x^0)$ with the real coefficient matrix $M \in \mathbb{R}^{n \times n}$ such that the differential equation (1) transformed due to (6) possesses the following coefficient matrix*

$$(A_{kl}(\xi))_{k,l=1,\ldots,n} = \begin{pmatrix} \varepsilon_1 & & 0 \\ & \ddots & \\ 0 & & \varepsilon_n \end{pmatrix} + o(1), \quad \xi \to x^0,$$

with $\varepsilon_k \in \{-1, 0, 1\}$ *and* $k = 1, \ldots, n$.

Definition 8.2. *The differential equation (1) is named* elliptic at the point $x^0 \in \Omega$, *if and only if all eigenvalues of the matrix* $(a_{ij}(x^0))_{i,j=1,\ldots,n}$ *do not vanish and have the same sign. If (1) is elliptic for all* $x^0 \in \Omega$, *we denote the differential equation being* elliptic in Ω.

Remarks: Eventually multiplying by (-1) we achieve that $(a_{ij}(x^0))_{i,j=1,\ldots,n}$ becomes positive-definite. Pointwise transformation into the normal form gives us the leading matrix

$$\begin{pmatrix} 1 & & 0 \\ & \ddots & \\ 0 & & 1 \end{pmatrix}.$$

The Laplace equation

$$\Delta u(x, \ldots, x_n) = 0$$

is the easiest and most important elliptic differential equation in $\mathbb{R}^n$. We have no characteristic surfaces for elliptic differential equations (compare Section 4).

Definition 8.3. *The differential equation (1) is named* hyperbolic at the point $x^0 \in \Omega$, *if and only if all eigenvalues of the matrix* $(a_{ij}(x^0))_{i,j=1,\ldots,n}$ *do not vanish and exactly one eigenvalue differs in its sign from the others. This being correct for all* $x^0 \in \Omega$, *we speak of a* hyperbolic differential equation in Ω.

Remarks: Eventually multiplying (1) by the factor (-1), the pointwise transformation into the normal form yields the leading matrix

$$\begin{pmatrix} 1 & & & 0 \\ & -1 & & \\ & & \ddots & \\ 0 & & & -1 \end{pmatrix}.$$

As the most important and easiest hyperbolic differential equation, we became familiar with the wave equation

$$\Box u(x_1, \ldots, x_n, t) = 0 \quad \text{in} \quad \mathbb{R}^n.$$

For hyperbolic equations, characteristic surfaces appear reducing to cones for the wave equation (see Section 4).

Definition 8.4. *The differential equation (8.2) is named* ultrahyperbolic at the point $x^0 \in \Omega$ *if and only if all eigenvalues of the matrix* $(a_{ij}(x^0))_{i,j=1,\ldots,n}$ *do not vanish, and at least two of them exist with a positive and a negative sign, respectively. This being correct for all* $x^0 \in \Omega$*, the differential equation (1) is called* ultrahyperbolic in Ω.

Remark: For $n \geq 2$ the differential equation

$$\Delta_x u(x_1,\ldots,x_n,y_1,\ldots,y_n) = \Delta_y u(x_1,\ldots,x_n;y_1,\ldots,y_n)$$

is ultrahyperbolic in $\mathbb{R}^{2n}$.

Definition 8.5. *The condition* $det(a_{ij}(x^0))_{i,j=1,\ldots,n} = 0$ *being fulfilled, we call (1)* parabolic at the point $x^0 \in \Omega$*; we name (1)* parabolic in Ω *if and only if* $det(a_{ij}(x))_{i,j=1,\ldots,n} = 0$ *for all* $x \in \Omega$ *holds true.*

Remarks: The equation (1) is exactly parabolic in Ω if and only if one eigenvalue of the matrix $(a_{ij}(x))_{i,j=1,\ldots,n} \neq 0$ vanishes for all $x \in \Omega$. The heat equation in $\mathbb{R}^n$ appears as the main example:

$$u_t(x_1,\ldots,x_n,t) = \Delta_x u(x_1,\ldots,x_n,t).$$

We shall now determine those affine-linear transformations leaving the wave equation in $\mathbb{R}^n$, $n \in \mathbb{N}$, invariant. In this context we consider the *transformation matrix*

$$F = (f_{kl})_{k,l=1,\ldots,n+1} \in \mathbb{R}^{(n+1)\times(n+1)}$$

and the *translation vector*

$$f = (f_1,\ldots,f_{n+1})^* \in \mathbb{R}^{n+1}.$$

We define the affine-linear, nonsingular, positive-oriented transformation $\varphi : \mathbb{R}^{n+1} \to \mathbb{R}^{n+1}$ by

$$(\xi,\tau) = (\xi_1,\ldots,\xi_n,\tau) = \varphi(x,t) = (\varphi_1(x_1,\ldots,x_n,t),\ldots,\varphi_{n+1}(x_1,\ldots,x_n,t))$$

with

$$\begin{aligned} \xi_k &= \sum_{l=1}^{n} f_{kl}x_l + f_{k,n+1}t + f_k, \qquad k = 1,\ldots,n, \\ \tau &= \sum_{l=1}^{n} f_{n+1,l}x_l + f_{n+1,n+1}t + f_{n+1} \end{aligned} \tag{13}$$

and equivalently

$$(\xi,\tau)^* = F \circ (x,t)^* + f, \qquad (x,t) \in \mathbb{R}^{n+1}. \tag{14}$$

Definition 8.6. *The transformation (13) or equivalently (14) is called* Lorentz transformation, *if and only if all* $u = u(\xi,\tau) = u(\xi_1,\ldots,\xi_n,\tau) \in C^2(\mathbb{R}^{n+1})$ *satisfy the invariance condition*

$$\Box_{(x,t)}\{u\circ\varphi\}\Big|_{(x,t)} = \{\Box_{(\xi,\tau)}u(\xi,\tau)\}\circ\varphi(x,t) \qquad \text{in} \quad \mathbb{R}^{n+1}. \tag{15}$$

Here we have used the d'Alembert operator

$$\Box_{(x,t)} := -c^2\Big(\frac{\partial^2}{\partial x_1^2}+\ldots+\frac{\partial^2}{\partial x_n^2}\Big)+\frac{\partial^2}{\partial t^2}$$

with the constant $c > 0$.

Remarks:

1. The relation (15) reveals that the set of Lorentz transformations $\varphi : \mathbb{R}^{n+1} \to \mathbb{R}^{n+1}$ is a group $\mathcal{G}$, with the composition of mappings as group operation and with the neutral element $\varphi = \mathrm{id}_{\mathbb{R}^{n+1}}$. We name $\mathcal{G}$ the *Lorentz group.*
2. The subgroup $\mathcal{G}_0 := \{\varphi\in\mathcal{G} : \varphi(0)=0\}$ of the *origin-preserving Lorentz transformations* consists of those mappings (14) with $f = 0$ satisfying the condition (15).

On account of the calculations at the beginning of this section the invariance condition (15) is equivalent to the matrix equation

$$F\circ\begin{pmatrix} -c^2 & & & 0\\ & \ddots & & \\ & & -c^2 & \\ 0 & & & 1\end{pmatrix}\circ F^* = \begin{pmatrix} -c^2 & & & 0\\ & \ddots & & \\ & & -c^2 & \\ 0 & & & 1\end{pmatrix}. \tag{16}$$

Definition 8.7. *Those Lorentz transformations* $\varphi = \varphi(x_1,\ldots,x_n,t)\in\mathcal{G}$ *allowing a time-independent measurement for the position in space, that means*

$$\frac{d}{dt}\varphi_k(x_1,\ldots,x_n,t)\equiv 0, \qquad k=1,\ldots,n, \tag{17}$$

and the time measurement being independent of the position in space, that means

$$\frac{d}{dx_k}\varphi_{n+1}(x_1,\ldots,x_n,t)\equiv 0, \qquad k=1,\ldots,n, \tag{18}$$

and which do not cause a time reversal, that means

$$\frac{d}{dt}\varphi_{n+1}(x_1,\ldots,x_n,t) > 0, \tag{19}$$

are called Galilei transformations.

Remark: The Galilei transformations $\mathcal{G}' \subset \mathcal{G}$ constitute a subgroup of the group of Lorentz transformations.

If (13) represents a Galilei transformation, we obtain the conditions

$$f_{k,n+1} = 0 = f_{n+1,k}, \quad k = 1, \ldots, n, \qquad f_{n+1,n+1} > 0.$$

Setting

$$F' := \begin{pmatrix} f_{11} & \cdots & f_{1n} \\ \vdots & & \vdots \\ f_{n1} & \cdots & f_{nn} \end{pmatrix} \in \mathbb{R}^{n \times n}, \qquad F = \begin{pmatrix} F' & 0 \\ 0 & f_{n+1,n+1} \end{pmatrix},$$

the relation (16) implies

$$F' \circ (F')^* = \begin{pmatrix} 1 & & 0 \\ & \ddots & \\ 0 & & 1 \end{pmatrix}, \qquad f_{n+1,n+1} = 1, \qquad \det F' > 0.$$

These considerations yield the following

Theorem 8.8. *In the class of Lorentz transformations (13) the Galilei transformations take on the form*

$$\xi^* = F' \circ x^* + f', \qquad t = t + f_{n+1} \tag{20}$$

with the positive-oriented, orthogonal $n \times n$-matrix

$$F' := \begin{pmatrix} f_{11} & \cdots & f_{1n} \\ \vdots & & \vdots \\ f_{n1} & \cdots & f_{nn} \end{pmatrix},$$

the translation vector $f' = (f_1, \ldots, f_n)^* \in \mathbb{R}^n$ and the time dilation $f_{n+1} \in \mathbb{R}$.

Let $\psi = \psi(x,t) = \psi(x_1, \ldots, x_n, t) \in \mathcal{G}$ be an arbitrary Lorentz transformation. Then we compose a Galilei transformation $\chi = \chi(x,t) \in \mathcal{G}'$ by a translation in (x,t)-space and a rotation in x-space, such that the Lorentz transformation

$$\varphi = \varphi(\xi, \tau) = \chi \circ \psi \circ \chi^{-1}(\xi, \tau) \tag{21}$$

satisfies the following conditions:

$$\varphi \in \mathcal{G}_0, \tag{22}$$

$$\varphi_k(\xi_1, \ldots, \xi_n, \tau) = \xi_k \quad \text{for} \quad k = 2, \ldots, n, \qquad (\xi, \tau) \in \mathbb{R}^{n+1}. \tag{23}$$

Therefore, we only have to study the origin-preserving Lorentz transformations in the case $n = 1$: For the real 2×2-matrix

$$F = \begin{pmatrix} \alpha & \beta \\ \gamma & \delta \end{pmatrix} \in \mathbb{R}^{2\times 2}$$

we read off the following condition from (16):

$$F \circ \begin{pmatrix} -c^2 & 0 \\ 0 & 1 \end{pmatrix} \circ F^* = \begin{pmatrix} -c^2 & 0 \\ 0 & 1 \end{pmatrix}. \tag{24}$$

Defining the symmetric matrix

$$\Lambda := \begin{pmatrix} ic & 0 \\ 0 & 1 \end{pmatrix} \qquad \text{with} \qquad \Lambda^{-1} = \begin{pmatrix} -\frac{i}{c} & 0 \\ 0 & 1 \end{pmatrix},$$

we rewrite (24) equivalently into the form

$$F \circ \Lambda \circ \Lambda^* \circ F^* = \Lambda \circ \Lambda^*,$$

and finally

$$(\Lambda^{-1} \circ F \circ \Lambda) \circ (\Lambda^{-1} \circ F \circ \Lambda)^* = E.$$

Here E denotes the unit matrix in $\mathbb{R}^2$. Consequently, the subsequent matrix $G := \Lambda^{-1} \circ F \circ \Lambda$ is orthogonal, and we have $\det G > 0$. We observe $G \in \mathrm{SO}(2)$ and obtain

$$G = \begin{pmatrix} \cos z & \sin z \\ -\sin z & \cos z \end{pmatrix} \qquad \text{with} \qquad z \in \mathbb{C}.$$

Respecting $z = i\vartheta$ and $\vartheta \in \mathbb{R}$, we now calculate

$$\begin{aligned} F = \Lambda \circ G \circ \Lambda^{-1} &= \begin{pmatrix} ic & 0 \\ 0 & 1 \end{pmatrix} \circ \begin{pmatrix} \cos z & \sin z \\ -\sin z & \cos z \end{pmatrix} \circ \begin{pmatrix} -\frac{i}{c} & 0 \\ 0 & 1 \end{pmatrix} \\ &= \begin{pmatrix} ic & 0 \\ 0 & 1 \end{pmatrix} \circ \begin{pmatrix} -\frac{i}{c}\cos z & \sin z \\ \frac{i}{c}\sin z & \cos z \end{pmatrix} = \begin{pmatrix} \cos z & ic\sin z \\ \frac{i}{c}\sin z & \cos z \end{pmatrix} \\ &= \begin{pmatrix} \cos(i\vartheta) & -c\frac{1}{i}\sin(i\vartheta) \\ -\frac{1}{c}\frac{1}{i}\sin(i\vartheta) & \cos(i\vartheta) \end{pmatrix} = \begin{pmatrix} \cosh\vartheta & -c\sinh\vartheta \\ -\frac{1}{c}\sinh\vartheta & \cosh\vartheta \end{pmatrix}. \end{aligned}$$

We combine our considerations to

Theorem 8.9. *For each Lorentz transformation $\psi \in \mathcal{G}$ there exists a Galilei transformation $\chi \in \mathcal{G}'$ and a special hyperbolic transformation*

$$\varphi(x_1, \ldots, x_n, t) := \begin{pmatrix} \cosh\vartheta & 0 \ldots 0 & -c\sinh\vartheta \\ 0 & 1 \quad 0 & 0 \\ \vdots & \ddots & \vdots \\ 0 & 0 \quad 1 & 0 \\ -\frac{1}{c}\sinh\vartheta & 0 \ldots 0 & \cosh\vartheta \end{pmatrix} \circ \begin{pmatrix} x_1 \\ \vdots \\ x_n \\ t \end{pmatrix}, \tag{25}$$

such that the following representation holds true:

$$\psi = \chi^{-1} \circ \varphi \circ \chi. \tag{26}$$

In classical physics those reference systems (x_1, x_2, x_3, t) and $(\xi_1, \xi_2, \xi_3, \tau)$ are equivalent, which refer to each other by a Galilei transformation. Due to Theorem 8.9, the origin of the first system is translated by such a motion from the origin of the second system, however, the time is simply transferred. On account of (20) the time measurement coincides in both systems, that means

$$d\tau = dt. \tag{27}$$

Additionally, the distance measurement is the same in both systems, more precisely

$$d\xi_1^2 + d\xi_2^2 + d\xi_3^2 = dx_1^2 + dx_2^2 + dx_3^2. \tag{28}$$

In *relativistic physics* of A. Einstein, the Galilei transformations $\mathcal{G}'$ are replaced by the larger group $\mathcal{G}$ of Lorentz transformations. Since they simultaneously transfer space and time coordinates, a separate time and space measurement is *not* possible any more. In special relativity theory we assume the speed of light having the same value c in all reference systems, if these systems move towards each other with a velocity smaller than c. Measuring the physical phenomena by the d'Alembert operator

$$\frac{1}{c^2}\Box = \frac{1}{c^2}\frac{\partial^2}{\partial t^2} - \Delta_x,$$

the Lorentz transformations appear as those mappings referring two equivalent reference systems to each other.

Considering the case $n = 1$, at first, we have the special hyperbolic transformation

$$\begin{pmatrix} \xi \\ \tau \end{pmatrix} = \begin{pmatrix} \cosh\vartheta & -c\sinh\vartheta \\ -\frac{1}{c}\sinh\vartheta & \cosh\vartheta \end{pmatrix} \circ \begin{pmatrix} x \\ t \end{pmatrix}. \tag{29}$$

This implies

$$\begin{pmatrix} d\xi \\ c\,d\tau \end{pmatrix} = \begin{pmatrix} \cosh\vartheta & -c\sinh\vartheta \\ -\sinh\vartheta & c\cosh\vartheta \end{pmatrix} \circ \begin{pmatrix} dx \\ dt \end{pmatrix}$$

and consequently

$$\begin{aligned} c^2\,d\tau^2 - d\xi^2 &= \sinh^2\vartheta\,dx^2 - 2c\sinh\vartheta\cosh\vartheta\,dx\,dt + c^2\cosh^2\vartheta\,dt^2 \\ &\quad -\Big(\cosh^2\vartheta\,dx^2 - 2c\cosh\vartheta\sinh\vartheta\,dx\,dt + c^2\sinh^2\vartheta\,dt^2\Big) \\ &= c^2\,dt^2 - dx^2. \end{aligned}$$

Combined with Theorem 8.9 and the equations (27) and (28) as well, we obtain the invariance property

$$c^2\,d\tau^2 - d\xi_1^2 - d\xi_2^2 - d\xi_3^2 = c^2\,dt^2 - dx_1^2 - dx_2^2 - dx_3^2. \tag{30}$$

Therefore, the Lorentz transformations leave the distance of two events (x_1, x_2, x_3, t) and $(\xi_1, \xi_2, \xi_3, \tau)$ invariant with respect to the *Minkowski metric*

$$d\sigma^2 := c^2\,d\tau^2 - d\xi_1^2 - d\xi_2^2 - d\xi_3^2. \tag{31}$$

The quantities $d\tau$ and $d\xi_1^2 + d\xi_2^2 + d\xi_3^2$ are *not* preserved under Lorentz transformations, in general.

We name a vector (x_1, x_2, x_3, t) time-like (or space-like) if and only if

$$c^2t^2 > x_1^2 + x_2^2 + x_3^2 \qquad (\text{or} \quad c^2t^2 < x_1^2 + x_2^2 + x_3^2\ )$$

is satisfied. Two events (x_1', x_2', x_3', t'), $(x_1'', x_2'', x_3'', t'')$ occur at different times (and at different positions in space), if and only if the vector

$$(x_1' - x_1'', x_2' - x_2'', x_3' - x_3'', t' - t'')$$

is time-like (and space-like, respectively). Then we find a Lorentz transformation such that both events occur at the same position in space (or at the same time). Finally, the surface

$$ct^2 = x_1^2 + x_2^2 + x_3^2$$

represents the characteristic light cone, on which each two events can be transferred into each other by a Lorentz transformation.

9 Some Historical Notices to the Chapters 5 and 6

Jean d'Alembert (1717–1783) may be seen as the founder of a great *school for mathematical physics* in France. We know his name from the solution of the one-dimensional wave equation. Furthermore, the corresponding differential operator in $\mathbb{R}^n$ is denoted in his honor. Directly and indirectly, he inspired most of the following mathematicians, creating in France after the Revolution a *golden era for mathematics*:
J.L. Lagrange (1736–1813), G. Monge (1746–1818), P.S. Laplace (1749–1827), A. Legendre (1752–1833), J. Fourier (1768–1830), and S. Poisson (1781–1840).

We got aquainted to Laplace and Poisson by their investigations of the homogeneous and inhomogeneous potential-equation, respectively. The names of Monge together with Ampère are connected with their nonlinear differential equation describing embedding problems for Riemannian metrics. In the theory of spherical harmonics, Laplace and Legendre have given essential contributions. Fourier is, besides *his* series, well-known for the *Théorie de la Chaleur*. Last but not least, Lagrange's name stands aequo loco to Euler's (1707–1783), when we are speaking of the variational equations in the *Calculus of Variations.*

In those times, even the more theoretical mathematicians were confronted with the difficult decision, wether to sympathize with the King, the Revolution, the Emperor, or the Republic. In the imperial times, some of the French mathematicians even joined the *Expedition to Egypt*, which was scientifically a success. Then applied mathematics enjoyed a high recognition and played a central role for the education of the youth in the Technical Sciences. The *École Polytechnique* of Paris was founded as the preimage for all modern Technical Universities – as well as the Research University *Scuola Normale Superiore* at Pisa. Simultaneously, classical Universities in the Rhine area were closed.

It was C.F. Gauss, who declined an offer for an attractive chair of mathematics in a Parisian University – the scientific metropolis. At home, Gauß successfully founded a tradition for mathematics in the University of Göttingen. Within his life-time, this prestigeous institution was administrated by the Kingdoms, then in a personal union, of Hannover and England.

The beginning of the modern theory of partial differential equations is marked by the seminal paper of E. Hopf, from 1927, on the maximum principle for linear elliptic differential equations. While the classical results are mostly derived by integral representations, the Hopf maximum principle is independent of these ingredients. This paved the way for J. Schauder, to start his ingenious treatment of elliptic equations in 1932/34 by functional analytic methods.

Figure 1.9 SCHWARZ-RIEMANN MINIMAL SURFACE
spanning a quadrilateral – taken from *H. A. Schwarz: Mathematische Abhandlungen I*, page 2, Springer-Verlag, Berlin... (1890).

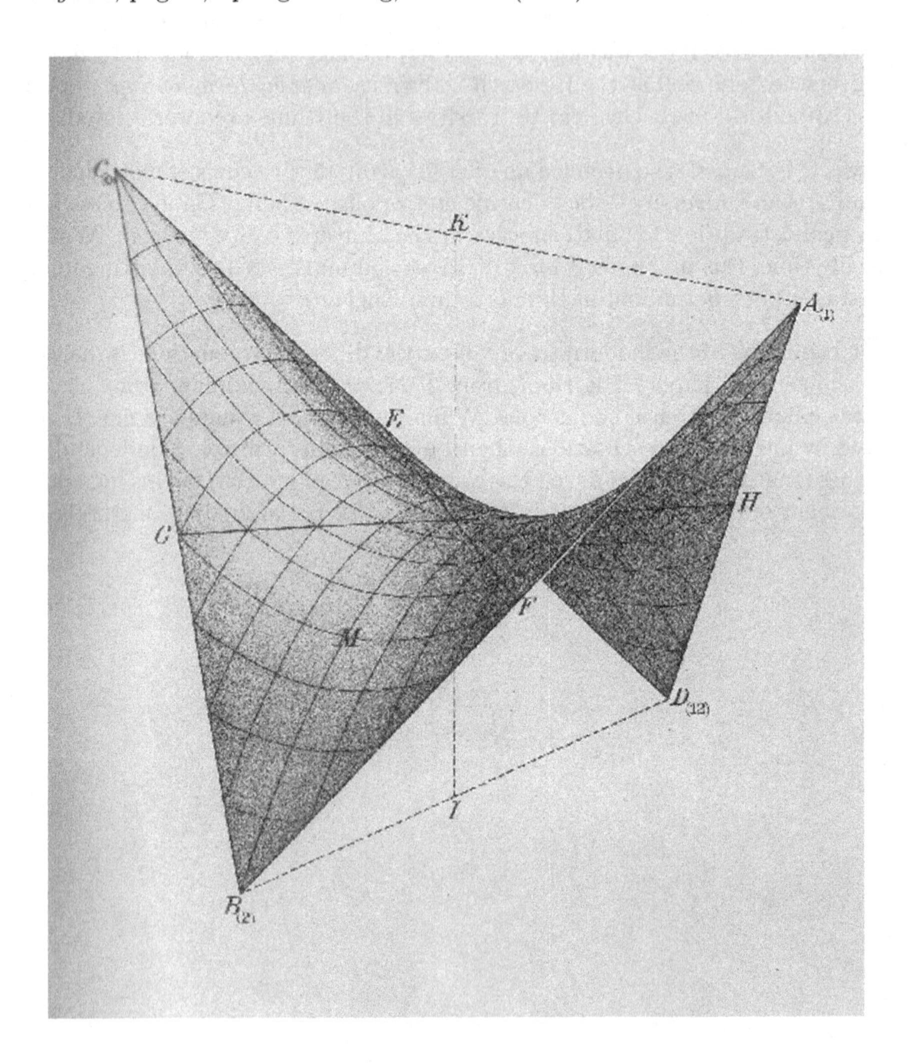

References

[BS] H. Behnke, F. Sommer: *Theorie der analytischen Funktionen einer komplexen Veränderlichen.* Grundlehren der math. Wissenschaften **77**, Springer-Verlag, Berlin ..., 1955.

[B] L. Bianci: *Vorlesungen über Differentialgeometrie. Dt. Übersetzung von Max Lukat.* Verlag B. G. Teubner, Leipzig, 1899.

[BL] W. Blaschke, K. Leichtweiss: *Elementare Differentialgeometrie.* Grundlehren der math. Wissenschaften **1**, 5. Auflage, Springer-Verlag, Berlin ..., 1973.

[CH] R. Courant, D. Hilbert: *Methoden der mathematischen Physik I, II.* Heidelberger Taschenbücher, Springer-Verlag, Berlin ..., 1968.

[D] G. Darboux: *Leçons sur la théorie générale des surfaces I – IV.* Première édition: Gauthier-Villars, Paris, 1887 – 1896. Reprint: Chelsea Publ. Co., New York, 1972.

[De] K. Deimling: *Nichtlineare Gleichungen und Abbildungsgrade.* Hochschultext, Springer-Verlag, Berlin ..., 1974.

[DHKW] U. Dierkes, S. Hildebrandt, A. Küster, O. Wohlrab: *Minimal surfaces I, II.* Grundlehren der math. Wissenschaften **295**, **296**, Springer-Verlag, Berlin ..., 1992.

[DHS] U. Dierkes, S. Hildebrandt, F. Sauvigny: *Minimal surfaces.* Grundlehren der math. Wissenschaften **339**, Springer-Verlag, Berlin ..., 2010.

[DHT1] U. Dierkes, S. Hildebrandt, A. Tromba: *Regularity of minimal surfaces.* Grundlehren der math. Wissenschaften **340**, Springer-Verlag, Berlin ..., 2010.

[DHT2] U. Dierkes, S. Hildebrandt, A. Tromba: *Global analysis of minimal surfaces.* Grundlehren der math. Wissenschaften **341**, Springer-Verlag, Berlin ..., 2010.

[E] L. C. Evans: *Partial Differential Equations.* AMS-Publication, Providence, RI., 1998.

[G] P. R. Garabedian: *Partial Differential Equations.* Chelsea, New York, 1986.

[Gi] M. Giaquinta: *Multiple integrals in the calculus of variations.* Annals Math. Stud. **105**. Princeton University Press, Princeton N.J., 1983.

[GT] D. Gilbarg, N. S. Trudinger: *Elliptic Partial Differential Equations of Second Order.* Grundlehren der math. Wissenschaften **224**, Springer-Verlag, Berlin ..., 1983.

F. Sauvigny, *Partial Differential Equations 1*, Universitext, DOI 10.1007/978-1-4471-2981-3, © Springer-Verlag London 2012

[Gr] H. Grauert: *Funktionentheorie I.* Vorlesungsskriptum an der Georg-August-Universität Göttingen im Wintersemester 1964/65.

[GF] H. Grauert, K. Fritzsche: *Einführung in die Funktionentheorie mehrerer Veränderlicher.* Hochschultext, Springer-Verlag, Berlin ..., 1974.

[GL] H. Grauert, I. Lieb: *Differential- und Integralrechnung III.* 1. Auflage, Heidelberger Taschenbücher, Springer-Verlag, Berlin ..., 1968.

[GuLe] R. B. Guenther, J. W. Lee: *Partial Differential Equations of Mathematical Physics and Integral Equations.* Prentice Hall, London, 1988.

[HW] P. Hartman, A. Wintner: *On the local behavior of solutions of nonparabolic partial differential equations.* American Journ. of Math. **75** (1953) 449–476.

[H1] E. Heinz: *Differential- und Integralrechnung III.* Ausarbeitung einer Vorlesung an der Georg-August-Universität Göttingen im Wintersemester 1986/87.

[H2] E. Heinz: *Partielle Differentialgleichungen.* Vorlesung an der Georg-August-Universität Göttingen im Sommersemester 1973.

[H3] E. Heinz: *Lineare Operatoren im Hilbertraum I.* Vorlesung an der Georg-August-Universität Göttingen im Wintersemester 1973/74.

[H4] E. Heinz: *Fixpunktsätze.* Vorlesung an der Georg-August-Universität Göttingen im Sommersemester 1975.

[H5] E. Heinz: *Hyperbolische Differentialgleichungen.* Vorlesung an der Georg-August-Universität Göttingen im Wintersemester 1975/76.

[H6] E. Heinz: *Elliptische Differentialgleichungen.* Vorlesung an der Georg-August-Universität Göttingen im Sommersemester 1976.

[H7] E. Heinz: *On certain nonlinear elliptic systems and univalent mappings.* Journal d'Analyse Math. **5** (1956/57), 197–272.

[H8] E. Heinz: *An elementary analytic theory of the degree of mapping.* Journal of Math. and Mechanics **8** (1959), 231–248.

[H9] E. Heinz: *Zur Abschätzung der Funktionaldeterminante bei einer Klasse topologischer Abbildungen.* Nachr. Akad. Wiss. in Göttingen, II. Math.-Phys. Klasse, Heft Nr.9, 1968.

[H10] E. Heinz: *Lokale Abschätzungen und Randverhalten von Lösungen elliptischer Monge-Ampèrescher Gleichungen.* Journ. reine angew. Math. **438** (1993), 1–29.

[H11] E. Heinz: *Monge-Ampèresche Gleichungen und elliptische Systeme.* Jahres-bericht der Deutschen Mathematiker-Vereinigung **98** (1996), 173–181.

[He1] G. Hellwig: *Partielle Differentialgleichungen.* Teubner-Verlag, Stuttgart, 1960.

[He2] G. Hellwig: *Differentialoperatoren der mathematischen Physik.* Springer-Verlag, Berlin ..., 1964.

[He3] G. Hellwig: *Höhere Mathematik I – IV.* Vorlesungen an der Rheinisch-Westfälischen Technischen Hochschule Aachen; Mitschrift aus dem Wintersemester 1978 bis zum Sommersemester 1980.

[Hi1] S. Hildebrandt: *Analysis 1.* Springer-Verlag, Berlin ..., 2002.

[Hi2] S. Hildebrandt: *Analysis 2.* Springer-Verlag, Berlin ..., 2003.

[HS] F. Hirzebruch und W. Scharlau: *Einführung in die Funktionalanalysis.* Bibl. Inst., Mannheim, 1971.

[HC] A. Hurwitz, R. Courant: *Funktionentheorie.* Grundlehren der math. Wissenschaften **3**, 4. Auflage, Springer-Verlag, Berlin ..., 1964.

[J] F. John: *Partial Differential Equations.* Springer-Verlag, New York ..., 1982.

[Jo] J. Jost: *Partielle Differentialgleichungen. Elliptische (und parabolische) Gleichungen.* Springer-Verlag, Berlin ..., 1998.

[K] W. Klingenberg: *Eine Vorlesung über Differentialgeometrie.* Heidelberger Taschenbücher, Springer-Verlag, Berlin ..., 1973.

[Kn] H. Kneser: *Lösung der Aufgabe 42.* Jahresber. Dt. Math.-Vereinigung **35** (1926), 123–124.

[M] C. Müller: *Spherical Harmonics.* Lecture Notes in Math. **17**, Springer-Verlag, Berlin ..., 1966.

[Mu] Frank Müller: *Funktionentheorie und Minimalflächen.* Vorlesung an der Universität Duisburg-Essen, Campus Duisburg, im Sommersemester 2011.

[N] J.C.C. Nitsche: *Vorlesungen über Minimalflächen.* Grundlehren der math. Wissenschaften **199**, Springer-Verlag, Berlin ..., 1975.

[Ni] L. Nirenberg: *On nonlinear elliptic partial differential equations and Hölder continuity.* Comm. Pure Appl. Math. **6** (1953), 103–156.

[R] F. Rellich: *Die Bestimmung einer Fläche durch ihre Gaußsche Krümmung.* Math. Zeitschrift **43** (1938), 618–627.

[Re] R. Remmert: *Funktionentheorie I.* Grundwissen Mathematik **5**, 2. Auflage, Springer-Verlag, Berlin ..., 1989.

[Ru] W. Rudin: *Principles of Mathematical Analysis.* McGraw Hill, New York, 1953.

[S1] F. Sauvigny: *Einführung in die reelle und komplexe Analysis mit ihren gewöhnlichen Differentialgleichungen 1.* Vorlesungsskriptum an der BTU Cottbus im Wintersemester 2006/07.

[S2] F. Sauvigny: *Einführung in die reelle und komplexe Analysis mit ihren gewöhnlichen Differentialgleichungen 2.* Vorlesungsskriptum an der BTU Cottbus im Sommersemester 2007.

[S3] F. Sauvigny: *Flächen vorgeschriebener mittlerer Krümmung mit eineindeutiger Projektion auf eine Ebene* Math. Zeitschrift **180** (1982), 41–67.

[S4] F. Sauvigny: *On immersions of constant mean curvature: Compactness results and finiteness theorems for Plateau's problem.* Archive for Rational Mechanics and Analysis **110** (1990), 125–140.

[S5] F. Sauvigny: *Deformation of boundary value problems for surfaces with prescribed mean curvature.* Analysis **21** (2001), 157–169.

[S6] F. Sauvigny: *Un problème aux limites mixte des surfaces minimales avec une multiple projection plane et le dessin optimal des escaliers tournants.* Annales de l'Institut Henri Poincaré – Analyse Non Linéaire **27** (2010), 1247–1270.

[Sc] F. Schulz: *Regularity theory for quasilinear elliptic systems and Monge-Ampère equations in two dimensions.* Lecture Notes in Math. **1445**, Springer-Verlag, Berlin ..., 1990.

[V] I. N. Vekua: *Verallgemeinerte analytische Funktionen.* Akademie-Verlag, Berlin, 1963.

Index

Almost everywhere (a.e.), 117
Analyticity theorem
  for Poisson's equation, 317
Approximation of L^p-functions, 159
Arc lenght, 13
Area
  of a classical surface in $\mathbb{R}^3$, 14
  of a hypersurface in $\mathbb{R}^n$, 16
  of an m-dimensional surface in $\mathbb{R}^n$, 17
Automorphism group, 271
Automorphism group of the unit disc, 277

Banach space, 140
  separable, 148
Beltrami operator of first order, 84
  in spherical coordinates, 87
Bessel's inequality, 150
Boundary behavior of conformal mappings
  $C^{1,1}$-regular, 295
  analytic, 288
Boundary condition
  of Dirichlet, 377
  of Neumann, 377
Boundary point lemma
  in $\mathbb{C}$, 289
  of E. Hopf, 368
Brouwer's fixed point theorem
  in $\mathbb{C}$, 180
  in $\mathbb{R}^n$, 194

Capacity zero of a set, 37
Cauchy's initial value problem, 396
Cauchy's integral formula, 222
  in $\mathbb{C}^n$, 227
Cauchy's principal value, 259
Cauchy-Riemann equations
  homogeneous, 225
  inhomogeneous, 242
Characteristic parameters, 403
Co-derivative of a 1-form, 82
Compactly contained sets $\Theta \subset\subset \Omega$, 222
Completeness relation, 150
Conformal equivalence, 271
Connected component, 205
Content, 113
Convergence
  almost everywhere, 124
  weak, 170
Convergence theorem
  of B.Levi, 105, 119
  of Fatou, 106, 120
  of Lebesgue, 108, 120
Courant-Lebesgue lemma, 50
Curve
  closed, 57
  piecewise continuously differentiable, 57
Curvilinear integral, 22, 58
Curvilinear integrals
  First theorem on, 58
  in simply connected domains, 66
  Second theorem on, 62

d'Alembert operator, 393

F. Sauvigny, *Partial Differential Equations 1*, Universitext,
DOI 10.1007/978-1-4471-2981-3,

Daniell's integral, 92, 151
  absolutely continuous, 164
  lower, 100
  upper, 100
Decomposition theorem of Jordan-Hahn, 166
Degree of mapping, 187, 191
  with respect to z, 206
Differential equation
  linear elliptic, 363, 430
  linear hyperbolic, 430
  linear parabolic, 431
  of Laplace, 310
  of Poisson, 311
  quasilinear elliptic, 373
  ultrahyperbolic, 411, 431
Differential form, 18
  *-operator
    parameter-invariant, 78
    properties, 78
  0-form, 19
  basic m-form, 19
  closed, 60
  exact, 68
  exterior derivative, 23
  exterior product, 20
    commutator relation, 21
  improper integral over a surface, 21
  in the class C^k, 18
  inner product, 72
    parameter-invariant, 73
  integration over an oriented manifold, 35
  of degree m, 18
  Pfaffian 1-form
    exact, 58
    Primitive of a, 58
  Pfaffian form (1-form), 22
  transformed, 26
Differential operator
  degenerate elliptic, 363
  elliptic, 363
  quasilinear, 373
  reduced, 364
  stable elliptic, 372
  uniformly elliptic, 363
Differential symbols
  equivalent, 19
Dirichlet domain, 337
Dirichlet problem
  for the Laplace equation, 329
    Uniqueness theorem, 329
  for the Laplacian, 338
  Uniqueness and stability, 365
Divergence of a vector-field, 26
Domain
  contractible, 68
  simply connected, 66
  star-shaped, 68
Domain integral, 22
Dual space, 147

Embedding
  continuous, 158
Equation of prescribed mean curvature, 373
Essential supremum, 155
Extension of linear functionals, 161

Fixed point, 180
Fourier coefficients, 150
Fourier expansion, 344
Fourier series, 150
Fourier transformation, 381
  inverse, 381
Fourier-Plancherel integral theorem, 380
Function
  p-times integrable, 152
  antiholomorphic, 237
  characteristic, 96
  differentiable in the sense of Pompeiu, 255
  Dini continuous, 254
  Gamma-, 309
  Green's, 318
  Hölder continuous, 254, 258
  harmonic, 310
  harmonically modified, 334
  holomorphic in several complex variables, 226
  Lebesgue integrable, 102
  Lipschitz continuous, 254
  measurable, 121
  pseudoholomorphic, 266
  simple, 127
  spherically harmonic of degree k, 347

subharmonic, 325
superharmonic, 325
weakharmonic, 325
Fundamental solution of the Laplace equation, 311
Fundamental theorem of Algebra, 180

Galilei transformation, 432
Gaussian integral theorem, 47
in the complex form, 245
Gramian determinant, 16
Green's formula, 49

Hölder's inequality, 152
Hadamard's estimate, 252
Hadamard's method of descent, 409, 420
Harnack's inequality, 323
Harnack's lemma, 335
Hausdorff null-set, 46
Heat conductivity coefficient, 378
Heat equation, 378
Initial value problem, 380
Initial-boundary-value problem, 390
Kernel function, 382
Uniqueness of the initial value problem, 388
Hilbert space, 142
Homotopy, 61, 65
Homotopy lemma
in $\mathbb{C}$, 178
in $\mathbb{R}^n$, 191
Hypersurfaces in $\mathbb{R}^n$, 14

Identity theorem
in $\mathbb{C}$, 224
in $\mathbb{C}^n$, 230
Index, 175, 182, 197
Index-sum formula, 175, 183
Inner product, 141
Integral equation
of Abel, 414
Integral operator
of Hadamard, 252
of Vekua, 261
Integral representation, 246, 311
of Poisson, 320
Integral-mean-value
spherical, 405

Jordan curve, 212
analytic, 286
of the class C^k, 286
Jordan domain, 212, 286

Laplace-Beltrami operator, 84
in cylindrical coordinates, 85
in polar coordinates, 85
in spherical coordinates, 88
Laurent expansion, 247
Lebesgue space
L^p-space, 152
Linear mapping
bounded, 146
continuous, 145
Linear space
of spherical harmonics of the order k, 354
of the complex potentials, 266
Liouville's theorem
for harmonic functions, 324
for holomorphic functions, 229
Lipschitz estimate for conformal mappings, 290
Lorentz transformation, 432

Manifold
atlas, 32
boundary of the, 33
chart of a, 32
closed, 33
compact, 33
differentiable, 32
induced atlas on the boundary, 34
induced orientation on the boundary, 34
oriented, 33
regular boundary, 33
singular boundary, 33
Mapping
conformal, 270
elementary, 272
Fixed point of a, 276
origin-preserving, 276
Maximum principle, 326
for holomorphic functions, 235
of E. Hopf, 370
strong, 370
Maximum-minimum principle

parabolic, 388
Mean value property, 323
Mean value theorem of Asgeirsson, 411
Measurable set, 109
Measure, 113
finite, 113
of a set, 109
Metric tensor, 16
Minimal surface equation, 374
Minimal surface operator, 374
Minimum principle, 326
Minkowski's inequality, 153
Mixed boundary value problem, 377
Monodromy theorem, 65
Multiplication functional, 161, 163

Norm, 140
L^∞-norm, 155
L^p-norm, 152
of a functional, 146
supremum-norm, 158
Normal domain, 245
in $\mathbb{R}^n$, 310
Normal vector, 14
exterior, 44
Normed space, 140
complete, 140
Null-set, 114

Order, 193
Orthogonal space, 143
Orthonormal system
complete, 150

Parallelogram identity, 144
Parameter domain, 13
Parametric representation, 13
equivalent, 13
Partial integration
in $\mathbb{R}^n$, 4
in arbitrary parameters, 82
Partition of unity, 11
Path
closed, 57
piecewise continuously differentiable, 57
Poincaré's condition, 340
Poincaré's lemma, 70
Poisson's kernel, 322
Polynomial
Gegenbaur's, 351
of Legendre, 351
Portrait of
Bernhard Riemann (1826–1866), 214
Carl Friedrich Gauß (1777–1855), 90
Hermann Amandus Schwarz (1843–1921), 304
Joseph A. F. Plateau (1801–1883), 361
Stefan Banach (1892–1945), 173
Pre-Hilbert-space, 141
Principle of the argument, 251
Product theorem for the degree of mapping, 209
Projection theorem in Hilbert spaces, 144
Proposition
of Hardy and Littlewood, 293

Regularity theorem
for L^p-functionals, 163
for the inhomogeneous Cauchy-Riemann equation, 264
for weakharmonic functions, 333
Representation theorem of Fréchet-Riesz, 147
Residue, 245
Residue theorem
general, 242
of Liouville, 245
Riemann's sphere, 239
Riemann's theorem on removable singularities, 247
Riemannian mapping theorem, 280
Riesz's representation theorem, 168
Root lemma, 281
Rotation of a vector-field, 25

Sard's lemma, 202
Schwarz-Riemann minimal suface, 438
Schwarzian lemma, 276
Schwarzian reflection principle, 237
Selection theorem of Lebesgue, 128
Separability of L^p-spaces, 160
Sigma-Additivity
σ-Additivity, 109
Sigma-Algebra
σ-Algebra, 113

Sigma-Subadditivity
σ-Subadditivity, 112
Similarity principle of Bers and Vekua, 267
Smoothing
of a closed curve, 62
of functions, 2
Spherical harmonic
n-dimensional, 347
Spherical harmonics
Addition theorem, 357
Completeness, 358
Stokes integral theorem
classical, 52
for manifolds, 38
local, 31
Support of a function, 206
Surface
characteristic, 392
m-dimensional in $\mathbb{R}^n$, 16
noncharacteristic, 392
parametrized, 13
regular and oriented, 13
Surface element
of a hypersurface, 15
of an m-dimensional surface in $\mathbb{R}^n$, 16
Surface integral, 22

Tangential space to a surface, 14
Theorem
about Fourier series, 342
of Arzelà-Ascoli, 280
of Carathéodory-Courant, 286
of Carleman, 269
of Casorati-Weierstraß, 249
of Cauchy-Riemann, 220
of Cauchy-Weierstraß, 222
of d'Alembert, 405
of Dini, 93
of Egorov, 131
of F. John, 406
of Fischer-Riesz, 154
of Fubini, 138
of Hurwitz, 282
of Jordan-Brouwer, 211
of Kirchhoff, 407
of Lusin, 133
of monodromy, 222
of Poincaré and Brouwer, 195
of Pompeiu-Vekua, 257
of Radon-Nikodym, 164
of Rouché, 179
on Cauchy's integral across the boundary, 300
on holomorphic parameter integrals, 231
on removable singularities, 255
on the harmonic extension, 301
on the invariance of domains in $\mathbb{C}$, 234
on the invariance of domains in $\mathbb{R}^n$, 213
Tietze's extension theorem, 7
Transformation
fractional linear, 271
of Möbius, 271

Uniqueness theorem of Vekua, 269
Unit normal vector, 14

Vector-potential, 72
Vekua's class of functions, 256
Volume form, 73

Wave equation
Cauchy's initial value problem, 403
for $n = 1$, 403
for $n = 2$, 409
for $n = 3$, 407
for even $n \geq 2$, 420
for odd $n \geq 3$, 417
Uniqueness, 396
homogeneous, 393
Energy estimate, 394
inhomogeneous, 422
Initial-boundary-value problem, 424
Weak compactness of $L^p(X)$, 170
Weierstraß approximation theorem, 6
Winding number, 175, 177, 181